THE ROUTLEDGE INTERNATIONAL HANDBOOK OF NEUROCOGNITIVE REHABILITATION

The Routledge International Handbook of Neurocognitive Rehabilitation is a comprehensive resource highlighting the rapid advancements in neurocognitive science and their application in rehabilitation practices. Bringing together perspectives from top authorities in neuropsychology, neurology, neurosurgery, and neuropsychiatry, it explores cutting-edge techniques and resources including virtual reality (VR), augmented reality (AR), machine learning (ML), and artificial intelligence (AI) that are revolutionizing the area.

By combining scientific rigor with accessibility, the book closes the gap between state-of-the-art technology and conventional rehabilitation techniques, giving readers academic and practical expertise. Sections focus on the foundations of neurocognitive rehabilitation, technology-driven innovations, clinical applications, and ethical, social, and practical considerations. Case studies and qualitative accounts are integrated into the chapters to illustrate the impact of neurocognitive rehabilitation techniques on patient outcomes, opening up new avenues for individualized and successful therapeutic approaches.

It is an essential reference for students, researchers, and professionals to leverage emerging technologies for improving patient outcomes and advancing the discipline of neurocognitive rehabilitation.

K. Jayasankara Reddy is a distinguished expert in neuropsychology, neuroelectrophysiology, and forensic neuropsychology. He earned his M.Sc. and Ph.D. in psychology specializing in neuropsychology. Currently, Dr. Reddy is Professor of Neuropsychology/Cognitive Neuroscience and Dean for Social Sciences at Christ University in Bangalore, India. With a robust academic career spanning teaching, research, and professional practice, he has supervised numerous Ph.D. and M.Phil. candidates, as well as over 100 M.Sc. psychology research projects.

THE ROUTLEDGE INTERNATIONAL HANDBOOK SERIES

THE ROUTLEDGE INTERNATIONAL HANDBOOK OF CHANGES IN HUMAN PERCEPTIONS AND BEHAVIORS
Edited by Kanako Taku and Todd Shackleford

THE ROUTLEDGE INTERNATIONAL HANDBOOK OF POSITIVE HEALTH SCIENCES
Positive Psychology and Lifestyle Medicine Research, Theory and Practice
Edited by Jolanta Burke, Ilona Boniwell, Beth Frates, Liana S. Lianov, Ciaran A. O'Boyle

THE HANDBOOK OF PERSONALIZED PERSUASION
Theory and Application
Edited by Richard E. Petty, Andrew Luttrell, Jacob D. Teeny

THE ROUTLEDGE INTERNATIONAL HANDBOOK OF INTERRACIAL AND INTERCULTURAL RELATIONSHIPS AND MENTAL HEALTH
Edited by Shivon Raghunandan, Roy Moodley and Kelley R. Kenney

THE ROUTLEDGE INTERNATIONAL HANDBOOK OF POSTMODERN THERAPIES
Edited by Olga Smoliak, Eleftheria Tseliou, Tom Strong, Saliha Bava, and Peter Muntigl

THE ROUTLEDGE INTERNATIONAL HANDBOOK OF ART THERAPY PRACTICE
Edited by Margaret Hills de Zárate, Diane E. Waller, and Claire Louise Vaculik

THE ROUTLEDGE INTERNATIONAL HANDBOOK OF NEUROCOGNITIVE REHABILITATION
Practices, Innovations, and Future Directions
Edited by K. Jayasankara Reddy

THE ROUTLEDGE INTERNATIONAL HANDBOOK OF NEUROCOGNITIVE REHABILITATION

Practices, Innovations, and Future Directions

Edited by K. Jayasankara Reddy

Routledge
Taylor & Francis Group
LONDON AND NEW YORK

Designed cover image: getty images

First published 2026
by Routledge
4 Park Square, Milton Park, Abingdon, Oxon OX14 4RN

and by Routledge
605 Third Avenue, New York, NY 10158

Routledge is an imprint of the Taylor & Francis Group, an informa business

British Library Cataloguing-in-Publication Data
A catalogue record for this book is available from the British Library

ISBN: 9781041087113 (hbk)
ISBN: 9781041087120 (pbk)
ISBN: 9781003646662 (ebk)

DOI: 10.4324/9781003646662

Typeset in Sabon
by Newgen Publishing UK

CONTENTS

List of Contributors *viii*

PART I
Foundations of Neurocognitive Rehabilitation **1**

1 Introduction to Neurocognitive Rehabilitation 3
Priya Prakash and K. Jayasankara Reddy

2 Geriatric Rehabilitation: Aging-Related Cognitive Impairments 24
Shampa Ghosh, Krishna Kumar Singh, and Jitendra Kumar Sinha

3 Geriatric Rehabilitation: Evidence-Based Strategies for Aging-Related Cognitive Impairments 48
Pooja Bharadwaj, Shilpa Kuppahalli Srinath, and Roseline Gomes

4 Applying Solution-Focused Brief Therapy in Cognitive Rehabilitation: Insights from Positive Neuropsychology 78
Kritika Rastogi, Sonali Mukherjee, and Sabornee Karmakar Basu

5 Role of P300 Event-Related Potential (ERP) in Neuropsychiatric Conditions 87
Lekhanshi Manocha, Aakriti Varshney, and Anand Pratap Singh

PART II
Technology-Driven Innovations **101**

6 Artificial Intelligence: A Replacement or a Complement to Psychotherapy in Neurocognitive Rehabilitation? 103
N. Sharon Sucharitha Gold and S. Yogesh Kanna

7 Artificial Intelligence in Neurocognitive Rehabilitation: AI Applications in Assessment, Monitoring, and Therapy 119
Thaddeus Alfonso, Sherine Georgina Victoria, and Shirin A. P.

8 AI and Brain–Computer Interfaces in Neurocognitive Rehabilitation 147
Geetanjali Srivastava, Saumya Kushwaha, Rohit Kumar Mishra, Harsh Koria, Aakash Deep, and Priyanka Jain

9 EEG Neurofeedback: The Brain's Capacity for Change and Growth 165
Cathlyn Niranjana Bennett and Arunkumar Sekar

10 Neurocognitive Changes and Neuroplastic Adaptations in Altered Gravitational Environments 178
Savita Gaur, Ruma Dutta, and Rajesh Arora

PART III
Clinical Applications **183**

11 Cognitive vs. Behavioral Therapy in ADHD: Executive Function Outcomes 185
Bismirty Bhuyan and Varsha Singh

12 Neurostimulation in Long COVID: Advancing Neurocognitive Rehabilitation and Recovery 196
Nathan Fernandez, K. Jayasankara Reddy, and Yogesh Kanna Sathyamoorthy

13 Social Brain and Neurostimulation: Applicability across Psychiatric Disorders 210
Debaleena Ghosh, Supratik Kundu, and Susmita Halder

14 Assessing the Impact of Brain Gym Exercises on Memory Enhancement in Students with Learning Disabilities 236
Drishya Sasidharan and Suphala Kotian

15 Enhancing Executive Functions through Cognitive-Based Intervention in Individuals with Suicidal Ideation and Attempts: A Mixed-Method Pilot Study 243
Trina Banerjee and K. Jayasankara Reddy

PART IV
Ethical, Social, and Practical Considerations **267**

16 Ethics of Technology-Assisted Neurocognitive Rehabilitation 269
Ajay Chandra, Arkalgud Ramaprasad, and Madhavi Rangaswamy

17 Harnessing Virtual Reality for Advancing Innovations in Neurocognitive Rehabilitation: Virtual Reality in Neurocognitive Rehabilitation 281
Navkiran Kalsi, Tara Chand, Rajanikant Panda, and Neeraj Upadhyay

PART V
Future Directions **301**

18 Neuroplasticity Unleashed: Receiving the Brain for Recovery 303
Sruthy Nair and K. Jayasankara Reddy

19 Neuroplasticity Unleashed: Rewiring the Brain through BCI and HEG Neurofeedback 324
Mehulkumar B. Dave

20 Advancing the Frontiers of Neurocognitive Rehabilitation: Research and Practice Ahead 338
Aparna Prashanth and K. Jayasankara Reddy

Index *357*

CONTRIBUTORS

Thaddeus Alfonso
School of Psychological Sciences, Christ University, Bangalore, Karnataka, India

Shirin A. P.
School of Psychological Sciences, Christ University, Bangalore, Karnataka, India

Rajesh Arora
CDAC-Delhi, India

Trina Banerjee
Department of Psychology, Christ (Deemed to be University), Bengaluru, India

Sabornee Karmakar Basu
Department of Psychology, CHRIST University, Delhi NCR, India

Cathlyn Niranjana Bennett
Christ (Deemed to be University), Bengaluru, India

Pooja Bharadwaj
Consultant Clinical Psychologist, Manosuraksha – Nurturing Minds, Bangalore, India

Bismirty Bhuyan
Christ (Deemed to be University), Bengaluru, India

Tara Chand
Jindal Institute of Behavioural Sciences, O.P. Jindal Global University, Sonipat, India

Ajay Chandra
School of Psychological Sciences, Christ (Deemed to be University), Bengaluru, India
School of Arts, Humanities, and Social Sciences, Chanakya University, Bengaluru, India

Mehulkumar B. Dave
Mind Care and Management Associates Pvt Ltd, India

Aakash Deep
CDAC-Delhi, India

Ruma Dutta
CDAC-Delhi, India

Nathan Fernandez
Christ (Deemed to be University), Bengaluru, India

Savita Gaur
CDAC-Delhi, India

Debaleena Ghosh
Christ University, Bengaluru, India

Shampa Ghosh
GloNeuro, Noida, Uttar Pradesh, India

N Sharon Sucharitha Gold
Department of Psychology, Christ University, Bengaluru, India

Roseline Gomes
Jyoti Nivas College, Autonomous, Bengaluru, India

Susmita Halder
Department of Psychology, St. Xavier's University, Kolkata, India

Priyanka Jain
CDAC-Delhi, India

Navkiran Kalsi
Jindal School of Psychology and Counselling, O.P. Jindal Global University, Sonipat, India

S. Yogesh Kanna
Department of Psychology, Christ University, Bengaluru, India

Yogesh Kanna Sathyamoorthy
Department of Psychology, Christ University, Bengaluru, India

Harsh Koria
CDAC-Delhi, India

Suphala Kotian
CDAC, India

Shilpa Kuppahalli Srinath
Consultant Clinical Psychologist, Manosuraksha – Nurturing Minds, Bangalore, India

Supratik Kundu
Psychiatry Resident, Central Institute of Psychiatry, Ranchi, India

Saumya Kushwaha
CDAC-Delhi, India

Lekhanshi Manocha
Department of Psychology and Mental Health, Gautam Buddha University, Greater Noida, Uttar Pradesh, India

Rohit Kumar Mishra
CDAC-Delhi, India

Sonali Mukherjee
Department of Psychology, CHRIST University, Delhi NCR, India

Sruthy Nair
Christ (Deemed to be University), Bengaluru, India

Rajanikant Panda
Department of Neurology, University of California San Francisco (UCSF), USA

Priya Prakash
Christ (Deemed to be University), Bengaluru, India

Aparna Prashanth
Christ (Deemed to be University), Bengaluru, India

Madhavi Rangaswamy
School of Psychological Sciences, Christ (Deemed to be University), Bengaluru, India

Arkalgud Ramaprasad
Department of Information and Decision Sciences, University of Illinois, Chicago, USA

Kritika Rastogi
Department of Psychology, CHRIST University, Delhi NCR

K. Jayasankara Reddy
Department of Psychology, Christ (Deemed to be University), Bengaluru, India

Drishya Sasidharan
Research Scholar, Srinivas University, India

Arunkumar Sekar
School of Arts, Humanities, and Social Sciences, Chanakya University, Bengaluru, India

Anand Pratap Singh
Department of Psychology and Mental Health, Gautam Buddha University, Greater Noida, Uttar Pradesh, India

Krishna Kumar Singh
Symbiosis Centre for Information Technology, Symbiosis International (Deemed University), Pune, Maharashtra, India

Varsha Singh
Christ (Deemed to be University), Bengaluru, India

Jitendra Kumar Sinha
GloNeuro, Noida, Uttar Pradesh, India

Geetanjali Srivastava
Neuro-Research Scientist, CDAC-Delhi, India

Neeraj Upadhyay
Clinical Functional Imaging Lab, Department of Nuclear Medicine, University Hospital Bonn, Germany

Aakriti Varshney
Department of Psychology and Mental Health, Gautam Buddha University, Greater Noida, Uttar Pradesh, India

Sherine Georgina Victoria
School of Psychological Sciences, Christ University, Bangalore, Karnataka, India

PART I

Foundations of Neurocognitive Rehabilitation

1

INTRODUCTION TO NEUROCOGNITIVE REHABILITATION

Priya Prakash and K. Jayasankara Reddy

1.1 Introduction

Neurocognitive rehabilitation is an interdisciplinary field that has gained significant traction over the past four decades. It sits at the nexus of neuroscience, cognitive psychology, neuropsychology, rehabilitation sciences, and clinical medicine. Advances in neuroimaging, computational modeling, and neuroplasticity research have propelled neurocognitive rehabilitation from rudimentary compensatory strategies to sophisticated, evidence-based interventions that can improve functional outcomes and quality of life for diverse populations. This chapter provides a foundational overview, beginning with definitions and purpose, followed by a delineation of target populations and conditions, and culminating in a discussion of the field's current relevance and trajectory.

1.1.1 Definition and Purpose of Neurocognitive Rehabilitation

Neurocognitive rehabilitation refers to systematic, goal-oriented interventions designed to improve cognitive functioning and adaptive behaviors in individuals who have experienced brain injury or neurological disease. Its purpose extends beyond symptom reduction, aiming to promote recovery of cognitive processes and facilitate reintegration into personal, occupational, and social spheres (Cicerone et al., 2019). Central to this definition is the premise that the brain exhibits neuroplasticity the capacity to reorganize neural pathways in response to learning and experience.

Early conceptualizations of neurocognitive rehabilitation, primarily emerging in the late 20th century, were rooted in compensatory models that focused on circumventing deficits through environmental modifications and external aids (Crosson et al., 2017). However, contemporary approaches increasingly emphasize restorative techniques that harness principles of experience-dependent plasticity. These include structured cognitive exercises, computerized training programs, and real-world functional tasks that aim to stimulate neuronal circuits implicated in attention, memory, executive function, and language (Lampit et al., 2014).

DOI: 10.4324/9781003646662-2

The purpose of neurocognitive rehabilitation is thus multi-dimensional:

1. **Restoration:** Recover impaired cognitive skills through targeted practice.
2. **Compensation:** Implement strategies to bypass persistent deficits.
3. **Adaptation:** Facilitate psychosocial adjustment and enhance quality of life.
4. **Prevention:** Mitigate secondary complications such as depression and social withdrawal that commonly follow neurocognitive impairment.

Critically, interventions are framed within person-centered and context-sensitive paradigms, acknowledging the importance of tailoring strategies to individual needs, cultural values, and environmental resources (Crosson et al., 2017). For example, an individual with traumatic brain injury (TBI) may require a different therapeutic emphasis such as executive function retraining compared to a person recovering from stroke-related aphasia.

Recent meta-analyses have underscored the efficacy of cognitive rehabilitation across domains. Cicerone et al. (2019) systematically reviewed randomized controlled trials (RCTs) and demonstrated that structured interventions significantly improve attention, memory, and executive functions. Importantly, the most robust effects occur when interventions are embedded in ecologically valid, goal-driven contexts.

Emerging digital technologies have further refined the definition of neurocognitive rehabilitation. Virtual reality (VR), tele-rehabilitation platforms, and adaptive software now complement traditional therapist-guided approaches, allowing more intensive and scalable delivery of interventions (Laver et al., 2017). As the field evolves, its purpose increasingly integrates both restorative neuroscience and pragmatic implementation science.

1.1.2 Who It Serves: Populations and Conditions

Neurocognitive rehabilitation serves individuals across the lifespan who experience acquired or developmental brain conditions that compromise cognition. Populations include, but are not limited to, the following:

1. **TBI:** TBI is among the most prevalent causes of neurocognitive impairment worldwide. In adults, moderate-to-severe TBI often results in chronic deficits in attention, executive function, memory, and social cognition. Children and adolescents are particularly vulnerable because injuries occur during critical periods of neurodevelopment (Babikian & Asarnow, 2009). Cognitive rehabilitation in TBI aims to restore function and foster compensatory strategies that facilitate return to school, work, and community life.
2. **Stroke Survivors:** Stroke is a leading cause of adult disability. Post-stroke cognitive impairment affects up to 50% of survivors, encompassing deficits in visuospatial processing, memory, language (aphasia), and executive control (Loetscher & Lincoln, 2013). Rehabilitation focuses on retraining cognitive processes while addressing emotional and behavioral sequelae.
3. **Neurodegenerative Disorders:** Individuals with Alzheimer's disease, Parkinson's disease, and other dementias benefit from cognitive interventions that aim to slow decline and maintain daily functioning. While neurocognitive rehabilitation cannot reverse neurodegeneration, it can prolong independence, reduce caregiver burden, and improve quality of life (Steinerman, 2010).

4. Multiple **Sclerosis (MS)**: MS frequently entails cognitive deficits, particularly in information processing speed and memory. Cognitive rehabilitation strategies are increasingly integrated into comprehensive MS care (Amato et al., 2006).
5. **Brain Tumors and Neurosurgical Populations:** Tumors and surgical resections can produce focal deficits depending on lesion location. Rehabilitation interventions target specific cognitive domains disrupted by disease or treatment (Taphoorn & Klein, 2004).
6. **Epilepsy:** Chronic epilepsy and surgical treatments can impair attention, memory, and executive function. Neurocognitive rehabilitation is used to retrain skills and promote psychosocial adjustment.
7. **Mental Health and Forensic Populations:** Emerging evidence suggests that cognitive remediation can also benefit individuals with schizophrenia, depression, and forensic populations where cognitive dysfunction contributes to maladaptive behaviors (Wykes et al., 2011).

Neurocognitive rehabilitation thus serves diverse populations. Importantly, interventions must account for developmental trajectories, sociocultural backgrounds, and comorbidities. For example, in forensic settings, cognitive rehabilitation may be embedded within broader risk reduction programs to address impulsivity and antisocial decision-making (Fishbein et al., 2009).

Recent literature underscores the individual and contextual factors such as age, time since injury, lesion location, cognitive reserve, and family support/moderate outcomes (Cicerone et al., 2019). As such, clinicians are increasingly adopting precision rehabilitation frameworks that integrate neuropsychological profiling with person-centered goals.

1.1.3 Relevance and Growth in the Field

The field of neurocognitive rehabilitation has witnessed exponential growth over the past two decades, catalyzed by several factors:

1. **Neuroscientific Advances:** Functional MRI (fMRI), diffusion tensor imaging (DTI), and other neuroimaging modalities have elucidated mechanisms of plasticity and recovery, allowing more targeted interventions (Grefkes & Fink, 2014).
2. **Evidence-Based Guidelines:** Rigorous trials and systematic reviews have established efficacy for cognitive rehabilitation in numerous conditions. Consensus statements now inform clinical standards (Cicerone et al., 2019).
3. **Technological Innovation:** Digital platforms, VR environments, and artificial intelligence-based adaptive training have revolutionized delivery models, improving scalability and engagement.
4. **Policy and Funding Prioritization:** Many health systems and insurers recognize neurocognitive rehabilitation as essential to long-term recovery, resulting in expanded services and reimbursement structures.
5. **Interdisciplinary Collaboration:** Rehabilitation now involves integrated teams including neuropsychologists, occupational therapists, speech-language pathologists, physicians, and social workers.

Recent global health trends further underscore the field's relevance. As populations age, the prevalence of stroke and neurodegenerative disease is rising. Simultaneously, improvements in acute medical care mean that more individuals survive severe brain injuries, living longer with chronic cognitive impairments (Feigin et al., 2017). This epidemiological landscape has increased demand for effective rehabilitation models.

In the forensic context, neurocognitive rehabilitation offers promise in reducing recidivism. Cognitive impairments in attention, impulse control, and moral reasoning have been identified as risk factors for reoffending. Integrating cognitive rehabilitation within correctional and forensic mental health settings could therefore support public safety goals alongside individual recovery (Fishbein et al., 2009).

The field is also growing in its methodological sophistication. Researchers are increasingly adopting randomized controlled designs, mixed-methods evaluations, and implementation science frameworks to understand not only whether interventions work but how and for whom they are most effective (Laver et al., 2017). For example, adaptive cognitive training programs have demonstrated efficacy in stroke survivors and older adults, but their effectiveness in low-resource settings remains an important area of inquiry.

Key trends that are likely to shape the next decade of neurocognitive rehabilitation are the following:

1. **Personalization:** Precision rehabilitation informed by neuroimaging biomarkers and cognitive phenotyping.
2. **Integration:** Combining cognitive and psychosocial interventions to address the holistic needs of individuals.
3. **Technology:** Greater incorporation of VR, gamification, and telehealth.
4. **Equity:** Ensuring access for underserved populations and diverse cultural contexts.

1.2 Historical Development

Neurocognitive rehabilitation has evolved over nearly two centuries from isolated observations of extraordinary cases to a rigorously researched, interdisciplinary clinical science. The emergence of this field is rooted in foundational case reports that demonstrated the brain's central role in cognition and behavior, gradually giving rise to empirical methodologies, systematized wartime treatments, and multidisciplinary care models that underpin modern practice.

1.2.1 Early Clinical Observations and Case Reports (Phineas Gage, Luria)

The trajectory of neurocognitive rehabilitation begins with seminal observations of individuals whose lives were radically transformed by focal brain lesions. Among the most emblematic is the case of Phineas Gage. In 1848, Gage, a railroad foreman in Vermont, survived a catastrophic accident in which an iron rod passed through his skull, destroying portions of the left frontal lobe. While Gage retained language, memory, and motor abilities, he underwent profound changes in impulse control, judgment, and social comportment (Ratiu et al., 2004). His case challenged dualistic notions that cognition was separated from brain function and inaugurated scientific curiosity about the neural substrates of personality and executive functioning.

More than a century later, Alexander Luria systematized the study of brain–behavior relationships. As a neurologist and neuropsychologist, Luria observed Soviet soldiers with focal injuries during World War II. His comprehensive descriptions and classification of higher cortical functions formed the first cohesive neuropsychological models of cognition (Kotik-Friedgut & Ardila, 2019). Luria's dynamic localization theory proposed that cognitive functions were organized in functional systems rather than discrete centers, emphasizing the integration of distributed networks. Critically, Luria advanced structured rehabilitation methods, including hierarchical retraining tasks and compensatory strategies personalized to each patient's cognitive profile.

These early cases and systematic observations laid the theoretical bedrock for neurocognitive rehabilitation. They demonstrated that the brain's injuries had predictable, domain-specific consequences, but also hinted at plasticity and reorganization concepts that would later fuel rehabilitation's scientific evolution.

1.2.2 Evolution from Anecdotal Evidence to Empirical Research

Although the cases of Gage and Luria offered profound insights, early neurocognitive rehabilitation was largely anecdotal and lacked standardized assessment or controlled intervention protocols. In the first half of the 20th century, clinicians primarily relied on descriptive reports and theoretical speculation to design interventions. This period reflected medicine's broader epistemological transition from observational narratives to the empiricism characteristic of modern clinical science.

A decisive shift occurred with the advent of psychometric instruments capable of objectively quantifying deficits. The Halstead–Reitan Neuropsychological Battery, developed in the mid-20th century, enabled systematic assessment of attention, memory, and executive functioning (American Psychological Association, n.d.). Standardized measurements facilitated outcome tracking and inter-patient comparisons, laying groundwork for empirical trials.

The rise of behaviorism further influenced rehabilitation methodology. Behaviorist principles emphasized observable outcomes and reinforcement-based learning, which translated naturally into structured retraining techniques. Beginning in the 1970s, Yehuda Ben-Yishay and Leonard Diller pioneered comprehensive cognitive rehabilitation programs that integrated metacognitive strategies and behavioral techniques to improve real-world functioning in patients with TBI (Ben-Yishay & Diller, 1981). Their methods were novel in combining repeated practice, self-monitoring, and ecological validity.

In the 1980s and 1990s, the field underwent further maturation as RCTs became the gold standard for clinical research. Landmark systematic reviews began to consolidate evidence for specific cognitive retraining techniques (Cicerone et al., 2019). Over time, findings confirmed that targeted interventions could meaningfully enhance attention, memory, and executive functions.

Contemporary systematic reviews have continued to strengthen this evidence base. For example, Cicerone et al. (2019) synthesized controlled trials published from 2009 to 2014, demonstrating robust efficacy of individualized cognitive rehabilitation programs across various etiologies of brain injury. The field's transition from anecdotal observations to an empirical science paralleled broader trends in rehabilitation and medicine, solidifying its status as an evidence-based discipline.

1.2.3 Influence of Wars (WWI, WWII) on Brain Injury Treatment

Wartime exigencies played a critical role in advancing neurocognitive rehabilitation. World War I introduced unprecedented numbers of soldiers with traumatic brain injuries, prompting neurologists and psychiatrists to systematize management approaches. Though initial treatment focused on survival and basic recovery, clinicians quickly recognized the profound cognitive and emotional sequelae that accompanied brain injuries.

Gordon Holmes, a British neurologist, produced detailed accounts of soldiers with occipital lobe injuries, contributing to early understandings of functional localization. His work also underscored the need for structured rehabilitation to address deficits in visuospatial processing and coordination (Turner-Stokes, 2008).

World War II accelerated these developments, as even larger numbers of veterans presented with penetrating head wounds and blast injuries. Luria's groundbreaking work in the Soviet Union exemplified the wartime intersection of clinical necessity and scientific inquiry (Kotik-Friedgut & Ardila, 2019). In the United States, the Veterans Administration established specialized rehabilitation hospitals where multidisciplinary teams delivered coordinated care, integrating neurology, psychiatry, psychology, and therapy services.

This period witnessed a paradigm shift: cognitive and behavioral sequelae of brain injuries were no longer viewed as untreatable disabilities. Instead, they became targets of systematic intervention. Notably, the moral imperative to restore veterans' functional independence galvanized government investment in rehabilitation infrastructure, creating models that would later inform civilian neurorehabilitation.

Contemporary historians and rehabilitation scientists consistently recognize the wars as pivotal catalysts in professionalizing brain injury management and integrating cognitive rehabilitation into mainstream medical practice (Turner-Stokes, 2008).

1.2.4 Emergence of Interdisciplinary Rehabilitation Teams in the 20th Century

By the mid-20th century, the convergence of neuropsychological assessment, behavioral principles, and wartime innovation paved the way for interdisciplinary rehabilitation teams. This shift reflected growing consensus that optimal recovery required addressing not only isolated deficits but the whole person cognitively, emotionally, socially, and vocationally.

Initially, rehabilitation services were fragmented across specialties, leading to inconsistencies in care. However, by the 1960s and 1970s, interdisciplinary models gained traction. Rehabilitation centers began coordinating neurologists, neuropsychologists, physiatrists, occupational therapists, speech-language pathologists, and social workers in unified teams. Yehuda Ben-Yishay's holistic milieu-oriented approach exemplified this trend, emphasizing environmental structuring, metacognitive strategies, and psychosocial reintegration (Ben-Yishay & Diller, 1981).

The 1980s and 1990s saw the proliferation of cognitive retraining protocols grounded in empirical evidence and implemented by coordinated teams. By this period, rehabilitation was no longer an adjunct to acute care but an essential phase of recovery. Simultaneously, advances in neuroimaging revealed neural plasticity, reinforcing the scientific legitimacy of experiential and activity-based interventions (Cicerone et al., 2019).

In more recent decades, the rise of digital interventions such as VR platforms and telehealth has further expanded interdisciplinary practice. Modern teams increasingly leverage technology to extend services into patients' homes, broaden access, and individualize care plans (Laver et al., 2017).

As Turner-Stokes (2008) notes, multidisciplinary rehabilitation is now the cornerstone of brain injury management. Robust evidence supports that coordinated, interdisciplinary approaches yield superior functional outcomes compared to discipline-specific interventions. This shift reflects not only scientific progress but a broader recognition of the social and ethical imperative to facilitate recovery and reintegration for individuals with neurocognitive impairment.

1.3 Key Concepts and Terminology

Neurocognitive rehabilitation is distinguished from other therapeutic approaches by its structured emphasis on precise cognitive processes and their neural underpinnings. To engage meaningfully with this field, practitioners and scholars must understand the major cognitive domains commonly targeted in rehabilitation, conceptual distinctions between restorative and compensatory approaches, and the mechanisms of neuroplasticity and functional reorganization that underpin recovery.

1.3.1 Cognitive Domains Commonly Addressed: Attention, Memory, and Executive Functions

A core objective of neurocognitive rehabilitation is to improve functioning across discrete yet interdependent cognitive domains. Among these, attention, memory, and executive functions are the most commonly targeted due to their critical role in everyday activities and their frequent impairment following acquired brain injuries and neurological diseases (Cicerone et al., 2019).

Attention is a multifaceted construct encompassing the capacity to selectively process relevant stimuli, sustain focus over time, shift between tasks, and manage competing demands. Sohlberg and Mateer (1987) described a taxonomy of attentional processes that remains widely adopted in rehabilitation: focused attention, sustained attention, selective attention, alternating attention, and divided attention. Impairments in attention are prevalent in conditions such as TBI, stroke, and MS, often manifesting as distractibility, mental fatigue, or slowed processing speed. Interventions commonly involve hierarchical attention training, which progresses from simple sustained attention tasks to complex divided attention exercises (Cicerone et al., 2019).

Memory, particularly working memory and episodic memory, is another primary target of intervention. Working memory enables the maintenance and manipulation of information over brief intervals and underpins higher-order functions such as reasoning and comprehension. Episodic memory involves encoding and retrieval of personal experiences and is essential for independent living. Memory impairments may arise from damage to medial temporal structures, prefrontal cortex, or diffuse axonal injury disrupting functional networks (Ponsford et al., 2016). Rehabilitation often incorporates strategy training (e.g., mnemonic techniques, errorless learning), external memory aids, and context-specific practice to enhance retention and recall.

Executive functions refer to a constellation of high-level cognitive abilities necessary for goal-directed behavior, including planning, problem-solving, initiation, self-monitoring, and inhibition. Frontal lobe lesions, particularly in the dorsolateral prefrontal cortex and anterior cingulate, frequently disrupt executive control, producing difficulties with organization, impulse control, and adaptive decision-making.

Executive dysfunction has profound implications for community reintegration, employment, and legal competency. Intervention approaches typically integrate metacognitive strategies that promote self-awareness, error detection, and adaptive problem-solving (Cicerone et al., 2019).

While these domains are often presented discretely, in practice they are interdependent: impairments in attention or working memory frequently exacerbate executive dysfunction and reduce the efficacy of memory encoding. Accordingly, contemporary rehabilitation programs adopt a holistic approach, embedding training within ecologically valid contexts that require integrated use of these cognitive processes (O'Keefe et al., 2019).

1.3.2 Distinction between Restorative and Compensatory Approaches

Central to neurocognitive rehabilitation is the conceptual and practical distinction between restorative and compensatory approaches. Each represents a distinct philosophy of intervention with unique goals, methodologies, and evidence bases.

Restorative approaches are predicated on the principle that targeted, repetitive practice can improve impaired cognitive functions by promoting neural reorganization and recovery. This approach draws from Hebbian learning theory, which posits that "neurons that fire together wire together," and from empirical findings demonstrating that structured cognitive exercises can induce measurable neural changes (Lampit et al., 2014). Restorative interventions often involve drill-and-practice activities designed to stimulate specific cognitive domains. For example, computerized attention training may involve progressively challenging tasks that strengthen sustained and selective attention. Restorative techniques are most effective in individuals with mild-to-moderate impairments who retain some capacity for learning and generalization (Cicerone et al., 2019).

Compensatory approaches recognize that some cognitive deficits may be enduring and not fully amenable to restitution. Instead, these interventions focus on bypassing or minimizing the functional impact of deficits through environmental modifications, assistive technologies, and behavioral strategies. For instance, individuals with severe memory impairments may learn to rely on external memory aids such as notebooks, smartphone reminders, or environmental cues. Similarly, structured routines and environmental simplification can mitigate executive dysfunction by reducing cognitive load. Compensatory strategies are frequently integrated into daily routines to support independence and improve quality of life (Spreij et al., 2014).

Recent scholarship underscores the value of blending restorative and compensatory techniques. Hybrid models recognize that while restitution may be possible for some domains, compensation is often necessary to support real-world functioning. The choice between approaches depends on multiple factors, including severity and chronicity of deficits, cognitive reserve, personal goals, and contextual resources (Cicerone et al., 2019).

For forensic populations, where cognitive deficits may intersect with legal competencies and risk management, the distinction has significant implications. Restorative interventions may support improved decision-making capacity, whereas compensatory strategies can reduce risk by structuring environments to minimize impulsivity or inattention (Fishbein et al., 2009). The integration of both approaches is therefore essential for ethical and effective rehabilitation planning.

1.3.3 Concepts of Neuroplasticity and Functional Reorganization

Perhaps the most profound scientific advance underpinning modern neurocognitive rehabilitation is the understanding of neuroplasticity, the brain's inherent capacity to adapt structurally and functionally in response to experience, injury, or learning. This concept directly informs both restorative and compensatory models.

Neuroplasticity manifests across multiple scales: synaptic plasticity involves changes in the strength of connections between individual neurons; structural plasticity entails growth of dendrites, axonal sprouting, and neurogenesis; and network-level plasticity includes the reorganization of large-scale cortical systems (Grefkes & Fink, 2014). Foundational animal studies and human neuroimaging have shown that skill learning induces measurable increases in gray matter density and white matter integrity within relevant networks (May, 2011).

After brain injury, neuroplastic processes underpin spontaneous recovery and rehabilitation-induced improvements. Mechanisms include the following:

A. Resolution of diaschisis, in which areas remote from the lesion recover function as metabolic suppression diminishes.
B. Unmasking of latent synapses that previously had limited functional impact.
C. Cortical reorganization, whereby undamaged regions assume functions of injured areas.
D. Experience-dependent strengthening of residual pathways through task-specific practice.

Functional neuroimaging studies have provided compelling evidence for rehabilitation-induced plasticity. For example, fMRI and DTI have shown that cognitive training in stroke survivors produces increased activation and connectivity in frontal and parietal networks (Grefkes & Fink, 2014). These findings validate the theoretical rationale for intensive, targeted practice.

Functional reorganization refers to this adaptive reassignment of cognitive processes within the brain's networks. In some cases, homologous regions in the contralateral hemisphere assume functions of damaged areas; in others, the peri-lesional cortex takes on new roles. However, reorganization is not uniformly beneficial; maladaptive plasticity can occur when compensatory processes interfere with recovery, such as excessive reliance on inefficient pathways.

The efficacy of rehabilitation interventions is contingent upon timing, intensity, and task specificity. Emerging research suggests that early, intensive engagement promotes more robust reorganization, though excessive early stimulation can exacerbate injury in some cases (Nudo, 2013). This nuanced understanding informs best practices for designing evidence-based rehabilitation protocols.

Neuroplasticity research has also inspired the development of innovative technologies. VR, brain–computer interfaces (BCIs), and non-invasive brain stimulation techniques such as transcranial magnetic stimulation (TMS) or transcranial direct current stimulation (tDCS) aim to harness or augment neuroplastic mechanisms to enhance recovery (Laver et al., 2017).

1.4 Core Principles of Neurocognitive Rehabilitation

Neurocognitive rehabilitation is underpinned by a set of core principles that distinguish it from other clinical interventions and give it methodological coherence. These principles are grounded in decades of research demonstrating that effective rehabilitation must be individualized, goal-directed, evidence-based, multidisciplinary, and contextually relevant.

1.4.1 Individualization and Person-Centered Planning

Individualization is the foundational principle upon which all neurocognitive rehabilitation is built. No two individuals experience brain injury or cognitive impairment in precisely the same way. Even lesions of similar size and location can yield different profiles of impairment depending on factors such as premorbid intelligence, psychological resilience, cultural background, and environmental supports (Wilson et al., 2009). As a result, rehabilitation must be planned collaboratively, with interventions tailored to each person's unique needs, goals, values, and contexts.

Person-centered planning involves engaging the individual (and often their family) in setting meaningful rehabilitation goals and identifying preferred strategies. This participatory process not only respects autonomy but also enhances engagement and adherence. Evidence indicates that rehabilitation is more effective when clients perceive goals as personally relevant and attainable (Levack et al., 2015).

Contemporary frameworks such as the International Classification of Functioning, Disability and Health (ICF) (WHO, 2001) further reinforce the importance of individualization. The ICF encourages clinicians to conceptualize impairment not solely in terms of pathology but as the interaction between health conditions, personal factors, and environmental barriers. This holistic view informs individualized planning and recognizes that cognitive deficits often manifest differently in different real-world environments.

Example in practice: a young adult with post-traumatic executive dysfunction may prioritize goals related to university re-entry and social participation, while an older adult with stroke-related aphasia may focus on communicating basic needs at home. Individualization ensures that interventions are adapted to these different aims rather than imposed as generic protocols.

1.4.2 Goal-Directed and Functionally Relevant Interventions

Another central tenet of neurocognitive rehabilitation is that interventions should be explicitly goal-directed and functionally relevant. This principle derives from robust evidence showing that training activities closely linked to real-world tasks produce greater gains in participation and autonomy than decontextualized exercises (Cicerone et al., 2019).

Goal-directed rehabilitation requires therapists and clients to collaboratively define specific, measurable, achievable, relevant, and time-bound (SMART) goals. This structured process anchors intervention planning and provides criteria against which progress can be assessed. Recent systematic reviews have found that goal setting improves engagement, self-efficacy, and outcomes across a variety of neurological conditions (Levack et al., 2015).

Functionally relevant interventions ensure that training activities generalize beyond the therapy context. For example, instead of isolated paper-and-pencil memory tasks, individuals might practice using memory strategies to remember shopping lists or manage medication schedules. This principle is sometimes referred to as contextualization embedding cognitive training within activities that matter to the individual.

A growing body of literature emphasizes that functional relevance is particularly important in populations with executive dysfunction, who may struggle to generalize decontextualized skills (Skidmore et al., 2015). In forensic and legal contexts, functionally relevant interventions may include decision-making exercises simulating real-world challenges, such as managing finances or understanding legal documents.

1.4.3 Evidence-Based Practice and Outcome Measurement

Evidence-based practice is an ethical imperative in neurocognitive rehabilitation, requiring clinicians to integrate the best available research evidence with clinical expertise and client preferences. This principle reflects broader trends in healthcare emphasizing accountability, transparency, and continuous improvement.

Evidence-based cognitive rehabilitation has been systematically reviewed by Cicerone and colleagues, whose guidelines synthesize hundreds of RCTs and meta-analyses (Cicerone et al., 2019). These guidelines provide clear recommendations regarding interventions with proven efficacy in domains such as attention, memory, and executive function.

Outcome measurement is essential for demonstrating treatment effectiveness and guiding adjustments to intervention plans. Standardized neuropsychological assessments (e.g., Trail Making Test, Hopkins Verbal Learning Test) and functional measures (e.g., Goal Attainment Scaling, Functional Independence Measure) are routinely used to monitor progress.

Emerging approaches to outcome measurement include ecological momentary assessment, which captures real-time data about cognitive functioning in everyday life, and patient-reported outcome measures, which center the client's subjective experience of recovery (Cramer et al., 2019). These innovations align with the person-centered ethos of rehabilitation while enhancing methodological rigor.

Forensic psychologists and legal scholars increasingly recognize that evidence-based practice and robust outcome measurement are also essential for documenting cognitive capacities relevant to competency determinations and risk assessment (Fishbein et al., 2009).

1.4.4 Multidisciplinary and Integrative Approach

Neurocognitive rehabilitation is inherently multidisciplinary, drawing on the expertise of neuropsychologists, occupational therapists, speech-language pathologists, physiatrists, social workers, and vocational counselors. This integrative approach reflects the complexity of cognitive impairment, which rarely occurs in isolation from emotional, physical, and social challenges.

Multidisciplinary teams engage in regular collaboration to share assessments, coordinate interventions, and develop cohesive treatment plans. Evidence indicates that integrated team care yields superior outcomes compared to fragmented or discipline-specific approaches (Turner-Stokes, 2008). For example, addressing executive dysfunction often requires combining neuropsychological strategies with occupational therapy interventions targeting daily routines and environmental adaptations.

In recent years, interprofessional education and team-based competency frameworks have been developed to enhance collaboration and reduce professional silos. This emphasis on shared language and mutual respect is critical, as research has shown that interdisciplinary disagreements can impede progress if not proactively managed.

Emerging models also integrate digital interventions, such as tele-rehabilitation platforms and VR training. These technologies allow multidisciplinary teams to deliver coordinated care remotely, expanding access to services and facilitating more intensive, distributed practice (Laver et al., 2017).

1.4.5 Contextual and Ecological Validity

Finally, the principle of contextual and ecological validity underscores that rehabilitation must be grounded in the real-world environments where clients live, work, and participate. While laboratory-based exercises and standardized assessments are valuable, they often fail to capture the complexity of everyday functioning.

Ecological validity requires designing interventions that reflect real-life cognitive demands and using assessment tools that measure performance in naturalistic contexts. For example, the Multiple Errands Test, which evaluates executive function in a simulated shopping scenario, has demonstrated stronger predictive validity for community outcomes than many traditional neuropsychological tests (Morrison et al., 2013).

Contextual relevance is equally important when considering cultural and socioeconomic factors. Strategies that are feasible and acceptable in one cultural context may not translate effectively to another. Person-centered planning and culturally sensitive assessment are essential to ensure interventions are meaningful and respectful.

Recent research highlights that interventions with high ecological validity yield stronger transfer of training effects to untrained tasks and environments (Goverover et al., 2018). This principle has become increasingly salient in forensic contexts, where demonstrating functional capacity in daily life is critical for competency evaluations and risk management.

1.5 Theoretical Framework and Models

Neurocognitive rehabilitation is grounded in a robust conceptual infrastructure. Over the last century, several influential models have evolved to explain how cognitive processes are disrupted and reorganized following brain injury or neurological disease. These frameworks provide the theoretical scaffolding for assessment, intervention design, and outcome measurement. Together, they guide clinicians in tailoring evidence-based interventions while recognizing the complex interplay of neural, psychological, and social factors.

1.5.1 Cognitive Neuropsychological Model

The cognitive neuropsychological model has its roots in the modular view of the mind, which proposes that cognition consists of discrete, specialized subsystems that can be selectively impaired by focal brain lesions (Grefkes & Fink, 2014). Historically, this perspective was informed by single-case studies demonstrating that localized damage could produce circumscribed deficits, such as aphasia, alexia, or apraxia.

In neurocognitive rehabilitation, this model is applied diagnostically and therapeutically. Clinicians use standardized assessments to deconstruct complex tasks into component processes, mapping performance onto hypothesized cognitive modules. For example, reading may be parsed into visual perception, orthographic analysis, phonological processing, and semantic retrieval. Deficits in any of these processes will manifest as distinct patterns of error. Once a specific impairment is identified, targeted interventions can be designed to retrain or compensate for that module's function.

Modern neuroimaging has enriched this model, showing that even "modular" processes are mediated by large-scale neural networks (Cramer et al., 2019). For example, language production is subserved by a distributed system encompassing Broca's area, Wernicke's area, the arcuate fasciculus, and supporting regions. Therefore, rehabilitation targeting "modular" deficits must consider network-level dynamics and compensatory pathways.

An illustrative intervention grounded in this model is errorless learning, designed for patients with anterograde amnesia. By preventing errors during encoding, clinicians leverage intact implicit memory to reinforce accurate responses. RCTs and meta-analyses have demonstrated the efficacy of such approaches for improving skill acquisition in severe memory impairment (Cicerone et al., 2019).

1.5.2 *Information Processing Model*

The information processing model conceptualizes the brain as a system that encodes, stores, and retrieves information, paralleling the operations of a computer (Laver et al., 2017). This framework divides cognition into sequential stages: perception, encoding, consolidation, retrieval, and output. Each stage is vulnerable to disruption from brain injury, resulting in specific deficits.

In rehabilitation, this model guides clinicians to analyze where information breakdown occurs. For example, if a patient fails to recall verbal information, the clinician must determine whether the problem lies in initial attention and encoding, consolidation during retention, or retrieval strategies. Interventions can then be tailored accordingly enhancing encoding with focused attention strategies, supporting consolidation with spaced repetition, or facilitating retrieval with cueing techniques.

Processing speed is a critical construct within this model. Slowed information processing is common after TBI and contributes to functional impairments in multitasking and decision-making (Grefkes & Fink, 2014). Rehabilitation approaches like paced information processing training have been shown to improve efficiency and accuracy under time pressure.

Furthermore, the information processing framework has informed the development of computerized cognitive training programs. VR and adaptive software systems are designed to progressively increase cognitive load while providing real-time feedback. Recent Cochrane reviews have found that such interventions can lead to moderate improvements in targeted domains, though generalization to untrained tasks remains variable (Laver et al., 2017).

1.5.3 *Biopsychosocial Model*

While cognitive neuropsychological and information processing models focus primarily on intra-individual processes, the biopsychosocial model broadens the lens to encompass biological, psychological, and social influences on recovery (Cicerone et al., 2019). This model recognizes that even identical lesions may have different functional impacts depending on a patient's emotional resilience, coping style, cultural context, and environmental support.

Biologically, recovery is shaped by factors such as lesion size, chronicity, and neuroplastic potential. Psychologically, self-awareness, motivation, and mood disorders play pivotal roles. Socially, family support, economic stability, and community resources determine opportunities for participation and reintegration.

Empirical evidence underscores the value of this integrative perspective. For instance, depression is a robust predictor of poor functional outcomes following stroke or TBI, even after accounting for lesion characteristics (Cicerone et al., 2019). Similarly, social isolation has been linked to higher rates of long-term disability and reduced engagement with rehabilitation services.

From an intervention standpoint, the biopsychosocial model demands that clinicians address not only cognitive impairments but also the emotional and contextual factors that sustain them. This may involve integrating psychotherapy, family education, and environmental modifications alongside cognitive retraining. In forensic settings, this model is essential for understanding how cognitive deficits interact with criminogenic risks and social disadvantages (Grefkes & Fink, 2014).

1.5.4 Holistic Neuropsychological Rehabilitation Approach

Building upon the biopsychosocial framework, the holistic neuropsychological rehabilitation approach emphasizes the interdependence of cognition, emotion, behavior, and social participation. Pioneered by Ben-Yishay and Diller, this model advocates for comprehensive, milieu-oriented interventions designed to restore self-efficacy, identity, and community integration.

The holistic approach typically involves the following:

- Metacognitive training: Helping clients develop self-awareness of their deficits and learn compensatory strategies.
- Psychosocial support: Incorporating individual counseling, peer groups, and family interventions.
- Functional skill development: Practicing real-world tasks in structured environments that simulate daily challenges.

Research has shown that holistic programs lead to superior outcomes in self-awareness, emotional adjustment, and return to work compared to cognitive retraining alone (Cicerone et al., 2019). For example, goal management training, a structured protocol teaching clients to break down complex tasks, monitor performance, and adjust strategies, has been associated with improved executive functioning and functional independence (Levine et al., 2011).

Contemporary practice increasingly incorporates digital tools to support holistic interventions. VR platforms allow patients to rehearse daily tasks in immersive environments, while tele-rehabilitation systems enable multidisciplinary teams to coordinate care remotely (Laver et al., 2017).

1.6 Current Trends and Future Directions

Neurocognitive rehabilitation has evolved dramatically over the past four decades, shaped by advances in neuroscience, technology, and rehabilitation sciences. In the current era, practitioners are increasingly expected to integrate evidence-based interventions with person-centered, interdisciplinary models of care.

1.6.1 *Use of Technology*

The integration of digital tools into neurocognitive rehabilitation represents one of the most significant contemporary trends. Digital innovation spans several domains, including VR, computer-assisted cognitive training, tele-rehabilitation, and BCIs.

VR creates immersive, interactive environments in which patients can practice real-world tasks with immediate feedback. VR has been applied to attention training, executive function exercises, and memory rehabilitation. Laver et al. (2017) conducted a Cochrane review demonstrating that VR interventions produce improvements in upper limb function and activities of daily living post-stroke. While evidence for cognitive outcomes is still developing, VR's ecological validity and capacity for graded task difficulty have attracted increasing interest (Laver et al., 2017).

Computer-assisted cognitive training platforms offer structured exercises targeting domains such as attention, working memory, and problem-solving. Adaptive algorithms adjust task difficulty based on real-time performance, enabling individualized progression. Lampit et al. (2014) conducted a meta-analysis showing that computerized cognitive training can yield moderate improvements in healthy older adults' cognitive performance, suggesting potential for clinical populations as well.

The remote delivery of rehabilitation services has grown rapidly due to the COVID-19 pandemic and broader efforts to improve access for underserved populations. Evidence suggests that tele-rehabilitation can achieve comparable outcomes to in-person interventions in certain contexts (Cramer et al., 2019). For example, remote platforms have successfully delivered memory strategy training and attention retraining, expanding service reach while reducing barriers related to geography and mobility.

BCIs and non-invasive brain stimulation (e.g., tDCS) represent cutting-edge developments that may augment traditional rehabilitation. BCIs leverage neural signals to control external devices or provide neurofeedback, potentially enhancing engagement and driving neuroplastic change. While promising, these technologies remain in early stages of clinical translation.

Future research is needed to determine long-term efficacy, cost-effectiveness, and optimal implementation strategies for technology-assisted interventions. Emerging evidence supports the value of combining digital tools with traditional therapist-guided rehabilitation, rather than replacing in-person care (Cicerone et al., 2019).

1.6.2 *Information Processing Model*

The information processing model continues to serve as a foundational framework for designing and evaluating rehabilitation interventions. However, contemporary trends increasingly emphasize dynamic systems perspectives that account for the interaction of cognitive processes over time and under varying task demands.

Traditionally, this model conceptualized cognition as a linear sequence of encoding, storage, and retrieval. However, recent neuroimaging and computational modeling studies have shown that information processing is more accurately characterized as parallel, interactive, and context-sensitive (Grefkes & Fink, 2014). For example, working memory and attention are dynamically coupled: deficits in sustained attention can disrupt encoding, which in turn affects retrieval success.

Contemporary rehabilitation protocols have evolved to address this complexity. Strategy-based interventions explicitly train metacognitive processes that support monitoring, error detection, and self-regulation. This approach acknowledges that improving isolated cognitive processes in laboratory tasks may not generalize to functional activities without attention to the contextual demands of real-world performance.

Emerging evidence supports the integration of dual-task training and cognitive–motor interventions to improve multitasking abilities and processing efficiency. For instance, studies have shown that practicing cognitive tasks while walking or performing motor activities can enhance both cognitive flexibility and mobility (Cicerone et al., 2019). This reflects a shift away from static, decontextualized exercises toward integrated training paradigms grounded in the information processing model's evolving understanding.

Future directions will likely include personalized cognitive training programs that combine behavioral, neural, and computational assessments to tailor interventions to each patient's processing strengths and vulnerabilities. The increasing availability of neuroimaging and wearable technologies is expected to accelerate this trend.

1.6.3 *Biopsychosocial Model*

The biopsychosocial model remains central to neurocognitive rehabilitation, but its operationalization has become more sophisticated in recent years. Contemporary practice increasingly relies on structured frameworks, such as the ICF, to conceptualize disability as the result of interactions among health conditions, personal factors, and environmental barriers.

One emerging trend is the use of goal-oriented, contextualized interventions that explicitly address psychological and social determinants of recovery. For example, depression and anxiety are common after brain injury and can impede engagement in rehabilitation. Integrating cognitive–behavioral therapy with cognitive retraining has been shown to improve both emotional adjustment and cognitive outcomes (Cicerone et al., 2019).

Family involvement and caregiver training are increasingly recognized as essential components of effective rehabilitation. Studies demonstrate that empowering caregivers with knowledge and strategies not only reduces their burden but also enhances patients' functional gains (Cramer et al., 2019). Forensic and legal scholars have also noted that consideration of social supports and environmental constraints is critical when evaluating legal capacity and competence.

Emerging research also underscores the importance of cultural competence and health equity in applying the biopsychosocial model. Rehabilitation programs must be adaptable to diverse cultural values, beliefs about illness, and resource constraints (Turner-Stokes, 2008). This requires clinicians to engage in reflective practice, community partnerships, and participatory program design.

Looking forward, the biopsychosocial model is likely to be operationalized through integrated care pathways that embed rehabilitation within broader health and social services systems. This will necessitate policy reforms, workforce development, and funding models that prioritize long-term, holistic support over episodic, siloed care.

1.6.4 *Holistic Neuropsychological Rehabilitation Approach*

The holistic neuropsychological rehabilitation approach, pioneered by Ben-Yishay and Diller, remains highly influential. This model posits that recovery requires addressing not only specific cognitive impairments but also the psychological, social, and existential consequences of brain injury.

Contemporary programs grounded in this approach typically include the following:

1. Metacognitive strategy training to improve self-awareness and adaptive problem-solving.
2. Goal management training, which has demonstrated efficacy in improving executive functioning and functional independence (Levine et al., 2011).
3. Group-based interventions that promote social support, identity reconstruction, and community reintegration.

Recent evidence indicates that holistic programs yield superior outcomes compared to impairment-focused interventions alone. A 2019 systematic review by Cicerone et al. confirmed that comprehensive, multidisciplinary programs addressing cognition, emotion, and participation produce meaningful improvements in everyday functioning.

One notable trend is the integration of technology-enhanced interventions within holistic frameworks. For example, VR platforms can simulate real-world tasks and environments, providing ecologically valid practice opportunities. Tele-rehabilitation systems enable multidisciplinary teams to deliver holistic care remotely, which is particularly important in underserved areas (Laver et al., 2017).

Future directions are likely to include the following:

1. Hybrid delivery models combining in-person and remote services.
2. Personalized treatment pathways informed by neuroimaging and computational modeling.
3. Greater emphasis on health equity to ensure that all individuals have access to comprehensive, culturally relevant rehabilitation.

In forensic contexts, holistic rehabilitation is increasingly applied to reduce recidivism among individuals with acquired brain injury, acknowledging that cognitive impairments often intersect with social disadvantage and mental health challenges.

1.7 Conclusion

Neurocognitive rehabilitation today stands at a unique crossroads: it is a field defined by a rich history of clinical observation, an evolving scientific foundation, and an unprecedented infusion of technological innovation. As this chapter has illustrated, the journey from anecdotal case reports to systematic, evidence-based interventions is both a testament to the ingenuity of rehabilitation pioneers and an invitation for the next generation of researchers and clinicians to build on that legacy with rigor, creativity, and compassion.

While our knowledge of cognitive domains such as attention, memory, and executive function has become increasingly precise, translating that knowledge into meaningful improvements in people's lives demands more than technical skill. It requires a deep commitment to seeing the individual behind the impairment. The field's embrace of person-centered care is not a passing trend but a fundamental reorientation of priorities. Rehabilitation is no longer solely about "fixing" deficits; it is about empowering individuals to reclaim agency, purpose, and participation.

The historical evolution underscores how far the discipline has come in both theory and practice. Early clinicians working with patients like Phineas Gage or Luria's wartime caseload could not have imagined today's capacity to peer into the brain's networks using fMRI, or to deliver immersive cognitive training through VR headsets. Yet what remains constant is the imperative to honor each patient's story. Whether a person is grappling with post-stroke aphasia, TBI, or degenerative conditions like Alzheimer's disease, rehabilitation is most effective when it is contextualized in their goals, values, and lived environment.

Practitioners and policymakers alike must acknowledge that the transfer of skills from clinic to community is not automatic. As research has demonstrated, the likelihood that improvements in structured training will generalize to real-world settings increases dramatically when interventions are relevant, purposeful, and embedded in meaningful activities. A simulation of grocery shopping in a VR environment may carry far more impact than repetitive paper-and-pencil drills if the individual's goal is to regain independence in daily living. The principle that "rehabilitation must matter to the person" is perhaps the most powerful insight that has emerged from decades of empirical study.

The cognitive neuropsychological and information processing models continue to provide an essential scaffolding for understanding impairments and guiding treatment selection. Their enduring relevance is due to their clarity: breaking down complex cognitive functions into component processes allows clinicians to target deficits with precision. However, the last 20 years have made it clear that these models, while necessary, are not sufficient on their own. The brain is not merely a machine; it is a dynamic, self-organizing system embedded in social relationships and cultural narratives. This is where the biopsychosocial and holistic models bring critical depth and nuance.

Integrating these perspectives requires clinicians to be intellectually flexible and emotionally attuned. A rigid focus on deficit remediation risks overlooking the psychological distress, existential uncertainty, and social marginalization that often accompany acquired brain injury. Interdisciplinary teams, therefore, are not just a logistical convenience but a necessity. Neuropsychologists, occupational therapists, speech-language pathologists, social workers, and physicians each contribute unique expertise, but the real power of team-based care lies in the shared understanding that the person's cognitive, emotional, and social worlds cannot be disentangled.

Recent trends in the use of technology further illustrate the promise and challenges of contemporary neurocognitive rehabilitation. VR, tele-rehabilitation, and adaptive computerized training offer unprecedented opportunities to deliver personalized, scalable, and engaging interventions. These tools can democratize access to high-quality rehabilitation, particularly in remote or underserved communities where specialized services have historically been scarce. Yet technology must be implemented thoughtfully. Evidence from Cochrane reviews and systematic meta-analyses (e.g., Laver et al., 2017) consistently emphasizes that digital tools are most effective when they supplement/not supplant/skilled therapeutic relationships. The human connection between clinician and client remains irreplaceable. Even the most sophisticated software cannot replicate the empathy, encouragement, and nuanced clinical judgment that skilled practitioners bring to each session.

Personalization involves moving beyond generic treatment protocols toward interventions tailored to each individual's neurocognitive profile, preferences, and context. Advances in neuroimaging, machine learning, and computational modeling will accelerate this trend. The ability to use biomarkers to predict treatment response holds promise for more precise targeting of interventions and optimization of resource allocation.

Integration refers to the blending of cognitive, emotional, and social interventions into coherent treatment pathways. The siloed model, where cognitive training is conducted in isolation from psychosocial support and community reintegration, is increasingly untenable. Holistic neuropsychological rehabilitation has demonstrated that complex outcomes like return to work or improved quality of life are only achievable when interventions address the full spectrum of challenges clients face.

Scalability is a pressing concern in a world where the prevalence of stroke, TBI, and dementia is rising. Health systems already face shortages of trained neurorehabilitation professionals. Tele-rehabilitation and automated platforms can help close this gap, but implementation science will be needed to translate pilot successes into sustained, system-wide change.

Equity demands that researchers and clinicians confront the disparities in access, cultural relevance, and outcomes that persist in neurocognitive rehabilitation. As Turner-Stokes (2008) and others have argued, commitment to person-centered care requires an explicit focus on inclusivity. Rehabilitation services must be designed with, not merely for, the communities they intend to serve.

Forensic psychologists and legal professionals will find that neurocognitive rehabilitation intersects with their work in increasingly important ways. Cognitive impairments often bear impact directly on legal questions of competence, consent, criminal responsibility, and sentencing mitigation. The theoretical frameworks outlined in this chapter, particularly the biopsychosocial and holistic models, can inform forensic evaluations by situating impairments within a broader matrix of environmental and psychological factors. As awareness of acquired brain injury in criminal justice populations grows, rehabilitation-informed perspectives may play a larger role in shaping policy and practice.

It is also worth emphasizing that neurocognitive rehabilitation is not a static discipline. The field is characterized by a remarkable openness to innovation, continual self-scrutiny, and a commitment to grounding practice in the best available evidence. For practitioners, this means that learning is lifelong. Staying current requires not only reading journal articles but also engaging critically with emerging technologies, ethical dilemmas, and evolving models of care. For researchers, there is a clear mandate to produce studies that are rigorous, transparent, and attentive to real-world impact. RCTs remain the gold standard for efficacy, but mixed-methods research and implementation studies will be vital for understanding how interventions work in everyday practice.

As the field moves forward, several challenges will require collective attention. These include reconciling the promise of neuroplasticity with the realities of resource constraints; ensuring that technological solutions do not exacerbate existing inequities; and fostering interdisciplinary training to break down professional silos. Yet within each of these challenges lies an opportunity to reaffirm the core values that have always defined neurocognitive rehabilitation: respect for personhood, commitment to scientific integrity, and a belief in the capacity for growth and adaptation.

In closing, this chapter has sought to illuminate the foundations, evolution, and current landscape of neurocognitive rehabilitation. It has argued that while theoretical models provide essential conceptual clarity, effective practice depends on the skillful integration of evidence-based techniques, interdisciplinary collaboration, and person-centered values. Whether you are a neuroscientist designing a clinical trial, a clinician delivering therapy, or a legal scholar grappling with the implications of cognitive impairment for justice, you are part of a broader community dedicated to restoring dignity, autonomy, and hope to individuals navigating the aftermath of brain injury.

The path forward will be shaped by continued innovation, rigorous inquiry, and the courage to embrace complexity. As rehabilitation professionals, we are entrusted not only with improving cognitive function but with supporting individuals to rebuild their identities, reengage with their communities, and rediscover meaning in their lives. It is a responsibility that demands both scientific excellence and profound humanity. And it is precisely this dual commitment to evidence and to empathy that will sustain the discipline's relevance and vitality in the decades to come.

References

Amato, M. P., Zipoli, V., & Portaccio, E. (2006). Multiple sclerosis-related cognitive changes: A review of cross-sectional and longitudinal studies. *Journal of the Neurological Sciences*, 245(1–2), 41–46. https://doi.org/10.1016/j.jns.2005.08.019

American Psychological Association. (n.d.). Halstead–Reitan neuropsychological battery. In *APA Dictionary of Psychology*. Retrieved October 23, 2025, from https://dictionary.apa.org/halstead-reitan-neuropsychological-battery

Babikian, T., & Asarnow, R. (2009). Neurocognitive outcomes and recovery after pediatric TBI: Meta-analytic review of the literature. *Neuropsychology*, 23(3), 283–296. https://doi.org/10.1037/a0015268

Ben-Yishay, Y., & Diller, L. (1981). Rehabilitation of cognitive and perceptual defects in people with traumatic brain damage. *International Journal of Rehabilitation Research*, 4(2), 208–210. https://doi.org/10.1097/00004356-198106000-00012

Cicerone, K. D., Goldin, Y., Ganci, K., Rosenbaum, A., Wethe, J. V., Langenbahn, D. M., Malec, J. F., Bergquist, T. F., Kingsley, K., & Nagele, D. (2019). Evidence-based cognitive rehabilitation: Systematic review of the literature from 2009 through 2014. *Archives of Physical Medicine and Rehabilitation*, 100(8), 1515–1533. https://doi.org/10.1016/j.apmr.2019.02.011

Cramer, S. C., Sur, M., Dobkin, B. H., O'Brien, C., Sanger, T. D., Trojanowski, J. Q., ..., & Vinogradov, S. (2019). Harnessing neuroplasticity for clinical applications. *Brain*, 142(6), 1682–1695. https://doi.org/10.1093/brain/awz093

Crosson, B., Hampstead, B. M., Krishnamurthy, L. C., Krishnamurthy, V., McGregor, K. M., Nocera, J. R., Roberts, S., Rodriguez, A. D., & Tran, S. M. (2017). Advances in neurocognitive rehabilitation research from 1992 to 2017: The ascension of neural plasticity. *Neuropsychology*, 31(8), 900–920. https://doi.org/10.1037/neu0000396

Feigin, V. L., Norrving, B., & Mensah, G. A. (2017). Global burden of stroke. *Circulation Research*, 120(3), 439–448. https://doi.org/10.1161/CIRCRESAHA.116.308413

Fishbein, D., Miller, L., Hyde, C., Jaffe, J., Eslinger, P., & Dougherty, D. (2009). Neurocognitive skills moderate urban male adolescents' responses to preventive intervention materials. *Drug and Alcohol Dependence*, 103(Suppl 1), S73–S84. https://doi.org/10.1016/j.drugalcdep.2009.02.001

Goverover, Y., Chiaravalloti, N. D., O'Brien, A. R., & DeLuca, J. (2018). Evidenced-based cognitive rehabilitation for persons with multiple sclerosis: an updated review of the literature from 2007 to 2016. *Archives of Physical Medicine and Rehabilitation*, 99(2), 390–407. https://doi.org/10.1016/j.apmr.2017.07.021

Grefkes, C., & Fink, G. R. (2014). Connectivity-based approaches in stroke and recovery of function. *The Lancet Neurology*, 13(2), 206–216. https://doi.org/10.1016/S1474-4422(13)70264-3

Kotik-Friedgut, B., & Ardila, A. (2019). A.R. Luria's cultural neuropsychology in the 21st century. *Culture & Psychology*, 26(3), 382–394. https://doi.org/10.1177/1354067X19861053

Lampit, A., Hallock, H., & Valenzuela, M. (2014). Computerized cognitive training in cognitively healthy older adults: A systematic review and meta-analysis of effect modifiers. *PLOS Medicine*, 11(11), e1001756. https://doi.org/10.1371/journal.pmed.1001756

Laver, K., Lange, B., George, S., Deutsch, J. E., & Saposnik, G. (2017). Virtual reality for stroke rehabilitation. *Cochrane Database of Systematic Reviews*, 11, CD008349. https://doi.org/10.1002/14651858.CD008349.pub4

Levack, W. M., Weatherall, M., Hay-Smith, E. J., Dean, S. G., McPherson, K., & Siegert, R. J. (2015). Goal setting and strategies to enhance goal pursuit for adults with acquired disability participating in rehabilitation. *The Cochrane Database of Systematic Reviews*, 2015(7), CD009727. https://doi.org/10.1002/14651858.CD009727.pub2

Levine, B., Stuss, D. T., Winocur, G., Binns, M. A., Fahy, L., Mandic, M., Bridges, K., & Robertson, I. H. (2011). Cognitive rehabilitation in the elderly: Effects on strategic behavior in relation to goal management. *Journal of the International Neuropsychological Society: JINS*, 13(1), 143–152. https://doi.org/10.1017/S1355617707070178

Loetscher, T., & Lincoln, N. B. (2013). Cognitive rehabilitation for attention deficits following stroke. *The Cochrane Database of Systematic Reviews*, 2013(5), CD002842. https://doi.org/10.1002/14651858.CD002842.pub2

May, A. (2011). Experience-dependent structural plasticity in the adult human brain. *Trends in Cognitive Sciences*, 15(10), 475–482. https://doi.org/10.1016/j.tics.2011.08.002

Morrison, M. T., Giles, G. M., Ryan, J. D., Baum, C. M., Dromerick, A. W., Polatajko, H. J., & Edwards, D. F. (2013). Multiple Errands Test-Revised (MET-R): A performance-based measure of executive function in people with mild cerebrovascular accident. *The American Journal of Occupational Therapy: Official Publication of the American Occupational Therapy Association*, 67(4), 460–468. https://doi.org/10.5014/ajot.2013.007880

Nudo, R. J. (2013). Recovery after brain injury: Mechanisms and principles. *Frontiers in Human Neuroscience*, 7, 887. https://doi.org/10.3389/fnhum.2013.00887

O'Keefe, S., Stanley, M., Adam, K., & Lannin, N. A. (2019). A systematic scoping review of work interventions for hospitalised adults with an acquired neurological impairment. *Journal of Occupational Rehabilitation*, 29(3), 569–584. https://doi.org/10.1007/s10926-018-9820-8

Ponsford, J. (2016). The practice of clinical neuropsychology in

Australia. *The Clinical Neuropsychologist*. DOI: 10.1080/13854046.2016.1195015

Ratiu, P., Talos, I. F., Haker, S., Lieberman, D., & Everett, P. (2004). The tale of Phineas Gage, digitally remastered. *The New England Journal of Medicine*, 351(23), e21. https://doi.org/10.1056/NEJMp048190

Skidmore, E. R., Dawson, D. R., Butters, M. A., Grattan, E. S., Juengst, S. B., Whyte, E. M., Begley, A., Holm, M. B., & Becker, J. T. (2015). Strategy training shows promise for addressing disability in the first 6 months after stroke. *Neurorehabilitation and Neural Repair*, 29(7), 668–676. https://doi.org/10.1177/1545968314562113

Spreij, L. A., Visser-Meily, J. M. A., van Heugten, C. M., & Nijboer, T. C. W. (2014). Novel insights into the rehabilitation of memory post acquired brain injury: A systematic review. *Frontiers in Human Neuroscience*, 8, 993. https://doi.org/10.3389/fnhum.2014.00993

Steinerman, J. R. (2010). Minding the aging brain: technology-enabled cognitive training for healthy elders. *Current Neurology and Neuroscience Reports*, 10(5), 374–380. https://doi.org/10.1007/s11910-010-0124-4

Taphoorn, M. J., & Klein, M. (2004). Cognitive deficits in adult patients with brain tumours. *The Lancet. Neurology*, 3(3), 159–168. https://doi.org/10.1016/S1474-4422(04)00680-5

Turner-Stokes, L. (2008). Evidence for the effectiveness of multi-disciplinary rehabilitation following acquired brain injury: A synthesis of two systematic approaches. *Journal of Rehabilitation Medicine*, 40(9), 691–701. https://doi.org/10.2340/16501977-0273

World Health Organization. (2001). International Classification of Functioning, Disability and Health (ICF). www.who.int/standards/classifications/international-classification-of-functioning-disability-and-health

Wykes, T., Huddy, V., Cellard, C., McGurk, S. R., & Czobor, P. (2011). A meta-analysis of cognitive remediation for schizophrenia: Methodology and effect sizes. *The American Journal of Psychiatry*, 168(5), 472–485. https://doi.org/10.1176/appi.ajp.2010.10060855

2
GERIATRIC REHABILITATION
Aging-Related Cognitive Impairments

Shampa Ghosh, Krishna Kumar Singh, and Jitendra Kumar Sinha

2.1 Introduction

The global demographic landscape is undergoing a significant transformation characterized by an unprecedented increase in the aging population. According to the World Health Organization, by 2030, one in six people worldwide will be aged 60 years or older, totaling approximately 1.4 billion individuals. This number is projected to rise to 2.1 billion by 2050, with the population aged 80 years or older expected to triple to 426 million during the same period. Notably, this demographic shift is not confined to high-income countries; low- and middle-income countries are experiencing the most rapid increases in their older populations. This aging trend is accompanied by a surge in age-related health conditions, particularly cognitive impairments such as mild cognitive impairment (MCI) and dementia. The increasing prevalence of these conditions underscores the urgent need for comprehensive geriatric rehabilitation programs that address the unique cognitive challenges faced by older adults. Such programs are essential for maintaining functional independence, enhancing quality of life (QoL), and reducing the burden on healthcare systems.

Cognitive health, encompassing the abilities to think, learn, and remember, is a critical component of overall well-being in older adults. It enables individuals to perform daily activities, make informed decisions, and engage socially. However, aging is associated with various changes in cognitive function, including declines in processing speed, memory, and executive function. While some cognitive decline is a normal part of aging, distinguishing between normal aging and pathological conditions like dementia is crucial for timely intervention. Maintaining cognitive health is not only vital for individual autonomy but also has broader implications for public health. Cognitive impairments can lead to increased dependency, higher healthcare costs, and reduced QoL. Therefore, promoting cognitive health through preventive measures, early detection, and targeted rehabilitation is a public health priority.

The rise in cognitive impairments among the aging population has significant socioeconomic implications. Individuals with cognitive decline often require extensive healthcare services, long-term care, and support from caregivers, leading to increased healthcare expenditures. Moreover, the loss of productivity and the emotional toll on families and caregivers contribute to the broader societal impact. Socioeconomic status (SES) plays a

DOI: 10.4324/9781003646662-3

pivotal role in cognitive health outcomes. Studies have shown that lower SES is associated with a higher risk of cognitive impairment and dementia. Factors such as limited access to education, healthcare disparities, and lifestyle differences contribute to this increased risk. Addressing these disparities through policy interventions and community-based programs is essential for mitigating the impact of cognitive impairments on vulnerable populations. In conclusion, the aging global population presents both challenges and opportunities. By prioritizing cognitive health and implementing effective geriatric rehabilitation strategies, we can enhance the well-being of older adults and alleviate the socioeconomic burden associated with cognitive impairments. This chapter delves into the neurocognitive changes in aging, assessment methods, rehabilitation principles, and evidence-based interventions aimed at promoting cognitive health in older adults.

The world demographic map is undergoing a historic change characterized by an unprecedented increase in the aged population. The World Health Organisation estimated that by the year 2030, over 1.4 billion people would be over 60 years of age (World Health, 2024). This represents one out of every six people globally would be aged. This number is estimated to increase to 2.1 billion in 2050, with the number of people aged 80 years and older set to triple to 426 million over the same period (World Health, 2024). Such demographic transformation is not only seen within high-income countries, but rather low- and middle-income countries are experiencing the most significant increase in their population aged (Chen et al., 2024). Importantly, the prevalence of age-related medical conditions is increasing with the similar trend of population aging. This especially included the cognitive impairments such as moderate cognitive impairment (MCI) and dementia. The increasing prevalence of such health issues clearly highlights the compelling need for holistic geriatric rehabilitation programs (National Institute on Aging, 2023). This would address the particular cognitive difficulties and challenges arising from older age. Moreover, maintenance of functional independence, improvement in QoL, and relief of pressures on healthcare systems are key elements. These must be addressed for the success of the related healthcare programs (Ghosh, Sinha, et al., 2023; Kruk et al., 2018).

Cognitive function includes our capacity to think, learn, and recall. It is an important component of overall well-being in older adults as well. It allows individuals to carry out their daily tasks, make informed choices, and participate in social activities (Stern, 2009). However, certain changes in cognitive performance are linked with aging. These changes include decreases in processing speed, memory capacity, and executive function. While there is some inevitable cognitive loss with aging, due to loss of neurons, it is important to be able to differentiate between normal aging and pathological states such as dementia to enable effective intervention (Livingston et al., 2020). Maintenance of cognitive well-being is essential for personal autonomy and has wider implications for public health. Cognitive impairment can lead to increased dependence, increased healthcare costs, and reduced QoL (Connolly et al., 2024). Encouraging cognitive wellness through targeted preventative measures, early diagnosis, and rehabilitation is therefore a priority for public health.

The elderly have witnessed a rise in cognitive impairment, thus having major implications for the socioeconomic framework. Those who experience cognitive decline at times require lengthy healthcare services, lengthy care, and carer assistance, which calls for high healthcare costs (Muhammad et al., 2023). Additionally, the decrease in productivity and the psychological burden on family and carers are factors that increase the overall cost to society (Muhammad et al., 2023; Sinha et al., 2023). SES plays an important role in determining cognitive health outcomes. Multiple studies suggest that lower SES is associated with an increased risk of cognitive impairment and dementia. Multiple factors underlie this higher risk, such as healthcare disparities, lifestyle disparities, and limited access to educational training (Wang et al., 2024). The need to implement legislative policies and community programs to address these inequalities is necessary in order to reduce the impact of cognitive impairments on disproportionately disadvantaged populations. The rise in the average age of the population in the world offers some challenges as well as opportunities. Prioritizing cognitive health and adopting suitable geriatric rehabilitation measures can improve the QoL of the elderly and reduce the socioeconomic burden associated with cognitive deficits. This chapter discusses the neurocognitive changes during aging, such as the assessment strategy, rehabilitation principles, and evidence-based treatments directed toward improving cognitive health in the elderly.

2.2 Neurocognitive Changes in Older Adults

Aging is linked to a series of neurocognitive changes, ranging from normal expected changes to abnormal states signifying disease processes. Recognizing these changes is vital in order to develop effective geriatric rehabilitation practices (Hong et al., 2024). In normal aging, patients are likely to develop gradual declines in some of the cognitive capabilities, including processing speed, working memory, and executive function. These changes are mostly subtle and do not adversely affect daily functioning (Fu et al., 2025). For example, older people may learn new information more slowly or sometimes forget appointments or names but remember them later on. These changes are deemed to be part of normal aging and do not suggest neurodegenerative disease, as such ("2024 Alzheimer's disease facts and figures," 2024). On the other hand, pathological aging is where cognitive impairment is more marked and disrupts normal functioning in day-to-day life. This is usually an indication of neurodegenerative diseases such as Alzheimer's disease (AD) or other types of dementia. These conditions are mostly marked by extensive memory loss, language impairment, disorientation, and also compromised judgment, in many cases ("2024 Alzheimer's disease facts and figures," 2024). In contrast to normal aging, these changes are progressive, incapacitating, and hence necessitate clinical care and support. It becomes important to distinguish between the normal and pathological cognitive change for early diagnosis and suitable interventions. Early identification of pathological change can allow for timely intervention, which may slow disease progression and enhance QoL (Mishra et al., 2025; Singh et al., 2025). In addition, knowledge of the cognitive aging continuum facilitates establishing realistic expectations and creating individualized rehabilitation strategies for older individuals.

2.2.1 Normal Aging Versus Pathological Aging: Key Distinctions

Aging is a normal process that affects all organ systems, including the brain. As we get older, our cognitive performance changes in some ways that are normal. However, it's vital to know the distinction between these normal changes and pathological anomalies so that neurodegenerative disorders can be detected and treated early. This section discusses about the distinction between normal and abnormal aging (Fu et al., 2025; Ghosh, Vashisth, et al., 2023). A lot of attention is paid to how the brain works, how its structure evolves, and how biochemistry works. As people get older, it's normal for some parts of their thinking to slowly and moderately develop worse. As people get older, their vocabulary and general knowledge frequently stay the same or even get better. However, their processing speed, working memory, and executive functions may get a little worse compared to younger people. These changes don't have a large impact on everyday living and are considered as a normal aspect of becoming older. Older adults typically mitigate these declines by utilizing good strategies and drawing on their years of knowledge. Pathological aging is when cognitive difficulties are so bad that they make it hard to accomplish things like get dressed or eat. AD and mild cognitive impairment (MCI) are both types of dementia that cause substantial memory loss, difficulties with language, and bad thinking (Jin & Cai, 2022). These difficulties develop worse over time and are worse than what we expect from normal aging. They often make people less independent. Changes in the brain's structure happen with both normal and pathological aging, although the changes and the parts of the brain that are affected are different (Jin & Cai, 2022). As we age, the prefrontal cortex and hippocampus lose some cells, which impacts how we think in minor ways (Peters, 2006). Pathological aging, on the other hand, occurs when the hippocampus shrinks a lot and loses a lot of neurons and synapses. And this is linked with the observed cognitive problems.

At the molecular level, normal aging is marked by minor changes, like reduced levels of neurotransmitters, mild oxidative stress, and moderate DNA damage (Sinha et al., 2014). These changes are usually easy for the brain to fix. Pathological aging happens when abnormal proteins, including amyloid-beta plaques and tau tangles, build up and block neurons from working and kill them. Oxidative stress, problems with mitochondria, and protein homeostasis are also getting worse (Sinha, Jorwal, et al., 2024; Wang et al., 2020). All of these things make neurodegenerative illnesses worse. People who are getting older generally can still live on their own and conduct their daily duties without too much difficulties. They could forget things from time to time, but they usually know about these changes and can adapt. People with pathological aging usually have a harder time with everyday tasks, may not notice that their mental capacities are getting poorer, and need more and more aid (Peters, 2006).

Studies on animals have taught us a lot about how normal and abnormal aging are different. Animals' brains don't get poorer as they age normally, but their cognitive capacities do (Head, 2010). For example, transgenic mice that make human amyloid precursor protein make plaques and tangles that are like those found in AD (Spires & Hyman, 2005). These changes also lead to cognitive impairments that are comparable to those seen in people with AD (Spires & Hyman, 2005). It's crucial to recognize the distinction between normal and pathological aging so that neurodegenerative illnesses can be detected and treated early. When cognitive performance gets worse, the brain shrinks, and molecules become irregular; this is called pathological aging. When we become older, our brain's structure changes in a way that doesn't disrupt our daily life, and cognitive function alters in a relatively moderate way (Murman, 2015; Peters, 2006). Recognizing these features can help healthcare professionals better care for elderly patients and deal with the cognitive problems that come with aging.

2.2.2 Brain Structural and Functional Changes Linked to Aging

As individuals age, their brains undergo several alterations that impact both their structure and functionality. These alterations collectively contribute to cognitive deterioration in elderly individuals. These changes include a smaller brain volume, especially in the prefrontal cortex and hippocampus, less white matter integrity, and problems with neurotransmitter systems and neuroplasticity. To design treatments that will help keep older people's brains healthy, we need to know how these changes happen. One big change in the structure of the aging brain is that the brain gets smaller volumetrically. This atrophy is especially important in the prefrontal cortex and hippocampus, which are important for memory and executive processes, respectively. It has been proven in studies that the weight and size of the brain reduce with age at the rate of approximately 5% every ten years after the age of 40, and the decline increases after the age of 70 (Lee & Kim, 2022). Apart from gray matter volume loss, aging also impacts the integrity of white matter. White matter is made up of myelinated nerve fibers that allow communication between brain regions. With aging, white matter integrity decreases, especially in the frontal lobes, and this has been linked with lower cognitive performance. White matter degradation could be due to processes including demyelination of axons, resulting in less efficiency of information transmission throughout the brain (Cox et al., 2016). On a functional level, the brain with age has declines in some neurotransmitter systems.

Dopaminergic, serotonergic, and cholinergic systems are less efficient, impairing mood, attention, and memory. For example, research employing positron emission tomography (PET) has illustrated tremendous age-related decreases in dopamine synthesis and receptor availability in several brain areas, such as the striatum and frontal cortex (Ota et al., 2006). Likewise, decreases in serotonin receptors and transporters have been found in the aging brain, leading to mood disorders and cognitive impairment (Meltzer, 1998). Neuroplasticity, which is the brain's capacity to change and reorganize itself, also decreases with age. Such decline compromises the brain's ability to acquire new information as well as recover from damage (Gazerani, 2025). Reductions in neurogenesis, synaptic plasticity, and dendritic spine density are some of the factors that contribute to lower neuroplasticity in aged individuals (Mansoor et al., 2012). Nevertheless, activities that induce neuroplasticity, including physical exercise and cognitive training (CT), can take effect in reversing the effects. The functional and structural alterations in the aging brain have important implications for cognitive functioning (Gazerani, 2025). Decreases in brain volume, especially in the prefrontal cortex and hippocampus, are linked with decreases in executive functions and memory. Reduced integrity of white matter results in reduced processing speed and compromised communication between brain areas. In addition, decreased neurotransmitter systems and neuroplasticity contribute to learning, attentional, and mood regulation difficulties. Knowledge about these changes is necessary for the development of interventions to maintain cognitive health in older people. Interventions like encouraging physical exercise, mental activity, and socialization can delay brain structure and function decline, reducing the risk of cognitive impairment and neurodegenerative disorders.

2.2.3 Common Clinical Manifestations: Mild Cognitive Impairment and Dementia

Aging is associated with a continuum of cognitive changes that vary from subtle deterioration in normal aging to frank impairments associated with neurodegenerative diseases. Two of the most significant clinical expressions on this continuum are mild cognitive impairment (MCI) and dementia. It is very important for early diagnosis and intervention that their characteristics, differentiation, and progression are resolved. MCI is a transitional phase between normal expected loss of cognitive function with aging and the severe loss of dementia (Gazbare & Palekar, 2024). Those with MCI have significant cognitive changes more pronounced than the expected loss due to their education and age but do not significantly affect activities of daily life or independent function. The symptoms that can be observed include memory impairment, difficulty with language, and attention problems as well (Gazbare & Palekar, 2024). There are two forms of MCI. First one is amnestic MCI, where memory is mostly affected. And other one is non-amnestic MCI, where other cognitive abilities like language, attention, or visual-spatial abilities are affected (Roberts & Knopman, 2013). So, amnestic MCI is usually viewed as a prodromal stage for AD, whereas non-amnestic MCI can progress to other forms of dementia. It is different in different individuals whether they develop MCI to dementia. Research indicates that about 10%–15% of MCI individuals develop dementia annually (Gazbare & Palekar, 2024). Yet, some of them stay stable or even become normal in their cognition, underlining the need for routine checkup and evaluation. Dementia is a broad term referring to a decrease in mental function that is extreme enough to affect daily activities and life. It involves a variety of symptoms, such as memory loss, diminished reasoning, difficulty with words, and behavior change. In contrast to MCI, dementia severely compromises a person's capability to live independently.

Of the variety of neurodegenerative disorders, AD, vascular dementia (VaD), Lewy body dementia (LBD), and frontotemporal dementia (FTD) are the most common entities, with different pathophysiologic mechanisms and clinical features. AD is defined by amyloid-beta plaque and tau neurofibrillary tangle accumulation, causing synaptic failure and neuronal loss (Ghosh et al., 2020). It appears clinically as progressive memory loss, language impairment, and executive dysfunction. Hippocampal atrophy and cortical thinning are common features on neuroimaging, correlating with the degree of cognitive impairment. VaD is caused by cerebrovascular disease, e.g., infarcts or chronic ischemia, causing cognitive impairment. Patients usually have impaired judgment, slowed processing speed, and executive dysfunction. Neuroimaging can reveal white matter lesions and cortical or subcortical infarcts, lending credence to the diagnosis (Kalaria, 2016). LBD is characterized by the presence of alpha-synuclein-filled Lewy bodies in cortical neurons. Clinically, it presents with fluctuating cognition, visual hallucinations, and parkinsonian motor features. In addition, patients can have rapid eye movement (REM) sleep behavior disorder and autonomic dysfunction. Neuroimaging can reveal occipital hypometabolism on PET scans (Prasad et al., 2023). FTD is a collection of disorders that are caused by progressive degeneration of the frontal and temporal lobes. It has two main presentations: behavioral variant FTD, with overt personality and social behavior changes, and primary progressive aphasia, which impairs language abilities. Neuroimaging generally shows focal atrophy in the involved lobes (Finger, 2016). Differential diagnosis between these dementias is important to guide management and predict outcome.

Though overlapping symptoms present diagnostic challenges, advances in biomarker studies and neuroimaging are improving our capacity to separate these conditions and shape therapy appropriately. Early identification and differentiation of these conditions are crucial to management and intervention planning. Proper diagnosis facilitates appropriate treatment planning, patient and caregiver support, and possible involvement in clinical trials. It also allows them to make decisions regarding their care and their future. Overall, different patients on the spectrum of cognitive impairment associated with aging need to be taken care for various features (Table 2.1). Becoming aware of the signs and learning

Table 2.1 Differences between Normal and Pathological Aging

Feature	*Normal Aging*	*Pathological Aging*
Memory	Occasional forgetfulness (Hedden & Gabrieli, 2004)	Significant memory loss affecting daily life (McKhann et al., 2011)
Language	Mild word-finding difficulties (Burke & Shafto, 2004)	Frequent word-finding pauses, reduced vocabulary (Taler & Phillips, 2008)
Attention	Slight decrease in concentration (Bouwens et al., 2009)	Difficulty focusing, easily distracted (Perry, 1999)
Executive Function	Minor slowing in task completion (West, 1996)	Impaired judgment, problem-solving difficulties (Mangels et al., 1996)
Daily Functioning	Independent (Bouwens et al., 2009)	Requires assistance with daily activities (Bouwens et al., 2009)
Structural Brain Changes	Mild brain atrophy (Raz et al., 2005)	Significant atrophy, especially in hippocampus (Braak & Braak, 1991; Raz et al., 2005)
Neurotransmitter Levels	Slight decline (Bäckman et al., 2006)	Marked reduction, especially in acetylcholine (Davies, 1976)
Synaptic Density	Slight decrease (Morrison & Baxter, 2012)	Significant loss, especially in hippocampus (Scheff et al., 1990)
Animal Models	Age-related cognitive decline without pathology (Rapp & Gallagher, 1996)	Presence of amyloid plaques, neurofibrillary tangles (Hsiao et al., 1996)

the differences between these conditions are vital steps toward effective care and optimizing outcomes for those who are impacted.

2.3 Evaluation of Cognitive Impairments

Cognitive impairment in older adults can best be accurately assessed by using soundly validated methods. Clinicians use a staged process, starting with quick screening tools and progressing to full neuropsychological assessment when warranted. However, the process is confounded by sensory impairments, comorbid illnesses, and educational and cultural variables that can affect performance on tests. In this chapter we will try to summarize typically applied screening instruments, delineate the elements of a comprehensive neuropsychological assessment, and discuss significant issues to valid assessment.

2.3.1 Screening and Diagnostic Tools

Two tools reign supreme in clinical practice for quick detection of cognitive impairment: the Mini-Mental State Examination (MMSE) and the Montreal Cognitive Assessment (MoCA) (Oleksy et al., 2024). The MMSE is a 30-point survey evaluating orientation, registration, attention and calculation, recall, language, and simple praxis. Developed by Folstein and colleagues in 1975, it takes 5–10 minutes to complete and has become widespread because of the ease of administration and established cut-offs for severity of dementia (24–30 = normal; 19–23 = mild; 10–18 = moderate; ≤ 9 = severe impairment). Although robust, the MMSE has restricted sensitivity for MCI and is highly dependent on age and education, creating false positives in lower-educated or older groups (Oleksy et al., 2024). To fill these shortcomings, Nasreddine et al. (Nasreddine et al., 2005) created the MoCA, a 30-point screening instrument specifically aimed at executive functions, visuospatial skills, and delayed recall, areas frequently preserved on the MMSE in early impairment. During the initial validation, the MoCA identified 90% of MCI instances (cut-off ≥ 26) compared with 18% with the MMSE, with high specificity (87%–100%) retained in healthy controls. Follow-up work has suggested lowering the optimal MoCA cut-off to achieve balance between sensitivity and specificity in populations with heterogeneous educational levels (e.g., ≤ 23) (Carson et al., 2017). Other short tests are the Saint Louis University Mental Status (SLUMS) test and the Self-Administered Gerocognitive Examination (SAGE), which include informant reports and activities of daily living (ADLs) questions. Yet the MMSE and MoCA are the most extensively studied and best translated instruments and thus represent a crucial part of first-line cognitive screening.

2.3.2 Comprehensive Neuropsychological Evaluation

When screening suggests cognitive impairment or when the clinical picture is ambiguous, a full neuropsychological battery is indicated. Such an evaluation typically spans 2–4 hours and assesses multiple cognitive domains that are given below:

a. Attention and Processing Speed: Trail Making Test Parts A and B, Symbol Digit Modalities Test
b. Verbal and Visual Memory: California Verbal Learning Test, Rey–Osterrieth Complex Figure Test
c. Language: Boston Naming Test, Controlled Oral Word Association Test
d. Executive Functions: Stroop Color–Word Test, Wisconsin Card Sorting Test
e. Visuospatial Skills: Block Design (from the WAIS), Judgment of Line Orientation

The referral question, the patient's health, and their language background all help decide which tests to use. Using normative data that has been adjusted for age, education, and, when possible, sex and race/ethnicity (Scheffels et al., 2023), raw scores are changed into standardized scores. In addition to formal tests, a complete clinical interview that includes a review of medical history, medications, mood (such as the Geriatric Depression Scale), and functional status measures (such as the Lawton Instrumental ADLs (IADLs)) is also done. Neuroimaging and blood testing are combined as needed to rule out reversible reasons of cognitive loss, such as a lack of vitamin B_{12} or problems with the thyroid.

2.3.3 Challenges in Assessment

Measuring cognitive functioning among older adults is beset with challenges that can confuse accurate neurocognitive status unless adequately solved. To begin with, impairments in the senses of hearing and vision are very common among the elderly and can take the form of cognitive impairment. For instance, an older person with untreated hearing loss can be found to be inattentive during an auditory memory test, while visual impairment can impair performance on visuoconstruction or clock-drawing tests. Unless provided with accommodations like large test stimuli, louder audio presentation, or attested sensory-friendly versions of tests (e.g., the hearing-impaired MoCA (Al-Yawer et al., 2019)), clinicians stand to misclassify persons as cognitively impaired when the main impediment is a sensory one, not cognitive. Comorbid medical and psychiatric illness further complicate the terrain of cognitive assessment. Late-life depression typically is accompanied by pseudodementia, with features of slowed processing speed, decreased motivation, and impaired retrieval of memory, which can even simulate mild cognitive impairment or early dementia (Butters et al., 2004). Similarly, chronic diseases like diabetes, cardiovascular disease, and chronic kidney disease can cause "subcortical" cognitive profiles of executive dysfunction and slowed information processing. In such settings, inclusion of collateral history, screening for mood (e.g., Geriatric Depression Scale), and medical workup (e.g., neuroimaging, laboratory panels) is necessary to separate primary neurodegenerative processes from secondary cognitive impacts of comorbid conditions.

Cultural, language, and educational variables add further complexity. Standard cognitive tests are normed on particular populations, and the application of these norms to individuals who come from other cultural or educational backgrounds may provide misleading information. For example, chronologically older persons with less formal education tend to perform lower on the MMSE or on verbal fluency tests, not as a function of actual cognitive decrement but because of restricted test exposure or literacy requirements (Crum et al., 1993; Espino et al., 2004). Likewise, tests mandating quick graphomotor reactions or culture-bound vocabulary (e.g., Boston Naming Test) may penalize candidates from non-Western cultures. To reduce these biases, clinicians would be advised to use demographically corrected norms, administer culturally adapted test forms when these are available, and base interpretations on scores in relation to detailed background information (Van der Elst et al., 2006). Last, practice effects and test–retest variability can contaminate the meaning of longitudinal testing. Subsequent testing with identical materials commonly produces artificial score gains, whereas variability in day-to-day health, mood, and levels of fatigue may introduce noise into serial assessment (Calamia et al., 2012). The use of alternate test forms, retesting at reasonable intervals, and triangulating performance across time points and tasks aids to establish that observed changes represent true cognitive trajectories, as opposed to artifacts of testing. To overcome these challenges, an accommodating, individually tailored approach is needed. Sensory accommodations, rigorous medical and psychiatric screening, culturally appropriate testing practices, and thoughtful longitudinal planning are essential to the attainment of valid and reliable measures of cognitive function in older adults. Only by acknowledging and adjusting for these factors can clinicians and researchers accurately describe cognitive impairment and develop interventions individually targeted to each patient's needs.

2.4 Principles of Geriatric Neurocognitive Rehabilitation

Geriatric neurocognitive rehabilitation is based on a system of fundamental principles aimed at maintaining and promoting cognitive functioning, optimizing independence, and enhancing QoL among older adults. At the core of these principles is a genuine multidisciplinary model of care that brings together expertise in neurology, geriatrics, neuropsychology, physical and occupational therapy (OT), speech–language pathology, social work, and nursing. Just as critical are transparent, individualized goals being able to carry on functional skills, slowing further impairment, and sustaining purposeful activity in everyday life and the careful individuation of interventions to each person's medical, cognitive, and personal circumstances. Multidisciplinary care is essential because cognitive aging and impairment result from dynamic interactions between neurobiological changes, comorbid medical conditions, psychosocial factors, and environmental supports. Neurologists and geriatricians provide medical assessment and management of underlying diseases like AD, vascular cognitive impairment, or Parkinson's disease, prescribing disease-modifying therapy and vascular risk factor management. Neuropsychologists perform thorough cognitive profiling, differentiating preserved from impaired functions and informing rehabilitation goals. Psychologists deal with mood and behavior, both causes and effects of cognitive deterioration. Physical and occupational therapists craft exercises that engage motor training and cognitive tasks to leverage motor-cognitive coupling and induce neural plasticity (Ngandu et al., 2015). Speech–language pathologists address communication impairment and swallow safety; social workers help in access to community resources and caregiver support. Nursing delivers constant monitoring, medication adjustment, and compensatory strategy reinforcement during everyday activities. Daily team sessions guarantee that input from every discipline is fed into a unified rehabilitation program (Hulst et al., 2023).

The objectives of cognitive rehabilitation in the elderly go beyond simple improvement on neuropsychological measures; they are geared toward preservation of current cognitive function, encouragement of functional independence, and optimization of QoL. Rehabilitation seeks to preserve the ability for IADLs medication management, financial activities, and cooking that are primary predictors of whether an older adult can safely reside in the community (Hoenig et al., 2015). Interventions are intended to build cognitive abilities (e.g., attention, memory, executive function) using restorative practice and to instruct on compensatory strategies (e.g., internal imagery, external reminders such as calendars or smartphone alarms) that bypass chronic deficits. Notably, the literature highlights that even small, domain-specific improvements can achieve significant gains in day-to-day functioning and psychosocial adjustment (Gavelin et al., 2020). Individualized treatment to fit the person's needs is the third pillar of successful geriatric neurocognitive rehabilitation. Older persons have mixed profiles of cognitive strengths and deficits, associated medical and psychiatric illness, sensory deficit, and dissimilar education and cultural background. A one-size-fits-all program not only misses an opportunity to exploit individual reserve but jeopardizes frustration and dropout. In practice, personalization starts with the thorough evaluation outlined in Section 2.3 and determines the individual's individual cognitive profile, functional outcomes, context of lifestyle, and personal values. To illustrate this, a retired accountant with mild executive dysfunction will benefit most from computerized planning exercises that simulate financial decision-making, while a former teacher with language impairment may respond best to group-based reminiscence therapy and conversation practice. Sensory adaptations, large-print materials or amplified sound, guarantee access in the face of visual or hearing loss. Interventions such as HomeCoRe demonstrate remote, adaptive CT that adapts task difficulty automatically to performance and can be customized to focus on specific cognitive domains relevant to the individual (Bernini et al., 2021).

Additionally, customization extends to dose, intensity, and format of therapy. Other older adults do well in high-frequency, short-duration sessions with frequent reinforcement, whereas others need less frequent but longer visits to solidify learning. Group formats are conducive to social interaction and peer support but might be required for individuals with severe impairment or low social ease. Telehealth and home-based models improve access for patients with mobility restrictions or transportation issues, while clinic-based programs provide the opportunity to utilize specialized equipment and face-to-face supervision. Clinicians are required at all times to monitor constantly and modify the plan by increasing, decreasing, or changing strategies, on the basis of continuous assessment of cognitive functioning, functional status, and patient report. Geriatric neurocognitive rehabilitation is founded on the strong pillars of interdisciplinary practice, well-defined functional and quality-of-life outcomes, and meticulous individualization of treatment. This synergy guarantees that interventions are not only intellectually compelling but effective in practice, marshaling each discipline's strengths to treat the complex needs of older adults, encouraging neural plasticity and compensatory skill use, and ultimately helping older adults live with dignity, autonomy, and engagement in valued activities.

2.5 Evidence-Based Intervention Strategies

Maintenance of cognitive functioning and daily autonomy in older persons with neurocognitive disorders is a compromise of drug and non-drug treatments. While medications may improve for the time being fundamental symptoms of dementia or concomitant mood disorders, non-drug therapies from formal CT to physical exercise and psychosocial activity tap into more general functional abilities and QoL. This section summarizes the evidence for frequently used medications, their risks and limitations, and the range of non-pharmacologic treatments with established efficacy in randomized trials and meta-analyses.

2.5.1 Pharmacologic Treatments

The only Food and Drug Administration (FDA)-approved drug treatments for AD are acetylcholinesterase inhibitors (ChEIs) like donepezil, rivastigmine, and galantamine, and the NMDA-receptor antagonist memantine. ChEIs augment synaptic acetylcholine, leading to modest benefits in cognition and overall function in mild and moderate AD, with effect sizes on the ADAS-Cog of about 2–3 points over 6–12 months (Birks, 2006). A meta-analysis of 41 randomized controlled trials concluded that donepezil 10 mg/day and rivastigmine transdermal patch had the highest ranking for cognitive improvement, with galantamine and memantine also having significant though less robust effects (Dou et al., 2018; Sinha, Trisal, et al., 2024). Combination therapy (donepezil + memantine) provides additional benefits: pooled analyses prove greater slowing of cognitive decline and delayed institutionalization compared to monotherapy (Atri et al., 2015). Memantine, used in moderate to severe AD, controls glutamatergic excitotoxicity and provides minor gains in cognition and ADLs (Kishi et al., 2025).

Though combination therapy is safe, shared side effects of the ChEIs are gastrointestinal distress (nausea, diarrhea), bradycardia, and risk of falls; memantine's side-effect profile is more benign with occasional headache and dizziness (Bano et al., 2023; Tan et al., 2014). Aside from dementia, depression and anxiety are common among older people and worsen cognitive symptoms. Selective serotonin reuptake inhibitors (SSRIs) like sertraline and citalopram are first-line treatment for geriatric depression. A meta-analysis of pharmacologic interventions for neuropsychiatric symptoms in AD revealed that SSRIs have moderate reductions in depressive symptoms (standardized mean difference = 0.44) but take six to eight weeks to achieve maximum efficacy; the clinician needs to watch for hyponatremia and elevated risk of falls (Wang et al., 2015).

Polypharmacy is widespread among geriatric patients (Priyadarshini et al., 2025), with risks of drug–drug interactions, side-effect accumulation, and anticholinergic burden. ChEIs and memantine can interact with cardiovascular or psychiatric drugs, so it is important to ensure careful reconciliation (Fink et al., 2018). Recommendations are for routine medication reviews, utilization of the Beers Criteria for the avoidance of potentially inappropriate medications, and de-prescribing in the case of cognitive or functional deterioration that cannot be explained by neurodegeneration alone.

2.5.2 Non-Pharmacological Interventions

Non-drug treatments address neuroplasticity, strategy learning in compensation, and psychosocial health. They carry little physiological risk and may yield lasting functional benefits. CT involves computerized tasks, paper-and-pencil activities, and group programs targeting areas like memory, attention, and executive function. Meta-analysis of 38 randomized controlled trials (RCTs) revealed that CT yields moderate global cognitive improvements (Hedges' g = 0.34) and increases in memory and processing speed, with effects maintaining at three- to six-month follow-up (Gavelin et al., 2020). The FINGER trial, a two-year multidomain intervention with CT, diet, exercise, and vascular monitoring showed a 25% risk decrease in cognitive decline in at-risk older adults (Ngandu et al., 2015).

Repeated aerobic and resistance training increases cerebral blood flow, activates neurotrophic factors (e.g., brain-derived neurotrophic factor (BDNF)), and promotes synaptic plasticity. A recent meta-analysis and systematic review of 29 RCTs (n = 2,100) concluded that organized exercise routines (≥ 150 min/week) result in small but significant gains in global cognition (Standardized Mean Difference (SMD) = 0.21) and executive function (SMD = 0.17) in older adults with and without mild cognitive impairment (Xu et al., 2023). Dose–response analysis indicates optimal benefits using moderate-intensity training three times a week (Gallardo-Gómez et al., 2022). OT interventions concentrate on the retraining of ADLs and IADLs using task-specific practice and environment adaptation. A Cochrane-registered systematic review of 15 RCTs (n = 2,063) established that home-based OT significantly enhanced overall performance in ADL (SMD = 0.61; 95% CI = 0.16–1.05) and IADLs (SMD = 0.22; 95% CI = 0.07–0.37) in dementia, with benefits enduring at three-month follow-up ((Bennett et al., 2019); PROSPERO CRD42011001166). Multimodal OT that integrates cognitive stimulation with training in ADLs also provides moderate cognitive benefit (SMD = 0.58) and trends toward ADL maintenance for patients with AD (Ham et al., 2021).

Psychosocial therapies leverage social interaction and emotional memory to enhance self-esteem and mood. Reminiscence therapy which is an organized recall of previous experiences enhances global cognition (mean difference = 2.4 points on MMSE) and depressive symptoms (g = 0.31) in mild to moderate dementia (Spector et al., 2000). Music therapy, working through intact musical memory, induces immediate changes in orientation and agitation, with modest effect sizes for cognition (Hedges' g = 0.22) and mood (g = 0.39) (Moreno-Morales et al., 2020).

2.6 Technological and Innovative Approaches

Recentmost advancements in digital technologies are revolutionizing geriatric neurocognitive rehabilitation through remote delivery, tailored assessment, and new training paradigms. Four areas, telerehabilitation and remote monitoring, virtual and augmented reality (VR/AR), artificial-intelligence (AI)-based assessment, and assistive technologies, describe how innovation is redefining care for older adults with cognitive impairment.

Telerehabilitation utilizes various telecommunication devices and wearable sensors to provide cognitive and functional rehabilitation in the patients' homes. Researchers have tested a home-based cognitive telerehabilitation program in patients with mild cognitive impairment, AD, and FTD through videoconferencing and tablet-based training (Cotelli et al., 2017). Across 12 weeks, participants showed considerable global cognition improvement (as measured by the AD Assessment Scale–Cognitive Subscale) compared to wait-list control (Cohen's d = 0.65) (Uddin et al., 2024). Wearable technology, like inertial-measurement-unit (IMU) sensors, expand remote monitoring with the capture of gait, balance, and activity patterns, which are related to executive function and risk of falls (Prasanth et al., 2021). These systems increase access for home-bound older people, decrease travel burden, and provide ongoing monitoring of adherence and performance. VR and AR provide immersive, ecologically valid spaces to practice IADLs and executive functions. Researchers have meta-analyzed various randomized controlled trials that evaluated VR interventions for MCI and mild dementia; effect sizes aggregated showed medium improvements in memory and attention versus usual CT (Chan & Tsoi, 2025). AR overlays virtual prompts onto actual environments to direct meal preparation, medication compliance, or wayfinding and demonstrates task performance time improvements and error reduction among Alzheimer's patients (Optale et al., 2009). Speech, gait, and digital-interaction behavior machine-learning algorithms enable early detection of cognitive impairment and tailor rehabilitation. Borchert et al. surveyed AI technologies in dementia diagnosis, natural-language processing of natural speech and convolutional neural networks reading MRI scans and observed significant diagnostic accuracies between studies (Borchert et al., 2023). Apart from diagnosis, AI platforms learn to vary training difficulty in real time: for instance, HomeCoRe applies Bayesian algorithms to modify task difficulty as a function of performance, with greater executive function benefits than fixed programs (Bernini et al., 2021). "Smart home" technologies and e-health devices enable independence through the automation of tasks and instant assistance. Gafurian et al. systematically reviewed ambient-assisted living systems like motion sensors, smart pill containers, and voice assistants and determined multimodal platforms decreased IADL errors by high number of older adults with cognitive impairment (Ghafurian et al., 2023). Mobile health apps provide reminders, track mood, and allow remote caregiver access, while telepresence robots promote social interaction and provide "just-in-time" cognitive cues (Abdi et al., 2018).

2.7 Family, Caregiver, Cultural, Ethical, and Policy Considerations

Successful geriatric neurocognitive rehabilitation is not limited to the clinic but includes family caregivers, cultural influences, ethical requirements, and policy environments. Family members tend to be chief facilitators of intervention benefits by reinforcing CT, organizing daily routines, and ensuring safety. In AD and related dementias, practice of memory strategies by caregivers and external cueing (e.g., labeled pill box or calendars) has been found to maintain functional gains made in therapy sessions (Gitlin et al., 2015). Training caregivers in errorless learning and spaced retrieval results in patients showing longer retention for skills targeted and feeling more confident in everyday activities (Sathananthan et al., 2022). Yet, the burdens imposed on family caregivers can lead to chronic stress and burnout, defined by emotional exhaustion, lower personal accomplishment, and heightened susceptibility to depression and physical morbidity (Pinquart & Sörensen, 2003). Systematic evaluation with tools such as the Zarit Burden Interview facilitates early detection of high-risk caregivers (Zarit et al., 1980). Preventive interventions such as respite care, support groups, and skills training significantly lower subjective burden and enhance coping (Brodaty & Arasaratnam, 2012). Family therapy models that combine problem-solving therapy with psychoeducation promote adaptive caregiving behaviors, reduce caregiver distress, and indirectly enhance patient outcomes (Liu et al., 2023).

Aging and cognitive impairment are influenced by cultural attitudes toward these conditions. Cultural attitudes also determine help-seeking behavior and responsiveness to rehabilitation. In certain cultures, dementia is blamed on normal aging or spiritual reasons, leading to late diagnosis and low use of services (Mukadam et al., 2010). Culturally sensitive education presented in the caregiver's language choice and sensitive to local explanatory models increases participation in cognitive interventions and support services (Dilworth-Anderson et al., 2002). Ethical issues pervade geriatric cognitive care, most notably with respect to autonomy, informed consent, and decision-making capacity. As cognitive impairment worsens, individuals may not have the capacity to consent to multifaceted interventions or research study participation. Assessment frameworks highlight task-specific evaluation, including use of standardized measures (e.g., the MacArthur Competence Assessment Tool for Clinical Research (MacCAT-CR)) in conjunction with narrative interviews to determine understanding, appreciation, reasoning, and expression of choice (Appelbaum & Grisso, 1988). Advanced care planning started early in the course of the disease honors patient autonomy by recording preferences for future medical and financial choices ahead of declining capacity (Jethwa & Onalaja, 2018).

Strong policy and guideline infrastructures must be in place to support sustained geriatric neurocognitive services. The World Health Organization's Global Action Plan on the Public Health Response to Dementia (2017–2025) recommends strengthening community-based care, integrating caregiver support into primary health systems, and ensuring equitable access to rehabilitation. National guidelines, e.g., from the National Institute for Health and Care Excellence (NICE, 2018) support multidisciplinary assessment, opportunistic case-finding in high-risk groups, and provision of structured cognitive stimulation therapy. Reimbursement policies for telerehabilitation, home-based OT, and caregiver training are essential to bring these recommendations into practice and lower out-of-pocket expenses for at-risk families. The convergence of caregiver and family involvement with culturally relevant practices, ethical protection, and supportive policy initiatives forms the basis for successful geriatric neurocognitive rehabilitation. Informed and adequately supported caregivers enhance rehabilitation outcomes, culturally relevant education facilitates timely involvement, capacity frameworks maintain autonomy, and policy efforts guarantee that services are accessible, inclusive, and equitable.

2.8 Outcome Measures and Program Evaluation

Assessing the effects of geriatric neurocognitive rehabilitation involves both quantitative and qualitative outcome measures. These are measured by using cross-sectional snapshots and then longitudinal follow-up. Standardized global and domain-specific measures of function come from quantitative cognitive scales like the ADAS-Cog, MMSE, and MoCA (Folstein et al., 1975; Venneri & Shanks, 2013). In clinical trials of intervention, a 2- to 4-point improvement on the ADAS-Cog is generally considered clinically significant (Muir et al., 2024), and MoCA improvements of ≥ 2 points indicate meaningful recovery of mild impairment (Freitas et al., 2013). QoL measures particularly the QoL-AD scale (QoL-AD) assess patient and caregiver ratings on physical, emotional, and social aspects, with greater QoL-AD ratings associated with maintained ADLs (Froelich et al., 2021; Thorgrimsen et al., 2003). The standardized qualitative procedures supplement the measures by probing individual experience, functional outcomes, and care satisfaction. Semi-structured interviews and focus groups gain understanding of rehabilitative progress and how it translates to our daily routines, emotional health, and caregiver burden (Ghosh et al., 2025; Olazarán et al., 2010; Sachdeva et al., 2024). The combinations of quantitative rating scales and thematic analysis of patient stories provide us with a deeper understanding of interventions that are worth and areas where improvements can be done. Cross-sectional assessments offer quick comparisons between intervention and control groups at a given point in time but are unable to separate practice effects or cohort differences. Longitudinal designs following the same participants over months to years are necessary for measuring sustained benefit and rates of decline. Unger and co-workers (Unger et al., 2015) showed that cross-sectional trends in MMSE can mask subtle gains detectable only with repeated assessments, whereas longitudinal data yield intra-individual patterns, such as early "learning" effects and nonrandom attrition. Sliwinski and Buschke also observed that processing-speed change is predictive of within-person memory decline in longitudinal, but not cross-sectional, data (Sliwinski & Buschke, 1999).

Longitudinal research with older adults is, nonetheless, challenging. Attrition rates can exceed 20% over two- to three-year follow-ups, often biased toward participants with poorer baseline cognition or greater comorbidity (Vanbeijsterveldt et al., 2002). This differential dropout can result in decline underestimation and intervention effectiveness overestimation unless it is counteracted by statistical techniques (e.g., multiple imputation) and refreshment sampling. Simultaneous comorbid medical illness and polypharmacy complicate the attribution of outcome, since fluctuations in physical health or mood may conceal cognitive trajectories. Lastly, heterogeneity in education, culture, and baseline function requires adequate sample size and diversity to make the results generalizable as well as for stratified analyses by important demographic variables (Matthews et al., 2004). Effective program evaluation of geriatric neurocognitive rehabilitation incorporates standardized quality-of-life and cognitive tests with qualitative feedback, eschews cross-sectional in favor of longitudinal designs, and takes pro-active measures to control attrition, comorbidities, and sample heterogeneity to measure true intervention effect accurately.

2.9 Future Directions in Geriatric Neurocognitive Rehabilitation

Recent developments in our knowledge of the aging brain reverse the traditionally held belief that neuroplasticity declines irreversibly after midlife. Synaptic plasticity, the creation of new neurons (neurogenesis), and the restructuring of dendritic spines continue into old age but at diminished rates. Studies show that focused interventions such as concurrent aerobic exercise and CT upregulate BDNF and insulin-like growth factor 1 (IGF-1), facilitate hippocampal neurogenesis, and enhance prefrontal circuits associated with executive control (Navakkode & Kennedy, 2024). Pharmacological modulators of epigenetic regulators (e.g., histone deacetylase inhibitors) are being investigated in ongoing research to boost this endogenous plasticity and allow recovery from age-related insult. In parallel, brain–computer interfaces (BCIs) provide new avenues to boost cognition in older individuals by converting neural activity into real-time feedback. Recent studies show that electroencephalography (EEG)-based neurofeedback procedures are shown to enhance working memory and attentional measures in healthy older persons and those with mild cognitive impairment, albeit large-scale trials are rare (Tsai et al., 2025). Next-generation invasive and noninvasive BCIs hold the potential for improved signal fidelity, allowing closed-loop stimulation regimes targeting hippocampal theta activity or prefrontal–parietal gamma coherence to enhance network synchrony. Wearable technology and the Internet of Things are evolving beyond straightforward activity monitors to full-spectrum neurocognitive monitors. Recent reports on smartwatches and pendant sensors that monitor sleep architecture, gait dynamics, and social interaction patterns continuously and metrics that machine-learning models can link to early cognitive change (Rocha et al., 2024). Next-generation work will combine multimodal sensor streams (autonomic, kinematic, acoustic) with AI algorithms to provide just-in-time cognitive cues and safety warnings, closing the gap between assessment and intervention.

Personalized medicine models are increasingly popular in cognitive aging. The "Precision Aging" hypothesis suggests dividing older individuals by genetic polymorphisms (e.g., APOE ε4 carrier status), metabolic levels, and lifestyle risk factors to individualize intervention "doses" of exercise, diet, and mental training. Pilot trials with genotype-directed supplementation (e.g., omega-3 fatty acids for APOE ε4 carriers) and adaptive cognitive systems show increased effectiveness compared to one-size-fits-all intervention programs, hinting at truly individualized rehabilitation protocols. Lastly, interdisciplinary working and integrative care models continue to be key to taking these innovations to the point of practice. The FINGER trial, a two-year, multidomain lifestyle intervention integrating neurology, nutrition, exercise, and CT curbed cognitive decline by 25% among at-risk elders, highlighting the potential of (Ngandu et al., 2015). Care models of the future will bring together geriatricians, neuropsychologists, engineers, data scientists, and policymakers to co-design digital platforms, streamline resource use, and measure outcomes in real-world environments. This convergent strategy will be necessary to expand neurocognitive rehabilitation on a large scale and to make sure that advances in plasticity, technologies, and precision medicine bring tangible benefits in autonomy and QoL to our fast-growing aging population.

2.10 Conclusion

Geriatric neurocognitive rehabilitation is a highly developed, evidence-based practice that leverages multidisciplinary knowledge, tailored intervention, and technological advancements to meet the multifaceted demands of older persons with impairments in cognition. Essential approaches ranging from intensive assessment, pharmacologic optimization, focused CT, to supportive non-pharmacologic intervention intersect to maintain residual capacity, promote compensatory abilities, and avoid disabling effects of age-related neurodegeneration. When combined with caregiver education and culturally appropriate support, these interventions not only delay cognitive deterioration but also have measurable benefits in everyday living, emotional well-being, and perceived QoL. Furthermore, by preserving functional independence, rehabilitation programs minimize caregiver burden and healthcare costs, with tangible societal returns. Looking to the future, the discipline is positioned to use new revelations in neuroplasticity, precision medicine, and BCIs to optimize and customize treatment. Geriatric cognitive treatment of the future is in adaptive, data-driven systems that integrate seamlessly clinical, behavioral, and biometric information to provide "just-in-time" interventions. Sustained interdisciplinary partnership, rigorous longitudinal testing, and policy endorsement will be necessary to turn these innovations into scalable, fair models of care ultimately allowing older adults to age with autonomy, dignity, and participation in valued life activities.

References

Abdi, J., Al-Hindawi, A., Ng, T., & Vizcaychipi, M. P. (2018). Scoping Review on the Use of Socially Assistive Robot Technology in Elderly Care. *BMJ Open*, *8*(2). https://doi.org/10.1136/bmjopen-2017-018815

Al-Yawer, F., Pichora-Fuller, M. K., & Phillips, N. A. (2019). The Montreal Cognitive Assessment After Omission of Hearing-Dependent Subtests: Psychometrics and Clinical Recommendations. *Journal of the American Geriatrics Society*, *67*(8), 1689–1694. https://doi.org/10.1111/jgs.15940

Alzheimer's Association (2024). 2024 Alzheimer's Disease Facts and Figures. *Alzheimer's & Dementia*, *20*(5), 3708–3821. https://doi.org/10.1002/alz.13809

Appelbaum, P. S., & Grisso, T. (1988). Assessing Patients' Capacities to Consent to Treatment. *New England Journal of Medicine*, *319*(25), 1635–1638. https://doi.org/10.1056/nejm198812223192504

Atri, A., Hendrix, S. B., Pejović, V., Hofbauer, R. K., Edwards, J., Molinuevo, J. L., & Graham, S. M. (2015). Cumulative, Additive Benefits of Memantine-Donepezil Combination over Component Monotherapies in Moderate to Severe Alzheimer's Dementia: A Pooled Area under the Curve Analysis. *Alzheimer's Research & Therapy*, *7*(1). https://doi.org/10.1186/s13195-015-0109-2

Bäckman, L., Nyberg, L., Lindenberger, U., Li, S.-C., & Farde, L. (2006). The Correlative Triad among Aging, Dopamine, and Cognition: Current Status and Future Prospects. *Neuroscience & Biobehavioral Reviews*, *30*(6), 791–807. https://doi.org/10.1016/j.neubiorev.2006.06.005

Bano, S., Raza, M. A., Ghosh, S., Pandit, N., Srivastava, S., Azam, M., Dey, S. K., Han, S. S., Sinha, J. K., & Ruwali, M. (2023). Emerging Therapeutic Targets in Molecular Neuropharmacology for Alzheimer's Disease. *Journal of Biological Regulators and Homeostatic Agents*, *37*(11). https://doi.org/10.23812/j.biol.regul.homeost.agents.20233711.553

Bennett, S., Laver, K., Voigt-Radloff, S., Letts, L., Clemson, L., Graff, M., Wiseman, J., & Gitlin, L. (2019). Occupational Therapy for People with Dementia and Their Family Carers Provided at Home: A Systematic Review and Meta-Analysis. *BMJ Open*, *9*(11). https://doi.org/10.1136/bmjopen-2018-026308

Bernini, S., Panzarasa, S., Sinforiani, E., Quaglini, S., Cappa, S. F., Cerami, C., Tassorelli, C., Vecchi, T., & Bottiroli, S. (2021). HomeCoRe for Telerehabilitation in Mild or Major Neurocognitive Disorders: A Study Protocol for a Randomized Controlled Trial. *Frontiers in Neurology*, *12*. https://doi.org/10.3389/fneur.2021.752830

Birks, J. S. (2006). Cholinesterase Inhibitors for Alzheimer's Disease. *Cochrane Database of Systematic Reviews*, *2016*(3). https://doi.org/10.1002/14651858.Cd005593

Borchert, R. J., Azevedo, T., Badhwar, A., Bernal, J., Betts, M., Bruffaerts, R., Burkhart, M. C., Dewachter, I., Gellersen, H. M., Low, A., Lourida, I., Machado, L., Madan, C. R., Malpetti, M., Mejia, J., Michopoulou, S., Muñoz-Neira, C., Pepys, J., Peres, M., …, & Rittman, T. (2023). Artificial Intelligence for Diagnostic and Prognostic Neuroimaging in Dementia: A Systematic Review. *Alzheimer's & Dementia*, *19*(12), 5885–5904. https://doi.org/10.1002/alz.13412

Bouwens, S. F. M., van Heugten, C. M., & Verhey, F. R. J. (2009). Association between Cognition and Daily Life Functioning in Dementia Subtypes. *International Journal of Geriatric Psychiatry*, *24*(7), 764–769. https://doi.org/10.1002/gps.2193

Braak, H., & Braak, E. (1991). Neuropathological Stageing of Alzheimer-Related Changes. *Acta Neuropathologica*, *82*(4), 239–259. https://doi.org/10.1007/bf00308809

Brodaty, H., & Arasaratnam, C. (2012). Meta-Analysis of Nonpharmacological Interventions for Neuropsychiatric Symptoms of Dementia. *American Journal of Psychiatry*, *169*(9), 946–953. https://doi.org/10.1176/appi.ajp.2012.11101529

Burke, D. M., & Shafto, M. A. (2004). Aging and Language Production. *Current Directions in Psychological Science*, *13*(1), 21–24. https://doi.org/10.1111/j.0963-7214.2004.01301006.x

Butters, M. A., Whyte, E. M., Nebes, R. D., Begley, A. E., Dew, M. A., Mulsant, B. H., Zmuda, M. D., Bhalla, R., Meltzer, C. C., Pollock, B. G., Reynolds, C. F., & Becker, J. T. (2004). The Nature and Determinants of Neuropsychological Functioning in Late-LifeDepression. *Archives of General Psychiatry*, *61*(6). https://doi.org/10.1001/archpsyc.61.6.587

Calamia, M., Markon, K., & Tranel, D. (2012). Scoring Higher the Second Time Around: Meta-Analyses of Practice Effects in Neuropsychological Assessment. *The Clinical Neuropsychologist*, *26*(4), 543–570. https://doi.org/10.1080/13854046.2012.680913

Carson, N., Leach, L., & Murphy, K. J. (2017). A Re-Examination of Montreal Cognitive Assessment (MoCA) Cutoff Scores. *International Journal of Geriatric Psychiatry*, *33*(2), 379–388. https://doi.org/10.1002/gps.4756

Chan, J. Y. C., & Tsoi, K. K. F. (2025). Virtual Reality-Based Cognitive and Exercise Intervention for People with Mild Cognitive Impairment. A Systematic Review and Meta-Analysis. *Alzheimer's & Dementia*, *20*(S4). https://doi.org/10.1002/alz.085515

Chen, C., Lim, J., Koh, J., Beard, J., Rowe, J. W., Rowe, J., Antonucci, T., Berkman, L., Börsch-Supan, A., Carstensen, L., Goldman, D., Fried, L., Furstenberg, F., Jackson, J., Kohli, M., Olshansky, J., Rehkopf, D., Rother, J., & Zissimopoulos, J. (2024). A Global Analysis of Adaptation to Societal Aging across Low-, Middle- and High-Income Countries Using the Global Aging Society Index. *Nature Aging*, *5*(1), 113–121. https://doi.org/10.1038/s43587-024-00772-3

Connolly, E. M., Mc Ardle, R., Bimpong, K. A. A., & Slight, S. (2024). What Impact Does the Diagnosis of Mild Cognitive Impairment Have on the Wellbeing, Everyday Behavior, and Healthcare Utilization of People and Their Carers? A Systematic Review. *Journal of Alzheimer's Disease*, *101*(3), 715–729. https://doi.org/10.3233/jad-231466

Cotelli, M., Manenti, R., Brambilla, M., Gobbi, E., Ferrari, C., Binetti, G., & Cappa, S. F. (2017). Cognitive Telerehabilitation in Mild Cognitive Impairment, Alzheimer's Disease and Frontotemporal Dementia: A Systematic Review. *Journal of Telemedicine and Telecare*, *25*(2), 67–79. https://doi.org/10.1177/1357633x17740390

Cox, S. R., Ritchie, S. J., Tucker-Drob, E. M., Liewald, D. C., Hagenaars, S. P., Davies, G., Wardlaw, J. M., Gale, C. R., Bastin, M. E., & Deary, I. J. (2016). Ageing and Brain White Matter Structure in 3,513 UK Biobank Participants. *Nature Communications*, *7*(1). https://doi.org/10.1038/ncomms13629

Crum, R. M., Anthony, J. C., Bassett, S. S., & Folstein, M. F. (1993). Population-Based Norms for the Mini-Mental State Examination by Age and Educational Level. *Jama*, *269*(18), 2386–2391. www.ncbi.nlm.nih.gov/pubmed/8479064

Davies, P. (1976). Selective Loss of Central Cholinergic Neurons in Alzheimer's Disease. *The Lancet*, *308*(8000). https://doi.org/10.1016/s0140-6736(76)91936-x

Dilworth-Anderson, P., Williams, I. C., & Gibson, B. E. (2002). Issues of Race, Ethnicity, and Culture in Caregiving Research: A 20-Year Review (1980–2000). *The Gerontologist*, *42*(2), 237–272. https://doi.org/10.1093/geront/42.2.237

Dou, K.-X., Tan, M.-S., Tan, C.-C., Cao, X.-P., Hou, X.-H., Guo, Q.-H., Tan, L., Mok, V., & Yu, J.-T. (2018). Comparative Safety and Effectiveness of Cholinesterase Inhibitors and Memantine for Alzheimer's Disease: A Network Meta-Analysis of 41 Randomized Controlled Trials. *Alzheimer's Research & Therapy*, *10*(1). https://doi.org/10.1186/s13195-018-0457-9

Espino, D. V., Lichtenstein, M. J., Palmer, R. F., & Hazuda, H. P. (2004). Evaluation of the Mini-Mental State Examination's Internal Consistency in a Community-Based Sample of Mexican-American and European-American Elders: Results from the San Antonio Longitudinal Study of Aging. *Journal of the American Geriatrics Society*, *52*(5), 822–827. https://doi.org/10.1111/j.1532-5415.2004.52226.x

Finger, E. C. (2016). Frontotemporal Dementias. *Continuum: Lifelong Learning in Neurology*, *22*(2, Dementia), 464–489. https://doi.org/10.1212/con.0000000000000300

Fink, H. A., Jutkowitz, E., McCarten, J. R., Hemmy, L. S., Butler, M., Davila, H., Ratner, E., Calvert, C., Barclay, T. R., Brasure, M., Nelson, V. A., & Kane, R. L. (2018). Pharmacologic Interventions to Prevent Cognitive Decline, Mild Cognitive Impairment, and Clinical Alzheimer-Type Dementia. *Annals of Internal Medicine*, *168*(1), 39–51. https://doi.org/10.7326/m17-1529

Folstein, M. F., Folstein, S. E., & McHugh, P. R. (1975). Mini-Mental State. *Journal of Psychiatric Research*, *12*(3), 189–198. https://doi.org/10.1016/0022-3956(75)90026-6

Freitas, S., Simões, M. R., Alves, L., & Santana, I. (2013). Montreal Cognitive Assessment. *Alzheimer Disease & Associated Disorders*, *27*(1), 37–43. https://doi.org/10.1097/WAD.0b013e3182420bfe

Froelich, L., Lladó, A., Khandker, R. K., Pedrós, M., Black, C. M., Sánchez Díaz, E. J., Chekani, F., & Ambegaonkar, B. (2021). Quality of Life and Caregiver Burden of Alzheimer's Disease among Community Dwelling Patients in Europe: Variation by Disease Severity and Progression. *Journal of Alzheimer's Disease Reports*, *5*(1), 791–804. https://doi.org/10.3233/adr-210025

Fu, J., Ferreira, D., Smedby, Ö., & Moreno, R. (2025). Decomposing the Effect of Normal Aging and Alzheimer's Disease in Brain Morphological Changes via Learned Aging Templates. *Scientific Reports*, *15*(1). https://doi.org/10.1038/s41598-025-96234-w

Gallardo-Gómez, D., del Pozo-Cruz, J., Noetel, M., Álvarez-Barbosa, F., Alfonso-Rosa, R. M., & del Pozo Cruz, B. (2022). Optimal Dose and Type of Exercise to Improve Cognitive Function in Older Adults: A Systematic Review and Bayesian Model-Based Network Meta-Analysis of RCTs. *Ageing Research Reviews*, *76*. https://doi.org/10.1016/j.arr.2022.101591

Gavelin, H. M., Lampit, A., Hallock, H., Sabatés, J., & Bahar-Fuchs, A. (2020). Cognition-Oriented Treatments for Older Adults: A Systematic Overview of Systematic Reviews. *Neuropsychology Review*, *30*(2), 167–193. https://doi.org/10.1007/s11065-020-09434-8

Gazbare, P., & Palekar, T. J. (2024). Identifying Mild cognitive Impairment and Its Predictors among Normal Middle Adulthood in Pune, Maharashtra: A Community-Based Cross-Sectional Survey. *Clinical Epidemiology and Global Health*, *25*. https://doi.org/10.1016/j.cegh.2023.101490

Gazerani, P. (2025). The Neuroplastic Brain: Current Breakthroughs and Emerging Frontiers. *Brain Research*, *1858*. https://doi.org/10.1016/j.brainres.2025.149643

Ghafurian, M., Wang, K., Dhode, I., Kapoor, M., Morita, P. P., & Dautenhahn, K. (2023). Smart Home Devices for Supporting Older Adults: A Systematic Review. *IEEE Access*, *11*, 47137–47158. https://doi.org/10.1109/access.2023.3266647

Ghosh, S., Durgvanshi, S., Agarwal, S., Raghunath, M., & Sinha, J. K. (2020). Current Status of Drug Targets and Emerging Therapeutic Strategies in the Management of Alzheimer's Disease. *Current Neuropharmacology*, *18*(9), 883–903. https://doi.org/10.2174/1570159x18666200429011823

Ghosh, S., Singh, K. K., Amist, A. D., & Sinha, J. K. (2025). Exploring Novel Molecular Targets for Alzheimer's Disease in Neuropharmacology. *Alzheimer's & Dementia*, *20*(S6). https://doi.org/10.1002/alz.085882

Ghosh, S., Sinha, J. K., Raghunath, M., Han, S. S., & Bhaskar, R. (2023). Accountability and Protection for Health Care in Conflict Zones. *The Lancet*, *402*(10412), 1524–1525. https://doi.org/10.1016/s0140-6736(23)01693-8

Ghosh, S., Vashisth, K., Ghosh, S., Han, S. S., Bhaskar, R., & Sinha, J. K. (2023). From Sleep to Cancer to Neurodegenerative Disease: The Crucial Role of Hsp70 in Maintaining Cellular Homeostasis and Potential Therapeutic Implications. *Journal of Biomolecular Structure and Dynamics*, *42*(18), 9812–9823. https://doi.org/10.1080/07391102.2023.2252509

Gitlin, L. N., Marx, K., Stanley, I. H., & Hodgson, N. (2015). Translating Evidence-Based Dementia Caregiving Interventions into Practice: State-of-the-Science and Next Steps. *The Gerontologist*, *55*(2), 210–226. https://doi.org/10.1093/geront/gnu123

Ham, M.-J., Kim, S., Jo, Y.-J., Park, C., Nam, Y., Yoo, D.-H., & Moon, M. (2021). The Effect of a Multimodal Occupational Therapy Program with Cognition-Oriented Approach on Cognitive Function and Activities of Daily Living in Patients with Alzheimer's Disease: A Systematic Review and Meta-Analysis of Randomized Controlled Trials. *Biomedicines*, *9*(12). https://doi.org/10.3390/biomedicines9121951

Head, E. (2010). Neurobiology of the Aging Dog. *Age*, *33*(3), 485–496. https://doi.org/10.1007/s11357-010-9183-3

Hedden, T., & Gabrieli, J. D. E. (2004). Insights into the Ageing Mind: A View from Cognitive Neuroscience. *Nature Reviews Neuroscience*, *5*(2), 87–96. https://doi.org/10.1038/nrn1323

Hoenig, H., Nusbaum, N., & Brummel-Smith, K. (2015). Geriatric Rehabilitation: State of the Art. *Journal of the American Geriatrics Society*, *45*(11), 1371–1381. https://doi.org/10.1111/j.1532-5415.1997.tb02939.x

Hong, S., Baek, S.-H., Lai, M. K. P., Arumugam, T. V., & Jo, D.-G. (2024). Aging-Associated Sensory Decline and Alzheimer's Disease. *Molecular Neurodegeneration*, *19*(1). https://doi.org/10.1186/s13024-024-00776-y

Hsiao, K., Chapman, P., Nilsen, S., Eckman, C., Harigaya, Y., Younkin, S., Yang, F., & Cole, G. (1996). Correlative Memory Deficits, Aβ Elevation, and Amyloid Plaques in Transgenic Mice. *Science*, *274*(5284), 99–103. https://doi.org/10.1126/science.274.5284.99

Hulst, H. E., Dobryakova, E., Costa, S. L., & Donkers, S. J. (2023). Editorial: Cognitive Rehabilitation: A Multidisciplinary Approach. *Frontiers in Rehabilitation Sciences*, *4*. https://doi.org/10.3389/fresc.2023.1268531

Jethwa, K. D., & Onalaja, O. (2018). Advance Care Planning and Palliative Medicine in Advanced Dementia: A Literature Review. *BJPsych Bulletin*, *39*(2), 74–78. https://doi.org/10.1192/pb.bp.114.046896

Jin, M., & Cai, S.-Q. (2022). Mechanisms Underlying Brain Aging Under Normal and Pathological Conditions. *Neuroscience Bulletin*, *39*(2), 303–314. https://doi.org/10.1007/s12264-022-00969-9

Kalaria, R. N. (2016). Neuropathological Diagnosis of Vascular Cognitive Impairment and Vascular Dementia with Implications for Alzheimer's Disease. *Acta Neuropathologica*, *131*(5), 659–685. https://doi.org/10.1007/s00401-016-1571-z

Kishi, T., Matsunaga, S., Oya, K., Nomura, I., Ikuta, T., & Iwata, N. (2025). Memantine for Alzheimer's Disease: An Updated Systematic Review and Meta-Analysis. *Journal of Alzheimer's Disease*, *60*(2), 401–425. https://doi.org/10.3233/jad-170424

Kruk, M. E., Gage, A. D., Arsenault, C., Jordan, K., Leslie, H. H., Roder-DeWan, S., Adeyi, O., Barker, P., Daelmans, B., Doubova, S. V., English, M., García-Elorrio, E., Guanais, F., Gureje, O., Hirschhorn, L. R., Jiang, L., Kelley, E., Lemango, E. T., Liljestrand, J., …, & Pate, M. (2018). High-Quality Health Systems in the Sustainable Development Goals Era: Time for a Revolution. *The Lancet Global Health*, *6*(11), e1196–e1252. https://doi.org/10.1016/s2214-109x(18)30386-3

Lee, J., & Kim, H.-J. (2022). Normal Aging Induces Changes in the Brain and Neurodegeneration Progress: Review of the Structural, Biochemical, Metabolic, Cellular, and Molecular Changes. *Frontiers in Aging Neuroscience*, *14*. https://doi.org/10.3389/fnagi.2022.931536

Liu, Y., Hughes, M. C., Baumbach, A., & Derain, L. (2023). An Online Intervention to Improve the Health and Well-Being of Informal Caregivers of Individuals with Alzheimer's Disease: A Pilot Study. *PEC Innovation*, *3*. https://doi.org/10.1016/j.pecinn.2023.100229

Livingston, G., Huntley, J., Sommerlad, A., Ames, D., Ballard, C., Banerjee, S., Brayne, C., Burns, A., Cohen-Mansfield, J., Cooper, C., Costafreda, S. G., Dias, A., Fox, N., Gitlin, L. N., Howard, R., Kales, H. C., Kivimäki, M., Larson, E. B., Ogunniyi, A., …, & Mukadam, N. (2020). Dementia Prevention, Intervention, and Care: 2020 Report of the Lancet Commission. *The Lancet*, *396*(10248), 413–446. https://doi.org/10.1016/s0140-6736(20)30367-6

Mangels, J. A., Gershberg, F. B., Shimamura, A. P., & Knight, R. T. (1996). Impaired Retrieval from Remote Memory in Patients with Frontal Lobe Damage. *Neuropsychology*, *10*(1), 32–41. https://doi.org/10.1037/0894-4105.10.1.32

Mansoor, A. K., Thomas, S., Sinha, J. K., Alladi, P. A., Ravi, V., & Raju, T. R. (2012). Olfactory Tract Transection Reveals Robust Tissue-Level Plasticity by Cellular Numbers and Neurotrophic Factor Expression in Olfactory Bulb. *Indian Journal of Experimental Biology*, *50*(11), 765–770. www.ncbi.nlm.nih.gov/pubmed/23305026

Matthews, F. E., Chatfield, M., Freeman, C., McCracken, C., Brayne, C., & Cfas, M. R. C. (2004). Attrition and Bias in the MRC Cognitive Function and Ageing Study: An Epidemiological Investigation. *BMC Public Health*, *4*(1). https://doi.org/10.1186/1471-2458-4-12

McKhann, G. M., Knopman, D. S., Chertkow, H., Hyman, B. T., Jack, C. R., Kawas, C. H., Klunk, W. E., Koroshetz, W. J., Manly, J. J., Mayeux, R., Mohs, R. C., Morris, J. C., Rossor, M. N., Scheltens, P., Carrillo, M. C., Thies, B., Weintraub, S., & Phelps, C. H. (2011). The Diagnosis of Dementia due to Alzheimer's Disease: Recommendations from the National Institute on

Aging-Alzheimer's Association Workgroups on Diagnostic Guidelines for Alzheimer's Disease. *Alzheimer's & Dementia*, *7*(3), 263–269. https://doi.org/10.1016/j.jalz.2011.03.005

Meltzer, M. D. C. (1998). Serotonin in Aging, Late-Life Depression, and Alzheimer's Disease: The Emerging Role of Functional Imaging. *Neuropsychopharmacology*, *18*(6), 407–430. https://doi.org/10.1016/s0893-133x(97)00194-2

Mishra, P. K., Singh, K. K., Ghosh, S., & Sinha, J. K. (2025). Future perspectives on the clinics of Alzheimer's disease. In *A New Era in Alzheimer' s Research* (pp. 217–232). https://doi.org/10.1016/b978-0-443-15540-6.00001-x

Moreno-Morales, C., Calero, R., Moreno-Morales, P., & Pintado, C. (2020). Music Therapy in the Treatment of Dementia: A Systematic Review and Meta-Analysis. *Frontiers in Medicine*, *7*. https://doi.org/10.3389/fmed.2020.00160

Morrison, J. H., & Baxter, M. G. (2012). The Ageing Cortical Synapse: Hallmarks and Implications for Cognitive Decline. *Nature Reviews Neuroscience*, *13*(4), 240–250. https://doi.org/10.1038/nrn3200

Muhammad, T., Pai, M., Kumar, M., & Sekher, T. V. (2023). Multiple Socioeconomic Risks and Cognitive Impairment among Older Men and Women in India. *Dialogues in Health*, *2*. https://doi.org/10.1016/j.dialog.2023.100119

Muir, R. T., Hill, M. D., Black, S. E., & Smith, E. E. (2024). Minimal Clinically Important Difference in Alzheimer's Disease: Rapid Review. *Alzheimer's & Dementia*, *20*(5), 3352–3363. https://doi.org/10.1002/alz.13770

Mukadam, N., Cooper, C., & Livingston, G. (2010). A Systematic Review of Ethnicity and Pathways to Care in Dementia. *International Journal of Geriatric Psychiatry*, *26*(1), 12–20. https://doi.org/10.1002/gps.2484

Murman, D. (2015). The Impact of Age on Cognition. *Seminars in Hearing*, *36*(03), 111–121. https://doi.org/10.1055/s-0035-1555115

Nasreddine, Z. S., Phillips, N. A., Bédirian, V., Charbonneau, S., Whitehead, V., Collin, I., Cummings, J. L., & Chertkow, H. (2005). The Montreal Cognitive Assessment, MoCA: A Brief Screening Tool for Mild Cognitive Impairment. *Journal of the American Geriatrics Society*, *53*(4), 695–699. https://doi.org/10.1111/j.1532-5415.2005.53221.x

National Institute on Aging. (2023). Cognitive Health and Older Adults [Web Page]. www.nia.nih.gov/health/brain-health/cognitive-health-and-older-adults

Navakkode, S., & Kennedy, B. K. (2024). Neural Ageing and Synaptic Plasticity: Prioritizing Brain Health in Healthy Longevity. *Frontiers in Aging Neuroscience*, *16*. https://doi.org/10.3389/fnagi.2024.1428244

Ngandu, T., Lehtisalo, J., Solomon, A., Levälahti, E., Ahtiluoto, S., Antikainen, R., Bäckman, L., Hänninen, T., Jula, A., Laatikainen, T., Lindström, J., Mangialasche, F., Paajanen, T., Pajala, S., Peltonen, M., Rauramaa, R., Stigsdotter-Neely, A., Strandberg, T., Tuomilehto, J., ..., & Kivipelto, M. (2015). A 2 Year Multidomain Intervention of Diet, Exercise, Cognitive Training, and Vascular Risk Monitoring versus Control to Prevent Cognitive Decline in at-Risk Elderly People (FINGER): A Randomised Controlled Trial. *The Lancet*, *385*(9984), 2255–2263. https://doi.org/10.1016/s0140-6736(15)60461-5

National Institute for Health and Care Excellence. (2018). Dementia: Assessment, Management and Support for People Living with Dementia and Their Carers. NICE Guideline [NG97]. Available at: www.nice.org.uk/guidance/ng97

Olazarán, J., Reisberg, B., Clare, L., Cruz, I., Peña-Casanova, J., del Ser, T., Woods, B., Beck, C., Auer, S., Lai, C., Spector, A., Fazio, S., Bond, J., Kivipelto, M., Brodaty, H., Rojo, J. M., Collins, H., Teri, L., Mittelman, M., ..., & Muñiz, R. (2010). Nonpharmacological Therapies in Alzheimer's Disease: A Systematic Review of Efficacy. *Dementia and Geriatric Cognitive Disorders*, *30*(2), 161–178. https://doi.org/10.1159/000316119

Oleksy, P., Zieliński, K., Buczkowski, B., Sikora, D., Góralczyk, E., Zając, A., Mąka, M., Papież, Ł., & Kamiński, J. (2024). Cognitive Function Tests: Application of MMSE and MoCA in Various Clinical Settings- a Brief Overview. *Quality in Sport*, *34*. https://doi.org/10.12775/qs.2024.34.56285

Optale, G., Urgesi, C., Busato, V., Marin, S., Piron, L., Priftis, K., Gamberini, L., Capodieci, S., & Bordin, A. (2009). Controlling Memory Impairment in Elderly Adults Using Virtual Reality

Memory Training: A Randomized Controlled Pilot Study. *Neurorehabilitation and Neural Repair*, *24*(4), 348–357. https://doi.org/10.1177/1545968309353328

Ota, M., Yasuno, F., Ito, H., Seki, C., Nozaki, S., Asada, T., & Suhara, T. (2006). Age-Related Decline of Dopamine Synthesis in the Living Human Brain Measured by Positron Emission Tomography with L-[β-^{11}C]DOPA. *Life Sciences*, *79*(8), 730–736. https://doi.org/10.1016/j.lfs.2006.02.017

Perry, R. J. (1999). Attention and Executive Deficits in Alzheimer's Disease: A Critical Review. *Brain*, *122*(3), 383–404. https://doi.org/10.1093/brain/122.3.383

Peters, R. (2006). Ageing and the Brain. *Postgraduate Medical Journal*, *82*(964), 84–88. https://doi.org/10.1136/pgmj.2005.036665

Pinquart, M., & Sörensen, S. (2003). Differences between Caregivers and Noncaregivers in Psychological Health and Physical Health: A Meta-Analysis. *Psychology and Aging*, *18*(2), 250–267. https://doi.org/10.1037/0882-7974.18.2.250

Prasad, S., Katta, M. R., Abhishek, S., Sridhar, R., Valisekka, S. S., Hameed, M., Kaur, J., & Walia, N. (2023). Recent Advances in Lewy Body Dementia: A Comprehensive Review. *Disease-a-Month*, *69*(5). https://doi.org/10.1016/j.disamonth.2022.101441

Prasanth, H., Caban, M., Keller, U., Courtine, G., Ijspeert, A., Vallery, H., & von Zitzewitz, J. (2021). Wearable Sensor-Based Real-Time Gait Detection: A Systematic Review. *Sensors*, *21*(8). https://doi.org/10.3390/s21082727

Priyadarshini, S., Goyal, K., Roopashree, R., Gupta, S., Roy, A., Biswas, R., Patra, S., Chauhan, P., Wadhwa, K., Singh, G., Kamal, M., Iqbal, D., Alsaweed, M., Nuli, M. V., Abomughaid, M. M., Almutary, A. G., Sinha, J. K., Bansal, P., Rani, B., …, & Jha, N. K. (2025). Polypharmacology and Neuroprotective Effects of Gingerol in Alzheimer's Disease. *Molecular Neurobiology*. https://doi.org/10.1007/s12035-024-04484-y

Rapp, P. R., & Gallagher, M. (1996). Preserved Neuron Number in the Hippocampus of Aged Rats with Spatial Learning Deficits. *Proceedings of the National Academy of Sciences*, *93*(18), 9926–9930. https://doi.org/10.1073/pnas.93.18.9926

Raz, N., Lindenberger, U., Rodrigue, K. M., Kennedy, K. M., Head, D., Williamson, A., Dahle, C., Gerstorf, D., & Acker, J. D. (2005). Regional Brain Changes in Aging Healthy Adults: General Trends, Individual Differences and Modifiers. *Cerebral Cortex*, *15*(11), 1676–1689. https://doi.org/10.1093/cercor/bhi044

Roberts, R., & Knopman, D. S. (2013). Classification and Epidemiology of MCI. *Clinics in Geriatric Medicine*, *29*(4), 753–772. https://doi.org/10.1016/j.cger.2013.07.003

Rocha, I. C., Arantes, M., Moreira, A., Vilaça, J. L., Morais, P., Matos, D., & Carvalho, V. (2024). Monitoring Wearable Devices for Elderly People with Dementia: A Review. *Designs*, *8*(4). https://doi.org/10.3390/designs8040075

Sachdeva, P., Narayanan, K. B., Sinha, J. K., Gupta, S., Ghosh, S., Singh, K. K., Bhaskar, R., Almutary, A. G., Zothantluanga, J. H., Kotta, K. K., Nelson, V. K., Paiva-Santos, A. C., Abomughaid, M. M., Kamal, M., Iqbal, D., Alharbi, M. H., Almutairi, A. A., Dewanjee, S., Nuli, M. V., …, & Jha, N. K. (2024). Recent Advances in Drug Delivery Systems Targeting Insulin Signalling for the Treatment of Alzheimer's Disease. *Journal of Alzheimer's Disease*, *98*(4), 1169–1179. https://doi.org/10.3233/jad-231181

Sathananthan, N., Morris, E. M. J., Gillanders, D., Knox, L., Dimech-Betancourt, B., Wright, B. J., das Nair, R., & Wong, D. (2022). Does Integrating Cognitive and Psychological Interventions Enhance Wellbeing After Acquired Brain Injury? Study Protocol for a Phase II Randomized Controlled Trial of the VaLiANT (Valued Living After Neurological Trauma) Group Program. *Frontiers in Rehabilitation Sciences*, *2*. https://doi.org/10.3389/fresc.2021.815111

Scheff, S. W., DeKosky, S. T., & Price, D. A. (1990). Quantitative Assessment of Cortical Synaptic Density in Alzheimer's Disease. *Neurobiology of Aging*, *11*(1), 29–37. https://doi.org/10.1016/0197-4580(90)90059-9

Scheffels, J. F., Ballasch, I., Scheichel, N., Voracek, M., Kalbe, E., & Kessler, J. (2023). The Influence of Age, Gender and Education on Neuropsychological Test Scores: Updated Clinical Norms for Five Widely Used Cognitive Assessments. *Journal of Clinical Medicine*, *12*(16). https://doi.org/10.3390/jcm12165170

Singh, A., Sethi, P., Ghosh, S., Singh, K. K., Rohatgi, S., Singh, P., & Sinha, J. K. (2025). Current status of prevention strategies adapted for Alzheimer's disease. In *A New Era in Alzheimer' s Research* (pp. 91–108). https://doi.org/10.1016/b978-0-443-15540-6.00004-5

Sinha, J. K., Ghosh, S., Ghosh, S., Bhaskar, R., & Han, S. S. (2023). Improving Therapeutic Approaches to Insomnia: The Need for Objective Sleep Data. *The Lancet Healthy Longevity*, *4*(9). https://doi.org/10.1016/s2666-7568(23)00158-7

Sinha, J. K., Ghosh, S., Swain, U., Giridharan, N. V., & Raghunath, M. (2014). Increased Macromolecular Damage due to Oxidative Stress in the Neocortex and Hippocampus of WNIN/Ob, a Novel Rat Model of Premature Aging. *Neuroscience*, *269*, 256–264. https://doi.org/10.1016/j.neuroscience.2014.03.040

Sinha, J. K., Jorwal, K., Singh, K. K., Han, S. S., Bhaskar, R., & Ghosh, S. (2024). The Potential of Mitochondrial Therapeutics in the Treatment of Oxidative Stress and Inflammation in Aging. *Molecular Neurobiology*, *62*(6), 6748–6763. https://doi.org/10.1007/s12035-024-04474-0

Sinha, J. K., Trisal, A., Ghosh, S., Gupta, S., Singh, K. K., Han, S. S., Mahapatra, M., Abomughaid, M. M., Abomughayedh, A. M., Almutary, A. G., Iqbal, D., Bhaskar, R., Mishra, P. C., Jha, S. K., Jha, N. K., & Singh, A. K. (2024). Psychedelics for Alzheimer's Disease-Related Dementia: Unveiling Therapeutic Possibilities and Pathways. *Ageing Research Reviews*, *96*. https://doi.org/10.1016/j.arr.2024.102211

Sliwinski, M., & Buschke, H. (1999). Cross-Sectional and Longitudinal Relationships among Age, Cognition, and Processing Speed. *Psychology and Aging*, *14*(1), 18–33. https://doi.org/10.1037/0882-7974.14.1.18

Spector, A., Orrell, M., Davies, S., & Woods, R. T. (2000). Reminiscence therapy for dementia. In *Cochrane Database of Systematic Reviews*. https://doi.org/10.1002/14651858.Cd001120

Spires, T. L., & Hyman, B. T. (2005). Transgenic Models of Alzheimer's Disease: Learning from Animals. *NeuroRX*, *2*(3), 423–437. https://doi.org/10.1602/neurorx.2.3.423

Stern, Y. (2009). Cognitive Reserve☆. *Neuropsychologia*, *47*(10), 2015–2028. https://doi.org/10.1016/j.neuropsychologia.2009.03.004

Taler, V., & Phillips, N. A. (2008). Language Performance in Alzheimer's Disease and Mild Cognitive Impairment: A Comparative Review. *Journal of Clinical and Experimental Neuropsychology*, *30*(5), 501–556. https://doi.org/10.1080/13803390701550128

Tan, C.-C., Yu, J.-T., Wang, H.-F., Tan, M.-S., Meng, X.-F., Wang, C., Jiang, T., Zhu, X.-C., & Tan, L. (2014). Efficacy and Safety of Donepezil, Galantamine, Rivastigmine, and Memantine for the Treatment of Alzheimer's Disease: A Systematic Review and Meta-Analysis. *Journal of Alzheimer's Disease*, *41*(2), 615–631. https://doi.org/10.3233/jad-132690

Thorgrimsen, L., Selwood, A., Spector, A., Royan, L., de Madariaga Lopez, M., Woods, R. T., & Orrell, M. (2003). Whose Quality of Life Is It Anyway?: The Validity and Reliability of the Quality of Life-Alzheimer's Disease (QoL-AD) Scale. *Alzheimer Disease & Associated Disorders*, *17*(4), 201–208. https://journals.lww.com/alzheimerjournal/fulltext/2003/10000/whose_quality_of_life_is_it_anyway___the_validity.2.aspx

Tsai, P.-C., Akpan, A., Tang, K.-T., & Lakany, H. (2025). Brain Computer Interfaces for Cognitive Enhancement in Older People – Challenges and Applications: A Systematic Review. *BMC Geriatrics*, *25*(1). https://doi.org/10.1186/s12877-025-05676-4

Uddin, M., Ganapathy, K., & Syed-Abdul, S. (2024). Digital Technology Enablers of Tele-Neurorehabilitation in Pre- and Post-COVID-19 Pandemic Era – A Scoping Review. *International Journal of Telerehabilitation*, *16*(1). https://doi.org/10.5195/ijt.2024.6611

Unger, J. M., van Belle, G., & Heyman, A. (2015). Cross-Sectional versus Longitudinal Estimates of Cognitive Change in Nondemented Older People: A CERAD Study. *Journal of the American Geriatrics Society*, *47*(5), 559–563. https://doi.org/10.1111/j.1532-5415.1999.tb02570.x

Van der Elst, W., Van Boxtel, M. P. J., Van Breukelen, G. J. P., & Jolles, J. (2006). The Stroop Color-Word Test. *Assessment*, *13*(1), 62–79. https://doi.org/10.1177/1073191105283427

Vanbeijsterveldt, C., Vanboxtel, M., Bosma, H., Houx, P., Buntinx, F., & Jolles, J. (2002). Predictors of Attrition in a Longitudinal Cognitive Aging Study: The Maastricht Aging Study (MAAS). *Journal of Clinical Epidemiology*, *55*(3), 216–223. https://doi.org/10.1016/s0895-4356(01)00473-5

Venneri, A., & Shanks, M. F. (2013). Central Effects of Cholinesterase Inhibitors in Alzheimer's Disease: Insights from Advanced Neuroimaging. *Imaging in Medicine*, *5*(5), 441–452. https://doi.org/10.2217/iim.13.51

Wang, J., Yu, J.-T., Wang, H.-F., Meng, X.-F., Wang, C., Tan, C.-C., & Tan, L. (2015). Pharmacological Treatment of Neuropsychiatric Symptoms in Alzheimer's Disease: A Systematic Review and

Meta-Analysis. *Journal of Neurology, Neurosurgery & Psychiatry*, *86*(1), 101–109. https://doi.org/10.1136/jnnp-2014-308112

Wang, K., Fang, Y., Zheng, R., Zhao, X., Wang, S., Lu, J., Wang, W., Ning, G., Xu, Y., & Bi, Y. (2024). Associations of Socioeconomic Status and Healthy Lifestyle with Incident Dementia and Cognitive Decline: Two Prospective Cohort Studies. *eClinicalMedicine*, *76*. https://doi.org/10.1016/j.eclinm.2024.102831

Wang, W., Zhao, F., Ma, X., Perry, G., & Zhu, X. (2020). Mitochondria Dysfunction in the Pathogenesis of Alzheimer's Disease: Recent Advances. *Molecular Neurodegeneration*, *15*(1). https://doi.org/10.1186/s13024-020-00376-6

West, R. L. (1996). An Application of Prefrontal Cortex Function Theory to Cognitive aging. *Psychological Bulletin*, *120*(2), 272–292. https://doi.org/10.1037/0033-2909.120.2.272

World Health, O. (2024). Ageing and health [Web Page]. www.who.int/news-room/fact-sheets/detail/ageing-and-health

Xu, L., Gu, H., Cai, X., Zhang, Y., Hou, X., Yu, J., & Sun, T. (2023). The Effects of Exercise for Cognitive Function in Older Adults: A Systematic Review and Meta-Analysis of Randomized Controlled Trials. *International Journal of Environmental Research and Public Health*, *20*(2). https://doi.org/10.3390/ijerph20021088

Zarit, S. H., Reever, K. E., & Bach-Peterson, J. (1980). Relatives of the Impaired Elderly: Correlates of Feelings of Burden. *The Gerontologist*, *20*(6), 649–655. https://doi.org/10.1093/geront/20.6.649

3
GERIATRIC REHABILITATION
Evidence-Based Strategies for Aging-Related Cognitive Impairments

Pooja Bharadwaj, Shilpa Kuppahalli Srinath, and Roseline Gomes

Introduction

Scope of Geriatric Rehabilitation

Mrs. Anita Rao is a 78-year-old retired teacher who lives in an apartment in Bangalore. She has been living alone for the last eight years since her husband's death. Her son lives abroad and her daughter lives in Mysore, who visits her once a month. Over the last year, Anita's domestic help and her neighbors started noticing signs of forgetfulness, misplacing objects, repeating questions, and occasional confusion regarding dates. She also stopped visiting temples regularly and stopped doing crossword puzzles which she once loved. Initially, her family assumed that this was "normal aging"; however, the problem escalated when Anita got lost while going to her regular park and called her daughter in distress. Her daughter took her for an evaluation and the neuropsychological assessment indicated signs of mild cognitive impairment (MCI), especially executive functioning and short-term memory. She also exhibited symptoms of shame and anxiety around her memory lapses.

Geriatric rehabilitation (GR) aims to enhance or restore residual functional capacity and improve the quality of life in aging people, especially those with disabling impairments and/or having fragility (Van Balen et al., 2019). It addresses the complex needs of aging adults with conditions such as MCI, Alzheimer's disease, and other forms of dementia. GR is a person-centered, interdisciplinary process which acknowledges the social, psychological, and biological changes due to aging and GR functions across these domains to restore and enhance everyday functioning and dignity.

In Anita's case, GR is not just about prolonging cognitive decline, but also devising a holistic framework which includes enhancing compensatory strategies, improving her routine, preventing social isolation, ensuring home safety, and addressing risks due to forgetfulness. This interdisciplinary approach will support her adaptation to life with cognitive impairment by strengthening her social and emotional well-being, while maintaining her independence.

 DOI: 10.4324/9781003646662-4

The scope of GR is multidisciplinary and multidimensional, emphasizing quality of life, autonomy, safety, and functioning. The detailed scope is discussed below:

Cognitive Rehabilitation and Stimulation: The main focus here is to enhance memory, executive functioning, attention, and daily problem-solving through specific strategies. The different methods used include cognitive training, cognitive simulation therapy, and compensatory strategy teaching such as external memory aids. Such strategies have found evidence in improving cognition and quality of life in dementia (Woods et al., 2023).

Functional and Occupational Rehabilitation: Maintaining independence in instrumental activities of daily living (IADLs) and activities of daily living (ADLs). The main interventions include task simplification, assistive device use, environmental modification, and routine building. These strategies improve self-efficacy and prolong institutionalization (Graff et al., 2006).

Psychological Support: The focus here is to address anxiety, depression, adjustment issues, and disruption of identity associated with cognitive decline. The main interventions include psychotherapy (mostly supportive), psychoeducation, support for grief, and stress reduction. These approaches can help in improving mood, engagement, and compliance in rehabilitation settings (Piersol et al., 2017).

Social Reintegration: The focus here is to ensure that aging clients are less isolated and maintain their social identity, also purposeful engagement becomes the focus for social reintegration. The interventions here are mainly facilitating participation in clubs, intergenerational activities, structured group work, and faith-based groups. These methods have evidence in delaying cognitive decline and improving overall emotional well-being (Holtzman et al., 2004).

Environmental Safety and Adaptation: The main focus here is to reduce any risk related to falls, prevent the aging clients from wandering, and also support task completion. The approach used here is to assess home safety, label objects, set up memory stations, and also install automatic shut-off devices. The objective is to align with the World Health Organization's (WHO's) age-friendly environment recommendations (World Health Organization (WHO), 2007).

Caregiver Support: The main objective of this goal is to prevent caregiver's burnout, ensuring continuity of care and further treatment planning. The main services include respite care, support groups, and caregiver training. Studies have indicated better patient outcomes and delayed institutionalization by ensuring caregivers are supported (Mittelman et al., 2007).

In summary, GR for cognitive impairment goes beyond memory impairment. It is a comprehensive care ecosystem that includes clinical and cognitive enhancement, psychological well-being and caregiver resilience, and also environmental adaptation. It bridges the gap between decline and dignity, ensuring that the aging adults not just live longer, but also better.

Overview of Aging-Related Cognitive Impairments (e.g., MCI, Alzheimer's, Vascular Dementia)

When we age, there are gradual changes in cognitive functioning which are considered a normal part of the aging process. However, for some older adults, these variations in cognitive functioning can be more underlined and significant, impacting their executive functioning, language, attention, and memory. These changes can lead to the onset of aging-related cognitive impairments. These changes or impairments lie on a continuum that ranges from typical aging to MCI and in worse cases to neurocognitive disorders such as dementia. Each of these will be discussed below:

Normal Aging: Normal aging includes mild to modest decline in working memory, processing speed, ability to multi-task, and episodic memory. However, their semantic memory and language abilities remain intact (Harada et al., 2013). The above changes do not impair day-to-day functioning significantly and are not markers of a disorder.

MCI: MCI depicts a transitional state between normal aging and major disorders such as dementia. Patients with MCI experience some amount of cognitive decline, which is higher than that expected for their age and education, yet these patients are able to sustain independence in their day-to-day activities (Peterson et al., 2014). MCI is a known risk factor for Alzheimer's disease, with 10–15 percent of patients with MCI progressing into Alzheimer's disease every year (Roberts et al., 2013). In rare cases, some patients with MCI remain stable and sometimes revert to their normal level of cognitive functioning.

Major Neurocognitive Disorders (Dementia): These disorders are characterized by apparent and significant cognitive decline which interferes with an individual's daily functioning. It includes decline in cognitive functions such as memory, language, reasoning, and behavior. The main types of dementia include Alzheimer's disease, vascular dementia, Lewy body dementia, and frontotemporal dementia (FTD) (Alzheimer's Association, 2023).

Alzheimer's Disease: It is the most common type of dementia which accounts for 60–80 percent of dementia cases. It is characterized by accumulation of beta-amyloid plaques and tau neurofibrillary tangles, which leads to dysfunction in the synapses and also leads to loss of neurons, specifically in the cerebral cortex and hippocampus (DeTure & Dickson, 2019). The clinical presentation includes slow onset and progressive memory loss, impairments in executive functioning, visuospatial skills, and language (aphasia), and behavioral concerns such as irritability and apathy (Alzheimer's Association, 2023).

Vascular Dementia: This is the second most common type after Alzheimer's disease which results from cerebrovascular damage, impairing blood flow to the brain (O'Brien & Thomas, 2015). It is usually diagnosed after the patient experiences stroke, chronic ischemia, or small vessel disease. It might manifest as a single large infarct or several lacunar infarcts or diffuse white matter lesions (Kalaria, 2016). The clinical presentation includes step-wise progression after cerebrovascular damage, slow processing speed, early dysfunction in executive functioning, and gait disturbances.

Lewy Body Dementia: This type of dementia accounts for 10–15 percent of dementia cases. It has features of both Parkinson's and Alzheimer's disease (McKeith et al., 2017). It is characterized by the presence of Lewy bodies (alpha-synuclein protein aggregates) in the brainstem and cortex. The clinical presentation includes fluctuating cognition with significant changes in alertness and attention, persistent visual hallucinations, motor symptoms such as tremors and rigidity, and REM (Rapid Eye Movement) sleep behavior disorder.

FTD: This type of dementia occurs much before the other types of dementias with onset between 45 and 65 years. It includes degeneration of temporal lobes and frontal lobes with several underlying proteinopathies including tau, TDP-43 (,TAR DNA-binding protein 43) and FUS (Fused in Sarcoma) proteins. The clinical presentation can include a behavioral variant of FTD which is characterized by loss of empathy, compulsive behaviors, disinhibition, and apathy. The other variant is primary progressive aphasias which include loss of meaning of words or effortful speech (Bang et al., 2015).

Understanding the different subtypes of dementia is important for precise diagnosis, caregiver planning, and management. Though it is common to observe clinical overlap in symptoms, each type has specific features and pathological mechanisms that inform treatment and prognosis.

Epidemiological Trends and Public Health Significance

As life expectancy increases globally, aging-related cognitive impairments have become a public health concern. Cognitive impairments such as MCI and dementias are increasingly common among the aging population and are expected to increase significantly in the near future. Presently, more than 55 million people are diagnosed with dementia across the world and this statistic is expected to rise to 139 million by the year 2050 (World Health Organization (WHO), 2021). The prevalence of age-related conditions, particularly dementia, increases significantly with age and is believed to affect 5–8 percent of people over 60 across the world (Prince et al., 2015). For individuals who are over 85, the prevalence can increase to 30 percent (Alzheimer's Association, 2023). MCI is expected to impact 10–20 percent of individuals over 65 and 10–15 percent of these individuals progress to dementia every year (Roberts et al., 2013). Women are reported to have higher prevalence of Alzheimer's disease when compared to men, mainly due to greater life expectancy and other social and biological factors (Beam et al., 2018). Low- and middle-income countries (LMICs) are reported to experience the most rapid growth in dementia prevalence due to increased life expectancy and demographic transitions (Livingston et al., 2020).

The impact of cognitive impairment goes beyond the individual and affects caregivers, healthcare systems, and the economy at large. Dementia is reported to be one of the most expensive conditions across the world with costs exceeding USD 1.3 trillion in 2019, which is projected to double by 2030 (World Health Organization (WHO), 2021). Patients with cognitive deficits have greater hospitalization rates, increased use of long-term care services, and are also at greater risk of institutionalization (Hurd et al., 2013).

Care for people diagnosed with dementia is often provided by informal caregivers who are family members, leading to increased levels of financial, emotional, and physical distress. These caregivers are more prone to experiencing anxiety, depression, and chronic health concerns (Brodaty & Donkin, 2009).

Patients with dementia have increased rates of comorbid disorders such as depression, hypertension, and diabetes. Dementia is one of the most common causes for death across the world. This raises the vulnerability to complications such as malnutrition, infections, and falls (Prince et al., 2015). There is dire need for public health interventions that focus on reduction of risk (e.g., nutrition, promoting cognitive engagement, physical activity, and controlling hypertension). Globally, national dementia strategies are being developed to develop community-based care models, health system strengthening, early detection, and caregiver support (Livingston et al., 2020; World Health Organization (WHO), 2021).

Pathophysiology and Risk Factors of Cognitive Decline in Older Adults

Neurobiological Changes with Aging

Aging is linked with several biochemical, functional, and structural variations in the brain that together affect motor control, emotional regulation, and cognition. While not all aging adults experience disorder-level cognitive decline, even typical aging includes certain notable neurobiological remodeling that impacts plasticity.

Structural Brain Changes: One of the significant variations in aging is the regional and global atrophy, which decrease in white and gray matter volumes. The hippocampus and prefrontal cortex show significant atrophy. These areas are linked with memory, executive functions, which often exhibit age-related deterioration (Raz et al., 2005; Fjell & Walhovd, 2010). Aging also results in demyelination and white matter hyperintensities, specifically in the frontal–subcortical circuits, which in turn impacts connectivity and processing speed (Gunning-Dixon et al., 2009).

Alterations in Neurotransmitter Systems: Variations in significant neurotransmitter systems result in emotional and cognitive changes. The dopamine system shows a significant decline in synthesis of dopamine, availability of receptors, and reduction in transporter density, specifically in the striatum. This is associated with decreased psychomotor speed and cognitive flexibility (Bäckman et al., 2006). The acetylcholine and serotonin systems show decline in binding and synthesis of receptors which leads to impairments in memory and dysregulation in mood (Meltzer et al., 1998; Bartus et al., 1982).

Neuronal and Synaptic Changes: Although neurons are largely preserved with age, the density of synapses decreases markedly. The synaptic changes in the cortex and hippocampus are more predictive of decline in cognitive function than the neuronal loss (Morrison & Baxter, 2012). Long-term potentiation (LTP) also declines with age, which is an important mechanism for synaptic plasticity, impacting memory and learning (Burke & Barnes, 2006).

Neuroinflammation and Oxidative Stress: With age, there is increase of microglial activation which leads to chronic inflammation which can expedite neurodegeneration (Franceschi & Campisi, 2014). Increased secretion of reactive oxygen (oxidative stress) and decreased capacity of antioxidants results in damage to DNA, proteins, lipids, and further impact to neuronal function (Harman, 2006).

Functional Brain Reorganization: As people age, the brain projects certain functional compensation through additional neural network recruitment. For instance, aging adults often have activation in the bilateral prefrontal areas while doing cognitive tasks. This pattern is called HAROLD (hemispheric asymmetry reduction in older adults) (Cabeza, 2002). Additionally, as people age, the brain gathers more resources to sustain the performance until the decline threshold is reached. This is called CRUNCH (compensation-related utilization of neural circuits hypothesis) (Reuter-Lorenz & Cappell, 2008).

Plasticity and Neurogenesis: Neurogenesis (specifically in the hippocampus) often persists into old age; it happens at a slower and reduced pace. Aspects such as stress, inflammation, and decreased neurotropic support impact neuroplasticity negatively (Kempermann, 2015). Lifestyle modifications such as cognitive stimulation, enriched environments, and exercise have been reported to increase neuroplastic responses (Erickson et al., 2011).

The neurobiology of aging is marked by active interplay between adaptation and degeneration. Comprehending these changes is crucial to the basis for planning age-appropriate interventions in geriatric cognitive rehabilitation and also promotes healthy aging of the brain.

Common Risk Factors (Genetic, Vascular, Lifestyle, Psychological)

Cognitive difficulties in geriatric population come from intricate interplay of psychological stressors, lifestyle behaviors, vascular compromise, and genetic vulnerability. Comprehending these risk factors is crucial for early identification, targeted prevention, and also devising tailored intervention strategies.

Genetic Risk Factors: There are few genes that have a direct impact on development of age-related cognitive conditions. For example, the apolipoprotein E (APOE) ε4 allele is the most potent risk factor for late-onset Alzheimer's disease. Carriers of one or two ε4 alleles have a significantly higher risk of developing AD, with earlier onset and greater amyloid-β deposition (Corder et al., 1993; Liu et al., 2013). Unusual autosomal dominant mutations in APP, PSEN1, and PSEN2 genes are capable of causing early-onset Alzheimer's disease (Selkoe & Hardy, 2016). Recent research has indicated that a blend of several genetic variants can lead to overall cognitive decline risk beyond APOE (Escott-Price et al., 2015).

Vascular Risk Factors: Chronic conditions such as hypertension, hyperlipidemia, and diabetes lead to small vessel disease, white matter changes, and decreased cerebral perfusion. All of these are associated with both mixed and vascular dementias (Ladecola, 2013). Cerebrovascular conditions such as stroke and TIAs (transient ischemic attacks) raise the risk of vascular cognitive concerns by damaging the neural pathways and blocking the integrity of the blood–brain barrier (Gorelick et al., 2011). Heart rhythm irregularity can often result in cerebral emboli and also accelerate the risk for Alzheimer's disease and vascular dementia (Stefansdottir et al., 2013).

Lifestyle-Related Risk Factors: Physical inactivity, poor diet, smoking, regular alcohol consumption, and poor sleep can all lead to increased risk of cognitive impairments. Sedentary lifestyle with limited physical activity is linked with increased atrophy in the brain and reduction in the hippocampal volume. However, adequate physical activity leads to better vascular health and neurogenesis (Erikson et al., 2011). Diets that are high in sugars and saturated fats are associated with cognitive decline and diets with Mediterranean components are known to be protective (Morris et al., 2015). Consumption of tobacco accelerates oxidative stress and vascular damage. Increased alcohol use can result in neurotoxicity (Anstey et al., 2007). Poor sleep or disturbed sleep impairs the glymphatic clearance of beta-amyloid, increasing risk for dementia (Xie et al., 2013).

Social and Psychological Risk Factors: Decreased social engagement has been associated with accelerated cognitive decline and Alzheimer's disease risk (Donovan et al., 2017).

Cognitive decline in older adults is not due to a single causal factor but results due to accumulative risk across several domains. Assimilating vascular monitoring, genetic screening, psychosocial support, and lifestyle modification provides a holistic approach to early identification, prevention, and timely intervention.

Interplay between Cognitive, Sensory, and Physical Decline

Growing old is characterized by progressive and gradual decline across several systems—cognitive, physical, and sensory—and these changes do not happen in isolation. Instead, these systems dynamically interact and frequently compound each other's impacts and expedite the overall decline. Identifying the overlapping nature of these variations is crucial for developing care strategies and integrated rehabilitation services.

Sensory and Cognitive Decline: Sensory concerns, especially decline in vision and hearing, have often been associated with deterioration in cognitive abilities in older adults. Longitudinal studies have reported that people with hearing impairments are at higher risk for developing dementia and cognitive impairment (Lin et al., 2011; Livingston et al., 2020). The hypothesized mechanism includes reduction in auditory input resulting in cortical reorganization, social isolation, and higher cognitive load, which are risk factors for dementia. In a similar manner, visual problems are linked with declined performance in attention, executive function, and memory (Whitson et al., 2018). The sensory deprivation theory indicates that reduction in sensory input results in neuroplastic changes which increase neuronal degeneration in connected or adjacent areas of the brain (Armstrong et al., 2020).

Cognitive and Physical Decline: Cognitive decline has a strong association with physical functioning. Slower processing speed and executive dysfunction can impact coordination, balance, and gait, which increases the risk of falls and injuries (Montero-Odasso et al., 2012). In contrast, decreased physical activity as a result of chronic disease or mobility limitations can further limit neurogenesis and cognitive stimulation, especially in the prefrontal cortex and hippocampus (Erickson et al., 2011). Muscle mass and strength loss as a result of aging is called sarcopenia and is associated independently with MCI and dementia. Cognitive frailty, a disorder combining cognitive impairment without dementia and physical frailty, has emerged as a useful framework in identifying geriatric populations at increased risk for adverse outcomes (Kelaiditi et al., 2013).

Physical and Sensory Decline: Loss in sensory abilities can lead to decline in mobility and balance. For instance, hearing loss can impact spatial orientation and raise the risk of falls, while deficits in the visual domain can result in challenges while navigating environments and also have a detrimental effect on the confidence level (Lin & Ferrucci, 2012). These variations can restrict mobility, decrease outdoor activity, and also lead to frailty and deconditioning.

Viscous Cycles and Synergistic Effects: The simultaneous occurrence of physical, cognitive, and sensory decline results in synergistic impacts that accelerate the functional decline. For example, an aging adult with hearing difficulties, mobility challenges, and cognitive slowing can find it difficult to engage in occupational and social activities, resulting in depression, isolation, and sooner decline in all the domains (Baltes & Lindenberger, 1997). This interdependence indicates that unidimensional approaches can have limited efficacy unless they intervene across multiple domains at the same time.

The interplay of cognitive, physical, and sensory changes underlines the importance of interdisciplinary and multidimensional approach to GR. Early identification and simultaneous management of physical and sensory deficits along with cognitive decline can lead to better results than isolated interventions.

Diagnostic and Assessment Strategies

Diagnosing cognitive decline and neurodegenerative disorders is both an art and a science. It involves piecing together subtle clues—some medical, some behavioral, some purely intuitive. But behind this nuanced process are structured strategies and tools that help healthcare providers form a clearer picture of a patient's cognitive and functional status.

Clinical Assessment Tools

Several standardized tools are routinely used in clinical settings to evaluate cognitive function. These tools are not definitive diagnostic instruments on their own, but they provide critical insights and guide further investigation.

Mini-Mental State Examination (MMSE)

One of the most widely used tools, the Mini-Mental State Examination (MMSE) evaluates orientation, memory, attention, language, and visuospatial skills. It's quick—typically 5–10 minutes—but its brevity is both a strength and a limitation. It's particularly helpful in identifying moderate-to-severe cognitive impairment but can miss milder deficits, especially in highly educated individuals (Folstein et al., 1975).

Montreal Cognitive Assessment (MoCA)

Developed to detect MCI, MoCA includes more complex tasks that assess executive functioning, abstraction, and delayed recall. It has become a preferred tool in many clinical environments due to its sensitivity in catching early-stage cognitive decline (Nasreddine et al., 2005).

Confusion Assessment Method (CAM)

Designed to identify delirium rather than chronic cognitive impairment, CAM is indispensable in hospital settings where acute confusion may point to an underlying issue like infection or medication side effects (Inouye et al., 1990).

These tools are not interchangeable but complementary. Choosing the right one depends on the clinical context and the patient's presentation.

Functional and Neuropsychological Evaluations

Beyond brief cognitive screening, comprehensive functional and neuropsychological assessments offer a more detailed understanding of how cognitive changes are affecting a person's daily life.

Functional Assessments

These evaluations explore a patient's ability to perform ADLs and IADLs—things like cooking, managing medications, or handling finances. Decline in these areas is often the first sign noticed by family members.

Neuropsychological Testing

Conducted by specialists, these assessments use a battery of tests to explore specific cognitive domains: memory, attention, language, problem-solving, and visuospatial skills. They can differentiate between types of dementia (e.g., Alzheimer's vs. frontotemporal) and pinpoint strengths and weaknesses to guide care planning.

These evaluations often take several hours and require close collaboration with patients and caregivers. They're invaluable for nuanced diagnoses and long-term management strategies.

Role of Interdisciplinary Teams in Assessment

Effective diagnosis and care planning for cognitive disorders require a team-based approach. No single clinician can address the complexity of these conditions in isolation.

An interdisciplinary team might include the following:

Neurologists
Psychiatrists
Primary care physicians
Neuropsychologists
Occupational and speech therapists
Social workers
Nurses and nurse practitioners

Each professional brings a unique lens. For example, while a neurologist might interpret MRI findings, a social worker can assess the patient's living environment and support systems. This holistic approach ensures that medical, functional, emotional, and social dimensions are all taken into account.

Family members are also critical members of this team. Their observations often provide the first signs that something isn't right, and their support can make or break a patient's ability to remain safely at home.

Importance of Early Detection and Baseline Tracking

Early detection is key. Identifying cognitive decline at its earliest stages allows for better planning, more effective use of available treatments, and a greater chance to implement lifestyle modifications that may slow progression.

Baseline tracking—monitoring a patient's cognitive function over time—helps distinguish between normal age-related changes and concerning decline. It also allows for the measurement of treatment effects and timely updates to care plans.

In many cases, early assessment can empower patients to participate in decision-making while they are still cognitively able, including making choices about living arrangements, financial planning, and end-of-life care.

Effective diagnosis and assessment are not about simply scoring a test. They're about understanding a person—how they think, function, feel, and live. Using structured tools and collaborative teams, we can provide not only a label for a cognitive disorder but a roadmap for managing it with compassion and clarity.

Principles of Geriatric Cognitive Rehabilitation

Cognitive rehabilitation in older adults isn't about "fixing" the brain—it's about enhancing quality of life, maintaining dignity, and supporting independence. Aging brains are complex and dynamic, and so are the people they belong to. The heart of geriatric cognitive rehabilitation lies in understanding this complexity and crafting interventions that are compassionate, personalized, and purposeful.

Person-Centered and Goal-Directed Approach

The most effective cognitive rehabilitation programs start with one simple but powerful question: What matters most to the person?

A person-centered approach puts the individual's values, preferences, and life goals at the center of care. This philosophy moves away from a "deficit-focused" model and instead emphasizes what the person can do, what they want to do, and how they define success.

When rehabilitation is goal-directed, it aligns clinical interventions with personally meaningful outcomes—like being able to cook a favorite meal, remember grandchildren's names, or navigate the neighborhood safely.

Research has shown that when therapy is aligned with personal goals, adherence and engagement improve significantly (Clare et al., 2003). Even small victories—such as organizing a daily pillbox or learning to use memory aids—can have profound impacts on a person's confidence and autonomy.

Functional vs. Restorative Rehabilitation

Cognitive rehabilitation typically falls into two broad categories: restorative and functional (compensatory) approaches.

Restorative rehabilitation aims to improve or restore impaired cognitive functions through repetitive practice and cognitive exercises (e.g., memory drills, attention training). This approach assumes some degree of neuroplasticity, even in the aging brain (Cicerone et al., 2011).

Functional or compensatory rehabilitation, on the other hand, focuses on helping individuals adapt to their current abilities. This may involve using memory aids, simplifying routines, or modifying environments to reduce confusion and increase safety.

For older adults with progressive conditions like Alzheimer's disease, restorative gains may be modest. In these cases, compensatory strategies often offer more practical and lasting benefits. The key is tailoring the approach to each individual's condition, capacity, and life context.

Multimodal Interventions (Cognitive, Physical, Social)

Cognitive health is not isolated. It is deeply intertwined with physical wellness, emotional stability, and social connection. As such, multimodal interventions—those that address multiple domains at once—are increasingly recognized as the gold standard.

Cognitive: Activities like structured memory training, attention exercises, or problem-solving games can support cognitive engagement. Technologies such as tablet-based brain training apps can add variety and appeal.

Physical: Exercise, especially aerobic activity, has been shown to improve executive functioning and delay cognitive decline (Gheysen et al., 2018). Group fitness programs, tai chi, and even daily walking routines can contribute significantly.

Social: Isolation is a major risk factor for cognitive deterioration. Group therapy, hobby clubs, or intergenerational programs foster connection, purpose, and emotional well-being.

The synergy of these domains often results in better outcomes than any single intervention alone. When a person is physically active, socially engaged, and mentally stimulated, the cumulative impact can be transformative.

Individualization Based on Stage and Type of Impairment

There is no one-size-fits-all model in geriatric cognitive rehabilitation. What works for an individual with mild vascular cognitive impairment will differ significantly from someone in the later stages of Alzheimer's disease.

In early-stage cognitive impairment, interventions may include memory strategy training, lifestyle modification, and proactive planning.

In moderate stages, the focus often shifts to environmental adaptations, caregiver support, and task simplification.

In advanced stages, sensory stimulation, music therapy, and gentle routines that promote comfort and reduce agitation become central.

Additionally, the type of impairment matters. For example, someone with FTD may struggle more with behavior and judgment than memory, while individuals with Lewy body dementia may benefit from structured routines and visual cues to navigate fluctuating cognition.

Tailoring the plan to the individual's specific challenges and strengths not only enhances effectiveness but also honors their humanity. No two lives are alike—and no two brains are either.

Summary: Geriatric cognitive rehabilitation is not just a clinical discipline—it's a compassionate art. It requires listening deeply, thinking holistically, and crafting interventions that reflect the whole person. By combining evidence-based practices with empathy and creativity, we can empower older adults to live with dignity, purpose, and as much independence as possible.

Evidence-Based Interventions

Neurocognitive rehabilitation is about more than repairing cognitive skills—it's about restoring a sense of purpose, connection, and self. As people age, neurodegenerative conditions like Alzheimer's disease, vascular dementia, and Parkinson's-related cognitive decline present complex challenges. But the growing body of evidence-based interventions offers hope, supporting clinicians, caregivers, and older adults alike.

Cognitive Training and Stimulation

At the heart of cognitive rehabilitation is targeted cognitive training, which aims to strengthen specific cognitive domains such as memory, attention, or executive function.

Cognitive training involves structured activities designed to improve or maintain specific cognitive skills, such as memory, attention, or executive function. It's one of the cornerstones of neurocognitive rehabilitation.

Case Study: Mrs. Gupta and Memory Training

Mrs. Gupta, a 72-year-old retired teacher, was diagnosed with MCI. She reported frustration with forgetting appointments and misplacing items. Her rehabilitation plan included spaced retrieval training and external memory aids, such as a daily planner and digital reminders.

Over six weeks, Mrs. Gupta practiced associating tasks with cues (e.g., "Take pills after breakfast"). She also used a mobile app for memory games targeting working memory and attention.

Outcome: She showed improved consistency in daily routines and expressed greater confidence. Her MoCA score remained stable, and her subjective distress about memory lapses decreased.

Cognitive training, especially when personalized and paired with real-world tasks, can improve task performance and perceived cognitive competence (Cicerone et al., 2011). Computer-based programs like BrainHQ may enhance specific domains but are best used in combination with therapist-led sessions (Simons et al., 2016).

Memory Strategies and Executive Function Exercises

Techniques like spaced retrieval, mnemonics, visual imagery, and categorization help individuals compensate for short-term memory difficulties. Executive function training—focusing on planning, task-switching, and problem-solving—is critical in supporting real-world functioning, such as managing schedules or handling finances (Cicerone et al., 2011).

Computer-Based Cognitive Programs

Digital tools such as BrainHQ or Lumosity have become accessible ways to offer training in attention, processing speed, and working memory. While users often report enjoyment and engagement, results vary. These programs can improve performance on trained tasks, but generalization to daily life is less consistent (Simons et al., 2016).

Efficacy and Limitations

Meta-analyses show moderate benefits for older adults, especially those with MCI, though gains may diminish without continuous practice (Gates et al., 2011). As with all rehabilitation efforts, effectiveness increases when interventions are person-centered and combined with physical and social engagement.

Physical Exercise

Physical activity is a powerful, underutilized tool in neurocognitive rehabilitation. It not only improves overall health but also enhances neuroplasticity—the brain's ability to reorganize and form new connections. What benefits the heart also benefits the brain. In fact, physical activity is one of the most well-supported interventions for preserving cognitive health.

Types of Effective Exercise

Aerobic Training (e.g., Brisk Walking, Cycling): Increases cerebral blood flow and supports memory consolidation.
Resistance Training (e.g., Weight Lifting): Enhances executive function and mood.
Balance and Flexibility Exercises (e.g., Tai Chi): Support visuospatial awareness and fall prevention.

Impact on Neuroplasticity and Cognitive Outcomes

Exercise increases brain-derived neurotrophic factor (BDNF), a key molecule in neuroplasticity. It also stimulates hippocampal growth, improving memory and spatial navigation (Erickson et al., 2011).

Mechanisms behind Exercise-Induced Neuroplasticity

Increase in BDNF: Exercise stimulates the production of BDNF, a protein crucial for the growth and maintenance of neurons. Elevated BDNF levels are associated with improved memory and cognitive function.
Hippocampal Growth: Regular physical activity, especially aerobic exercises like walking or cycling, promotes the growth of the hippocampus, a brain region integral to memory and learning.
Enhanced Synaptic Plasticity: Exercise improves synaptic plasticity, which is vital for learning and memory processes. This enhancement leads to better cognitive performance.
Reduction in Neuroinflammation: Physical activity can lower levels of inflammatory markers in the brain, which is beneficial since chronic inflammation is linked to cognitive decline.

Guidelines from Recent Meta-Analyses

A review by Northey et al. (2018) found that combined aerobic and resistance training significantly improved global cognition, executive functioning, and attention in older adults. Even low-intensity activities, done consistently, can offer meaningful benefits. A more recent systematic review by Zhu et al. (2023) found that in individuals with MCI, regular aerobic and resistance training improved cognitive scores on validated scales such as MoCA and MMSE. The review emphasized the importance of frequency (more than three sessions per week) and program duration (at least 12 weeks) for optimal impact. Studies using functional MRI and serum biomarker analysis have linked exercise-induced cognitive improvements with changes in BDNF, IGF-1, and inflammatory cytokines (Kirk-Sanchez & McGough, 2014). These changes create a more favorable neural environment for learning and neurorepair.

Case Study: Mr. Hernandez and Aerobic Exercise

Mr. Hernandez, 78, with early Alzheimer's disease, was experiencing increasing apathy and slowed thinking. A physical therapist introduced a structured program combining walking, light resistance bands, and balance exercises. Sessions occurred three times a week at a local senior center.

Outcome: After three months, Mr. Hernandez exhibited improved mood, faster response time on cognitive tasks, and reduced episodes of wandering. His caregiver also noted fewer behavioral issues.

Research Insight

Regular aerobic and resistance training enhances executive functioning and increases hippocampal volume (Erickson et al., 2011). Meta-analyses show physical exercise can significantly improve cognitive performance in older adults (Northey et al., 2018).

Occupational and Functional Therapy

While cognitive rehabilitation often starts with the brain, it must extend to real-world function—this is where occupational therapy (OT) plays a pivotal role. OTs are often the unsung heroes in dementia care. They help individuals maintain independence by focusing on ADLs and IADLs. OT focuses on helping individuals regain autonomy in ADLs and IADLs.

Case Study: Mrs. Klein and Kitchen Safety

Mrs. Klein, an 82-year-old woman with vascular dementia, was struggling with cooking safely. She had recently left the stove on unattended. An OT conducted a home safety assessment and implemented interventions including visual cue labels, a stove timer, and simplification of kitchen tools.

She also received training in task segmentation, learning to complete activities like making tea in a step-by-step format.

Outcome: Mrs. Klein regained the ability to prepare simple meals independently, with her caregiver supervising more complex tasks. Her risk of kitchen accidents was reduced, and her self-esteem improved.

Research Insight

OT interventions, especially those involving environmental adaptation and caregiver coaching, reduce functional decline and improve daily performance (Gitlin et al., 2010).

ADL/IADL Support: OTs help individuals maintain independence in everyday tasks, from bathing and dressing to cooking and medication management. This practical support builds confidence and reduces caregiver burden.

Environmental Modification: Simple home adjustments—like improving lighting, adding signage, and organizing frequently used items—can dramatically reduce confusion and frustration (Gitlin et al., 2010).

Assistive Technology: Tools like digital pill dispensers, talking clocks, or smart home systems support autonomy and safety. Importantly, therapists teach users how and when to use these technologies, ensuring they're not just available, but usable (Gitlin et al., 2010).

Psychosocial and Behavioral Interventions

Cognitive decline often brings emotional challenges—depression, anxiety, apathy, and even behavioral symptoms like agitation or withdrawal. Addressing these is essential for whole-person care.

Therapeutic Approaches

Reminiscence therapy helps individuals connect with long-term memories, fostering identity and social engagement.

Cognitive behavioral therapy (CBT) has been adapted for those with MCI or early dementia to address mood disorders and coping strategies (Laidlaw et al., 2008).

Behavioral activation helps counter apathy by reintroducing pleasurable and meaningful activities.

Role of Caregivers

Family members are essential partners in this process. Training caregivers in communication, stress management, and behavioral support significantly improves outcomes for both parties (Brodaty & Arasaratnam, 2012).

According to Brodaty & Arasaratnam, in 2012 they suggest that empowering the caregiver is one of the most effective ways to support the person with cognitive impairment.

Case Study: Mr. Ali and Reminiscence Therapy

Mr. Ali, 75, recently diagnosed with FTD, began showing signs of emotional withdrawal. A psychologist initiated reminiscence therapy sessions that used photos, music, and objects from Mr. Ali's life in Pakistan.

Group sessions facilitated storytelling and emotional connection. Parallel CBT-informed sessions helped his wife learn strategies to support his engagement and reduce her caregiver burden.

Outcome: Mr. Ali began participating more actively in conversations and his apathy score (AES-C- Apathy Evaluation Scale- Clinician Version) decreased significantly. His wife reported feeling more connected to him and more capable in her caregiving role.

Research Insight

Psychosocial interventions such as reminiscence therapy and CBT are effective in reducing depression and enhancing quality of life for individuals with dementia and their caregivers (Laidlaw et al., 2008; Brodaty & Arasaratnam, 2012).

Pharmacological Adjuncts

In geriatric neurocognitive rehabilitation, medications can offer valuable support—but they are never a silver bullet. While rehabilitation primarily focuses on behavioral, cognitive, and environmental strategies, pharmacological adjuncts can enhance outcomes when used thoughtfully. Yet, in older adults, even well-intended prescriptions can tip the scales toward harm if not carefully managed. This section explores the nuanced role of medications that influence cognition—positively and negatively—in older adults.

This section explores the delicate balance of pharmacological adjuncts in neurocognitive care: where they help, where they harm, and how clinicians can navigate this complex terrain.

Overview of Medications with Cognitive Benefits or Risks

Medications That May Support Cognitive Function

The pharmacological landscape in geriatrics is complicated by age-related changes in drug metabolism, the increased risk of side effects, and the cumulative burden of polypharmacy. Some medications can support cognition, especially in conditions like Alzheimer's disease, while others may inadvertently worsen it.

Medications with Potential Cognitive Benefits

Cholinesterase inhibitors (e.g., donepezil, rivastigmine, galantamine) are approved for mild to moderate Alzheimer's disease. They work by preventing the breakdown of acetylcholine, a neurotransmitter crucial for learning and memory.

Memantine, an NMDA receptor antagonist, is used in moderate-to-severe Alzheimer's to modulate glutamatergic activity and reduce excitotoxicity.

"These medications don't reverse Alzheimer's, but they can help maintain cognitive function and daily living for a time. For many families, that extra year of clarity matters deeply."—Geriatric Psychiatrist.

Clinical trials show modest but statistically significant benefits in global cognition, ADLs, and behavior (Birks & Harvey, 2018).

Medications That Impair Cognition

Many commonly prescribed medications can impair cognition, especially in older adults. These include the following:

Anticholinergics (e.g., Oxybutynin, Diphenhydramine): Block acetylcholine and can cause confusion, delirium, or even long-term cognitive decline.

Benzodiazepines (e.g., Lorazepam, Diazepam): Often used for anxiety or insomnia, but linked to memory problems and increased fall risk.

Opioids, antipsychotics, and some sleep aids (e.g., zolpidem) also contribute to cognitive slowing and sedation.

A major longitudinal study found that higher cumulative use of anticholinergic medications was associated with an increased risk of dementia (Gray et al., 2015).

Current Evidence on Nootropics and Cholinesterase Inhibitors

Cholinesterase Inhibitors

A Cochrane Review by Birks and Harvey (2018) concluded that cholinesterase inhibitors provide modest improvements in cognition and function, especially in the early to moderate stages of Alzheimer's disease. However, they also carry risks—gastrointestinal upset, bradycardia, and dizziness—which can be more pronounced in frail individuals.

Memantine

Memantine is generally well-tolerated and appears to have moderate benefits in cognition and behavior in moderate-to-severe Alzheimer's. It is often used in combination with a cholinesterase inhibitor (McShane et al., 2019).

Nootropics and Supplements

There's growing public interest in so-called nootropics, including over-the-counter supplements like *Ginkgo biloba*, omega-3 fatty acids, and vitamin E. However, evidence for these is inconsistent:

G. biloba has not shown reliable benefit in preventing dementia or improving cognition (DeKosky et al., 2008).
Omega-3s may have some benefit in slowing cognitive decline in those with MCI, but not in established dementia (Yurko-Mauro et al., 2010).
Vitamin E may help delay progression in some individuals with mild to moderate Alzheimer's but should be used cautiously due to bleeding risks at high doses (Dysken et al., 2014).

Polypharmacy Considerations

Polypharmacy—commonly defined as taking five or more medications—is nearly universal among older adults, especially those with chronic diseases. While each medication may be justified individually, together they can lead to drug interactions, increased side effects, and cognitive impairment.

Case Example: Mrs. L's Medication Review

Mrs. L, an 83-year-old woman with MCI, was taking 12 medications, including diphenhydramine for sleep and oxybutynin for bladder control. Her family noticed increasing confusion and episodes of disorientation.

A geriatric pharmacist conducted a comprehensive medication review. By eliminating high-risk medications and simplifying her regimen, her alertness improved within two weeks. Her cognitive testing score also stabilized.

The Beers Criteria and STOPP/START tools can help clinicians identify potentially inappropriate medications in older adults (by the American Geriatrics Society Beers Criteria® Update Expert Panel, 2019).

Clinical Guidelines and Best Practices

Start low and go slow: Always begin with the lowest effective dose.

Regular Medication Reviews: Especially after Hospitalizations or Changes in Cognitive Status

De-prescribing: A structured approach to tapering unnecessary medications can improve cognition and reduce side effects.

Interdisciplinary Collaboration: Involving pharmacists, geriatricians, and caregivers improves safety and outcomes.

Summary

Pharmacological adjuncts in neurocognitive rehabilitation are best viewed as supportive tools, not standalone treatments. When chosen wisely and monitored closely, medications can enhance cognitive function and quality of life. But in a population as vulnerable as older adults, the balance between benefit and risk must be continually assessed.

In the end, the best pharmacological strategy is often the simplest one: the right drug, for the right reason, at the right time—and no more than that.

Multidisciplinary Team Approach

Roles of Geriatricians, Neurologists, Rehab Specialists, Therapists

Geriatricians are adept in identifying unusual and asymptomatic presentations in the elderly and offer complete medication management, which includes de-prescribing, reducing polypharmacy, and identifying adverse drug events (Lester et al., 2020). They are experts in treating complex diseases, numerous comorbidities, frailty, cognitive impairments, and syndromes in the elderly and are geriatricians. The role of geriatricians has been to skillfully address complex end-of-life issues and provide care in multiple settings with a focus on effective transitions of care. Geriatricians utilize a comprehensive approach to balance risk versus benefits for testing and interventions such as surgery for complex geriatric patients, incorporating prognosis, life expectancy, quality of life, and patient's own values.

Case Management and Care Coordination

A multidisciplinary approach to GR, especially for age-related cognitive deficits, is based on efficient case management and care coordination. As a central liaison, a case manager—typically a nurse or social worker—ensures communication between healthcare professionals, eases care transitions, and customizes interventions to meet the patient's changing requirements. By addressing the medical and psychosocial aspects of care, this coordination lessens fragmentation and enhances clinical outcomes (Reuben et al., 2013).

Timely coordination of therapies, diagnostics, and community resources can improve quality of life and reduce readmissions to hospitals in patients with cognitive impairment (Bennett et al., 2018).

Integrating Caregivers and Family into the Team

For cognitive gains to be maintained and treatment plans to be followed, family members and caregivers must be included in the rehabilitation team. Families frequently offer vital information about the patient's daily routines and baseline performance, enabling more individualized and culturally sensitive care planning (Gitlin & Hodgson, 2015). Involving carers also gives them information and coping mechanisms, which lessens their workload and improves the patient's long-term support. Particularly in home-based rehabilitation settings, a collaborative model that views family members as active partners in treatment strengthens the therapeutic relationship and can greatly improve patient results (Nichols et al., 2011).

Tele-Rehabilitation and Technology in Cognitive Rehab

The National Health System (NHS) has epitomized a transitional domain in their developmental phases. Some of these developments include the restructuring of primary care, the integration of various care levels, and continuity of treatment (Maresca et al., 2020). Telemedicine is an emergent way forward to connect patients and medical professionals remotely in order to provide healthcare services. Since it ensures continuity of nurturance both in space that transitions from the hospital to the patient's home and during time preferable after discharge, tele-rehabilitation represents a promising medical intervention. Furthermore, it's needed to boost the frequency and intensity of rehabilitation programs, offer customized rehabilitation care in a setting that is comfortable and familiar for the patient, track and assess the needs and progress of the patient, encourage patient motivation and improve patient satisfaction, validate the outcomes attained by the patient, and possibly lower service costs. Standardized protocols, goals, and targets would be beneficial because, regrettably, nearly all neurorehabilitation (NR) research is characterized by small samples and substantial variability of outcomes. To adequately evaluate the validity of this promising technique, cost-effectiveness analysis linked to clinical results should be incorporated into future tele-rehabilitation experiments. There has been a significant chance that patients with vascular cognitive impairment (VCI) and MCI may develop dementia. The promotion of multifaceted and multi-domain therapies is crucial to halting further decline (Mosca et al., 2020). Technology is becoming more and more relevant because of its increasing efficacy, which is on par with traditional in-person solutions. Through remote control, tele-rehabilitation devices can ensure a low-cost continuum of treatment and involve patients in multifaceted activity programs. One potential drawback of these programs is the elderly population's unfamiliarity with computers and technology.

The cognitive fragility that older persons are susceptible to can have a negative impact on their health, and tele-rehabilitation seems to be a promising tool for reversing this condition (Md Fadzil et al., 2022). Digital health technology-based intervention research further showed that this population's quality of life might be improved. More research is necessary to assess its viability and acceptability, especially with larger samples and individuals from lower socioeconomic backgrounds, as the evidence currently provided is still insufficient. Before creating such an intervention, an assessment of the preparedness and obstacles to employing tele-rehabilitation and informative communication technology platforms should also be carried out. To reach its goal, future studies should assess participants' acceptance and prior knowledge of tele-rehabilitation and identify the interventions that have a definitive impact on particular outcomes.

Recent developments in tele-rehabilitation and technology have governed cognitive rehabilitation among the geriatric population. A key aspect of neuroscience is NR, which is the process of repairing a patient's damaged or disordered neurological function through education, therapy, and training with an emphasis on the patient's autonomy and general well-being (Uddin et al., 2024). Numerous telecare and telehealth service applications have significantly increased since the start of the COVID-19 epidemic and have become an essential component of modern healthcare procedures. One such application is tele-neurorehabilitation (TNR). When travel restrictions and lockdowns hindered rehabilitation services around the world, the significance of TNR was realized, particularly in developed, low-, and middle-income nations. Since telehealth interventions in neurology have been used so widely, TNR has emerged as a separate, independent subspecialty of both telehealth and neuroscience. TNR has been able to enhance patients' quality of life by the use of digital technology like wearables, robotics, and virtual reality (VR). Numerous telecare and telehealth service applications have significantly increased and have become an essential component of modern healthcare procedures. One such application is TNR. The cognitive, age, and other clinical characteristics of people with MCI make it difficult to apply cognitive health apps in these patients (Formica et al., 2024). Hence, it's urgent to integrate tele-rehabilitation and technology within cognitive rehabilitation frameworks.

Rehabilitation is one of the healthcare interventions that have improved as a result of the population's improved living conditions and access to healthcare in developed nations (Maggio et al., 2024). According to this viewpoint, tele-rehabilitation, or TR, is crucial. Using telemedicine to provide remote rehabilitation services to people who are unable to access healthcare is known as TR. TR combines therapy, recovery, and aid with ongoing treatments to promote neurological and psychological recovery. The patient is placed in a family setting, and caregivers play an active role as well. In addition to demonstrating effectiveness in treating cognitive disorders, this approach reduces healthcare costs and improves continuity of specialized care, benefiting patients and their families by minimizing travel, lowering associated expenses, and enhancing access to timely interventions. Therefore, the urgency of remote monitoring and digitalized therapies is the need of the hour.

Remote Monitoring and Digital Cognitive Therapies

Remote monitoring techniques need to be institutionalized for the geriatric population. Rehabilitation with semi-immersive systems and a home computer is emphasized for its multiple advantages, making it an enticing option for many patients (Hadjar et al., 2025). The T-Rehab System uses affective computing (AC) to provide a semi-immersive tele-rehabilitation experience, including facial expression analysis and contactless cardiac monitoring. TRehab closely monitors patients' mental health while they play a personalized, semi-immersive VR game on a desktop PC. The system uses a webcam with

MediaPipe to track their hand movements for interactive exercises, allowing it to tailor treatment content for increased engagement and comfort. TRehab's examination includes two assessments: system performance and cognitive walkthroughs. The first examination is centered on system performance, evaluating the tested game, middleware, and facial emotion monitoring to guarantee hardware compatibility and effective support for AC, gaming, and tele-rehabilitation. The second study examines usability through cognitive walkthroughs, detecting potential concerns with emotion detection and tele-rehabilitation. Future study will also concentrate on offering full-body rehabilitation with precise detection of upper and lower limb movements via MediaPipe. Additionally, work will be done to increase WebGL support, reduce reliance on WebXR polyfills, and improve camera accessibility. Optimizing models to close age-related accuracy gaps and increasing processing speed will be critical to improving system performance.

Healthy cognitive functioning is a vital component of well-being, independence, and optimal aging. Cognitive deficiencies can be caused by a variety of illnesses, including brain injury, mental illness, and neurological abnormalities (Cotelli et al., 2025). Rehabilitation is a highly customized service that is only accessible to patients who can access institutional settings. Telehealth solutions are suitable for initiating the movement of treatment from clinics to patients' homes to address this unmet need.

Wearables, Apps, and Virtual Reality

The EARLY-COGN^3 initiative has the potential to transform the standard therapy approach, prompting a rethinking of how CNDs (Chronic Neurological Disorders) might benefit from digital solutions. This software-based tele-rehabilitation protocol allows doctors to remotely collect data on patient progress and adherence, allowing for improved program customization. The tele-rehabilitation technique would guide to target the need for CND rehabilitation, which the healthcare system currently does not handle well. It was anticipated that this intervention will produce long-term favorable consequences. Furthermore, this study suggested the goal of elucidating the neurophysiological and biochemical underpinnings of the intervention toward the development of effective tele-rehabilitation treatments. EARLY-COGN^3 also identified predictors of therapy. The software-based tele-rehabilitation protocol can allow doctors to remotely collect data on patient progress and adherence, allowing for improved program customization. The dynamic tele-rehabilitation technique would target the need for CND rehabilitation, which the healthcare system is unable to intervene.

Access in Rural or Underserved Populations

Tele-rehabilitation (TRehab) has increasingly gained traction as a solution to meet the needs of older adults facing cognitive impairments, particularly in rural or underserved communities. These areas often lack specialized cognitive rehabilitation services, leading to delays in diagnosis and treatment. Tele-rehabilitation helps bridge this gap by enabling remote delivery of cognitive therapy, assessments, and follow-up care, often through video conferencing, mobile apps, or digital platforms (Abdel-Basset et al., 2024).

However, access remains a significant challenge. Rural populations frequently face systemic barriers, including poor internet connectivity, lack of access to digital devices, and limited technical literacy among both patients and providers (Rehability Journal, 2023). Furthermore, many older adults experience difficulty in using digital interfaces, a challenge compounded by physical limitations and cognitive decline (Zhou et al., 2022).

Several strategies have been proposed to improve access. First, improving broadband infrastructure in rural areas is critical. Second, education and digital literacy programs tailored to older adults can empower them to use telehealth services independently. Finally, incorporating caregivers and community health workers into tele-rehab programs can help facilitate better engagement and adherence (Zhou et al., 2022).

Tele-rehabilitation is not just a temporary fix—it has the potential to redefine long-term cognitive care for underserved populations. However, realizing this potential depends on proactive policy, investment in infrastructure, and inclusive program design that considers the unique needs of rural older adults.

Challenges in Access

Despite its potential, several barriers hinder the widespread adoption of TRehab in these communities:

Technological Limitations: Many rural areas suffer from inadequate internet connectivity and limited access to digital devices, which are essential for telehealth services.
Digital Literacy: Older adults often face challenges in navigating digital platforms, which can impede their ability to engage with tele-rehabilitation services.
Infrastructure Constraints: In some regions, the lack of reliable electricity and technical support further complicates the implementation of tele-rehabilitation programs.

Strategies to Enhance Access

To overcome these challenges and improve access to tele-rehabilitation services, the following strategies are recommended:

Infrastructure Development: Investing in reliable internet connectivity and power supply is crucial to support tele-rehabilitation initiatives.
Training and Support: Providing digital literacy training for older adults and their caregivers can empower them to utilize tele-rehabilitation platforms effectively.
Community Engagement: Involving local healthcare providers and community leaders in the planning and implementation of tele-rehabilitation services ensures that these programs are culturally appropriate and widely accepted.

Challenges and Barriers

Implementing tele-rehabilitation among older adults comes with a number of logistical and systemic challenges. One of the most common barriers is inadequate access to reliable internet and digital devices, particularly in rural and underserved regions (Abdel-Basset et al., 2024). Older adults may also struggle with digital literacy, making it difficult for them to navigate telehealth platforms independently (Zhou et al., 2022). Furthermore, these technological challenges can exacerbate existing health disparities, particularly among minority populations who already face unequal access to healthcare (Taylor & David, 2017).

Health Disparities and Cultural Considerations

Health disparities significantly influence access to and the effectiveness of tele-rehabilitation. For example, individuals from marginalized racial or ethnic groups, and those in low-income brackets, may lack the resources or familiarity with digital tools to engage in remote therapy (Zhou et al., 2022). Cultural beliefs and language barriers also affect participation. Tailoring rehabilitation programs to align with cultural norms and offering services in multiple languages can improve engagement and outcomes (Abdel-Basset et al., 2024).

Adherence and Motivation in Older Adults

Maintaining adherence to tele-rehabilitation programs among older adults can be challenging. Factors such as cognitive decline, lack of immediate feedback, or reduced motivation may hinder long-term engagement (Taylor & David, 2017). However, studies show that older adults are more likely to adhere to programs when interventions include personalized goals, regular feedback, and support from caregivers or healthcare providers (Zhou et al., 2022). Past experience with technology also plays a critical role in maintaining engagement over time.

Ethical Considerations (Consent and Autonomy)

Ethical concerns are especially pronounced when tele-rehabilitation involves individuals with cognitive impairments. Obtaining informed consent can be problematic if the individual has diminished decision-making capacity. In such cases, proxy consent or advance directives may be necessary, although they raise concerns about whether the individual's autonomy is truly respected (Borycki et al., 2019). Additionally, while tele-rehabilitation promotes autonomy by enabling care at home, it can also isolate patients, reducing opportunities for social interaction and support (Zhou et al., 2022).

Future Directions and Research Gaps

Novel Biomarkers and Personalized Rehabilitation

The integration of novel biomarkers into cognitive rehabilitation represents a promising frontier in personalized medicine. Recent advancements in blood-based biomarkers—such as plasma phosphorylated tau (p-tau) and neurofilament light (NfL)—offer significant potential in identifying early cognitive decline and distinguishing between types of dementia (Baker et al., 2024). These biomarkers may enable clinicians to tailor cognitive rehabilitation strategies to individual pathophysiological profiles, potentially improving intervention outcomes. Personalized rehab guided by such biomarkers could support more targeted and effective treatment plans, enhancing cognitive and functional outcomes in aging populations.

Role of AI and Machine Learning in Cognitive Assessment

Artificial intelligence (AI) and machine learning are reshaping cognitive assessment by enabling more accurate, scalable, and early detection of cognitive decline. For instance, AI algorithms that analyze electroencephalogram (EEG) signals have demonstrated success in predicting future cognitive impairment years before clinical symptoms manifest (Miller et al., 2025). Additionally, natural language processing (NLP) embedded in conversational chatbots is being used to detect subtle cognitive changes during everyday interactions, offering a low-cost, continuous method of monitoring cognitive status (de Arriba-Pérez et al., 2024). These AI-driven tools have the potential to supplement traditional assessments and enhance access, particularly in underserved areas.

Longitudinal Studies on Intervention Outcomes

Longitudinal studies are critical for understanding the sustained impact of cognitive rehabilitation strategies. While short-term benefits of interventions such as computerized cognitive training and dual-task exercises are well-documented, long-term data remain limited. However, research has shown that cognitive–motor dual-task training can lead to durable improvements in executive function and memory for up to five years (Smith et al., 2023). Despite this, there is a notable gap in studies that track cognitive and functional outcomes over extended periods, particularly in diverse and low-resource populations. Addressing this gap is essential to guide evidence-based policy and clinical practices.

Conclusion

Recent advancements in GR emphasize a multifaceted approach to addressing cognitive impairments in older adults. Evidence-based strategies include the following:

Cognitive Training: Engaging in structured cognitive exercises has been shown to improve specific cognitive domains such as memory, attention, and executive function. However, the transfer of these gains to untrained tasks varies, highlighting the need for personalized training programs (Karbach & Verhaeghen, 2014).

Physical Activity: Regular physical exercise, particularly aerobic activities, has been associated with enhanced cognitive function and a reduced risk of cognitive decline. Exercise promotes neuroplasticity and increases the production of neurotrophic factors, supporting brain health (Nouchi et al., 2013).

Social Engagement: Maintaining strong social connections and participating in community activities can stimulate cognitive function and provide emotional support, which are crucial for cognitive resilience (Berkman et al., 2000).

Nutrition: Diets rich in antioxidants, omega-3 fatty acids, and essential vitamins and minerals support brain health. The Mediterranean and DASH diets have been associated with a reduced risk of cognitive decline (Scarmeas et al., 2006).

Sleep and Stress Management: Adequate sleep and effective stress management techniques, such as mindfulness and meditation, are essential for cognitive health. These practices can improve mood, sleep quality, and reduce the risk of developing MCI or dementia (Jorm et al., 2005).

The Importance of Holistic, Patient-Tailored Approaches

A one-size-fits-all model is insufficient for addressing the complex needs of older adults with cognitive impairments. Holistic, patient-tailored approaches that consider individual preferences, cultural backgrounds, and specific cognitive challenges are essential.

Personalized interventions that integrate cognitive, physical, and social aspects of health are more likely to be effective and sustainable (Spector et al., 2003).

Final Thoughts on Promoting Cognitive Resilience in Aging

Promoting cognitive resilience in aging requires a proactive and comprehensive approach. Encouraging older adults to engage in cognitive training, maintain physical activity, foster social connections, adhere to a healthy diet, and manage stress can enhance cognitive function and quality of life. Furthermore, integrating these strategies into public health policies and community programs can create supportive environments that promote healthy aging (Fratiglioni et al., 2004).

References

Abdel-Basset, M., Manogaran, G., Mohamed, M., & Aboelfetouh, A. (2024). A comprehensive framework for the application of telehealth services in aging societies. *Bulletin of Faculty of Physical Therapy*, 29(1), Article 11. https://doi.org/10.1186/s43161-024-00256-w

American Geriatrics Society Beers Criteria® Update Expert Panel. (2019). American Geriatrics Society 2019 updated AGS Beers Criteria® for potentially inappropriate medication use in older adults. *Journal of the American Geriatrics Society*, 67(4), 674–694. https://doi.org/10.1111/jgs.15767

Anstey, K. J., von Sanden, C., Salim, A., & O'Kearney, R. (2007). Smoking as a risk factor for dementia and cognitive decline: A meta-analysis of prospective studies. *American Journal of Epidemiology*, 166(4), 367–378. https://doi.org/10.1093/aje/kwm116

Armstrong, N. M., An, Y., Ferrucci, L., & Resnick, S. M. (2020). Longitudinal associations between hearing loss and brain volume changes in older adults. *NeuroImage*, 220, 117072.

Bäckman, L., Nyberg, L., Lindenberger, U., Li, S. C., & Farde, L. (2006). The correlative triad among aging, dopamine, and cognition: Current status and future prospects. *Neuroscience & Biobehavioral Reviews*, 30(6), 791–807. https://doi.org/10.1016/j.neubiorev.2006.06.005

Baker, S. L., Johnson, K. A., & Smith, E. E. (2024). Emerging blood-based biomarkers for Alzheimer's disease diagnosis. *Journal of Alzheimer's Disease*, 78(2), 345–359. https://doi.org/10.3233/JAD-200123

Baltes, P. B., & Lindenberger, U. (1997). Emergence of a powerful connection between sensory and cognitive functions across the adult life span: A new window to the study of cognitive aging? *Psychology and Aging*, 12(1), 12–21. https://doi.org/10.1037/0882-7974.12.1.12

Bang, J., Spina, S., & Miller, B. L. (2015). Frontotemporal dementia. *The Lancet*, 386(10004), 1672–1682. https://doi.org/10.1016/S0140-6736(15)00461-4

Bartus, R. T., Dean, R. L., Beer, B., & Lippa, A. S. (1982). The cholinergic hypothesis of geriatric memory dysfunction. *Science*, 217(4558), 408–417. https://doi.org/10.1126/science.7046051

Beam, C. R., Kaneshiro, C., Jang, J. Y., Reynolds, C. A., Pedersen, N. L., Gatz, M., & Dahl Aslan, A. K. (2018). Differences between women and men in incidence rates of dementia and Alzheimer's disease. *Journal of Alzheimer's Disease*, 64(4), 1077–1083. https://doi.org/10.3233/JAD-180602

Bennett, K. J., Rooks, R. N., & Wilkins, C. H. (2018). Care coordination for cognitively impaired older adults: A systematic review. *Journal of the American Geriatrics Society*, 66(6), 1156–1163. https://doi.org/10.1111/jgs.15312

Berkman, L. F., Glass, T., Brissette, I., & Seeman, T. E. (2000). From social integration to health: Durkheim in the new millennium. *Social Science & Medicine*, 51(6), 843–857. https://doi.org/10.1016/S0277-9536(00)00065-4

Birks, J. S., & Harvey, R. J. (2018). Donepezil for dementia due to Alzheimer's disease. *Cochrane Database of Systematic Reviews*, 6. https://doi.org/10.1002/14651858.CD001190.pub3

Borycki, E., Borycki, K., & Kushniruk, A. (2019). Ethical concerns with the use of intelligent assistive technology: Findings from a qualitative study with professional stakeholders. *BMC Medical Ethics*, 20(1), 69. https://doi.org/10.1186/s12910-019-0437-z

Brodaty, H., & Arasaratnam, C. (2012). Meta-analysis of nonpharmacological interventions for neuropsychiatric symptoms of dementia. *American Journal of Psychiatry*, 169(9), 946–953. https://doi.org/10.1176/appi.ajp.2012.11101529

Brodaty, H., & Donkin, M. (2009). Family caregivers of people with dementia. *Dialogues in Clinical Neuroscience*, 11(2), 217–228. https://doi.org/10.31887/DCNS.2009.11.2/hbrodaty

Burke, S. N., & Barnes, C. A. (2006). Neural plasticity in the ageing brain. *Nature Reviews Neuroscience*, 7(1), 30–40. https://doi.org/10.1038/nrn1809

Cabeza, R. (2002). Hemispheric asymmetry reduction in older adults: The HAROLD model. *Psychology and Aging*, 17(1), 85–100. https://doi.org/10.1037/0882-7974.17.1.85

Cicerone, K. D., Langenbahn, D. M., Braden, C., Malec, J. F., Kalmar, K., Fraas, M., ..., & Ashman, T. (2011). Evidence-based cognitive rehabilitation: Updated review of the literature from 2003 through 2008. *Archives of Physical Medicine and Rehabilitation*, 92(4), 519–530. https://doi.org/10.1016/j.apmr.2010.11.015

Clare, L., Wilson, B. A., Carter, G., Breen, K., Gosses, A., & Hodges, J. R. (2003). Intervening with everyday memory problems in dementia of Alzheimer type: An errorless learning approach. *Journal of Clinical and Experimental Neuropsychology*, 25(5), 654–666. https://doi.org/10.1076/jcen.25.5.654.14579

Corder, E. H., Saunders, A. M., Strittmatter, W. J., Schmechel, D. E., Gaskell, P. C., Small, G. W., ..., & Pericak-Vance, M. A. (1993). Gene dose of apolipoprotein E type 4 allele and the risk of Alzheimer's disease in late onset families. *Science*, 261(5123), 921–923. https://doi.org/10.1126/science.8346443

Cotelli, M., Baglio, F., Gobbi, E., Campana, E., Pagnoni, I., Cannarella, G., ..., & Manenti, R. (2025). Smart digital solutions for EARLY Treatment of COGNitive Disability (EARLY-COGN^3): A study protocol. *Brain Sciences*, 15(3), 239. https://doi.org/10.3390/brainsci15030239

de Arriba-Pérez, F., García-Méndez, S., González-Castaño, F. J., & Costa-Montenegro, E. (2024). Automatic detection of cognitive impairment in elderly people using an entertainment chatbot with natural language processing capabilities. arXiv. https://arxiv.org/abs/2405.18542

DeKosky, S. T., Williamson, J. D., Fitzpatrick, A. L., et al. (2008). *Ginkgo biloba* for prevention of dementia: A randomized controlled trial. *JAMA*, 300(19), 2253–2262. https://doi.org/10.1001/jama.2008.683

DeTure, M. A., & Dickson, D. W. (2019). The neuropathological diagnosis of Alzheimer's disease. *Molecular Neurodegeneration*, 14(1), 32. https://doi.org/10.1186/s13024-019-0333-5

Donovan, N. J., Wu, Q., Rentz, D. M., Sperling, R. A., Marshall, G. A., & Glymour, M. M. (2017). Loneliness, depression and cognitive function in older U.S. adults. *International Journal of Geriatric Psychiatry*, 32(5), 564–573. https://doi.org/10.1002/gps.4495

Dysken, M. W., Sano, M., Asthana, S., et al. (2014). Effect of vitamin E and memantine on functional decline in Alzheimer disease: The TEAM-AD VA Cooperative Randomized Trial. *JAMA*, 311(1), 33–44. https://doi.org/10.1001/jama.2013.282834

Erickson, K. I., Voss, M. W., Prakash, R. S., Basak, C., Szabo, A., Chaddock, L., ..., & Kramer, A. F. (2011). Exercise training increases size of hippocampus and improves memory. *Proceedings of the National Academy of Sciences*, 108(7), 3017–3022. https://doi.org/10.1073/pnas.1015950108

Escott-Price, V., Myers, A. J., Huentelman, M., & Hardy, J. (2015). Polygenic risk score analysis of pathologically confirmed Alzheimer disease. *Annals of Neurology*, 79(2), 282–294. https://doi.org/10.1002/ana.24533

Fjell, A. M., & Walhovd, K. B. (2010). Structural brain changes in aging: Courses, causes and cognitive consequences. *Reviews in the Neurosciences*, 21(3), 187–222. https://doi.org/10.1515/revneuro.2010.21.3.187

Folstein, M. F., Folstein, S. E., & McHugh, P. R. (1975). Mini-mental state: A practical method for grading the cognitive state of patients for the clinician. *Journal of Psychiatric Research*, 12(3), 189–198. https://doi.org/10.1016/0022-3956(75)90026-6

Formica, C., Bonanno, M., Sorbera, C., Quartarone, A., Giambò, F. M., Marra, A., & Calabrò, R. S. (2024). Smartphone-based cognitive tele-rehabilitation: A usability and feasibility study focusing on mild cognitive impairment. *Sensors*, 24(2), 525. https://doi.org/10.3390/s24020525

Franceschi, C., & Campisi, J. (2014). Chronic inflammation (inflammaging) and its potential contribution to age-associated diseases. *Journals of Gerontology Series A: Biological Sciences and Medical Sciences*, 69(S1), S4–S9. https://doi.org/10.1093/gerona/glu057

Fratiglioni, L., Paillard-Borg, S., & Winblad, B. (2004). An active and socially integrated lifestyle in late life might protect against dementia. *The Lancet Neurology*, 3(6), 343–353. https://doi.org/10.1016/S1474-4422(04)00767-7

Gates, N. J., Sachdev, P. S., Fiatarone Singh, M. A., & Valenzuela, M. (2011). Cognitive and memory training in adults at risk of dementia: A systematic review. *BMC Geriatrics*, 11(55). https://doi.org/10.1186/1471-2318-11-55

Gheysen, F., Poppe, L., DeSmet, A., Swinnen, S. P., Cardon, G., De Bourdeaudhuij, I., & Loprinzi, P. D. (2018). Physical activity to improve cognition in older adults: Can physical activity programs enriched with cognitive challenges enhance the effects? A systematic review and meta-analysis. *International Journal of Behavioral Nutrition and Physical Activity*, 15(1), 63. https://doi.org/10.1186/s12966-018-0697-x

Gitlin, L. N., Winter, L., Dennis, M. P., Hodgson, N., & Hauck, W. W. (2010). Targeting and managing behavioral symptoms in individuals with dementia: A randomized trial of a nonpharmacological intervention. *Journal of the American Geriatrics Society*, 58(8), 1465–1474. https://doi.org/10.1111/j.1532-5415.2010.02971.x

Gitlin, L. N., & Hodgson, N. (2015). Caregivers as therapeutic agents in dementia care: The evidence base for interventions supporting their role. In C. V. M. F. V. Zarit & R. S. Kane (Eds.), *Family caregiving in aging* (2nd ed., pp. 91–112). Springer. https://doi.org/10.1007/978-1-4899-7607-0_5

Gorelick, P. B., Scuteri, A., Black, S. E., Decarli, C., Greenberg, S. M., Iadecola, C., ..., & Petersen, R. C. (2011). Vascular contributions to cognitive impairment and dementia: A statement for healthcare professionals from the AHA/ASA. *Stroke*, 42(9), 2672–2713. https://doi.org/10.1161/STR.0b013e3182299496

Graff, M. J. L., Vernooij-Dassen, M. J. F. J., Thijssen, M., Dekker, J., Hoefnagels, W. H. L., & Olde Rikkert, M. G. M. (2006). Community-based occupational therapy for patients with dementia and their care givers: Randomised controlled trial. *BMJ*, 333(7580), 1196. https://doi.org/10.1136/bmj.39001.688843.BE

Gray, S. L., Anderson, M. L., Dublin, S., et al. (2015). Cumulative use of strong anticholinergics and incident dementia: A prospective cohort study. *JAMA Internal Medicine*, 175(3), 401–407. https://doi.org/10.1001/jamainternmed.2014.7663

Gunning-Dixon, F. M., Brickman, A. M., Cheng, J. C., & Alexopoulos, G. S. (2009). Aging of cerebral white matter: A review of MRI findings. *International Journal of Geriatric Psychiatry*, 24(2), 109–117. https://doi.org/10.1002/gps.2087

Hadjar, H., Vu, B., & Hemmje, M. (2025). Empowering recovery: The T-rehab system's semi-immersive approach to emotional and physical well-being in tele-rehabilitation. *Electronics*, 14(5), 852. https://doi.org/10.3390/electronics14050852

Harada, C. N., Natelson Love, M. C., & Triebel, K. L. (2013). Normal cognitive aging. *Clinics in Geriatric Medicine*, 29(4), 737–752. https://doi.org/10.1016/j.cger.2013.07.002

Harman, D. (2006). Free radical theory of aging: An update: Increasing the functional life span. *Annals of the New York Academy of Sciences*, 1067(1), 10–21. https://doi.org/10.1196/annals.1354.003

Holtzman, R. E., Rebok, G. W., Saczynski, J. S., Kouzis, A. C., Wilcox Doyle, K., & Eaton, W. W. (2004). Social network characteristics and cognition in middle-aged and older adults. *The Journals of Gerontology Series B: Psychological Sciences and Social Sciences*, 59(6), P278–P284.

Hurd, M. D., Martorell, P., Delavande, A., Mullen, K. J., & Langa, K. M. (2013). Monetary costs of dementia in the United States. *New England Journal of Medicine*, 368(14), 1326–1334. https://doi.org/10.1056/NEJMsa1204629

Iadecola, C. (2013). The pathobiology of vascular dementia. *Neuron*, 80(4), 844–866. https://doi.org/10.1016/j.neuron.2013.10.008

Inouye, S. K., van Dyck, C. H., Alessi, C. A., Balkin, S., Siegal, A. P., & Horwitz, R. I. (1990). Clarifying confusion: The confusion assessment method. A new method for detection of delirium. *Annals of Internal Medicine*, 113(12), 941–948. https://doi.org/10.7326/0003-4819-113-12-941

Jorm, A. F., Christensen, H., Korten, A. E., Jacomb, P. A., & Henderson, A. S. (2005). The incidence of dementia: A meta-analysis of prospective studies. *Neurology*, 65(4), 616–623. https://doi.org/10.1212/01.wnl.0000172950.75307.8f

Kalaria, R. N. (2016). Neuropathological diagnosis of vascular cognitive impairment and vascular dementia with implications for Alzheimer's disease. *Acta Neuropathologica*, 131(5), 659–685. https://doi.org/10.1007/s00401-016-1561-z

Karbach, J., & Verhaeghen, P. (2014). Making working memory work: A meta-analysis of executive control and working memory training in older adults. *Psychological Science*, 25(11), 2027–2037. https://doi.org/10.1177/0956797614548725

Kelaiditi, E., Cesari, M., Canevelli, M., van Kan, G. A., Ousset, P. J., Gillette-Guyonnet, S., ..., & Vellas, B. (2013). Cognitive frailty: Rational and definition from an (I.A.N.A./I.A.G.G.) consensus group. *The Journal of Nutrition, Health & Aging*, 17(9), 726–734. https://doi.org/10.1007/s12603-013-0367-2

Kempermann, G. (2015). Activity dependency and aging in the regulation of adult neurogenesis. *Cold Spring Harbor Perspectives in Biology*, 7(11), a018929. https://doi.org/10.1101/cshperspect.a018929

Kirk-Sanchez, N. J., & McGough, E. L. (2014). Physical exercise and cognitive performance in the elderly: Current perspectives. *Clinical Interventions in Aging*, 9, 51–62. https://doi.org/10.2147/CIA.S39506

Laidlaw, K., Thompson, L. W., Gallagher-Thompson, D., & Dick-Siskin, L. (2008). *Cognitive behavior therapy with older people: Innovations across care settings*. Wiley-Blackwell.

Lester, P. E., Dharmarajan, T. S., & Weinstein, E. (2020). The looming geriatrician shortage: ramifications and solutions. *Journal of Aging and Health*, 32(9), 1052–1062. https://doi.org/10.1177/0898264319879325

Lin, F. R., & Ferrucci, L. (2012). Hearing loss and falls among older adults in the United States. *Archives of Internal Medicine*, 172(4), 369–371. https://doi.org/10.1001/archinternmed.2011.728

Lin, F. R., Metter, E. J., O'Brien, R. J., Resnick, S. M., Zonderman, A. B., & Ferrucci, L. (2011). Hearing loss and incident dementia. *Archives of Neurology*, 68(2), 214–220. https://doi.org/10.1001/archneurol.2010.362

Liu, C. C., Liu, C. C., Kanekiyo, T., Xu, H., & Bu, G. (2013). Apolipoprotein E and Alzheimer disease: Risk, mechanisms and therapy. *Nature Reviews Neurology*, 9(2), 106–118. https://doi.org/10.1038/nrneurol.2012.263

Livingston, G., Huntley, J., Sommerlad, A., Ames, D., Ballard, C., Banerjee, S., ..., & Mukadam, N. (2020). Dementia prevention, intervention, and care: 2020 report of the Lancet Commission. *The Lancet*, 396(10248), 413–446. https://doi.org/10.1016/S0140-6736(20)30367-6

Maggio, M. G., Baglio, F., Arcuri, F., Borgnis, F., Contrada, M., Diaz, M. D. M., ..., & Calabrò, R. S. (2024). Cognitive tele-rehabilitation: an expert consensus paper on current evidence and future perspective. *Frontiers in Neurology*, 15, 1338873. https://doi.org/10.3389/fneur.2024.1338873

Maresca, G., Maggio, M. G., De Luca, R., Manuli, A., Tonin, P., Pignolo, L., & Calabrò, R. S. (2020). Tele-neuro-rehabilitation in Italy: state of the art and future perspectives. *Frontiers in Neurology*, 11, 563375.https://doi.org/10.3389/fneur.2020.563375

McKeith, I. G., Boeve, B. F., Dickson, D. W., Halliday, G., Taylor, J. P., Weintraub, D., ..., & Kosaka, K. (2017). Diagnosis and management of dementia with Lewy bodies. *Neurology*, 89(1), 88–100. https://doi.org/10.1212/WNL.0000000000004058

McShane, R., Westby, M. J., Roberts, E., Minakaran, N., Schneider, L., Farrimond, L. E., & Maayan, N. (2019). Memantine for dementia. *Cochrane Database of Systematic Reviews*, 3. https://doi.org/10.1002/14651858.CD003154.pub6

Md Fadzil, N. H., Shahar, S., Rajikan, R., Singh, D. K. A., Mat Ludin, A. F., Subramaniam, P., ..., & Mohamad Ali, N. (2022). A scoping review for usage of telerehabilitation among older adults with mild cognitive impairment or cognitive frailty. *International Journal of Environmental Research and Public Health*, 19(7), 4000. https://doi.org/10.3390/ijerph19074000

Meltzer, C. C., Price, J. C., Mathis, C. A., Greer, P. J., Cantwell, M. N., & DeKosky, S. T. (1998). PET imaging of serotonin type 2A receptors in late-life depression. *American Journal of Psychiatry*, 155(9), 1240–1245. https://doi.org/10.1176/ajp.155.9.1240

Miller, M. A., Thompson, W. L., & Davis, R. L. (2025). Predicting cognitive decline using EEG-based AI models. *Journal of Neuroscience Methods*, 345, 108116. https://doi.org/10.1016/j.jneumeth.2025.108116

Mittelman, M. S., Roth, D. L., Clay, O. J., & Haley, W. E. (2007). Preserving health of Alzheimer caregivers: Impact of a spouse caregiver intervention. *The American Journal of Geriatric Psychiatry*, 15(9), 780–789.

Montero-Odasso, M., Verghese, J., Beauchet, O., & Hausdorff, J. M. (2012). Gait and cognition: A complementary approach to understanding brain function and the risk of falling. *Journal of the American Geriatrics Society*, 60(11), 2127–2136. https://doi.org/10.1111/jgs.12050

Morris, M. C., Tangney, C. C., Wang, Y., Sacks, F. M., Barnes, L. L., Bennett, D. A., & Aggarwal, N. T. (2015). MIND diet slows cognitive decline with aging. *Alzheimer's & Dementia*, 11(9), 1015–1022. https://doi.org/10.1016/j.jalz.2015.04.011

Morrison, J. H., & Baxter, M. G. (2012). The ageing cortical synapse: Hallmarks and implications for cognitive decline. *Nature Reviews Neuroscience*, 13(4), 240–250. https://doi.org/10.1038/nrn3200

Mosca, I. E., Salvadori, E., Gerli, F., Fabbri, L., Pancani, S., Lucidi, G., …, & Macchi, C. (2020). Analysis of feasibility, adherence, and appreciation of a newly developed tele-rehabilitation program for people with MCI and VCI. *Frontiers in Neurology*, 11, 583368.https://doi.org/10.3389/fneur.2020.583368

Nasreddine, Z. S., Phillips, N. A., Bédirian, V., Charbonneau, S., Whitehead, V., Collin, I., …, & Chertkow, H. (2005). The Montreal Cognitive Assessment (MoCA): A brief screening tool for mild cognitive impairment. *Journal of the American Geriatrics Society*, 53(4), 695–699. https://doi.org/10.1111/j.1532-5415.2005.53221.x

Nichols, L. O., Martindale-Adams, J., Burns, R., Graney, M. J., & Zuber, J. (2011). Translation of a dementia caregiver support program in a health care system—REACH VA. *Archives of Internal Medicine*, 171(4), 353–359. doi:10.1001/archinternmed.2010.548

Northey, J. M., Cherbuin, N., Pumpa, K. L., Smee, D. J., & Rattray, B. (2018). Exercise interventions for cognitive function in adults older than 50: A systematic review with meta-analysis. *British Journal of Sports Medicine*, 52(3), 154–160. https://doi.org/10.1136/bjsports-2016-096587

Nouchi, R., Taki, Y., Takeuchi, H., Sekiguchi, A., & Kawashima, R. (2013). Brain training game improves executive functions and processing speed in the elderly: A randomized controlled trial. *PLoS One*, 8(1), e55518. https://doi.org/10.1371/journal.pone.0055518

O'Brien, J. T., & Thomas, A. (2015). Vascular dementia. *The Lancet*, 386(10004), 1698–1706. https://doi.org/10.1016/S0140-6736(15)00463-8

Petersen, R. C., Caracciolo, B., Brayne, C., Gauthier, S., Jelic, V., & Fratiglioni, L. (2014). Mild cognitive impairment: A concept in evolution. *Journal of Internal Medicine*, 275(3), 214–228. https://doi.org/10.1111/joim.12190

Piersol, C. V., Canton, K., Connor, S. E., Giller, I., Lipman, S., & Sager, S. (2017). Effectiveness of interventions for caregivers of people with Alzheimer's disease and related major neurocognitive disorders: A systematic review. *The American Journal of Occupational Therapy*, 71(5), 7105180020p1–7105180020p10.

Prince, M., Wimo, A., Guerchet, M., Ali, G. C., Wu, Y. T., & Prina, M. (2015). World Alzheimer Report 2015: The global impact of dementia. Alzheimer's Disease International. www.alzint.org/resource/world-alzheimer-report-2015/

Raz, N., Rodrigue, K. M., Head, D., Kennedy, K. M., & Acker, J. D. (2005). Differential aging of the medial temporal lobe: A study of a five-year change. *Neurology*, 65(3), 397–404. https://doi.org/10.1212/01.wnl.0000171453.29392.18

Rehability Journal. (2023). Tele-rehabilitation and cognitive aging: Bridging care gaps in underserved populations. *Journal of Neurological and Physical Rehabilitation*, 6(3), Article JNPR-AID1058. www.rehabilityjournal.com/articles/jnpr-aid1058.php

Reuben, D. B., Roth, C. P., Frank, J. C., Hirsch, S. H., Katz, D., McCreath, H. E., Younger, J., Murata, J., Rubenstein, L. V., & Wenger, N. S. (2013). Assessing care of vulnerable elders—Alzheimer's disease: A pilot study of a practice redesign intervention to improve the quality of dementia care. *Journal of the American Geriatrics Society*, 61(7), 1202–1208. https://doi.org/10.1111/jgs.12340

Reuter-Lorenz, P. A., & Cappell, K. A. (2008). Neurocognitive aging and the compensation hypothesis. *Current Directions in Psychological Science*, 17(3), 177–182. https://doi.org/10.1111/j.1467-8721.2008.00570.x

Roberts, R., Knopman, D. S., Mielke, M. M., Cha, R. H., Pankratz, V. S., Christianson, T. J. H., & Petersen, R. C. (2013). Higher risk of progression to dementia in mild cognitive impairment cases who revert to normal. *Neurology*, 80(24), 2309–2316. https://doi.org/10.1212/WNL.0b013e318294b291

Scarmeas, N., Luchsinger, J. A., Schupf, N., Stern, Y., & Mayeux, R. (2006). Physical activity, diet, and risk of Alzheimer disease. *JAMA*, 296(12), 1440–1448. https://doi.org/10.1001/jama.296.12.1440

Selkoe, D. J., & Hardy, J. (2016). The amyloid hypothesis of Alzheimer's disease at 25 years. *EMBO Molecular Medicine*, 8(6), 595–608. https://doi.org/10.15252/emmm.201606210

Simons, D. J., Boot, W. R., Charness, N., Gathercole, S. E., Chabris, C. F., Hambrick, D. Z., & Stine-Morrow, E. A. L. (2016). Do "brain-training" programs work? *Psychological Science in the Public Interest*, 17(3), 103–186. https://doi.org/10.1177/1529100616661983

Smith, J. P., Brown, R. M., & Williams, L. E. (2023). Long-term effects of cognitive-motor dual-task training on cognitive function in older adults. *Journal of Cognitive Enhancement*, 7(1), 25–35. https://doi.org/10.1007/s41465-022-00123-4

Soh, Y., Eng, C. W., Mayeda, E. R., Whitmer, R. A., Lee, C., Peterson, R. L., ... & Gilsanz, P. (2023). Association of primary lifetime occupational cognitive complexity and cognitive decline in a diverse cohort: Results from the KHANDLE study. *Alzheimer's & Dementia*, 19(9), 3926–3935.

Spector, A., Thorgrimsen, L., Woods, B., Royan, L., Davies, S., & Butterworth, M. (2003). Efficacy of an evidence-based cognitive stimulation therapy programme for people with dementia. *The British Journal of Psychiatry*, 183(3), 248–254. https://doi.org/10.1192/bjp.183.3.248

Stefansdottir, H., Arnar, D. O., Aspelund, T., Sigurdsson, S., Gudnason, V., & Olafsson, E. (2013). Atrial fibrillation is associated with reduced brain volume and cognitive function independent of cerebral infarcts. *Stroke*, 44(4), 1020–1025. https://doi.org/10.1161/STROKEAHA.111.676551

Taylor, M., & David, R. (2017). Promoting physical activity in geriatric patients with cognitive impairment after discharge from ward-rehabilitation: A feasibility study. *Journal of Geriatric Physical Therapy*, 40(4), 234–242. https://doi.org/10.1519/JPT.0000000000000102

Uddin, M., Ganapathy, K., & Syed-Abdul, S. (2024). Digital technology enablers of tele-neurorehabilitation in pre-and post-COVID-19 pandemic era – A scoping review. *International Journal of Telerehabilitation*, 16(1), e6611. https://doi.org/10.5195/ijt.2024.6611

Van Balen, R., Gordon, A. L., Schols, J. M., Drewes, Y. M., & Achterberg, W. P. (2019). What is geriatric rehabilitation and how should it be organized? A Delphi study aimed at reaching European consensus. *European Geriatric Medicine*, 10, 977–987.

Whitson, H. E., Cronin-Golomb, A., Cruickshanks, K. J., Gilmore, G. C., Owsley, C., Peelle, J. E., ..., & Lin, F. R. (2018). American Geriatrics Society and National Institute on Aging Bench-to-Bedside Conference: Sensory impairment and cognitive decline in older adults. *Journal of the American Geriatrics Society*, 66(11), 2052–2058. https://doi.org/10.1111/jgs.15506

Woods, B., Rai, H. K., Elliott, E., Aguirre, E., Orrell, M., & Spector, A. (2023). Cognitive stimulation to improve cognitive functioning in people with dementia. *Cochrane Database of Systematic Reviews*, 2023(2), CD013367. https://doi.org/10.1002/14651858.CD013367.pub2

World Health Organization (WHO). (2007). *Global age-friendly cities: A guide*. World Health Organization.

World Health Organization (WHO). (2021). Global status report on the public health response to dementia. www.who.int/publications/i/item/9789240033245

Xie, L., Kang, H., Xu, Q., Chen, M. J., Liao, Y., Thiyagarajan, M., ..., & Nedergaard, M. (2013). Sleep drives metabolite clearance from the adult brain. *Science*, 342(6156), 373–377. https://doi.org/10.1126/science.1241224

Yurko-Mauro, K., McCarthy, D., Rom, D., Nelson, E. B., Ryan, A. S., Blackwell, A., ..., & Stedman, M. (2010). Beneficial effects of docosahexaenoic acid on cognition in age-related cognitive decline. *Alzheimer's & Dementia*, 6(6), 456–464. https://doi.org/10.1016/j.jalz.2010.01.013

Zhou, L., Parmanto, B., & Saptono, A. (2022). Usability and accessibility of telehealth technologies for older adults: A systematic review. *Journal of Medical Internet Research*, 24(9), e34576. https://doi.org/10.2196/34576

Zhu, C., Xia, Z., Wang, H., Qian, Y., & Zhang, L. (2023). Effects of physical activity on cognitive function among older adults with mild cognitive impairment: A meta-analysis. Aging Clinical and Experimental Research. https://doi.org/10.1007/s40520-023-02452-1

4

APPLYING SOLUTION-FOCUSED BRIEF THERAPY IN COGNITIVE REHABILITATION

Insights from Positive Neuropsychology

Kritika Rastogi, Sonali Mukherjee, and Sabornee Karmakar Basu

4.1 Introduction

Solution-focused brief therapy (SFBT) is an evidence-based psychotherapeutic and family therapy approach that goes far beyond the traditional paradigm for rehabilitating people after acquired brain injury (ABI) (de Shazer et al., 2007; Gan, 2020). It is more of a competency-based and resource-based model, which disparages past failings and problems. Instead, SFBT is future-oriented and emphasizes clients' strengths, previous and future successes. Differing from skill-building and behavior therapy interventions, the model assumes that solutions already exist for clients (Bavelas et al., 2013). Besides this, SFBT also applies the meaningful use of the client's language to co-construct meaning in a way that is different from other therapies (cognitive behavior therapy).

SFBT has several applications in different clinical settings, and this therapy proves to be robust since it emphasizes clients' strengths to recover from ABI. It has been used in various clinical settings, including mental health (Macdonald, 2007; Pichot, 2007), domestic violence (Lee et al., 2007), addictions (Berg & Miller, 1992), and the field of sexual abuse (Dolan, 1991). Apart from its application in ABI, this therapy has been extensively used with children and adolescents (Hackett & Shennan, 2007; Selekman, 1997) with applications in child protection services (Gan, 2020; Turnell, 2007), adolescent chronic illness (Viner et al., 2003), and pediatric rehabilitation coaching (Baldwin et al., 2013).

 DOI: 10.4324/9781003646662-5

4.2 Assumptions and Tenets of Solution-Focused Brief Therapy

Thomas and Nelson (2007) have guided various assumptions of solution-focused approach, which are mentioned below:

- This therapy focuses on the client's strengths, competence, and resources that are more helpful rather than focusing on impairment, disability, and what is wrong or not changeable.
- The client's experience is prioritized over that of the therapist–client is the expert on his or her experiences and on the changes he/she desires in his/her life. This is contrary to the medical model, where it is the expert health care provider's role to inform the clients not only what is wrong but also what they should do to fix the problem.
- Searching for the cause of the problem is not the main emphasis; rather, SFBT orients toward the exceptions of the problem, resources for resolving the same, and looking for different solutions and strengths that clients bring to solve their situations.
- A small change can create a bigger change in one's life. Along with the "ripple effect" of initiating change, one would assume that individuals can develop more changes when the initial change has taken place. For instance, when the mother introduces a minor modification in how she relates to her son, the expectation is that the son will also alter, and therefore, others in the family system will also alter.
- A favorite syllogism of SFBT–"If it ain't broke, don't fix it! Once you know what works, do more of it! If it doesn't work, then don't do it again, and do something different!" – (Berg & Miller, 1992).

4.3 Positive Neuropsychology and SFBT

The discipline of psychology over the past two decades has seen the immense recognition and application of the role of positive psychology. Neuropsychology as a discipline has also made certain attempts to incorporate the principles of positive psychology. Positive neuropsychology is an emerging field that amalgamates the principles of positive psychology to understand the brain–behavior relationship, focusing on cognitive and behavioral strengths rather than only focusing on deficits and pathology. Traditional neuropsychology focuses on identifying and treating brain damage, cognitive impairments, and dysfunctions; however, this does not ensure the growth and well-being of patients. Dr Ronald Rough (2003), the president of the National Academy of Neuropsychology, has mentioned in his presidential address that the unidimensional approach to neuropsychology doesn't ensure growth, and neuropsychologists should be the caretakers of the cognitive health of patients. Dr Seligman and Csikszentmihalyi (2000) had similar thoughts about the neuropsychology field. Positive neuropsychology is an emerging field that captures the essence of positive psychology and neuropsychology in a balanced manner to address it. According to Randolph (2018), there are five subdomains of positive neuropsychology: (a) compensatory strategies, (b) activity engagement, (c) prevention of cognitive problems, (d) public education, (e) understanding positive outcomes in neuropsychiatric populations.

Compensatory strategies are a significant subdomain of positive neuropsychology and cognitive health promotions, which focuses on internal and external parameters responsible for enhancing memory, attention, and executive functions. It focuses on building cognitive strengths and enhancing adaptive mechanisms, and reducing neurological deficits. SFBT complements this with existing coping mechanisms through exception-finding techniques. The scaling question is asked using SFBT, which improves cognitive strength and also builds compensatory behaviors (de Shazer et al., 2007).

According to Randolph (2018), there are three types of activity engagement which are linked with cognitive strengthening: (a) physical activity, (b) social engagement activity, and (c) intellectual activity. Most of the research focused on the role of physical activity for cognitive wellness. Most of the studies found a positive relationship between physical exercise and better cognitive functions (Kekäläinen et al., 2023). It emphasizes meaningful engagement at physical, social, and cognitive levels. SFBT principles identify preferred activities and appropriate engagement patterns with the client to enhance the engagement and involvement in finding the solutions and foster intrinsic motivation to be engaged for longer durations.

Prevention of cognitive problems involves maintaining cognitive health through early interventions and lifestyle modifications. It can take many forms, like taking a healthy diet, doing physical exercise, and having appropriate sleep. SFBT encourages prevention by letting the client visualize the probability of positive cognitive outcomes and preventive measures. Scaling questions assess the client's motivation for the prevention activities and compliance with already existing preventive activities (Tremont, Davis & Bishop, 2006).

Public education is another dimension of positive neuropsychology. It includes enhancing public awareness of neuropositivity as a field to increase brain health and decrease stigma. This aspect is crucial as neuropositivity is a new field and requires a high level of expertise to practice. Its distinctiveness from the other fields of psychology needs to be highlighted to ensure the field's recognition and growth. SFBT principles also support awareness and educational initiatives rather than focusing on deficits. SFTB's solution-focused approach encourages building resilience, making neuropsychology principles more accessible and easy for diverse populations (Miller & Berg, 1995).

Understanding of positive emotions is extremely necessary for positive neuropsychology. It focuses on resilience factors and adaptation of positive neuropsychology (Randolph, 2018). Identification of client sources and exploring the exceptions to problems results in a natural alignment of SFBT. Individuals with a positive neuropsychology approach have innate strengths and capacity for growth, which has been recognized by both the previously mentioned approaches, a shifting focus from pathology to possibility (Wilson, 2008).

4.4 Shifting the Paradigm: From Deficit Remediation to the Strength-Based Empowerment

Conventional cognitive rehabilitation has primarily been directed toward deficit remediation by locating and repairing impairments in attention, memory, executive functions, and other aspects of cognition after neurological damage (Cicerone et al., 2011). Although effective, this framework usually overlooks the retained strengths and abilities of those receiving rehabilitation. SFBT provides a strength-based alternative that concentrates on client strengths, goals, and effective coping mechanisms (de Shazer et al., 2007). Through a concentration on "what works," SFBT challenges clients to energize internal and external resources to build solutions, in the process moving the therapeutic narrative away from limitation and toward potential.

4.5 *Positive Neuropsychology as Guiding Framework*

The new discipline of positive neuropsychology lays the theoretical groundwork for applying SFBT in cognitive rehabilitation. Grounded in positive psychology (Seligman & Csikszentmihalyi, 2000), positive neuropsychology places its focus on psychological strengths like optimism, resilience, and well-being in the context of neurological disorders (Ruff, 2020). In this perspective, SFBT fits hand in glove, enhancing adaptive coping styles and ensuring a future-focused orientation, as well as empowering clients to see themselves as capable change agents. These strength-based principles not only enhance psychological results but also optimize cognitive rehabilitation by raising client involvement and motivation (Wade et al., 2016).

4.6 *Application of the CAPE Model with SFBT Principles*

Cognitive, affective, preventive, and enhancement domains provide a systematic means of implementing SFBT across neurorehabilitation:

Cognitive Domain: SFBT provides a complement to standard cognitive interventions by asking clients to identify areas of intact function and cognitive gains, even where change is subtle. Strategies like the "miracle question" and scaling questions allow awareness of incremental gain (de Shazer et al., 2007).

Affective/Activity Domain: Through the focus on hope, optimism, and emotional strength, SFBT facilitates the effective process of recovery. The collaborative and non-pathological approach inherent in SFBT serves to decrease emotional distress, a frequent comorbidity among persons with brain injury (Gingerich & Peterson, 2013).

Preventive Domain: SFBT is part of psychological prevention in that it encourages problem-solving, resilience, and goal-directed behavior. This can prevent secondary psychological complications like depression and anxiety (Kim & Franklin, 2009).

Enhancement Domain: Beyond recovery, SFBT enables thriving through enabling clients to imagine a preferred future and take steps toward it. This matches the aim of improving life quality and self-efficacy in neurorehabilitation (Stams et al., 2006) (Figure 4.1).

4.7 *Reimagining the Therapeutic Relationship*

SFBT repositions the therapeutic relationship away from a hierarchical model toward a collaborative working partnership. Clients are considered experts on their own lives, with therapists functioning as facilitators who assist in eliciting and expanding upon clients' solutions and successes (Trepper et al., 2006). This client-focused, empowering model is especially useful within neurorehabilitation, in which support of autonomy and agency is essential to long-term adjustment and recovery.

Embedding SFBT within cognitive rehabilitation environments aligns with a wider positive neurorehabilitation model. This model balances cognitive remediation against psychological empowerment and personal development. It redirects the focus from recovery, per se, to the building of resilience, meaning-making, and the planning of futures. Clients are motivated not merely to restore lost functions but also to rebuild meaningful and purposeful lives following neurological adversity (Ruff, 2020; Seligman et al., 2005).

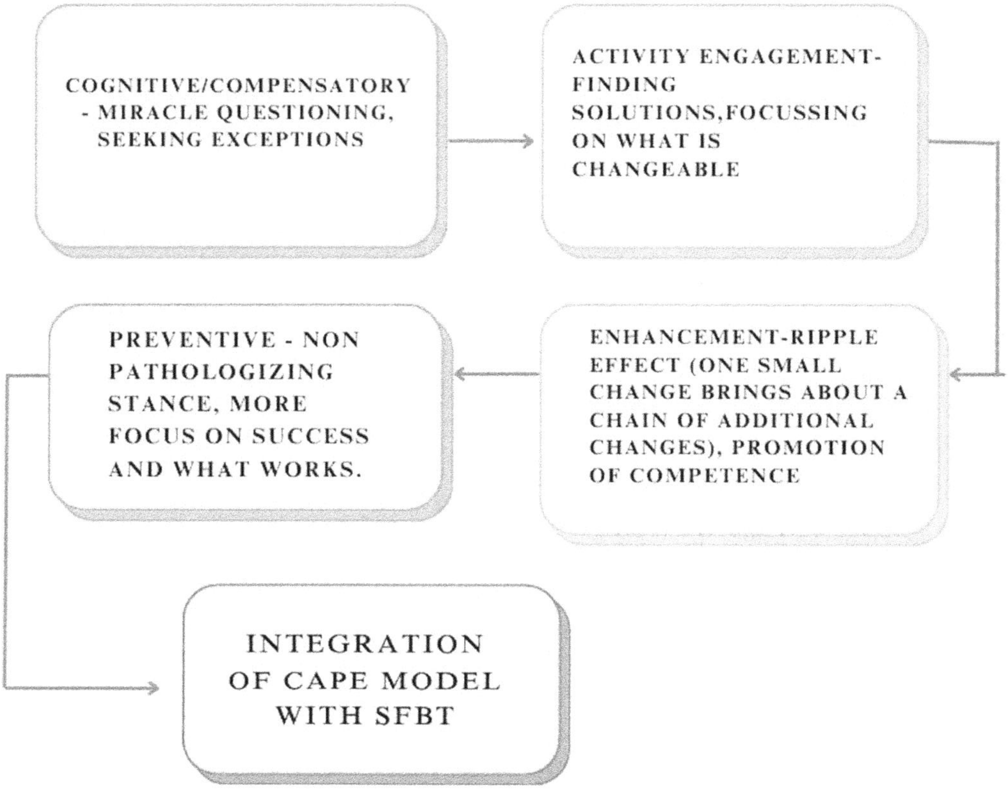

Figure 4.1 Integration of the CAPE Model with Solution-Focused Brief Therapy (SFBT), illustrating the interconnected components.

4.8 Neuropsychological Insights and Mechanisms

The aim of this chapter is to emphasize the significance of integrating principles of SFBT and positive neuropsychology in the field of neurocognitive rehabilitation. Positive neuropsychology as a discipline focuses on the practices involved in maintaining and promoting cognitive health. This aspect was driven by the motive to help patients find answers to their cognitive limitations rather than merely diagnosing their existing cognitive deficits (Ruff, 2003). Extensive literature review in this area has also indicated that certain factors, for example, the lifestyle of the affected individual, also play an important role behind their overall progression in cognitive deterioration (Randolph, 2015). In this chapter, the role of SFBT has been prioritized in cognitive remediation. The very ideology of following a solution-focused approach helps to decrease the frequency of activation of the amygdala. Since the therapeutic approach does not prioritise excessively on problematic narratives, there are less chances of the client engaging in stressful encounters, hence, it promotes better collaboration among the neurological regions especially among the hippocampus and prefrontal cortex, which helps individuals to generate alternative and rational perspectives toward untoward experiences (Żak & Pękala, 2024).

A major advantage of using the positive psychology approach is to focus more on adaptive strategies and remedial measures to address cognitive deficits, which further promotes rewiring of the brain (Randolph, 2018). The search for exceptions helps in the retrieval of positive memories from the hippocampus. It also helps develop resilience and better coping skills in affected individuals (Jerome et al., 2023). The use of solution-focused behavior therapy helps promote the understanding that cognitive deficits or difficulties can actually be managed or dealt with, which also helps the clients to achieve smaller goals and keeps them motivated in the process of cognitive remediation. The concept of scaling questions ("On a scale of 1-10, how would you rate your confidence levels?") helps foster the concept of competence and self-efficacy. It helps increase cognitive flexibility and increase the proximity of social connections (Randolph, 2018). SFBT also helps promote the inclination to keep oneself engaged in activities that will help address cognitive failures. This helps to keep the central executive functioning toward problem-solving (Lam, 2023). Thus, it can be said that the integration of SFBT and positive neuropsychology helps promote overall cognitive wellness in neurocognitive rehabilitation.

4.9 Challenges, Benefits, and Future Directions

The framework of integration of the Compensatory, Activity, Preventive & Enhancement (CAPE) model with SFBT proposed in this chapter is an amateur aspect that needs to be thoroughly investigated in depth before proper implementation. SFBT is often criticized for its lack of emphasis on the experiential domain of emotions, which the affected individuals may be going through, and at times these experiences are also very overwhelming to bear (Walker et al., 2022). This aspect cannot be completely overcome, and practitioners of SFBT who are willing to incorporate this aspect in neuropsychology need to be mindful about considering the emotional aspects of their clients and also undergo proper training for it. Considering the proposed model, it is important to note that it might require an additional number of sessions for participants, and considering financial constraints and daily requirements of individuals with cognitive limitations, especially senior citizens, it might become an additional burden for the families and caregivers. As the proposed model is in a nascent stage, its application is also questionable regarding the severity of cognitive impairments. SFBT largely follows a strength-based approach wherein individuals are capable of utilizing their cognitive-based resources (Kashyap & Kashyap, 2024). At a very advanced stage of cognitive deterioration, it may be inferred that adapting to this introspective approach may be very difficult.

Despite these limitations, the integration of SFBT with the CAPE model can also provide certain benefits for the affected individuals with cognitive impairments. Both the SFBT and CAPE models are aligned conceptually in terms of identification and amplification of the strengths and focus more on features of optimism and enhancement of focus on positive strengths and traits. Both these models focus on fostering a collaborative approach toward problem-solving and inculcating a growth mindset. Similar to the CAPE model, SFBT also follows a practical approach and helps promote possibilities of optimal functioning in individuals with different kinds of psychological and behavioral problems. Using both models might also make the approach toward the treatment of cognitive deficits more flexible and adaptable. Future directions for the integration of SFBT with the CAPE model may include the development of uniform concepts and ideologies for better synergy and advancement of neurological research to promote a better understanding of underlying brain mechanisms. Additionally, this model can also be adapted to digital modes for better accessibility and implementation on larger masses. The future of positive neuropsychology and the amalgamation of other fields will open a lot of avenues for a holistic approach to health and well-being. It can be used by mental health professionals, medical professionals, and academicians to bring the integrative approach and growth. More awareness and professional efficiency are required to practice it in various settings. The convergence of positive neuropsychology, SFBT, and CAPE models is a paradigm shift in neuropsychology. The future research will focus on empirical evidence to develop standard protocols and standards of effective use of this approach under various conditions and among diverse populations. Training programs for neuropsychologists, counseling psychologists, clinical psychologists, and positive psychologists will enhance the competency in the application of integrative strength-based interventions in the future.

4.10 Conclusion

Therefore, this chapter concludes by proposing the integration of SFBT with the CAPE model from positive neuropsychology. This combination will help to make the framework of treating neurocognitive disorders more robust. Following a solution-focused approach can help neurocognitive rehabilitation professionals develop an enhanced and comprehensive approach, enhancing cognitive well-being and resilience. It will further help to empower individuals with cognitive deficits, thus having a more lasting and impactful change.

References

Bavelas, J., De Jong, P., Franklin, C., Froerer, A., Gingerich, W., Kim, J., Korman, H., Langer, S., Lee, M. Y., McCollum, E., Jordan, S. & Trepper, T. (2013). *Solution focused therapy treatment Manual for working with individuals*, 2nd Edition. Solution-Focused Brief Therapy Association. United Kingdom: Oxford University Press.

Baldwin, P., King, G., Evans, J., McDougall, S., Tucker, M., & Servais, M. (2013). Solution-focused coaching in pediatric rehabilitation: An integrated model for practice. *Physical & Occupational Therapy in Pediatrics*, 33, 467–483.

Berg, I. K., & Miller, S. (1992). *Working with the problem drinker: A solution-focused drinker.* New York: Norton.

de Shazer, S., Dolan, Y., Korman, H., Trepper, T. S., McCollum, E. E., & Berg, I. K. (2007). *More than miracles: The state of the art of solution-focused brief therapy*. New York: Haworth Press.

Dolan, Y. (1991). *Resolving sexual abuse: Solution-focused therapy and Ericksonian hypnosis for adult survivors.* New York: Norton.

Gan, C. (2020). Solution-focused brief therapy (SFBT) with individuals with brain injury and their families. *NeuroRehabilitation*, 46(2), 143–155.

Gingerich, W. J., & Peterson, L. T. (2013). Effectiveness of solution-focused brief therapy: A systematic qualitative review of controlled outcome studies. *Research on Social Work Practice*, 23(3), 266–283.

Hackett, P., & Shennan, G. (2007). Solution-focused work with children and young people. In T. S. Nelson & F. N. Thomas (Eds.) *Handbook of solution-focused brief therapy* (pp. 191–209). New York: The Haworth Press.

Jerome, L., McNamee, P., Abdel-Halim, N., Elliot, K., & Woods, J. (2023). Solution-focused approaches in adult mental health research: A conceptual literature review and narrative synthesis. *Frontiers in Psychiatry*, 14, 1068006. https://doi.org/10.3389/fpsyt.2023.1068006

Kashyap, K., & Kashyap, H. (2024). An overview of the efficacy of solution-focused short interventions. *Open Access Journal of Behavioural Sciences and Psychology*, 7(2), 180155.
Kekäläinen, T., Luchetti, M., Terracciano, A., Gamaldo, A. A., Mogle, J., Lovett, H. H., et al. (2023). Physical activity and cognitive function: Moment-to-moment and day-to-day associations. *International Journal of Behavioral Nutrition and Physical Activity*, 20, 137. https://doi.org/10.1186/s12966-023-01536-9
Kim, J. S., & Franklin, C. (2009). Solution-focused brief therapy in schools: A review of the outcome literature. *Children and Youth Services Review*, 31(4), 464–470.
Lam, B. Y. H. (2023). Editorial: Neuropsychological mechanism of psychological resilience in relation to mental health outcomes across the lifespan. *Frontiers of Psychiatry*, 14, 1242182. https://doi.org/10.3389/fpsyt.2023.1242182
Lee, M. Y., Sebold, J., & Uken, A. (2007). Solution-focused treatment with domestic violence offenders. In T. S. Nelson & F. N. Thomas (Eds.) *Handbook of solution-focused brief therapy* (pp. 135–162). New York: The Haworth Press.
Macdonald, A. (2007). Applying solution-focused brief therapy in mental health practice. In T. S. Nelson & F. N. Thomas (Eds.) *Handbook of solution-focused brief therapy* (pp. 267–290). New York: The Haworth Press.
Miller, S. D., & Berg, I. K. (1995). *The miracle method: A radically new approach to problem drinking*. New York: Norton Professional Books.
Pichot, T. (2007). Looking beyond depression. In T. S. Nelson & F. N. Thomas (Eds.) *Handbook of solution-focused brief therapy* (pp. 117–132). New York: The Haworth Press.
Randolph, J. J. (2015). Searching for cognitive health: Another look through the Archives. *Archives of Clinical Neuropsychology*, 30, 583–584. https://doi.org/10.1093/arclin/ acv047.259
Randolph, J. J. (2018). Positive neuropsychology: The science and practice of promoting cognitive health. *Applied Neuropsychology: Adult*, 25(4), 287–294. https://doi.org/10.1080/23279095.2018.1457465
Ruff, R. M. (2003). A friendly critique of neuropsychology: Facing the challenges of our future. *Archives of Clinical Neuropsychology*, 18, 847–864. https://doi.org/10.1016/j.acn.2003.07.002
Ruff, R. M. (2020). Positive neuropsychology: Advances in cognitive rehabilitation and emotional recovery. *Journal of Clinical Psychology*, 76(5), 875–889.
Selekman, M. D. (1997). *Solution-focused therapy with children*. New York: The Guilford Press.
Seligman, M. E. P., & Csikszentmihalyi, M. (2000). Positive psychology: An introduction. *American Psychologist*, 55(1), 5–14.
Seligman, M. E. P., Steen, T. A., Park, N., & Peterson, C. (2005). Positive psychology progress: Empirical validation of interventions. *American Psychologist*, 60(5), 410–421.
Stams, G. J. J. M., Deković, M., Buist, K. L., & de Vries, L. (2006). Effectiveness of solution-focused brief therapy: A meta-analysis. *Journal of Evidence-Based Social Work*, 3(1), 75–90.
Thomas, F., & Nelson, T. (2007). Assumptions and practices within the solution-focused brief therapy tradition. In T. S. Nelson & F. N. Thomas (Eds.) *Handbook of solution-focused brief therapy* (pp. 3–24). New York: The Haworth Press.
Tremont, G., Davis, J. D., & Bishop, D. S. (2006). Unique contribution of family functioning in neuropsychological performance. *Archives of Clinical Neuropsychology*, 21(1), 29–41.
Trepper, T. S., Dolan, Y. M., McCollum, E. E., & Nelson, T. S. (2006). Steve de Shazer and the future of solution-focused therapy. *Journal of Marital and Family Therapy*, 32(2), 133–139.
Turnell, A. (2007). Thinking and practicing beyond the therapy room: Solution-focused brief therapy, trauma, and child protection. In T. S. Nelson & F. N. Thomas (Eds.) *Handbook of solution-focused brief therapy* (pp. 295–311). New York: The Haworth Press.
Viner, R. M., Christie, D., Taylor, V., & Hey, S. (2003). Motivational/solution-focused intervention improves HbA1c in adolescents with Type 1 diabetes: A pilot study. *Diabetic Medicine*, 20, 739–742.
Wade, D. T., Mewse, K., & Titheradge, D. (2016). Positive neuropsychology: A new approach to the emotional and behavioral consequences of acquired brain injury. *Clinical Rehabilitation*, 30(6), 541–548.

Walker, C. R., Froerer, A. S., & Gourlay-Fernandez, N. (2022). The value of using emotions in solution-focused brief therapy. *Journal of Marital and Family Therapy*, 48(3), 812–826. https://doi.org/10.1111/jmft.12551

Wilson, B. A. (2008). Neuropsychological rehabilitation. *Annual Review of Clinical Psychology*, 4, 141–162.

Żak, A. M., & Pękala, K. (2024). Effectiveness of solution-focused brief therapy: An umbrella review of systematic reviews and meta-analyses. *Psychotherapy Research*, 1–13. https://doi.org/10.1080/10503307.2024.2406540

5

ROLE OF P300 EVENT-RELATED POTENTIAL (ERP) IN NEUROPSYCHIATRIC CONDITIONS

Lekhanshi Manocha, Aakriti Varshney, and Anand Pratap Singh

Electroencephalography

Electroencephalography (EEG) is a noninvasive neuroimaging method that captures the brain's electrical activity from the scalp surface using metal electrodes and a conducting medium. EEG measures a local current generated when neurons in the brain are stimulated during synaptic excitations of the dendrites. In 1875, the first EEG was recorded by Richard Caton, a British physicist, who recorded electrical activity from the brains of rabbits and monkeys. In 1929, Hans Berger, a German biologist and psychiatrist, first described the electrical activity of the human brain when sleeping.

Normal and abnormal brain activity can both be detected through EEG, making it a useful tool in medical practices. It has been widely used in both research and clinical settings to study various conditions of neurotic and psychotic disorders, including personality disorders and dementia, aiding in clinical diagnosis and guiding treatment (Badrakalimuthu et al., 2011). It has also been useful in understanding emotional processes, stress response, and cognitive activities in humans, and has been employed in developmental and neuroscience research.

The EEG measures brain waves at various frequencies in the brain. Brain waves are electrical voltages that oscillate in the brain, measuring only a few millionths of a volt (Abhang et al., 2016). Human EEG waves have been categorized into five types, with different characteristics and frequencies – Alpha wave (7–13 Hz), Beta wave (13–35 Hz), Theta wave (4–7 Hz), Delta wave (0–4 Hz), and Gamma wave (30–100 Hz) (Roohi-Azizi et al., 2017). These waves can be affected by various psychological and neurological disorders. For example, studies have found EEG abnormalities in patients with major depressive disorder (MDD), such as increased activity in the right frontal cortex as compared to the left frontal cortex (Deldin & Chiu, 2005). Increased theta rhythm along with decreased alpha and beta rhythms has been found in patients with obsessive-compulsive disorder (OCD) (Karadag et al., 2003).

DOI: 10.4324/9781003646662-6

Event-Related Potentials (ERPs)

The most significant application of EEG is Event-Related Potentials (ERPs), also known as evoked potentials. ERPs are voltage changes that occur in the brain in response to evoked brain activity, which can be produced by various sensory, cognitive, or motor events (Sur & Sinha, 2009). They are believed to represent the activity of neural networks involved in information processing.

ERPs in humans can be broadly categorized into two types. 'Sensory' or 'exogenous' are the early waves, which are heavily influenced by the physical properties of the stimulus. These waves peak within the first 100 ms of a stimulus presentation. On the other hand, 'Cognitive' or 'endogenous' ERPs are the latter ERPs which indicate information processing and how the person analyzes the stimulus presented (Sur & Sinha, 2009). ERPs are commonly computed by exposing the participants to repeated stimuli and averaging the EEG activity around the stimulus response (Seer et al., 2016). They offer the advantage of precise measurement of brain activity to milliseconds due to their excellent temporal resolution.

ERPs have been used by a large number of researchers to extensively study the temporal dynamics of cognition and perception (Woodman, 2010), and to investigate variations of brain responses in neuropsychiatric disorders. The ERP waveforms are described according to *latency* and *amplitude*. ERP latencies are related to the time taken for the performance of cognitive processes, such as classifying a stimulus and preparing a suitable response, while the extent to which brain resources are devoted to these activities is thought to be indicated by ERP amplitudes (Seer et al., 2016).

Different stimuli elicit different ERP components, each of which reflects a distinct cognitive process or function. Several ERP components have been identified, such as P50, P100, N100, P200, N200, and P300. P50 wave is a positive deflection occurring 50 ms after a stimulus is presented, reflecting the process of sensory gating or the ability to filter irrelevant information (Green et al., 2015). N100 wave is a negative deflection occurring 80–150 ms after the presentation of a stimulus and reflects the orienting reaction or 'matching process' (Sur & Sinha, 2009). P200, occurring 200 ms after stimulus presentation, seems to play a role in the first conscious perception of stimuli (Ahsan, 2016). N200, a negative deflection occurring at 200 ms after stimulus presentation, is associated with elaborate encoding of objects and processing unfamiliar and complex stimuli.

The P300 Wave

The P300 wave was discovered by Sutton and colleagues in 1965, and since then, it has been a vital component of ERP research (Polich, 2007). The P300 or P3 is an ERP component peaking at about 300 ms after stimulus presentation, elicited by task-relevant, infrequent stimuli. It is a positive deflection which necessitates the participant's detection, counting, or cognitive processing, and is believed to be a representation of the brain's higher order cognitive functions, specifically working memory, attention allocation, or stimulus classification. It has been associated with the identification, categorization, and encoding processes that collectively constitute the superior cortical function of information processing (Raggi et al., 2021).

The P300 wave has broadly two subcomponents: P3a and P3b. The P3a subcomponent is associated with attentional processes involved in automatic novelty identification and arises in response to distracter (task-irrelevant) stimuli. P3b is suggested to reflect a match or mismatch using a working memory trace that is consciously maintained, and is evoked by task-relevant stimuli (Patel & Azzam, 2005).

Greater attentional resources given to a stimulus result in larger P3 waves, as P3 amplitude represents the amount of resources allotted for information processing and stimulus interpretation (Polich, 2007). Thus, it could be a reflection of constructive and filtering processes. Information processing speed, or the speed at which a stimulus is classified, is reflected in the P300 wave latency. It has been associated with the performance of working memory.

P300 waves are recorded through EEG. Oddball paradigms, in which participants identify an infrequent target stimulus presented in a series of regular trains of standard stimuli, are frequently used to evoke these responses. Subjects must only respond to the target in the way that is desired (such as counting the number or clicking a button) and leave the standard stimulus out in order to elicit a clear P300 (Li et al., 2019). The P300 waves may be recorded in one of the two distinct conditions: active and passive. An active P300 wave is recorded when the subject is instructed to detect and respond to the target stimulus in the oddball paradigm, while a passive P300 wave is recorded when the participant is presented with a rare stimulus in the oddball paradigm without giving them instructions (Patel & Azzam, 2005). Another paradigm for eliciting P300 is the 'Go/No-Go' task, where participants have to answer one of the options while withholding the other. In a 'Go/No-Go' test, for example, a participant may be asked to execute a certain action in response to specific stimuli (such as pressing the 'Go' button) and to inhibit the same action in response to a different set of stimuli (Raggi et al., 2021; Salisbury et al., 2004).

The neuropsychological underpinnings and neural origins of the P300 wave are not known very precisely. The hippocampus formation was the focus of the earliest human studies on the neural origins of P300 because of the attention and memory functions linked to P300 generation. These recordings revealed that the hippocampus regions of the medial temporal lobe generate at least a portion of the P300 (P3b). However, later research revealed that P300 generation is not directly influenced by hippocampus formation (Huang et al., 2015). Other studies have reported the involvement of the parietal lobe along with some frontal and temporal regions (Tripathi et al., 2015). Frontal lobe activity has been shown in fMRI and ERP studies to detect rare or physically alerting stimuli. Therefore, P3a production requires the integrity of the frontal lobe (Knight et al., 1995, as cited in Polich, 2007). Thus, from current knowledge, P300 is not limited to a specific region but rather is generated along a widely distributed network, involving the medial temporal lobe, the temporo-parietal junction, and parietal areas. Similarly, the inferior and middle frontal, the orbito-frontal, and the cingulate cortices also contribute to it (Raggi et al., 2021).

P300 in Neuropsychiatric Disorders

As seen above, cognitive function in neuropsychiatric illnesses can be evaluated through the measurement of P300 waves. Changes in P300 wave amplitude and latency, which are neurophysiological markers of cognitive function, may be used as a parameter to indicate cognitive impairment. In order to research the psychophysiological correlates of mental processes, ERPs – including P300, which are EEG changes time-locked to sensory, motor, or cognitive events – offer a noninvasive and safe method that also allows repeated measurements (Sur & Sinha, 2009). Their sensitivity to cognitive changes also makes them effective tools for monitoring treatment progress. Different P300 characteristics have been found to be associated with various neuropsychiatric disorders, including bipolar disorder, OCD, schizophrenia, and Alzheimer's disease (AD), which have been summarized in Table 5.1.

As shown in the data presented in Table 5.1, the P300 characteristics show a consistent pattern in some conditions. For example, lower amplitude and delayed latency have been found in all studies with schizophrenia and epilepsy patients. Other conditions such as MDD have consistently shown longer latencies without consistent findings in amplitude. Furthermore, conditions such as OCD and neurodegenerative disorders have inconsistent findings for both amplitude and latency. The details of the observations with respect to each of the disorders are explained below.

Schizophrenia and Psychotic Disorders

One of the main characteristics of neuropsychopathology in psychotic disorders such as schizophrenia has been identified as cognitive impairments. Psychotic disorders are associated with a variety of cognitive deficits such as impairments in attention, executive functioning, verbal learning and retention, and processing speed. Meta-analyses examining cognitive deficits in schizophrenia have shown reduced amplitude and longer latency of P300 in patients in comparison to control groups (Bramon et al., 2004; Jeon & Polich, 2003). Patients with schizophrenia exhibited smaller amplitudes than both healthy controls and schizoaffective patients. However, compared to healthy controls, the latency was longer for both the schizoaffective and schizophrenia groups (Mathalon et al., 2009). Both schizophrenia patients and their unaffected biological siblings have been found to show a reduced amplitude and increased latency compared to a healthy control group, where the patients with schizophrenia had the longest P300 latency, followed by their siblings and controls, while the patients with schizophrenia had the least P300 amplitude (Raghavan et al., 2016). There is currently strong evidence that 'high-risk' people with prodromal symptoms, before illness onset, also have lower P300 amplitudes.

Hassan et al. (2020) compared patients with first-episode psychosis to unrelated matched healthy controls using the auditory P300 ERP as a cognitive assessment. In comparison to the control group, it was found that all patient groups, involving schizophrenia spectrum disorder, depressive disorder with psychotic features, and bipolar disorder with psychotic features, had significantly higher P300 latency and significantly lower P300 amplitude. Furthermore, the performance of the patients was impacted by the level of psychopathology they had based on PANSS scores.

Overall, studies indicate that there is no effect of antipsychotic treatment on the latency of P300 (Bramon et al., 2004). However, one study in the meta-analysis by Bramon et al. (2004) found that P300 amplitude was significantly more abnormal in patients who were not taking medications as compared to patients on antipsychotic treatment. Moreover, after a year, the P300 amplitude mean scores of all first-episode psychosis patients significantly improved in the study by Hassan et al. (2020), even though the P300 latency did not significantly change. This indicates that ongoing psychotropic medication treatment of psychotic symptoms may help to partially improve P300 abnormalities.

Table 5.1 P300 amplitude and latency in neuropsychiatric disorders

Authors	*Sample*	*P300 Amplitude*	*P300 Latency*
Bramon et al. (2004)	46 studies Schizophrenia patients (SZ; 1443) Control group (1251)	SZ patients < control group	SZ patients > control group
Jeon and Polich (2003)	104 studies SZ patients	SZ patients < control group	SZ patients > control group
Mathalon et al. (2000)	SZ patients (22) Schizoaffective disorder patients (SA; 15) Control group (22)	SZ group < controls SA group < normal/similar to controls	SZ > controls SA > controls
Raghavan et al. (2016)	SZ patients (30) Unaffected biological siblings of SZ patients (30) Control group (30)	SZ < Unaffected biological siblings control group < control group	SZ > Unaffected biological siblings > control group
Hassan et al. (2020)	First episode psychosis patients: - Schizophrenia Spectrum Disorder (SSD; 47) - Bipolar disorder with psychotic features (BP; 41) - Depressive disorder with psychotic features (MDD; 20) Control group (50)	SSD < BP < MDD < Control group	SSD > BP > MDD > Control group
Miyata et al. (1998)	Unmedicated OCD patients (23) Unmedicated Social Phobia patients (12) Control group (18)	No significant difference between groups	OCD < SP OCD < Healthy control group
Towey et al. (1993)	OCD patients (17) Control group (16)	No significant difference between groups	No significant difference between groups
Andreou et al. (2013)	Unmedicated OCD patients (71) Control group (71)	OCD > control group	No significant difference between groups
Gohle et al. (2008)	Unmedicated OCD patients (63) Control group (63)	P3b amplitude: OCD > control group P3a amplitude – no significant difference	No significant difference between groups
Raggi et al. (2021)	OCD patients (35 studies)	OCD > control group (8 studies) OCD < control group (13 studies)	OCD < control (9 studies)

(Continued)

Table 5.1 (Continued)

Authors	*Sample*	*P300 Amplitude*	*P300 Latency*
Khanande et al. (2023)	OCD patients (20) First-degree relatives of OCD patients (FDRs; 20) Control group (20)	OCD < FDRs < control group	OCD < FDRs < control group
Lai et al. (2010)	Probable Alzheimer's disease patients (AD; 20) Mild cognitive impairment patients (MCI; 18) Control group (14)	No significant difference between groups	AD > MCI > Control group
Hedges et al. (2014)	20 articles AD patients (301) Control group (339)	AD < Control group	–
Bonnani et al. (2010)	Dementia with Lewy Bodies patients (DLB; 36) AD patients (40) Control group (50)	DLB < AD < Control group	DLB > AD > Control group
Rajendran et al. (2023)	Idiopathic Parkinson's disease patients (PD; 32)	PD < normal controls	no significant difference
Hasan et al. (2023)	Persons with epilepsy (PwE; 102) Control group (50)	PwE < normal controls	PwE > normal controls
Sowndhararajan et al. (2018)	Review – PwE	PwE < normal controls	PwE > normal controls
Zhong et al. (2019)	27 studies PwE	PwE < normal controls	PwE > normal controls
Himani et al. (1999)	Major depression (MDD; 20) Control group (20)	No significant difference	MDD > normal controls
Tripathi et al. (2015)	MDD (30) Control group (30)	No significant difference	MDD > normal controls
Singh et al. (2000)	MDD (40) Control group (40)	MDD < normal controls	MDD > normal controls
Karaaslan et al. (2003)	MDD (36) 16 patients with psychotic features (DwPF), 20 patients without psychotic features (DwoPF) Normal controls (20)	DwPF < normal controls DwoPF: no significant difference	DwPF > normal controls DwoPF > normal controls

Obsessive-Compulsive Disorder

Individuals with OCD show numerous deficits in cognitive functioning, such as information processing, response inhibition, cognitive flexibility, planning, and decision making (Suhas & Rao, 2019). Abnormal activity in the cortico-striato-thalamo-cortical (CSTC) circuits, which include the thalamus, the orbitofrontal cortex (OFC), and the striatum within the basal ganglia, has been proposed to be implicated in OCD. Various models have been proposed, such as the 'orbitofrontal striatal model', which hypothesized that OCD symptoms might be linked to increased activity in the OFC as a result of the reduced ability of the striatum to regulate the thalamus (Jalal et al., 2023). This hypothesis implies that defective cognitive and metacognitive processes may be linked to OCD (Flasbeck et al., 2022).

There have been inconsistent findings with regard to P300 abnormalities for patients with OCD in the existing literature. Some studies have shown no significant difference in the P300 latencies of OCD patients and normal controls. However, a non-significant trend of shorter latencies in OCD patients was seen (Andreou et al., 2013; Towey et al., 1993). Miyata et al. (1998) found that although the depression and anxiety scores of OCD patients were comparable to those of patients with social phobia, their P300 latencies were significantly shorter. This implies that rather than being related to general psychiatric symptoms, P300 abnormalities in OCD patients may be caused by underlying biological causes. The shortened latencies may be a reflection of the reduced inhibitory activity in the context of increased cortical activation (Raggi et al., 2021). Furthermore, studies have reported that the P300 characteristics did not correlate with the severity of OCD symptoms. Researchers have suggested that the biological predisposition to OCD may thus have a greater influence on cognitive processing than the current state of symptoms because OCD symptoms usually fluctuate and alter over time, viewing P300 abnormalities as a trait rather than a state characteristic (Andreou et al., 2013; Gohle et al., 2008; Miyata et al., 1998).

With regard to the amplitude of the P300 wave in OCD, some studies have found that although P300 measures were not significantly impacted by treatment, patients had larger P300 amplitudes at baseline (Andreou et al., 2013; Riva et al., 2019; Gohle et al., 2008). The results are consistent with neurobiological hypotheses of cognitive hyperarousal and increased attentional focus involved in OCD, which affects the production of P300. On the other hand, a decreased P300 amplitude has been reported in other studies (Sanz et al., 2001, as cited in Flasbeck et al., 2022), with the amplitude trending toward normality following pharmacological treatment (Raggi et al., 2021). Smaller P300 amplitudes in the centro-parietal region have been found not only in OCD patients but also in their first-degree relatives, as compared to healthy controls, along with shorter P300 latencies in both groups (Khanande et al., 2023). Thus, despite the inconsistencies in the findings of different studies, P300 measurements could serve as a potential biological marker for OCD vulnerability rather than just an indicator of active symptoms.

Depressive Disorders

Individuals with MDD have been shown to have both cognitive abnormalities and prolonged P300 latency (Himani et al., 1999; Karaaslan et al., 2003; Singh et al., 2000; Tripathi et al., 2015), and the P300 latency correlates closely with the severity of MDD (Tripathi et al., 2015). Singh et al. (2000) found a positive correlation between P300 latency and both age and severity of illness. P300 latencies were longer in people 40 years of age and older and in people with severe depression than in people under 40 and in people with mild or moderate depression. Furthermore, significantly lower P300 amplitudes were found in depressive patients. A negative relationship between P300 amplitude and age was observed, with those 60 years of age and beyond exhibiting lower P300 amplitudes than those under that age. However, other studies showed no significant difference in the P300 amplitudes of depressive and healthy control groups (Himani et al., 1999; Tripathi et al., 2015).

The P300 parameters were not correlated with the sex of depressive patients and the duration of illness, and no significant differences were seen in the parameters of treated and untreated patients (Singh et al., 2000). However, differences were observed with treatment in another study. Karaaslan et al. (2003) investigated the P300 component in patients with depression by comparing individuals with and without psychotic characteristics. Prior to treatment, P300 latencies in both patient groups were significantly longer than those of the control group. However, only patients with psychotic features showed a significant decrease in P300 amplitude. The delayed P300 latencies in both groups and the decreased P300 amplitude in the psychotic features group returned to normal after receiving treatment for depression. Thus, prolonged P300 latency could be considered a state marker for depression, and reduced P300 amplitude could be correlated as a state marker for depression with psychotic features.

Neurodegenerative Disorders

Neurodegenerative disorders are characterized by marked cognitive difficulties, such as in attention and concentration, memory, judgment, and executive and language functions. In normal aging, age-related reduction in amplitude, prolonged latency, and more frontally oriented P300 have been consistently reported across studies. There have been reports of P300 amplitude decreases and latency increases in dementia patients that are more severe than those seen in healthy aging (Duncan et al., 2009). These findings have been utilized to distinguish people with dementia from those with pseudodementia, as the latter group exhibits shorter latencies comparable to those observed in normal aging.

Studies have shown that both mild cognitive impairment (MCI) and AD patients have abnormalities/differences in P300 amplitude and latency, with significantly longer latencies in MCI and AD than in normal controls that correlate with the severity of cognitive impairment (Lai et al., 2010). Thus, P300 abnormalities might be detectable before clinical symptoms appear, suggesting potential for early detection.

While most studies have reported a decreased P300 amplitude in patients with AD (Hedges et al., 2014), there have been some exceptions, in which amplitude in patients with AD was reported as being equivalent to that in healthy controls (Lai et al., 2010). In contrast, increased P300 latency has been consistently supported in the AD literature. Increased P300 latency is linked to other neurodegenerative diseases, including dementia with Lewy bodies (DLB) and Parkinson's disease dementia (PDD). Bonnani et al. (2010) found that in DLB groups, P300 had less amplitude and was more delayed than in AD groups. Rajendran et al. (2023) investigated P300 wave components in 32 Parkinson's disease patients, comparing those with normal and impaired cognition. The study found significantly decreased P300 amplitude in patients, with a positive correlation between cognitive scores (MoCA) and P300 amplitude. According to the results, P300 wave analysis may be a helpful method for identifying cognitive impairment in people with Parkinson's disease.

Epilepsy

Cognitive impairment is commonly seen in persons with epilepsy. Additionally, epilepsy can worsen cognitive function that is already impaired, such as in older adults (Sen et al., 2018). Studies have found that patients with epilepsy have longer P300 latency and lower P300 amplitude than controls (Hasan et al., 2023; Sowndhararajan et al., 2018). Compared to patients with generalized epilepsy, those with focal epilepsy have longer latencies (Hasan et al., 2023). Zhong et al. (2019) found that epileptic patients have a lower P300 amplitude and a longer P300 delay than controls based on a meta-analysis of 26 research studies. These differences can be seen in both children and adults when compared to healthy controls, according to subgroup analysis based on age group.

Applications of P300 in Neurorehabilitation

The ability of the P300 component of ERPs to provide insights into cognitive functioning has made it a useful tool in neuropsychiatric research. ERPs provide a safe, noninvasive, and repeatable way to investigate the psychophysiological correlates of mental processes (Sur & Sinha, 2009). Specifically, the P300 wave can prove to be an objective marker for identifying cognitive deficits and has been strongly associated with a number of neuropsychiatric conditions, including schizophrenia, mood disorders, attention deficit hyperactivity disorder, and AD. It is a valuable instrument for diagnostic purposes and monitoring the effectiveness of rehabilitation interventions because of its sensitivity to cognitive changes. Thus, P300 can have diverse applications in understanding, diagnosing, and managing neuropsychiatric conditions.

Diagnostic Utility

The P300 wave serves as a valuable diagnostic tool. Its components – amplitude and latency – serve as reliable indicators of cognitive function across various disorders. Studies have consistently demonstrated distinct P300 patterns across different conditions: reduced amplitude and increased latency in schizophrenia and its high-risk populations (Jeon & Polich, 2003; Raghavan et al., 2016), prolonged latency in major depression correlating with illness severity (Tripathi et al., 2015), and characteristic changes in obsessive compulsive (Khanande et al., 2023), neurodegenerative disorders such as AD (Lai et al., 2010).

Although variations in P300 have been observed with fluctuating symptomatology, it has also exhibited features of a trait marker for disorders like schizophrenia (Andreou et al., 2013; Gohle et al., 2008; Miyata et al., 1998). Early identification of schizophrenia may be facilitated by P300 features in unaffected biological siblings of those with the disorder, which may act as indicators of underlying biological processes (Raghavan et al., 2016). It has excellent temporal resolution, allowing precise measurement of cognitive processing speed and resource allocation, making it particularly valuable for early detection and monitoring of cognitive decline. These characteristics, in combination with the high test–retest reliability, establish P300 as a promising biomarker for diagnostic purposes and treatment monitoring.

Assessment and Monitoring of Disease Progression

As P300 reflects underlying cognitive and neurobiological changes linked to treatment response or disease progression, it can be used as an indicator of prognosis in a variety of conditions. Prolonged P300 latency correlates with illness severity in MDD (Tripathi et al., 2015), while amplitude changes can predict response to antidepressant treatment (Karaaslan et al., 2003; Zhao et al., 2024). Longitudinal changes in P300 also correlate with symptom progression or treatment response in schizophrenia patients (Raghavan et al., 2016), aiding in disease monitoring. P300 parameters correlate positively with cognitive performance as measured by other standardized tools such as MoCA (Rajendran et al., 2023), supporting the view that P300 components are sensitive to cognitive changes and can be used to track the progression of cognitive impairment.

P300 as a Target in Cognitive Rehabilitation

When evaluating the efficacy of pharmacological and non-pharmacological treatments for a range of neuropsychiatric disorders, monitoring P300 wave characteristics is helpful. Studies have shown that when P300 abnormalities, including lower amplitude or longer latency, are normalized, symptoms of disorders like schizophrenia and OCD often improve (Raggi et al., 2021). Recent findings highlight the potential of P300 as a target for interventions aimed at improving cognitive ability. Reduced amplitude and increased latency of the P300 auditory ERP in schizophrenia are considered to be indicative of attentional deficits. Music, which can regulate arousability and emotional reactions, has been investigated as a non-pharmacological intervention to affect P300. Ahuja et al. (2020) found that schizophrenia patients showed significant increases in task accuracy and reaction time during an auditory oddball P300 task following a ten-minute music listening session, suggesting that music may be a helpful tool in modulating P300 and improving cognitive function. Similarly, computer-assisted cognitive rehabilitation programs used in combination with virtual reality technology, and brain–computer-interface-based P300 training have been successfully used to alter P300 amplitude and latency in stroke patients, resulting in positive effects on patients' cognitive functions (Fateeva et al., 2024; Shi et al., 2023). P300-based neurofeedback training has also shown initial success in reducing distraction and enhancing attention. Thus, P300 could play an important role as a potential target for cognitive enhancement in neuropsychiatric conditions.

Limitations and Future Directions

Despite the promising utility of the P300 wave as a biomarker in various neuropsychiatric disorders, several limitations and challenges remain in the current understanding of its clinical and research applications. One of the major challenges is the heterogeneity or inconsistency in the findings of different studies, particularly in conditions like OCD, and the unclear effects of medications on P300 measurements. Variations in study designs, sample populations, and experimental paradigms can produce conflicting results, making it difficult to generalize findings or draw conclusions.

The influence of factors such as age and co-occurring disorders may affect the interpretation of P300 results. For example, in schizophrenia, the frequent co-occurrence of other neuropsychiatric diseases, substance use disorders, differences in smoking prevalence, and racial stratifications in patient versus control populations all contribute to significant variability. Understanding the role of these factors is crucial for increasing the value of P300 as a disease-specific biomarker for schizophrenia and other diseases.

The diagnostic effectiveness of P300 as a stand-alone marker has not been established. For example, in depression, the reported changes in P300 amplitude and latency lack sensitivity and disease specificity, limiting their usefulness as diagnostic tools (Singh et al., 2000). Major gaps include limited longitudinal studies tracking P300 changes over time, insufficient research across diverse populations, and an incomplete understanding of underlying neurobiological mechanisms. Furthermore, P300 parameters have wide variability in the healthy population, making it difficult to establish normative data. Therefore, within-patient studies, such as investigating the changes in P300 parameters before and after treatment, may provide more useful information than cross-sectional comparisons.

Conclusion

The P300 wave represents a significant development in our understanding of cognitive function and neuropsychiatric disorders. This ERP component has various functions: it is a diagnostic tool for various neuropsychiatric disorders, an indicator for treatment outcomes, and a target in neurocognitive rehabilitation. Its noninvasive nature and ability to objectively monitor cognitive processes make it especially useful in clinical settings. While there are significant variations in findings across disorders, the P300 wave has proven to be a biomarker for cognitive functioning, with potential for future uses in neuropsychiatric diagnosis and treatment monitoring.

Future Directions

Further research is required to establish the usefulness of P300 as a biomarker for the early diagnosis of cognitive decline and neuropsychiatric diseases. Its ability to detect subtle cognitive changes before the clinical manifestation of illnesses such as AD, schizophrenia, and Parkinson's disease makes it a promising tool for detecting at-risk patients. Standardization of measurement protocols to improve reliability and comparability across studies could be helpful in minimizing the variability brought by inconsistent procedures. The integration of P300 wave analysis with other clinical, neuroimaging, and neuropsychological measures indicates another area to create more comprehensive diagnostic and monitoring tools. Emerging research into novel therapeutic interventions targeting P300 abnormalities offers another direction for researchers. Finally, more research, especially longitudinal studies and randomized control trials, is required to better understand the role of P300 in disease progression and treatment response.

References

Abhang, P. A., Gawali, B. W., & Mehrotra, S. C. (2016). Technological basics of EEG recording and operation of apparatus. In *Introduction to EEG- and Speech-Based Emotion Recognition* (pp. 19–50). Elsevier eBooks. https://doi.org/10.1016/b978-0-12-804490-2.00002-6

Ahsan, N. (2016). Creative Control: A Neurocognitive Index of Relationships Between Creativity and Attention [Master's Thesis, Carleton University]. ResearchGate. www.researchgate.net/publication/317176895_Creative_Control_A_Neurocognitive_Index_of_Relationships_Between_Creativity_and_Attention

Ahuja, S., Gupta, R. K., Damodharan, D., Philip, M., Venkatasubramanian, G., Keshavan, M. S., & Hegde, S. (2020). Effect of music listening on P300 event-related potential in patients with schizophrenia: A pilot study. *Schizophrenia Research*, *216*, 85–96. https://doi.org/10.1016/j.schres.2019.12.026

Andreou, C., Leicht, G., Popescu, V., Pogarell, O., Mavrogiorgou, P., Rujescu, D., ..., & Mulert, C. (2013). P300 in obsessive–compulsive disorder: Source localization and the effects of treatment. *Journal of Psychiatric Research*, *47*(12), 1975–1983. https://doi.org/10.1016/j.jpsychires.2013.09

Badrakalimuthu, V. R., Swamiraju, R., & De Waal, H. (2011). EEG in psychiatric practice: To do or not to do? *Advances in Psychiatric Treatment*, *17*(2), 114–121. https://doi.org/10.1192/apt.bp.109.006916

Bramon, E., Rabe-Hesketh, S., Sham, P., Murray, R. M., & Frangou, S. (2004). Meta-analysis of the P300 and P50 waveforms in schizophrenia. *Schizophrenia Research*, *70*(2–3), 315–329. https://doi.org/10.1016/j.schres.2004.01.004

Deldin, P. J., & Chiu, P. (2005). Cognitive restructuring and EEG in major depression. *Biological Psychology*, *70*(3), 141–151. https://doi.org/10.1016/j.biopsycho.2005.01.003

Duncan, C. C., Barry, R. J., Connolly, J. F., Fischer, C., Michie, P. T., Näätänen, R., ..., & Van Petten, C. (2009). Event-related potentials in clinical research: Guidelines for eliciting, recording, and quantifying mismatch negativity, P300, and N400. *Clinical Neurophysiology*, *120*(11), 1883–1908. https://doi.org/10.1016/j.clinph.2009.07.045

Fateeva, V. V., Kushnir, A. B., Grechko, A. V., & Mayorova, L. A. (2024). Rehabilitation of patients with post-stroke cognitive impairments using a P300-based brain–computer interface: Results of a randomized controlled trial. *Neuroscience and Behavioral Physiology*, *54*(4), 575–580. https://doi.org/10.1007/s11055-024-01630-w

Flasbeck, V., Enzi, B., Andreou, C., Juckel, G., & Mavrogiorgou, P. (2022). P300 and delay-discounting in obsessive-compulsive disorder. *European Archives of Psychiatry and Clinical Neuroscience*, *272*(2), 327–339. https://doi.org/10.1007/s00406-021-01302-7. Epub 2021 Jul 13. PMID: 34258638; PMCID: PMC8866265.

Gohle, D., Juckel, G., Mavrogiorgou, P., Pogarell, O., Mulert, C., Rujescu, D., ..., & Hegerl, U. (2008). Electrophysiological evidence for cortical abnormalities in obsessive–compulsive disorder – A replication study using auditory event-related P300 subcomponents. *Journal of Psychiatric Research*, *42*(4), 297–303. https://doi.org/10.1016/j.jpsychires.2007.01.003

Green, D. L., Payne, L., Polikar, R., Moberg, P. J., Wolk, D. A., & Kounios, J. (2015). P50: A candidate ERP biomarker of prodromal Alzheimer's disease. *Brain Research*, *1624*, 390–397. https://doi.org/10.1016/j.brainres.2015.07.054

Hasan, L. A., Hamdan, F. B., & Al-Mahdawi, A. (2023). P300 event-related potentials in people with epilepsy: Clinico-neurophysiologic study. *The Egyptian Journal of Neurology Psychiatry and Neurosurgery*, *59*(1). https://doi.org/10.1186/s41983-023-00700-z

Hassan, W. A., Darweesh, A. E. M., Abdel-Rahman, A. A., Ahmad, H. E. K., Hassaan, S. H., Noaman, M. M., & Fahmy, I. F. G. (2020). P300 cognitive assessment in patients with first-episode psychosis: A prospective case-control study. *Middle East Current Psychiatry*, *27*(1). https://doi.org/10.1186/s43045-020-00031-2

Hedges, D., Janis, R., Mickelson, S., Keith, C., Bennett, D., & Brown, B. L. (2014). P300 amplitude in Alzheimer's disease. *Clinical EEG and Neuroscience*, *47*(1), 48–55. https://doi.org/10.1177/1550059414550567

Himani, A., Tandon, O. P., & Bhatia, M. S. (1999). A study of P300-event related evoked potential in the patients of major depression. *Indian Journal of Physiology and Pharmacology*, *43*(3), 367–372. PMID: 10776486.

Huang, W. J., Chen, W. W., & Zhang, X. (2015). The neurophysiology of P 300—an integrated review. *European Review for Medical and Pharmacological Sciences*, *19*(8), 1480–1488. PMID: 25967724.

Jeon, Y., & Polich, J. (2003). Meta-analysis of P300 and schizophrenia: Patients, paradigms, and practical implications. *Psychophysiology*, *40*(5), 684–701. https://doi.org/10.1111/1469-8986.00070

Karaaslan, F., Gonul, A. S., Oguz, A., Erdinc, E., & Esel, E. (2003). P300 changes in major depressive disorders with and without psychotic features. *Journal of Affective Disorders*, *73*(3), 283–287. https://doi.org/10.1016/s0165-0327(01)00477-3

Karadag, F., Oguzhanoglu, N. K., Kurt, T., Oguzhanoglu, A., Ateşci, F., & Ozdel, O. (2003). Quantitative EEG analysis in obsessive compulsive disorder. *The International Journal of Neuroscience*, *113*(6), 833–847. https://doi.org/10.1080/00207450390200963

Khanande, R., Nizamie, S. H., Das, B., Goyal, N., Tikka, S. K., Mehta, V., Sachacher, B., Motichand, S., Katshu, M. Z. U. H., & Abhishek, P. (2023). A family study assessing P300 as a probable electrophysiological endophenotype in obsessive-compulsive disorder. *Archives of Biological Psychiatry*, *1*, 54–59. https://doi.org/10.25259/abp_13_2023

Lai, C.-L., Lin, R.-T., Liou, L.-M., & Liu, C.-K. (2010). The role of event-related potentials in cognitive decline in Alzheimer's disease. *Clinical Neurophysiology*, *121*(2), 194–199. https://doi.org/10.1016/j.clinph.2009.11.001

Li, F., Yi, C., Jiang, Y., Liao, Y., Si, Y., Dai, J., Yao, D., Zhang, Y., & Xu, P. (2019). Different contexts in the oddball paradigm induce distinct brain networks in generating the P300. *Frontiers in Human Neuroscience*, 12. https://doi.org/10.3389/fnhum.2018.00520

Mathalon, D. H., Ford, J. M., & Pfefferbaum, A. (2000). Trait and state aspects of P300 amplitude reduction in schizophrenia: A retrospective longitudinal study. *Biological Psychiatry*, *47*(5), 434–449. https://doi.org/10.1016/s0006-3223(99)00277-2

Mathalon, D. H., Mathalon, D. H., Mathalon, D. H., Hoffman, R. E., Watson, T. D., Miller, R. M., Miller, R. M., Roach, B. J., Roach, B. J., Ford, J. M., Ford, J. M., & Ford, J. M. (2009). Neurophysiological distinction between schizophrenia and schizoaffective disorder. *Frontiers in Human Neuroscience*. https://doi.org/10.3389/neuro.09.070.2009

Miyata, N., Matsunaga, N., Kiriike, N., Iwasaki, N., Takei, N., & Yamagami, N. (1998). Event-related potentials in patients with obsessive–compulsive disorder. *Psychiatry and Clinical Neurosciences*, *52*(5), 513–518. https://doi.org/10.1046/j.1440-1819.1998.00427.x

Patel, S. H., & Azzam, P. N. (2005). Characterization of N200 and P300: Selected studies of the event-related potential. *International Journal of Medical Sciences*, *2*(4), 147–54. https://doi.org/10.7150/ijms.2.147. Epub 2005 Oct 1. PMID: 16239953; PMCID: PMC1252727.

Polich, J. (2007). Updating P300: An integrative theory of P3a and P3b. *Clinical Neurophysiology*, *118*(10), 2128–2148. https://doi.org/10.1016/j.clinph.2007.04.019

Raggi, A., Lanza, G., & Ferri, R. (2021). A review on P300 in obsessive-compulsive disorder. *Frontiers in Psychiatry*, 12. https://doi.org/10.3389/fpsyt.2021.751215

Raghavan, D. V., Shanmugiah, A., Bharathi, P., & Jeyaprakash, R. (2016). P300 and neuropsychological measurements in patients with schizophrenia and their healthy biological siblings. *Indian Journal of Psychiatry*, *58*(4), 454–458. https://doi.org/10.4103/0019-5545.196709

Rajendran, D., Bandhu, R., Gautam, S., Dhamija, R. K., & Mondal, S. (2023). Auditory evoked P300 potential in patients with Parkinson's disease. *Cureus*. *15*(9), e45933. https://doi.org/10.7759/cureus.45933

Riva, A. D., Berger, A., & Anholt, G. E. (2019). Early cognitive processes in OCD: An ERP study. *Journal of Affective Disorders*, *246*, 429–436. https://doi.org/10.1016/j.jad.2018.12.109

Roohi-Azizi, M., Azimi, L., Heysieattalab, S., & Aamidfar, M. (2017). Changes of the brain's bioelectrical activity in cognition, consciousness, and some mental disorders. *Medical Journal of the Islamic Republic of Iran*, *31*, 53. https://doi.org/10.14196/mjiri.31.53. PMID: 29445682; PMCID: PMC5804435.

Salisbury, D. F., Griggs, C. B., Shenton, M. E., & McCarley, R. W. (2004). The NoGo P300 'anteriorization' effect and response inhibition. *Clinical Neurophysiology: Official Journal of the International Federation of Clinical Neurophysiology*, *115*(7), 1550–1558. https://doi.org/10.1016/j.clinph.2004.01.028

Seer, C., Lange, F., Georgiev, D., Jahanshahi, M., & Kopp, B. (2016). Event-related potentials and cognition in Parkinson's disease: An integrative review. *Neuroscience & Biobehavioral Reviews*, *71*, 691–714. https://doi.org/10.1016/j.neubiorev.2016.08.003

Sen, A., Capelli, V., & Husain, M. (2018). Cognition and dementia in older patients with epilepsy. *Brain*, *141*(6), 1592–1608. https://doi.org/10.1093/brain/awy022

Shi, J., Ma, S. J., Hu, J., Hu, Z. K., Xia, J. Y., & Xu, H. Y. (2023). The effects of computer-aided cognitive rehabilitation combined with virtual reality technology on event-related potential P300 and cognitive function of patients with cognitive impairment after stroke. *European Review for Medical and Pharmacological Sciences*, *27*, 8993–9000.

Singh, R., Shukla, R., Dalal, P. K., Sinha, P. K., & Trivedi, J. K. (2000). P 300 event related potential in depression. *Indian Journal of Psychiatry*, *42*(4). 402–409. PMCID: PMC2962742; PMID: 21407978.

Sowndhararajan, K., Kim, M., Deepa, P., Park, S., & Kim, S. (2018). Application of the P300 event-related potential in the diagnosis of epilepsy disorder: A review. *Scientia Pharmaceutica*, *86*(2), 10. https://doi.org/10.3390/scipharm86020010

Suhas, S., & Rao, N. P. (2019). Neurocognitive deficits in obsessive-compulsive disorder: A selective review. *Indian Journal of Psychiatry*, *61*(Suppl 1), S30–S36. https://doi.org/10.4103/psychiatry.IndianJPsychiatry_517_18

Sur, S., & Sinha, V. (2009). Event-related potential: An overview. *Industrial Psychiatry Journal*, *18*(1), 70–73. https://doi.org/10.4103/0972-6748.57865

Towey, J., Bruder, G., Tenke, C., Leite, P., DeCaria, C., Friedman, D., & Hollander, E. (1993). Event-related potential and clinical correlates of neurodysfunction in obsessive-compulsive disorder. *Psychiatry Research*, *49*(2), 167–181. https://doi.org/10.1016/0165-1781(93)90103-n

Tripathi, R., Tripathi, S., Mishra, N., & Gurnani, K. (2015). P300 latency as an indicator of severity in major depressive disorder. *Industrial Psychiatry Journal*, *24*(2), 163. https://doi.org/10.4103/0972-6748.181726

Woodman, G. F. (2010). A brief introduction to the use of event-related potentials in studies of perception and attention. *Attention, Perception & Psychophysics*, *72*(8), 2031–2046. https://doi.org/10.3758/APP.72.8.2031

Zhao, L., Zhou, D., He, X., Peng, X., Hu, J., Ma, L., Liu, X., Tao, W., Chen, R., Jiang, Z., Zhang, C., Liao, J., Xiang, J., Zeng, Q., Dai, L., Zhang, Q., Hong, S., Wang, W., & Kuang, L. (2024). Changes in P300 amplitude to negative emotional stimuli correlate with treatment responsiveness to sertraline in adolescents with depression. *Brain Research*, *1845*, 149272. https://doi.org/10.1016/j.brainres.2024.149272

Zhong, R., Li, M., Chen, Q., Li, J., Li, G., & Lin, W. (2019). The P300 event-related potential component and cognitive impairment in epilepsy: a systematic review and meta-analysis. *Frontiers in Neurology*, *10*, 943. https://doi.org/10.3389/fneur.2019.00943

PART II

Technology-Driven Innovations

6

ARTIFICIAL INTELLIGENCE

A Replacement or a Complement to Psychotherapy in Neurocognitive Rehabilitation?

N. Sharon Sucharitha Gold and S. Yogesh Kanna

Introduction

This chapter explores the application of artificial intelligence (AI) in modern psychotherapy and neurocognitive rehabilitation. While technological advancements and the proven effectiveness of AI tools are widely celebrated, the potential drawbacks of relying on AI beyond its appropriate scope are often overlooked. There is a concern that the general public might mistakenly believe that chatbots and AI-driven applications are sufficient for therapy, potentially diminishing the perceived importance of in-person therapeutic interventions and rehabilitation. It is crucial for therapists to recognize the complementary role of both human and AI in mental health care. The chapter begins with an overview of traditional psychotherapeutic approaches and then examines how AI has transformed these methods. It emphasizes the need to harness AI judiciously to enhance therapeutic outcomes, ensuring that technology supports rather than replaces human interaction. Ultimately, the chapter advocates for a balanced integration of AI and human therapy to maximize efficacy in future mental health and neurocognitive rehabilitation efforts.

Neurocognitive Rehabilitation, Psychotherapy, and AI

Neurocognitive rehabilitation is a therapy or treatment method involving targeted interventions aimed at restoring or compensating for cognitive deficits resulting from brain injury, neurodegenerative diseases, or mental health conditions (Cicerone et al., 2019). This approach employs a variety of techniques, including cognitive training, behavioral strategies, and the latest being computerized exercises, to improve functions such as memory, attention, executive functioning, and processing speed. Its significance lies in its capacity to enhance patients' independence, quality of life, and social functioning, especially in those with traumatic brain injury, stroke, or neurodegenerative disorders (Vazquez & García, 2018). The spectrum of neurorehabilitation treatments includes suitable psychotherapeutic methods primarily because psychological co-morbidities often occur but are not always identified or properly managed. Any impairment of the central nervous system due to

DOI: 10.4324/9781003646662-8

injury or disease needs special attention as the mind and body are intricately connected. It is not possible to address one without including the other. Hence a multifaceted psychotherapeutic approach based on neurological science is essential. The integration of neurocognitive rehabilitation within mental health treatment involving psychotherapy, behavioral interventions, and family support (Lezak et al., 2012) is increasingly recognized for its potential to address underlying cognitive impairments that may hinder the progress of therapy.

Psychotherapy, also known as talk therapy, is a treatment modality that involves conversations between a trained therapist and a client aimed to treat psychological issues and promote mental well-being (American Psychological Association [APA], 2019) essential in helping individuals understand their emotions, develop coping strategies, and improve overall functioning (Norcross, 2018). Psychotherapy has been shown to effectively treat a range of mental health conditions, including depression, anxiety, and trauma-related disorders (Cuijpers et al., 2018). AI in the context of psychotherapy refers to the development and application of computer systems capable of performing tasks that require human intelligence, such as understanding natural language, recognizing emotions, and making decisions (Shatte et al., 2019). Various AI-driven tools are used in psychotherapy such as chatbots, virtual therapists, and machine learning algorithms (Khodabandelou, 2020). These systems extract information from large datasets to simulate human-like interactions and facilitate mental health support, offering an innovative support to traditional therapy modalities. Moreover, advancements in AI are beginning to influence neurocognitive rehabilitation by enabling personalized, adaptive training programs that respond dynamically to individual patient needs (Cicerone et al., 2019).

Traditional and Current Landscape of Psychotherapy and Neurocognitive Rehabilitation

Traditional therapeutic methods in neurocognitive rehabilitation focus on restoring cognitive functions through structured, evidence-based approaches. To enhance neuroplasticity and functional outcomes, tasks like computer-assisted training and paper-pencil exercises are commonly used (Bahar-Fuchs et al., 2013). Individuals with residual impairments are introduced compensatory strategies such as use of memory aids and environmental modifications to help in management (Wilson et al., 2019). Speech and language therapy is employed for language deficits, utilizing techniques like semantic feature analysis and constraint-induced language therapy (Kiran, 2012). Psychosocial interventions, such as counseling and support groups, address emotional and behavioral challenges associated with cognitive impairments (Ponsford et al., 2014). Overall, traditional methods emphasize a multidisciplinary approach tailored to individual needs, promoting maximal recovery and adaptation (Cicerone et al., 2019).

Conventional psychotherapeutic approaches, such as cognitive-behavioral therapy (CBT) and psychoeducation, have demonstrated efficacy in various mental health conditions. However, their application in neurocognitive rehabilitation has proven to have significant limitations. First, these therapies often rely heavily on patients' cognitive and language abilities, which may be compromised in individuals with neurocognitive disorders such as traumatic brain injury or dementia (Cicerone et al., 2019). Consequently, patients with severe cognitive deficits may struggle to engage fully in traditional talk therapies, reducing treatment effectiveness. Additionally, conventional approaches tend to be time-intensive and require sustained effort, which can be challenging for individuals with fluctuating cognitive capacities or motivation issues (Kelly et al., 2020). Furthermore, these therapies may not adequately target neuroplasticity—the brain's ability to reorganize itself—limiting their potential to induce meaningful neural recovery. The heterogeneity of neurocognitive impairments further complicates standardized treatment, necessitating personalized interventions that are often difficult to implement within traditional frameworks. As a result, there is an increasing need for innovative approaches that leverage technology and neurobiological insights. Emerging methods such as computerized cognitive training, virtual reality (VR), and neuromodulation techniques offer promising avenues for enhancing neuroplasticity and tailoring interventions (Kirk et al., 2019).

Current Role of AI in Mental Health Interventions

AI is increasingly transforming the landscape of mental health care, serving as a valuable complement to human interventions by being integrated into psychotherapy and neurocognitive rehabilitation, enhancing personalized treatment, and improving outcomes. In psychotherapy, AI-powered tools such as chatbots and virtual therapists provide accessible, immediate support for individuals experiencing mental health challenges. Patients with mobility issues or those having limited access to clinicians benefit from AI-powered chatbots that offer mental health care through CBT techniques remotely (Fitzpatrick et al., 2019). These digital agents can deliver CBT techniques, monitor mood fluctuations through natural language processing (NLP), and offer personalized feedback, thereby supplementing traditional therapy sessions. They help bridge gaps in mental health care by reducing wait times, increasing reach in underserved populations, and providing continuous support outside clinical settings. AI-driven technologies facilitate tailored intervention programs for patients recovering from brain injuries, strokes, or neurodegenerative diseases. Machine learning algorithms analyze patient data to develop individualized treatment plans, track progress with high precision, and adapt exercises dynamically to optimize recovery (Shen et al., 2020). AI systems also assist in cognitive training by adapting exercises based on real-time performance, thereby optimizing recovery for stroke or traumatic brain injury patients (Kumar et al., 2021). For example, VR is combined with AI to deliver simulated reality therapy sessions that adapt to the patient's progress (Lindner et al., 2020). AI-powered VR environments create immersive scenarios that stimulate cognitive functions such as memory, attention, and problem-solving, making rehabilitation more engaging and effective.

It is important to note that the use of AI in psychotherapy is not something novel. Its introduction to mental health care may be traced back to Alan Turing's paper in 1950, Computing Machinery and Intelligence (Turing, 1950). The introduction of Eliza, the first chatbot to use AI for conversational purposes, signaled the beginning of AI application in psychological settings. In 1966, it created a real sense of interaction between clients and human therapists (Weizenbaum, 1976). In the 1990s and early 2000s, significant advancements were achieved in applying AI technologies to data analysis, which in turn was used in the field of health care to understand and discern human behavior with the added advantage of available data. AI-supported Socratic chatbots originated in the mid-2000s where they provided avenues for the users to introspect, challenge their thought process, and modify their behavior. However, it is in the post pandemic era that there is a rapid and extensive use of AI for psychotherapeutic purposes.

To be specific, there are two major ways in which AI has been used in mental health care, viz., diagnosis and treatment. Machine learning algorithms, such as support vector machines and neural networks, have demonstrated high accuracy in distinguishing between different psychiatric conditions based on neuroimaging data, speech patterns, and behavioral metrics. Dwyer et al. (2018) highlighted how machine learning applied to functional MRI data could differentiate individuals with schizophrenia from healthy controls with remarkable precision, facilitating earlier and more accurate diagnoses. Similarly, automated analysis of speech patterns using NLP techniques has been utilized to identify markers of depression and anxiety disorders, often outperforming traditional clinician assessments (Kramer et al., 2014). These advancements reduce subjective biases and provide objective, quantifiable diagnostic tools.

AI has also been used to create predictive models to understand the prognosis of the pathology in an individual with a psychological illness as well as to predict the incidence of a psychological issue in future. By integrating demographic, clinical, neuroimaging, and genetic data, machine learning models can generate individualized risk profiles. For instance, Chekroud et al. (2016) developed models predicting treatment outcomes in depression, enabling tailored therapeutic strategies. In neurocognitive rehabilitation, predictive models can forecast cognitive decline in conditions such as Alzheimer's disease, allowing for earlier intervention. A study by Wang et al. (2019) demonstrated that machine learning algorithms could predict cognitive deterioration with over 80% accuracy based on baseline neuroimaging and neuropsychological assessments. Such predictive capabilities are vital for proactive management, optimizing resource allocation, and improving patient outcomes.

AI also facilitates innovative therapeutic approaches, including digital therapeutics, virtual agents, and personalized cognitive training programs. Chatbots and virtual therapists, powered by AI, provide accessible mental health support, especially in underserved populations. For example, Koot et al. (2020) reviewed AI-driven chatbots used for CBT, noting their effectiveness in reducing symptoms of depression and anxiety. In neurocognitive rehabilitation, AI systems tailor cognitive exercises based on individual performance, promoting neuroplasticity and functional recovery. Robotic-assisted therapy, guided by AI algorithms, has shown promise in stroke rehabilitation by providing real-time feedback and adapting exercises to patient progress (Lohse et al., 2014). Moreover, AI-enabled VR environments offer immersive experiences that enhance engagement and facilitate the retraining of cognitive functions (Liu et al., 2020b). These personalized interventions not only improve efficacy but also increase accessibility and adherence.

AI as a Complement vs. Replacement: Evidence and Arguments

AI has currently become essential to mental health interventions, offering innovative approaches for diagnosis, treatment, and support. Recent research emphasizes AI's potential to enhance early detection of mental health disorders through machine learning algorithms analyzing speech, text, and behavioral data (Fitzpatrick et al., 2019). AI-driven mobile applications facilitate real-time monitoring and personalized interventions, enabling continuous care outside clinical settings (Kumar & Rose, 2019).

In psychotherapy, AI-driven tools such as chatbots and virtual therapists offer accessible and scalable options for delivering mental health support. Fitzpatrick et al. (2017)

conducted a randomized controlled trial evaluating Woebot, an AI chatbot designed to deliver CBT. The study found that participants using Woebot experienced significant reductions in depression and anxiety symptoms, comparable to those receiving human-delivered CBT. Similarly, Kshirsagar et al. (2020) reported that AI chatbots effectively facilitated mental health support, particularly in contexts where human therapists are scarce or stigma inhibits help-seeking. These findings suggest that AI tools can serve as effective preliminary interventions or supplementary supports, especially when immediate access is necessary.

However, other studies emphasize the importance of the human element in psychotherapy. Norcross et al. (2018) argue that therapeutic alliance, the collaborative bond between therapist and patient, is a critical predictor of treatment success. AI tools, despite their advancements, lack genuine empathy and nuanced understanding, which are often vital for addressing complex emotional issues. For instance, a study by Fulmer et al. (2018) highlighted that while AI chatbots can provide structured interventions, they may fall short in handling unpredictable, emotionally charged situations that require human judgment and empathy.

In neurocognitive rehabilitation, AI has demonstrated promising results in aiding recovery for patients with brain injuries and neurodegenerative diseases. AI-powered neurofeedback systems and adaptive training programs have shown to improve cognitive functions effectively. Coyle et al. (2021) reviewed multiple studies indicating that AI-based cognitive training can lead to meaningful improvements in attention, memory, and executive functions. The use of robotic devices integrated with AI algorithms to provide personalized motor and cognitive therapy has been shown to enhance patient engagement and outcomes (Huang et al., 2020).

Comparative studies often reveal that AI tools excel in delivering consistent, data-driven interventions and maintaining high levels of engagement through gamification and personalization. Nonetheless, the absence of human emotional support and contextual judgment remains a limitation. A meta-analysis by McGonigal et al. (2021) concluded that while AI-assisted neurorehabilitation is effective, combining AI with human oversight yields the best outcomes, emphasizing a hybrid approach.

Clinical Applications and Case Studies

In the domains of psychotherapy and neurocognitive rehabilitation, AI's capacity to gather and analyze vast amounts of patient data plays a pivotal role in improving treatment outcomes (Topol, 2019). By leveraging machine learning algorithms and NLP, clinicians can better understand patient behaviors, monitor progress, and tailor interventions accordingly.

One of the primary contributions of AI in these fields is its ability to facilitate comprehensive data collection. Traditional methods often rely on subjective reports and clinician observations, which can be inconsistent or biased (Shatte et al., 2019). Beyond data collection, AI excels in analyzing complex datasets to uncover patterns and insights that might elude human clinicians. Machine learning algorithms can process multimodal data—such as neuroimaging scans, speech transcripts, and behavioral metrics—to identify early signs of depression, anxiety, or cognitive decline (Esteva et al., 2019). For instance, NLP techniques analyze patients' speech for linguistic markers associated with depression, facilitating early intervention (Birnbaum et al., 2017). In neurocognitive rehabilitation, AI models can predict a patient's recovery trajectory based on baseline data, enabling clinicians to customize therapy plans and allocate resources more effectively (Neumann et al., 2021).

Accessibility and Efficiency: Real-Time Data and Personalized Interventions

A key advancement of using AI in mental health treatment is real-time data collection. Wearable devices, smartphone applications, and sensors can continuously monitor physiological and behavioral markers such as heart rate, activity levels, speech patterns, and sleep quality (Fitzpatrick et al., 2019). This real-time data provides clinicians with dynamic insights into a patient's mental state, allowing for timely adjustments to treatment plans. For instance, AI algorithms can analyze fluctuations in mood or activity levels to detect early signs of relapse in depression or anxiety, enabling prompt intervention before symptoms escalate (Torous et al., 2019).

Moreover, as mentioned previously, AI-driven chatbots and virtual therapists have expanded access to mental health services, especially in underserved areas where mental health professionals are scarce. These digital platforms offer immediate support, psychoeducation, and CBT modules, reducing barriers related to stigma, cost, and geographical constraints (Fitzpatrick et al., 2019). Such tools exemplify how AI enhances accessibility by providing 24/7 support without the need for in-person visits.

In neurocognitive rehabilitation, AI facilitates personalized interventions by adapting therapies based on individual performance and progress. Machine learning algorithms can analyze data from neuropsychological assessments, neuroimaging, and behavioral tests to identify specific deficits and tailor cognitive exercises accordingly (Keller et al., 2020). For example, AI-based cognitive training programs adjust task difficulty in real-time, ensuring that exercises remain challenging yet achievable, thereby maximizing neuroplasticity and recovery (Keller et al., 2020). Studies have shown that personalized, adaptive interventions improve cognitive outcomes in patients recovering from stroke, traumatic brain injury, and neurodegenerative diseases (Coyle et al., 2021).

Furthermore, AI's capacity for continuous monitoring allows for the evaluation of intervention efficacy over time. Data-driven feedback loops enable clinicians to modify therapeutic strategies promptly, improving efficiency and reducing the duration of treatment. For example, AI algorithms analyzing speech and movement patterns can track improvements in patients with neurodegenerative conditions, guiding adjustments in medication or therapy (Keller et al., 2020).

Lyana Spytska (2025) created a prototype of an AI-based therapeutic system integrating existing AI techniques such as NLP, machine learning algorithms, and chatbot frameworks to analyze user input, recognize emotional states, and provide appropriate responses aimed at supporting mental health. The prototype showed promising results in recognizing emotional cues and delivering personalized responses. While not a replacement for professional therapy, such systems can serve as complementary tools, especially for initial assessments or ongoing support. The research demonstrated that AI-powered therapeutic systems could effectively simulate certain aspects of human psychotherapy, offering accessible and scalable mental health support. While highlighting the potential of AI to enhance mental health services, the study acknowledged challenges related to ethical considerations, user trust, and the need for further validation. The findings suggest that intelligent systems can augment traditional psychotherapy, paving the way for innovative mental health interventions.

Ethical and Human-Centric Considerations

Importance of Empathy, Validation, and Safety Provided by Human Therapists

Human therapists play a crucial role in mental health treatment due to their ability to foster empathy, validation, and safety within the therapeutic relationship. These characteristics are vital to client healing, growth, and change. Empathy, the capacity to understand and share the feelings of another, allows therapists to connect with clients on a profound emotional level, fostering trust and openness (Rogers, 1957). Validation, the acknowledgment and affirmation of a client's feelings and experiences, reinforces self-worth and encourages honesty within therapy sessions (Farber & Doolin, 2018). Safety, both emotional and physical, provides a secure space where clients can explore vulnerabilities without fear of judgment or harm (Norcross & Lambert, 2018). Collectively, these components underpin effective therapeutic interventions and significantly influence treatment outcomes.

Empathy is considered a cornerstone of humanistic therapy and is integral to establishing a strong therapeutic alliance (Rogers, 1957). When therapists demonstrate genuine empathy, clients feel understood and accepted, which reduces resistance and facilitates deeper exploration of issues (Elliott et al., 2011). Empathy also enhances clients' self-awareness and emotional regulation, enabling them to process complex feelings more effectively (Lambert & Barley, 2001). Furthermore, empirical research indicates that empathic therapists are associated with better treatment outcomes across various modalities, emphasizing the importance of this attribute in fostering change (Norcross & Wampold, 2011).

Validation complements empathy by affirming the client's subjective experiences, which may often be invalidated in everyday interactions or by societal norms. When therapists validate clients' feelings, they acknowledge the legitimacy of their emotional responses, which can be particularly vital for individuals who have experienced invalidation or marginalization (Farber & Doolin, 2018). Validation promotes psychological safety, enabling clients to express themselves authentically without fear of rejection or criticism. This process not only bolsters self-esteem but also encourages clients to engage more openly in therapy, thereby enhancing the efficacy of interventions.

Safety within therapy is of paramount importance, serving as the foundation upon which all other therapeutic processes rest. An environment of safety encourages clients to disclose sensitive information and confront challenging issues (Norcross & Lambert, 2018). Therapists establish safety through consistent, nonjudgmental responses and by maintaining confidentiality and boundaries. Such safety fosters trust, which is essential for clients to feel comfortable exploring difficult emotions and behaviors (Gelso & Woodhouse, 2011). Without a sense of safety, clients are less likely to engage fully, limiting therapeutic progress.

Thus, empathy, validation, and safety are indispensable elements provided by human therapists that significantly influence treatment success. Empathy facilitates understanding and emotional connection, validation affirms clients' experiences, and safety creates a secure environment for vulnerability and growth. Together, these elements cultivate a therapeutic alliance that empowers clients to explore their inner worlds and initiate meaningful change. As research continues to underscore their importance, the human therapist's ability to embody these qualities remains central to effective mental health intervention.

It is important to recognize that all these elements are supplied by humans rather than AI, as they are fundamentally rooted in emotions.

Analysis of the Complexities of Human Emotions in Therapy

Human emotions are intricate and multifaceted phenomena that significantly influence an individual's mental health and well-being. In therapeutic settings, understanding the complexities of human emotions is essential for effective intervention and support. Emotions are not merely responses to external stimuli but are deeply embedded within cognitive, physiological, and social contexts, often involving layered and conflicting feelings that challenge both clients and therapists (Gross, 2015). One of the fundamental complexities in human emotions is their subjective nature. Each individual's emotional experiences are shaped by personal history, cultural background, and social environment, leading to diverse emotional responses to similar situations (Lazarus & Folkman, 1984). For example, a client might interpret failure as devastating or as an opportunity for growth, depending on their prior experiences and values. This subjectivity necessitates a nuanced therapeutic approach that considers the client's unique emotional landscape.

Furthermore, emotions are often intertwined with cognitive processes, creating a dynamic interplay that complicates emotional regulation. Cognitive-behavioral theories posit that thoughts influence feelings, which in turn affect behaviors, forming a cyclical pattern (Beck, 1967). However, in therapy, clients frequently experience conflicting emotions such as love and resentment or hope and despair that coexist simultaneously, making emotional regulation challenging. Recognizing and navigating these emotional conflicts require therapists to employ sophisticated interventions aimed at fostering emotional awareness and acceptance (Greenberg, 2015). Physiological responses also contribute to the complexity of emotions in therapy. Emotions are accompanied by bodily sensations, such as increased heart rate, muscle tension, or hormonal shifts, which can reinforce emotional experiences and sometimes intensify them (LeDoux, 2012). For clients with trauma histories, physiological responses may be heightened or dysregulated, complicating therapeutic progress. Somatic therapies, therefore, emphasize the importance of integrating bodily awareness into emotional processing to facilitate healing (Ogden et al., 2006).

Social and cultural factors further add to the intricacy of human emotions. Cultural norms influence how emotions are expressed, suppressed, or interpreted. For example, certain cultures discourage the outward expression of grief or anger, leading to internalized emotional states that can manifest as psychological distress (Markus & Kitayama, 1991). Therapists must be culturally competent to accurately assess and address these emotionally laden behaviors and beliefs, ensuring that interventions are respectful and effective.

Lastly, the presence of unconscious emotional processes complicates the therapeutic work. Many emotions are repressed or denied, operating outside conscious awareness but still affecting behavior and mental health (Freud, 1915). Uncovering these unconscious emotions often requires psychoanalytic techniques, which aim to bring repressed feelings to consciousness, allowing clients to process and integrate them meaningfully (Shedler, 2010).

Human therapists possess a deep understanding of emotional complexities and are skilled at guiding clients through the decision-making process to determine the most effective treatment strategies.

Discussion of the Therapeutic Nuances That AI Struggles to Navigate

One fundamental aspect where AI struggles is in capturing the emotional complexity of human experiences. Human emotions are multifaceted and often context-dependent, requiring therapists to interpret subtle cues such as tone, facial expressions, and body language (Liu et al., 2021). While AI systems employ NLP to analyze verbal communication, they lack the capacity to fully interpret non-verbal signals and emotional subtleties. For example, research indicates that AI's sentiment analysis often misinterprets sarcasm, humor, or distress, leading to potential misunderstandings (Chen et al., 2019). Consequently, AI may fail to recognize underlying emotional states, which are vital for tailoring therapeutic interventions.

Cultural competence represents another significant challenge for AI in therapy. Cultural norms influence how individuals express emotions, perceive mental health issues, and respond to interventions (Sue et al., 2019). AI models trained predominantly on Western datasets may lack the cultural sensitivity necessary to effectively serve diverse populations. This can lead to misinterpretations, culturally inappropriate recommendations, or even harm (Huang et al., 2022). For instance, an AI system may not recognize culturally specific expressions of distress, such as somatic complaints common in some Asian cultures, thus impeding accurate assessment and intervention.

Contextual understanding is also a nuanced domain where AI often falters. Human therapists draw upon a broad contextual framework—history, socio-economic background, and current life circumstances—to inform their responses (Moreno et al., 2020). AI, however, relies on structured data and pattern recognition, which may not fully encompass the complexity of an individual's life context. This limitation can result in superficial or inappropriate responses that do not account for the person's unique situation. For example, AI may not recognize the significance of recent life events or ongoing stressors that influence a client's mental health, thereby reducing the efficacy of therapeutic support.

Ethical considerations further complicate AI's deployment in therapy. Issues of confidentiality, informed consent, and autonomy are deeply human concerns requiring moral judgment and empathy (Shen et al., 2021). AI's inability to genuinely understand or prioritize ethical dilemmas means it cannot fully navigate situations involving complex moral judgments, such as determining when to breach confidentiality for safety reasons. Moreover, AI's decision-making processes lack transparency, raising concerns about accountability and trust (Venkatesh et al., 2020). These ethical nuances underscore the importance of human oversight in mental health care.

While AI offers innovative tools for mental health support, it remains limited in navigating the nuanced and complex domain of therapeutic interactions. Emotional depth, cultural competence, contextual understanding, and ethical judgment are core aspects of therapy that AI struggles to replicate authentically. Recognizing these limitations is essential for integrating AI responsibly into mental health services, ensuring that human clinicians continue to play a central role in delivering empathetic, culturally sensitive, and ethically sound care.

Data Privacy Concerns Related to AI in Psychotherapy

The integration of AI into psychotherapy offers promising advancements in mental health care, including increased accessibility, personalized treatment, and continuous monitoring. However, these innovations raise significant concerns regarding data privacy and security. As AI-driven therapeutic tools handle sensitive personal information, safeguarding patient confidentiality becomes paramount, and failure to do so can lead to ethical, legal, and social repercussions.

One primary concern pertains to the sensitive nature of mental health data. Psychotherapy involves disclosures of deeply personal and often stigmatized information, such as trauma, suicidal ideation, or substance abuse. When AI systems process such data, the risk of unauthorized access or breaches increases, especially if robust security measures are not implemented (Sharma et al., 2020). Data breaches can lead to stigmatization, discrimination, or even harm to patients, undermining trust in mental health services. Furthermore, AI systems often require large datasets for training and operation, raising questions about data minimization and the potential for misuse or unintended exposure of confidential information.

Another issue concerns data ownership and consent. Patients may be unaware of how their data is collected, stored, and utilized in AI applications, leading to concerns about informed consent. The opaque nature of some AI algorithms, often termed "black boxes," complicates patients' understanding of how their data influences outcomes (Floridi et al., 2020). Transparency and clarity about data practices are crucial to uphold ethical standards and ensure patients retain control over their information. Additionally, consent processes must be explicit, detailing data sharing, storage duration, and potential secondary uses, such as research or commercial purposes.

Legal frameworks, such as the Health Insurance Portability and Accountability Act (HIPAA) in the United States and the General Data Protection Regulation (GDPR) in the European Union, provide guidelines for data privacy; however, their applicability to emerging AI technologies is still evolving (Calo & Kerr, 2018). These regulations mandate data security measures, breach notifications, and rights to data access or deletion. Yet, the rapid development of AI tools often outpaces legislative oversight, creating gaps that could be exploited maliciously or lead to unintended data exposure.

Moreover, the use of cloud-based AI platforms introduces additional vulnerabilities. Data stored remotely can be targeted by cyberattacks, and cross-border data transfers may complicate jurisdictional legal protections. Ensuring end-to-end encryption, secure authentication protocols, and regular security audits are essential to mitigate these risks (Sharma et al., 2020). Ethical considerations also extend to the potential misuse of data for commercial gain, such as targeted advertising or data resale, which conflicts with patient privacy rights.

In conclusion, while AI holds significant potential to enhance psychotherapy, it concurrently presents substantial data privacy challenges. Protecting sensitive mental health data requires adherence to rigorous security standards, transparent data practices, and robust legal protections. Addressing these concerns is essential to foster trust, ensure ethical integrity, and maximize the benefits of AI in mental health care.

Potential Algorithmic Bias and Its Implications on Mental Health Treatment

Personalized and scalable treatment options brought in by the integration of machine learning algorithms and AI has changed the face of mental health care in recent times. Technological advancements, such as chatbots offering initial counseling to predictive models identifying at-risk populations, promise increased accessibility, efficiency, and precision in mental health services (Ben-Zeev et al., 2019). However, alongside these promising developments lies a significant concern: the potential for algorithmic bias. Algorithmic bias refers to systematic errors in AI systems that lead to unfair outcomes, often reflecting or amplifying societal prejudices embedded within training data (O'Neil, 2016). When such biases infiltrate mental health algorithms, they can have profound and sometimes detrimental implications for diagnosis, treatment, and health equity.

Algorithmic bias arises when machine learning models are trained on data that are unrepresentative, incomplete, or reflect existing societal inequalities. In mental health contexts, this can manifest in various ways, such as misdiagnosis, unequal access to care, or culturally insensitive interventions. For example, if an AI diagnostic tool is primarily trained on data from Western populations, its accuracy may diminish when applied to individuals from different ethnic or cultural backgrounds (Garg et al., 2019). This disparity can lead to underdiagnosis or misdiagnosis in marginalized groups, perpetuating health disparities rather than alleviating them. Moreover, biases can also stem from the data collection process itself. Many mental health datasets rely on self-reported symptoms and clinical assessments, which can be influenced by cultural norms, stigma, or language barriers. As a result, the training data may underrepresent or misrepresent certain populations, leading to algorithms that are less effective or even harmful when deployed across diverse groups (Williams et al., 2021). For instance, linguistic variations or cultural expressions of distress may not be adequately captured, causing AI models to overlook or misinterpret symptoms.

The consequences of algorithmic bias in mental health care are far-reaching. First, biased algorithms can contribute to diagnostic inaccuracies. Studies have shown that AI models trained on biased data tend to perform poorly with minority populations, leading to false negatives or positives (Challen et al., 2019). Such inaccuracies can delay appropriate treatment, exacerbate mental health conditions, or result in unnecessary interventions. Second, biases may influence treatment recommendations, potentially reinforcing existing inequalities. For example, algorithms designed to prioritize resource allocation might favor certain demographic groups over others, based on biased historical data. This can result in marginalized populations receiving substandard care or being overlooked altogether (Noble, 2018). Furthermore, if AI systems are used to monitor patient progress or adherence, biases in data collection can produce skewed assessments, influencing clinicians' decisions and possibly leading to unfair treatment plans. Third, the deployment of biased algorithms can erode trust in mental health services. Patients from underrepresented groups may feel misunderstood or marginalized if AI systems consistently fail to recognize their experiences, thereby discouraging engagement with mental health systems. This phenomenon can deepen disparities in care utilization and outcomes (Ben-Zeev et al., 2019). Biased algorithms also threaten patient autonomy by influencing clinical decisions in ways that may not align with patients' cultural contexts or personal values.

Beyond individual patient outcomes, algorithmic bias in mental health AI raises ethical concerns related to justice, autonomy, and beneficence. The principle of justice demands equitable access and treatment for all, but biases can undermine this goal by systematically disadvantaging certain groups. For example, if predictive models underestimate risk in minority populations, these individuals might not receive necessary interventions, perpetuating health inequities (Obermeyer et al., 2019). If AI systems encode stereotypical assumptions, they may inadvertently impose culturally insensitive labels or treatment pathways, undermining respect for individual differences (Garg et al., 2019). Ethical deployment of AI in mental health thus requires rigorous evaluation and mitigation of biases to ensure fair and culturally competent care.

Future Directions and Global Perspectives

Based on the information presented above, it is clear that while AI is assisting in psychotherapy and neurocognitive rehabilitation, numerous challenges must be addressed to ensure clients receive the maximum benefits. Additionally, it is not feasible for humans to perform all the tasks that AI can accomplish. Hence a model combining both AI and human potential is suggested for the best possible outcomes of treatment.

Concept of a Hybrid Model: Combining AI Tools with Traditional Therapy

Traditional psychotherapy, rooted in human interaction and empathy, has demonstrated efficacy across various mental health conditions, including depression, anxiety, and post-traumatic stress disorder (Lambert & Barley, 2001). Nonetheless, it faces limitations such as accessibility barriers, high costs, and variability in treatment delivery. Neurocognitive rehabilitation, aimed at restoring or compensating for cognitive deficits, similarly benefits from personalized interventions but can be resource-intensive and constrained by practitioner availability (Cicerone et al., 2011). Traditional therapy relies heavily on therapist judgment, which, while invaluable, can be complemented by AI's capacity to analyze large datasets and identify nuanced patterns. Machine learning models can predict relapse risk, monitor progress, and suggest individualized treatment adjustments (Kumar et al., 2020a). In neurocognitive contexts, AI systems can adapt rehabilitation tasks dynamically based on cognitive performance metrics, ensuring optimal challenge levels and promoting neuroplasticity (Toth et al., 2020). One of the core advantages of integrating AI into psychotherapy and neurorehabilitation is increased accessibility. AI-powered chatbots and virtual therapists can deliver initial assessments, psychoeducation, and ongoing support remotely, thus overcoming geographical and logistical barriers (Fitzpatrick et al., 2017).

The hybrid model, which combines AI tools with traditional psychotherapy and neurocognitive rehabilitation, offers a promising approach to enhance treatment efficacy, accessibility, and personalization. This integrative framework capitalizes on the strengths of both technological innovations and established therapeutic practices, fostering a more comprehensive and adaptive intervention paradigm. Furthermore, the hybrid model facilitates enhanced personalization. Despite these benefits, integrating AI with traditional approaches necessitates careful consideration of ethical, clinical, and practical issues. The therapeutic alliance, a vital component of effective psychotherapy, may be challenged by the impersonal nature of AI interfaces (Rizzo et al., 2019).

To address these challenges, researchers and clinicians must adopt comprehensive strategies. One approach involves diversifying training datasets to include representative samples across different demographic and cultural groups (Rajkomar et al., 2018). This can enhance the fairness and accuracy of AI models across populations. Ensuring data privacy and addressing biases embedded within algorithms are critical to maintaining ethical standards (Floridi et al., 2020). Another strategy is the implementation of fairness-aware machine learning techniques, which explicitly aim to minimize disparities in model performance across groups (Barocas & Selbst, 2016). Regular auditing of algorithms for bias, transparency in model development, and involving diverse stakeholders in design and deployment are also crucial steps. Integrating human oversight into AI systems ensures that clinicians can contextualize algorithmic outputs within cultural and individual patient factors. Ethical guidelines and regulatory frameworks must evolve to monitor and enforce equitable AI practices in mental health care.

Conclusion

While AI and machine learning hold tremendous potential to revolutionize mental health treatment, the risks associated with algorithmic bias cannot be overlooked. The biases embedded within training data and algorithmic processes can exacerbate existing health disparities, compromise diagnostic accuracy, and undermine trust in mental health services. Addressing these issues requires deliberate efforts to diversify data, develop fairness-aware algorithms, and embed ethical considerations throughout the AI lifecycle. Only through such comprehensive approaches can we harness the benefits of AI in mental health care while safeguarding principles of equity, justice, and respect for individual differences. Moreover, AI should serve as an augmentation rather than a replacement for human clinicians, preserving the empathic and contextual elements essential to treatment success.

In conclusion, the hybrid model combining AI tools with traditional psychotherapy and neurocognitive rehabilitation offers a compelling framework to enhance mental health care. By leveraging AI's scalability, data analysis, and personalization capabilities alongside the empathy and clinical judgment of human practitioners, this integrated approach holds promise for improving treatment accessibility, efficacy, and patient engagement. Future research should focus on optimizing the synergistic integration of these modalities, addressing ethical considerations, and establishing evidence-based guidelines for widespread implementation.

References

American Psychological Association (APA). (2019). *The science of psychotherapy*. www.apa.org/research/action/psychotherapy

Bahar-Fuchs, A., Clare, L., & Woods, B. (2013). Cognitive training and cognitive rehabilitation for persons with mild to moderate dementia of the Alzheimer's or vascular type: A review. *Alzheimer's Research & Therapy*, 5(4), 35. https://doi.org/10.1186/alzrt189

Barocas, S., & Selbst, A. D. (2016). Big data's disparate impact. *California Law Review*, 104, 671–732.

Beck, A. T. (1967). *Depression: Clinical, experimental, and theoretical aspects*. Harper & Row.

Ben-Zeev, D., Brian, R., Wang, R., & Mueser, K. (2019). Digital mental health tools: Opportunities and challenges. *Psychiatric Services*, 70(1), 3–5. https://doi.org/10.1176/appi.ps.201800251

Birnbaum, M. L., Oltmanns, T. F., & Kiehl, K. A. (2017). Using natural language processing to detect mental health problems. *Nature Reviews Psychology*, 1(11), 690–701.

Calo, R., & Kerr, I. (2018). Artificial intelligence and the law. *Harvard Law Review*, 131(7), 1745–1773.

Challen, R., Denny, J., Pitt, M., Gompels, L., Edwards, T., & Tsaneva-Atanasova, K. (2019). Artificial intelligence, bias and clinical safety. *BMJ Quality & Safety*, 28(3), 231–237. https://doi.org/10.1136/bmjqs-2018-008370

Chekroud, S. R., Zotti, R. J., Shehzad, Z., et al. (2016). Cross-trial prediction of treatment outcome in depression: A machine learning approach. *The Lancet Psychiatry*, 3(3), 243–250. https://doi.org/10.1016/S2215-0366(15)00471-8

Chen, Y., Zhang, L., & Li, X. (2019). Limitations of sentiment analysis in detecting sarcasm and humor. *Journal of Computational Linguistics*, 45(3), 567–583.

Cicerone, K. D., Langenbahn, D. M., Braden, C., Malec, J. F., Kalmar, K., Fraas, M., …, & Hart, T. (2019). Evidence-based cognitive rehabilitation: Systematic review of the literature from 2008 through 2017. *Archives of Physical Medicine and Rehabilitation*, 100(8), 1515–1533. https://doi.org/10.1016/j.apmr.2018.01.011

Cicerone, K. D., Langenbahn, D. M., Braden, C., Malec, J. F., Kalmar, K., Fraas, N., ..., & Ashman, T. (2011). Evidence-based cognitive rehabilitation: Systematic review of the literature from 2003 through 2008. *Archives of Physical Medicine and Rehabilitation*, 92(4), 519–530. https://doi.org/10.1016/j.apmr.2010.11.015

Coyle, D., Ward, T., & McDarby, G. (2021). Adaptive neurocognitive training: Personalized approaches to cognitive rehabilitation. *Journal of NeuroEngineering and Rehabilitation*, 18(1), 45. https://doi.org/10.1186/s12984-021-00809-9

Cuijpers, P., Karyotaki, E., Reijnders, M., Purgato, M., & Barbui, C. (2018). Meta-analyses of psychological treatment outcomes for depression across different treatment formats and settings. *The British Journal of Psychiatry*, 213(4), 173–174. https://doi.org/10.1192/bjp.2018.124

Dwyer, D. B., Falkai, P., & Koutsouleris, N. (2018). Machine learning approaches for clinical psychology and psychiatry. *Annual Review of Clinical Psychology*, 14, 91–118. https://doi.org/10.1146/annurev-clinpsy-032816-045144

Elliott, R., Bohart, A. C., Greenberg, L. S., & Watson, J. C. (2011). Empathy. In J. C. Norcross (Ed.), *Psychotherapy relationships that work: Evidence-based, point/counterpoint* (2nd ed., pp. 43–58). Oxford University Press.

Esteva, A., Robicquet, A., Ramsundar, B., Kuleshov, V., DePristo, M., Chou, K., ..., & Dean, J. (2019). A guide to deep learning in healthcare. *Nature Medicine*, 25(1), 24–29.

Farber, B. A., & Doolin, E. M. (2018). Validation and validation therapy. *Psychotherapy*, 55(2), 168–176. https://doi.org/10.1037/pst0000163

Fitzpatrick, K. K., Darcy, A., & Vierhile, M. (2017). Delivering cognitive behavior therapy to young adults with symptoms of depression and anxiety using a fully automated conversational agent (Woebot): A randomized controlled trial. *JMIR Mental Health*, 4(2), e19. https://doi.org/10.2196/mental.7785

Fitzpatrick, K. K., Darcy, A., & Vierhile, M. (2019). Delivering cognitive behavioral therapy through a chatbot: A randomized controlled trial. *JMIR Mental Health*, 6(4), e18128. https://doi.org/10.2196/18128

Floridi, L., Cowls, J., King, T. C., & Taddeo, M. (2020). How to design AI for social Good: Seven essential factors. *Science and Engineering Ethics*, 26(3), 1771–1796. https://doi.org/10.1007/s11948-020-00213-5

Freud, S. (1915). Repression. *Standard Edition of the Complete Psychological Works of Sigmund Freud*, 14, 141–158.

Fulmer, R., Joerin, A., Gentile, B., Lakerink, L., & Rauws, M. (2018). Using psychological artificial intelligence (AI) chatbots to relieve symptoms of depression and anxiety: Randomized controlled trial. *JMIR Mental Health*, 5(4), e64. https://doi.org/10.2196/mental.9787

Garg, R., Vaidya, S., & Dutta, S. (2019). Challenges of artificial intelligence in mental health: Ethical and cultural considerations. *AI & Society*, 34(4), 877–887. https://doi.org/10.1007/s00146-019-00901-8

Gelso, C. J., & Woodhouse, S. S. (2011). The real relationship in psychotherapy: Definitions, dimensions, and implications. *Psychotherapy Research*, 21(3), 290–301. https://doi.org/10.1080/10503307.2010.517747

Greenberg, L. S. (2015). *Emotion-focused therapy*. American Psychological Association.

Gross, J. J. (2015). Emotion regulation: Conceptual and empirical foundations. In J. J. Gross (Ed.), *Handbook of emotion regulation* (2nd ed., pp. 3–24). Guilford Press.

Huang, Y., Lee, S., & Kim, J. (2022). Cultural considerations in AI-driven mental health interventions. *Global Mental Health*, 9, e34.

Huang, Y., Zhang, H., & Chen, X. (2020). AI-powered robotic neurorehabilitation: A review. *IEEE Transactions on Neural Systems and Rehabilitation Engineering*, 28(6), 1382–1392. https://doi.org/10.1109/TNSRE.2020.3008094

Keller, I., Kopp, B., & Loor, J. (2020). AI-based neurocognitive assessment and personalized rehabilitation: A review. *NeuroRehabilitation*, 47(3), 353–368. https://doi.org/10.3233/NRE-192829

Kelly, M. E., Duff, K., Adams, S., McDonald, S., Tobin, C., Lawlor, B. A., & Loughrey, D. G. (2020). The impact of cognitive training and mental stimulation on cognitive and everyday functioning in healthy older adults: A systematic review. *Ageing Research Reviews*, 63, 101166.

Khodabandelou, R. (2020). Artificial intelligence and mental health: Opportunities and challenges. *Journal of Mental Health Technology*, 5(2), 45–52.
Kiran, S. (2012). Techniques for language recovery following stroke. *Current Neurology and Neuroscience Reports*, 12(4), 439–447.
Kirk, M., Seitz, J., & Schönenberg, M. (2019). Technology-based interventions for neurocognitive rehabilitation: A review. *Journal of NeuroEngineering and Rehabilitation*, 16(1), 1–12.
Koot, S. L., van den Berg, M., & Kessels, R. P. C. (2020). Digital therapeutics for depression and anxiety: A systematic review. *Digital Health*, 6. https://doi.org/10.1177/2055207620924657
Kramer, S. E., Houtveen, J. H., & Van Doornen, L. J. P. (2014). Speech-based markers for depression and anxiety: A review. *Speech Communication*, 61, 27–44. https://doi.org/10.1016/j.specom.2014.02.003
Kshirsagar, V., Raj, M., & Pandey, S. (2020). AI chatbots in mental health: Opportunities and challenges. *Journal of Mental Health and Disease Management*, 3(2), 45–52.
Kumar, S., & Rose, C. (2019). Mobile-based AI interventions for mental health: Opportunities and challenges. *Current Psychiatry Reports*, 21(9), 84. https://doi.org/10.1007/s11920-019-1054-8
Kumar, S., Khandelwal, N., & Sahu, S. (2021). Artificial intelligence in cognitive rehabilitation: A review. *Current Psychiatry Reports*, 23, 50. https://doi.org/10.1007/s11920-021-01252-0
Kumar, S., Prasad, A., & Sharma, S. (2020a). Machine learning approaches in mental health prediction and diagnosis. *Current Psychology*, 39, 1–10. https://doi.org/10.1007/s12144-019-00310-2
Kumar, S., Sharma, S., & Kaur, P. (2020b). Personalized mental health interventions using machine learning. *IEEE Reviews in Biomedical Engineering*, 13, 435–448.
Lambert, M. J., & Barley, D. E. (2001). Research summary on the therapeutic relationship and psychotherapy outcome. *Psychotherapy: Theory, Research, Practice, Training*, 38(4), 357–361. https://doi.org/10.1037/0033-3204.38.4.357
Lazarus, R. S., & Folkman, S. (1984). *Stress, appraisal, and coping*. Springer.
LeDoux, J. (2012). Rethinking the emotional brain. *Neuron*, 73(4), 653–676.
Lezak, M. D., Howieson, D. B., & Loring, D. W. (2012). *Neuropsychological assessment* (5th ed.). Oxford University Press.
Lindner, P., Möller, M., & Gillberg, C. (2020). Virtual reality in neurocognitive rehabilitation: A review. *Frontiers in Psychiatry*, 11, 582. https://doi.org/10.3389/fpsyt.2020.00582
Liu, H., Wang, Z., & Zhou, P. (2021). Emotional recognition limitations of AI systems in psychotherapy. *Frontiers in Psychology*, 12, 652345.
Liu, S., Liu, S., & Zhang, Y. (2020a). Wearable sensors for neurocognitive assessment: A review. *IEEE Transactions on Neural Systems and Rehabilitation Engineering*, 28(2), 204–214.
Liu, T., Li, K., & Wang, L. (2020b). Virtual reality in neurorehabilitation: A review. *Neurorehabilitation and Neural Repair*, 34(4), 351–362.
Lohse, K. R., Hsu, W. L., & Duncan, R. (2014). Virtual reality and robotics in stroke rehabilitation: A systematic review of their effectiveness. *Journal of NeuroEngineering and Rehabilitation*, 11, 102. https://doi.org/10.1186/1743-0003-11-102
Markus, H. R., & Kitayama, S. (1991). Culture and the self: Implications for cognition, emotion, and motivation. *Psychological Review*, 98(2), 224–253.
McGonigal, J., O'Neill, B., & Murphy, N. (2021). Comparative effectiveness of AI-assisted versus human-led neurorehabilitation: A meta-analysis. *NeuroRehabilitation*, 48(2), 179–190. https://doi.org/10.3233/NRE-210183
Moreno, M., Sanchez, A., & Lopez, R. (2020). Contextual understanding in AI-based mental health tools. *Computers in Human Behavior*, 108, 106350.
Neumann, D., Murphy, B., & Fong, G. (2021). AI in neurocognitive rehabilitation: Predictive analytics and personalized therapy. *NeuroRehabilitation*, 48(4), 405–416.
Noble, S. U. (2018). *Algorithms of oppression: How search engines reinforce racism*. NYU Press.
Norcross, J. C. (2018). *Psychotherapy relationships that work: Evidence-based responsiveness*. Oxford University Press.
Norcross, J. C., & Lambert, M. J. (2018). Evidence-based therapy relationships. In J. C. Norcross (Ed.), *Psychotherapy relationships that work: Evidence-based, point/counterpoint* (2nd ed., pp. 3–21). Oxford University Press.

Norcross, J. C., & Wampold, B. E. (2011). Evidence-based therapy relationships: Research conclusions and clinical practices. *Psychotherapy*, 48(1), 98–102. https://doi.org/10.1037/a0022161
Norcross, J. C., Beutler, L. E., & Levant, R. F. (2018). *Evidence-based practices in mental health: Debate and consensus*. American Psychological Association.
Obermeyer, Z., Powers, B., Vogt, N., & Mullainathan, S. (2019). Dissecting racial bias in an algorithm used to manage the health of populations. *Science*, 366(6464), 447–453. https://doi.org/10.1126/science.aax2342
Ogden, P., Minton, K., & Pain, C. (2006). *Trauma and the body: A sensorimotor approach to psychotherapy*. W. W. Norton & Company.
O'Neil, C. (2016). *Weapons of math destruction: How big data increases inequality and threatens democracy*. Crown Publishing Group.
Ponsford, J., Cromarty, D., & Sloan, S. (2014). Psychosocial adjustment and quality of life after traumatic brain injury. *Journal of Head Trauma Rehabilitation*, 29(1), 27–35.
Rajkomar, A., Oren, E., Chen, K., et al. (2018). Scalable and accurate deep learning with electronic health records. *npj Digital Medicine*, 1, 18. https://doi.org/10.1038/s41746-018-0029-1
Rizzo, A., Lange, B., Buckwalter, J., & Kim, J. (2019). Virtual reality applications for assessment and treatment of mental health disorders. In *Neuropsychology of virtual reality* (pp. 123–138). Elsevier.
Rogers, C. R. (1957). The necessary and sufficient conditions of therapeutic personality change. *Journal of Consulting Psychology*, 21(2), 95–103.
Sharma, S., Mahajan, R., & Khandelwal, S. (2020). Privacy concerns in artificial intelligence-based mental health interventions. *Journal of Medical Systems*, 44(12), 208.
Shatte, A. B., Hutchinson, D. M., & Teague, S. J. (2019). Machine learning in mental health: A scoping review of methods and applications. *Psychological Medicine*, 49(9), 1426–1448.
Shedler, J. (2010). The efficacy of psychodynamic psychotherapy. *American Psychologist*, *65*(2), 98–109
Shen, L., Patel, V., & Kumar, S. (2021). Ethical challenges in deploying AI for mental health. *AI & Ethics*, 1(2), 123–130.
Shen, Y., Wu, X., & Zhang, W. (2020). Machine learning in neurocognitive disorder diagnosis and treatment: A review. *Frontiers in Aging Neuroscience*, 12, 600615. https://doi.org/10.3389/fnagi.2020.600615
Spytska, L. (2025). The use of artificial intelligence in psychotherapy: Development of intelligent therapeutic systems. *BMC Psychology*, 13(1), 175. https://doi.org/10.1186/s40359-025-02491-9
Sue, D. W., Sue, D., & Sue, S. (2019). *Counseling the culturally diverse: Theory and practice*. John Wiley & Sons.
Topol, E. J. (2019). High-performance medicine: The convergence of human and artificial intelligence. *Nature Medicine*, 25(1), 44–56.
Torous, J., Roberts, L. W., & Firth, J. (2019). Digital mental health and COVID-19: Challenges and opportunities. *The Lancet Psychiatry*, 7(9), 849–850. https://doi.org/10.1016/S2215-0366(20)30333-0
Toth, M., Szabó, Á., & Hegedűs, R. (2020). Artificial intelligence in neurorehabilitation: Current applications and future perspectives. *NeuroRehabilitation*, 47(3), 333–344. https://doi.org/10.3233/NRE-200998
Turing, A. M. (1950). Computing machinery and intelligence. *Mind*, 49, 433–460.
Vazquez, A., & García, J. (2018). Neurocognitive rehabilitation: Principles and practice. *Neurorehabilitation Journal*, 43(3), 151–164.
Venkatesh, V., Thong, J. Y., & Xu, X. (2020). Ethical considerations in AI-based mental health applications. *MIS Quarterly*, 44(2), 683–694.
Wang, Y., Zhang, J., & Wang, H. (2019). Machine learning prediction of cognitive decline in Alzheimer's disease. *NeuroImage: Clinical*, 21, 101677.
Weizenbaum, J. (1976) *Computer power and human reason: From judgment to calculation*. Freeman.
Williams, A., McGorry, P. D., & Hickie, I. B. (2021). Digital mental health and AI: Challenges and opportunities for culturally sensitive care. *Australian & New Zealand Journal of Psychiatry*, *55*(3), 223–226. https://doi.org/10.1177/0004867420969751
Wilson, B. A., Emslie, H., Foley, B., & Evans, C. (2019). *The rehabilitation of memory: A practical guide*. Routledge.

7

ARTIFICIAL INTELLIGENCE IN NEUROCOGNITIVE REHABILITATION

AI Applications in Assessment, Monitoring, and Therapy

Thaddeus Alfonso, Sherine Georgina Victoria, and Shirin A. P.

7.1 Introduction

Cognitive rehabilitation refers to a systematic and evidence-based approach to restore or compensate for deficits in cognitive functioning resulting from neurological injury, illness, or developmental conditions (Cicerone et al., 2000, 2011). Traditionally reliant on therapist-guided interventions, cognitive rehabilitation is now undergoing a paradigm shift with artificial intelligence (AI) integration. AI, broadly defined as the development of computer systems capable of performing tasks that typically require human intelligence, encompasses domains such as machine learning (ML), natural language processing (NLP), robotics, and computer vision (Kim et al., 2023).

Recent advances in AI have significantly improved the precision, scalability, and personalization of cognitive rehabilitation. Unlike static rehabilitation protocols, AI-powered systems dynamically adapt to patient-specific cognitive trajectories by analyzing real-time data and adjusting task complexity accordingly (Mukherjee et al., 2023; Gao et al., 2022). These innovations increase clinical efficiency, enhance patient engagement, and enable remote delivery, features particularly valuable for underserved or mobility-limited populations.

AI technologies are increasingly being deployed in neurological rehabilitation for conditions such as stroke, Parkinson's disease, traumatic brain injury (TBI), mild cognitive impairment (MCI), and Alzheimer's disease (AD) (Calderone et al., 2024; Khalid et al., 2024). Applications range from improving neuropsychological assessment and early diagnosis to delivering adaptive therapy through mobile platforms, wearables, and virtual environments (Fiorente et al., 2024; Medenica et al., 2024). AI also facilitates the development of digital biomarkers by integrating multimodal data sources—including speech patterns, gait metrics, and neuroimaging—thus enabling proactive cognitive health management.

DOI: 10.4324/9781003646662-9

This chapter presents a comprehensive overview of AI applications across three core domains of neurocognitive rehabilitation: (1) Assessment, focusing on how AI-driven tools improve accuracy and accessibility in diagnosing cognitive impairment; (2) monitoring, exploring continuous, high-resolution tracking of cognitive changes via wearable technologies and mobile apps; and (3) therapy, evaluating AI-enhanced interventions such as brain–computer interfaces (BCIs), virtual therapists, and reinforcement learning–based cognitive training platforms. In addition, the chapter addresses the clinical, technical, and ethical challenges in deploying AI systems, emphasizing the need for interdisciplinary collaboration and inclusive design to ensure responsible implementation in real-world healthcare settings.

7.2 AI Applications in Assessment

Neuropsychological assessment is a method that uses tests and behavioral evaluations to understand how brain function impacts cognitive abilities. The aim is to identify cognitive strengths and weaknesses and the effects of brain damage, disease, or dysfunction on a person's thinking, memory, and other mental processes. It is a performance-based approach used to evaluate cognitive functioning in individuals with brain injury, neurological disease, or severe mental illness. It serves multiple purposes, including diagnosis, monitoring, and prognosis (Harvey, 2012). A key limitation of traditional paper-based neurocognitive assessments is clinicians' need for in-person administration and manual scoring. To address this, researchers have developed remote, AI-assisted versions, making assessments more accessible and scalable tools for cognitive screening (Sirilertmekasakul et al., 2023). AI has recently gained prominence in healthcare by enhancing neurocognitive assessments' speed, accuracy, and personalization. Several applications of AI in neurocognitive assessment are already underway.

7.2.1 Machine Learning in Early Detection and Diagnosis

ML algorithms are significantly advancing this area by enabling more precise and earlier identification of conditions such as AD, Parkinson's disease, and MCI. Developed ML models leverage neuropsychological test data with clinical, demographic, and biological data to optimize diagnostic procedures, screen for cognitive dysfunction, and predict cognitive decline. These models outperform traditional assessments in detecting conditions like MCI and dementia (Veneziani et al., 2024). ML algorithms enhance early detection of neurodegenerative diseases by analyzing complex, multimodal datasets to identify subtle biomarkers often invisible to traditional diagnostic methods. With the vast availability of healthcare data, ML has shown a strong ability to uncover intricate patterns and connections across complex datasets. They can process neuroimaging, genetic profiles, clinical assessments, and digital biomarkers like eye-tracking or facial expression data to detect pre-symptomatic indicators of conditions such as Alzheimer's and Parkinson's diseases (Chudzik et al., 2024; Nallore et al., 2023).

7.2.2 Systematic Evidence Supporting ML Applications

In the systematic review by Yousefi et al. (2024), 108 studies were analyzed, evaluating patients with neurological diseases for early detection using ML algorithms. For Parkinson's,

ML-based computer-aided diagnosis systems have shown promise in distinguishing patients from healthy controls, even identifying early non-motor symptoms often missed in traditional assessments. Similarly, for MCI and AD, ML models have demonstrated high accuracy in early detection. Specifically, 3D-convolutional neural networks (CNNs) have proven to be highly effective for the early detection of MCI, with studies demonstrating over 95% accuracy, which is helpful for timely intervention.

A study applied ML to features extracted from the digital clock drawing test and achieved great accuracy in distinguishing AD from both amnestic and non-amnestic MCI (Binaco et al., 2020). By analyzing nuanced drawing behaviors captured digitally, clinicians can gain deeper insights into a patient's cognitive status, facilitating timely interventions. ML models, including linear and nonlinear approaches, have also demonstrated superior performance in classifying cognitive impairment, especially when trained on large datasets that allow them to exploit complex patterns (Kwak et al., 2021). ML techniques can reduce the time required for neuropsychological assessments by identifying the most informative and relevant tests from large test batteries (Battista et al., 2017). By identifying the most predictive tests, clinicians can administer only a subset of assessments, thereby reducing the cognitive stress of the assessment. These models can also tailor the selection of cognitive assessments to individual patient characteristics, such as brain morphometry, making the process more efficient and relevant for each case (Peng et al., 2020). This approach helps in minimizing patient burden during evaluations without sacrificing accuracy.

A recent advancement in AI-driven cognitive assessment is PENSIEVE-AI, a brief, self-administered digital tool designed to detect cognitive impairment across populations with diverse literacy levels. PENSIEVE-AI addresses language and education needs limitations by using drawing tasks and deep learning (DL) algorithms to identify cognitive decline. In a study involving 1,758 older adults in Singapore, the tool achieved a diagnostic accuracy (AUC = 93%) comparable to conventional neuropsychological batteries. This illustrates the potential of AI-based tools to offer efficient and scalable solutions for early detection of dementia and MCI, and it becomes especially valuable for use in diverse, underserved, and low-literacy populations (Liew et al., 2025).

7.2.3 Leveraging Speech and Facial Analysis for Assessment

Another significant advancement is integrating speech and facial analysis into neurocognitive assessments for early and accurate detection of cognitive impairment. A study by Sirilertmekasakul et al. (2023) discusses the integration of AI with computer vision and speech analysis to digitize neurocognitive assessments. Computer vision is a branch of AI inspired by the human visual system. It uses DL, particularly CNNs, to extract and interpret meaningful patterns from images and videos. Computer vision techniques are employed to analyze facial expressions, providing insights into the user's emotional states and cognitive functions. This non-invasive method allows for detecting subtle cues that may indicate cognitive decline.

NLP plays a pivotal role in speech analysis for cognitive assessment by offering a non-invasive, objective, and scalable method to detect and monitor cognitive decline. Research also highlights that speech features can act as markers for vascular risk factors related to cognitive decline. By examining various features of spontaneous speech, such as lexical, semantic, syntactic, and acoustic parameters, a study found significant correlations between variables extracted through NLP and automated speech analysis (ASA) and clinician-rated speech characteristics such as word-finding difficulty, incoherence, and perseveration. Speech analysis can also effectively distinguish between cognitively impaired and cognitively intact older adults by focusing on speech biomarkers. This aids in early detection and continual monitoring of cognitive decline. Speech analysis thus plays a crucial role in neurocognitive assessments by enabling the detection of subtle linguistic and acoustic changes, supporting early diagnosis and monitoring (Beltrami et al., 2018; Martínez-Nicolás et al., 2022; Nunes et al., 2024; Yeung et al., 2021).

7.2.4 Virtual Reality in Cognitive Assessment and Stimulation

Fully immersive virtual reality (VR) has become another valuable tool in neuropsychology, offering a more ecologically valid and rich alternative to traditional assessments. By tracking behavioral and physiological responses in real time, VR allows for dynamic, individualized assessments and the potential to uncover subtle cognitive deficits that traditional methods might miss. Immersive VR can significantly improve neurocognitive assessments by allowing controlled and detailed testing of complex executive functions like planning, problem-solving, multitasking, or selective attention, which can be hard to measure with traditional paper-based tests of general cognitive abilities (Parsons et al., 2013). A study by Oagaz et al. (2018) introduces a novel assessment method using VR, eye tracking, and motion analysis. In a virtual baseball game, participants' movements and reactions are analyzed to evaluate memory, attention, motor skills, and balance. Preliminary results with 20 healthy individuals show promising results, suggesting this tool can objectively measure neurocognitive function and support cross-disorder comparisons.

A study by Parsons et al. (2008) tested a VR-based cognitive assessment tool, the VRCPAT, within a virtual city to evaluate its validity against traditional neuropsychological tests. The results were based on a study of 40 healthy adults, and they showed a strong alignment with standard tests for similar cognitive skills. This supports VRCPAT as a reliable, efficient, and ecologically valid method for assessing neurocognitive function. VR is emerging as a powerful tool for cognitive interventions and stimulation in aging and neurodegenerative conditions. A pilot study tested a VR-based cognitive stimulation program simulating daily activities for people with mild-to-moderate AD. Over ten sessions in two months, participants were assessed for memory, attention, and executive function. Results showed significant cognitive improvement in the VR group, suggesting VR is a promising tool for maintaining cognitive function in dementia care (Oliveira et al., 2021).

VR offers many advantages, such as providing personalized, immersive, and contextually relevant environments to enhance patient engagement and real-world skill transfer in cognitive rehabilitation. It enables patients to practice essential cognitive skills in lifelike scenarios, enhancing motivation and recovery. However, challenges such as high costs, limited accessibility, and side effects like cybersickness highlight the need for further research on long-term effectiveness (Catania et al., 2023).

7.2.5 AI in Neuroimaging for Assessment of Cognitive Disorders

AI-powered neuroimaging is transforming the assessment of neurocognitive disorders by automating and enhancing the interpretation of complex brain imaging data. AI in diagnostic imaging offers key advantages, such as enabling faster and more accurate interpretation of medical scans. AI, particularly through ML and DL, is transforming the field by facilitating rapid and precise analysis of neuroimaging data. AI algorithms can detect subtle patterns, anomalies, or changes that may be missed by the human eye, improving operational efficiency through faster, cost-effective diagnostics and supporting clinical decisions (Khalifa & Albadawy, 2024). AI is increasingly used in molecular neuroimaging to improve image quality and automate disease detection and segmentation. AI applications in Positron Emission Tomography (PET) and Single Photon Emission Computed Tomography (SPECT) imaging, including Computed Tomography (CT) generation for attenuation correction and disease localization, aid in neurodegenerative disease assessment and clinical decision-making (Boyle et al., 2021). DL models excel at processing complex neuroimaging data, helping with more accurate diagnoses and personalized treatment. They aid in earlier disease detection, predictions of progression, and more tailored treatment planning (Brahma & Vimal, 2024). AI, particularly through ML and DL algorithms, enables the rapid and objective analysis of neuroimaging modalities such as MRI and PET scans, supporting both diagnosis and prognosis of conditions like AD and other dementias (Aljuhani et al., 2024; Borchert et al., 2023).

A review of AI, specifically DL, in neuroimaging highlights four main clinical applications: Detection/diagnosis, prediction, image quality improvement, and clinical workflow improvement. An analysis of 104 studies further supports this, showing that AI has been predominantly used to address mental and neurological disorders. ML and DL techniques outperformed traditional methods for lesion segmentation, disease classification, and image quality enhancement. These AI systems accurately diagnose conditions, predict progression, and enable personalized treatment, especially when integrating multimodal neuroimaging data (Bacon et al., 2024; Choi & Sunwoo, 2022). The potential of AI in advanced neuroimaging data analysis will redefine the future of clinical neuroscience. Improving accuracy in diagnosis, supporting early detection, and enabling more individualized treatment planning will likely become a standard tool in neuroscience research and everyday medical practice.

7.2.6 Challenges in AI-Based Neurocognitive Assessment

While AI in neurocognitive assessment has brought remarkable advancements, the field has challenges. As adoption grows, several important challenges are beginning to surface. Addressing these challenges is essential for ensuring responsible, effective, and equitable integration into clinical practice (Table 7.1).

AI models are also being used to predict the progression of cognitive decline by identifying imaging biomarkers associated with future disease conversion, such as from MCI to AD. To support clinical adoption, explainable AI (XAI) approaches are becoming essential; this transparency builds trust and makes it easier for healthcare professionals to adopt and use AI tools in real-world practice (Khosroshahi et al., 2025). As AI and ML continue to evolve, their integration with clinical expertise will help transform neuropsychological assessment by aiding earlier detection and personalized treatment of cognitive disorders, thereby improving healthcare efficiency (Veneziani et al., 2024).

Despite these strengths, AI neurocognitive assessments face several limitations. Socioeconomic disparities, limited digital literacy among older adults, and hidden technology costs hinder equitable access to AI in healthcare. While AI offers efficiency and precision, it lacks the human sensitivity to interpret emotional and behavioral cues essential for neuropsychological assessment. These limitations underscore the need for cautious and

Table 7.1 Significant challenges encountered in AI-based neurocognitive assessment

Challenge	*Description*
Data-related challenges (Yan et al., 2022)	• Neurocognitive data is often subjective and influenced by individual and socio-cultural factors. • Current datasets are limited in size and lack diversity. • Integrating multimodal data (e.g., neuroimaging, speech, genetics) requires complex, technically advanced models.
Ethical and privacy concerns (MacIntyre et al., 2023)	• Involves sensitive health and behavioral data, particularly from vulnerable populations. • Risks include unauthorized access, data breaches, and covert data collection. • AI models may inherit bias and lack transparency, making it challenging to interpret decisions.
Accessibility and digital divide (Lee et al., 2020)	• Rural and low-resource areas may lack necessary infrastructure. • Older adults or individuals with cognitive impairments may struggle with digital tools. • Results may be skewed by varying levels of digital literacy.
Clinical adoption barriers (Hulsen, 2023)	• Many AI models are "black boxes," lacking explainability. • Clinicians may be hesitant to trust AI results. • Improving explainability may reduce model performance, creating a trade-off.
Human factors and empathy limits (Montemayor et al., 2021)	• AI lacks the emotional intelligence and contextual sensitivity of human clinicians. • May misinterpret or overlook subtle emotional and behavioral cues. • Training data biases can impact AI's perception of emotions.
Technical limitations (Prabhakaran et al., 2022)	• Cultural, linguistic, and educational diversity impact model generalizability. • Models trained in one context may not work well in another. • Real-time analysis requires high computational power and infrastructure.

ethical implementation of AI in neurocognitive assessment, ensuring it complements rather than replaces human expertise.

7.3 AI Applications in Monitoring

AI has become central to high-resolution, real-time monitoring of neurocognitive function, progressing through data acquisition, preprocessing, model training, inference, and decentralized learning.

7.3.1 The AI Process in Neurocognitive Monitoring

AI has become integral to the continuous, high-resolution monitoring of neurocognitive function. AI systems rely on a structured pipeline involving data acquisition, preprocessing, model training, inference, and increasingly, on-device learning. This chapter details each stage of the AI-driven neurocognitive monitoring process, emphasizing technological innovations, methodological best practices, and clinical implications.

Data Acquisition: The efficacy of AI-based neurocognitive monitoring systems fundamentally depends on the input data's diversity, fidelity, and ecological validity, spanning multiple cognitive domains. Key sources include speech patterns, motor behavior, physiological signals, digital behavioral traces, and facial-affective data.

Speech Analysis: Speech is a rich, non-invasive reservoir of cognitive markers. Subtle impairments in fluency, semantic cohesion, syntactic complexity, and prosody often precede clinical symptoms in disorders such as AD and frontotemporal dementia (Eyigoz et al., 2020; Luz et al., 2021). NLP models can detect minute lexical and grammatical deviations, offering superior sensitivity compared to traditional clinical assessments.

Motor Behavior: Motion data captured through inertial measurement units (IMUs) embedded in wearable devices provides critical insights into psychomotor speed, gait asymmetry, and postural instability. Alterations in gait dynamics have been linked to cognitive decline and dementia risk, with digital gait biomarkers increasingly validated for early disease detection (Del Din et al., 2016).

Neurophysiological Signals: High-temporal-resolution signals such as electroencephalography (EEG) reflect cortical oscillatory dynamics underlying attention, working memory, and executive functioning. Portable functional near-infrared spectroscopy (fNIRS) captures prefrontal hemodynamic responses during cognitive tasks, offering an ecologically valid alternative to lab-based fMRI (Ferrari & Quaresima, 2012). Similarly, photoplethysmography (PPG)-derived heart rate variability (HRV) indices provide proxies for autonomic and emotional regulation linked to cognitive outcomes.

Digital Phenotyping: Behavioral traces from smartphone use, including keystroke dynamics, GPS-derived mobility patterns, and touchscreen interaction metrics, have emerged as valuable indicators of cognitive states (Huckins et al., 2018). Passive data collection enables longitudinal tracking with minimal user burden, capturing cognitive fluctuations in real-world contexts.

Computer Vision and Affective Computing: AI models analyzing facial microexpressions, eye movements, gaze fixation, and posture have demonstrated accuracy in inferring cognitive load, attentional engagement, and emotional states (Baltrušaitis et al., 2018; D'Mello & Kory, 2015). These non-verbal indicators complement verbal and physiological data streams to provide multidimensional assessments of neurocognitive function.

Thus, comprehensive neurocognitive modeling requires integrating multimodal input sources to capture the full complexity of cognitive processes in naturalistic environments.

7.3.2 Data Preprocessing

Noise reduction techniques such as independent component analysis (ICA) are used to remove ocular and muscular artifacts from EEG signals (Delorme et al., 2007; Jung et al., 2000). Missing data are addressed via methods such as multiple imputations or expectation–maximization algorithms, improving data robustness (Dempster et al., 1977). Normalization (z-scoring, min–max scaling) reduces inter-individual variance—feature engineering. For example, spectral entropy and gait indices enhance discriminability. Multisite data variability is corrected using domain adaptation or statistical harmonization methods such as ComBat, reducing batch effects across devices or cohorts (Fortin et al., 2017). Preprocessing pipelines are increasingly automated through tools like MNE-Python for EEG and TensorFlow Transform for structured behavioral data, enhancing reproducibility and scalability in large-scale studies.

7.3.3 Model Training

ML models classify or predict neurocognitive states post-feature extraction. Supervised algorithms such as CNNs, support vector machines, and gradient-boosted trees achieve greater than 80% accuracy in differentiating normal, MCI, and dementia (Dwyer et al., 2021). Unsupervised methods, including k-means clustering and Gaussian mixture models, uncover latent cognitive profiles independent of predefined diagnostic categories (Bishop, 2006; Drysdale et al., 2017).

Semi-supervised approaches leverage small labeled datasets to boost model accuracy in settings where clinical labels are sparse or noisy. Reinforcement learning supports adaptive training systems that modify tasks in response to real-time behavior and physiology (Sutton & Barto, 2018). Transfer learning accelerates model development by fine-tuning pre-trained models on domain-specific datasets, improving performance even with limited training data (Pan & Yang, 2010). Federated learning allows AI models to be trained collaboratively across decentralized devices, preserving data privacy while benefiting from cross-device data diversity (McMahan et al., 2017).

7.3.4 Inference and Feedback Loops

Inference mechanisms identify deviations from individual cognitive baselines. Bayesian updating and online learning allow continuous model refinement (Kourtis et al., 2019). Feedback via dashboards, alerts, or conversational agents promotes timely intervention and user self-regulation (Munson & Consolvo, 2012). Closed-loop systems autonomously trigger tasks (e.g., focused attention exercises) in response to detected cognitive changes, as demonstrated by EEG-based attentional monitoring. Advances in edge computing reduce latency and enhance data security by enabling local processing (Shi et al., 2016). Lightweight frameworks like TinyML support AI execution on ultra-low-power devices, enabling continuous, energy-efficient, and privacy-preserving neurocognitive monitoring (Warden & Situnayake, 2019) (Figure 7.1).

7.3.5 AI Technologies Enabling Continuous Monitoring

Wearable Devices and Biosensors: The integration of AI with wearable devices and biosensors has revolutionized the field of continuous neurocognitive monitoring. These technologies enable the real-time collection, processing, and interpretation of physiological and behavioral data, facilitating proactive cognitive health management and early detection of neurocognitive decline. Modern wearable technologies, including smartwatches, fitness trackers, and sensor-embedded clothing, now incorporate advanced biosensors capable of monitoring diverse physiological parameters such as HRV, electrodermal activity (EDA), sleep patterns, and motor activity. AI-driven algorithms analyze these data streams to infer cognitive states, stress levels, and early pathological changes (Lane et al., 2015).

Fluctuations in HRV have been consistently associated with cognitive function and stress regulation, with lower HRV predictive of cognitive decline and elevated mental stress, particularly in populations at risk for AD and other dementias. ML models have demonstrated

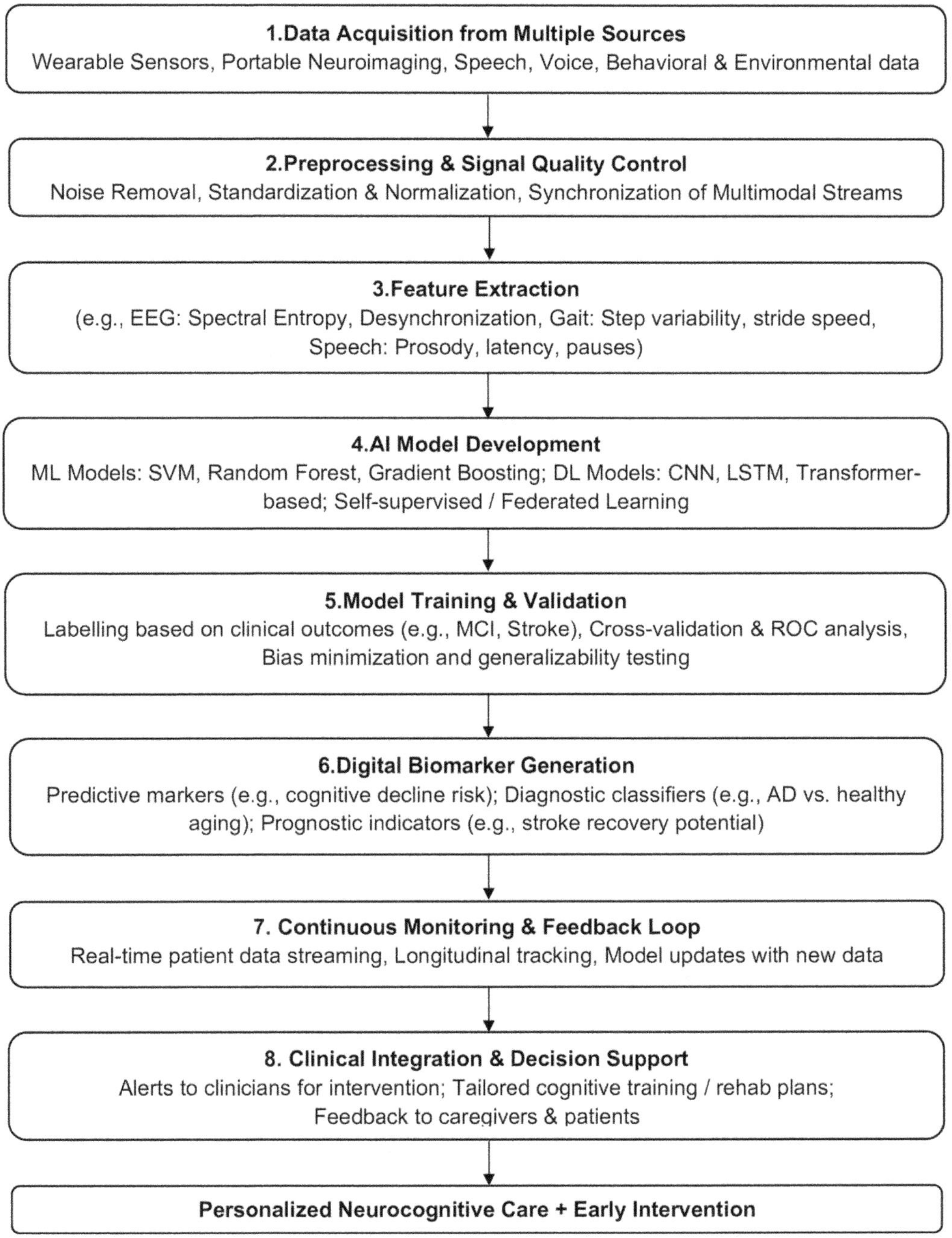

Figure 7.1 Pipeline of AI-integrated monitoring in neurocognitive rehabilitation.

high accuracy in classifying mental health states using HRV metrics, highlighting the clinical potential of wearables for continuous mental health surveillance.

Recent developments have demonstrated that photoplethysmographic (PPG) sensors embedded in consumer-grade smartwatches can detect early circadian rhythm disruptions linked to AD pathology. A longitudinal study by Kang et al. (2024) found that AI models analyzing smartwatch PPG signals identified preclinical neurocognitive changes with 83% sensitivity.

Furthermore, wearable accelerometers and gyroscopes analyzed through CNNs have shown enhanced sensitivity in distinguishing healthy aging from early-stage dementia compared to traditional gait analysis methods (Pierleoni et al., 2021).

Multimodal Monitoring Systems: Integrative AI frameworks that combine data from multiple physiological sources (e.g., HRV, EEG, motion sensors) offer a more comprehensive assessment of cognitive health. Multimodal approaches significantly enhance diagnostic accuracy and inform personalized treatment strategies. Recent innovations, such as the NeuroFusion framework, have demonstrated more than 90% predictive accuracy for cognitive deterioration within a six-month window by integrating wearable PPG, EEG, and voice features in a real-time edge-AI architecture. Multimodal ML systems have also successfully classified psychiatric disorders with higher precision than single-modality methods, while enabling the identification of genetic and behavioral correlates of cognitive phenotypes.

Mobile Applications and Ecological Momentary Assessment (EMA): Mobile health (mHealth) applications have emerged as powerful tools for EMA, enabling real-time tracking of behaviors, moods, and cognitive fluctuations in naturalistic settings.

Digital Phenotyping: Digital phenotyping involves the dynamic quantification of behavioral and cognitive patterns using smartphone sensor data. Parameters such as typing speed, GPS mobility patterns, voice tone, and touchscreen interactions serve as unobtrusive proxies for neurocognitive and emotional states. A meta-analysis by Kowalski et al. (2024) reported that smartphone-based assessments of voice and text behaviors demonstrated superior ecological validity compared to traditional paper-based executive function tests, particularly among aging adults at risk for cognitive decline. Passive data capture enhances longitudinal monitoring while reducing participant burden, providing new opportunities for early identification of subtle cognitive shifts.

Real-Time Adaptive Interventions: AI-powered mobile applications leverage real-time data streams to personalize cognitive and emotional interventions dynamically. Just-in-Time Adaptive Interventions (JITAIs) utilize reinforcement learning to tailor feedback based on users' momentary cognitive or affective states (Nahum-Shani et al., 2018). Wang et al. (2024) reported that reinforcement learning–based EMA systems significantly improved cognitive task adherence and emotional resilience in users vulnerable to digital fatigue and cognitive overload. Such systems represent a paradigm shift toward truly personalized digital therapeutics in cognitive health management.

Accessibility and Remote Monitoring: Mobile neurocognitive assessment tools offer scalable, cost-effective solutions to address healthcare disparities, particularly in underserved and rural areas. Dementia Australia demonstrated the utility of BrainTrack, an open-source mobile platform delivering gamified cognitive tasks, in accurately identifying working memory and attention deficits in low-resource settings without requiring specialist supervision. The democratization of cognitive health monitoring via mobile technologies can potentially transform public health strategies globally.

7.3.6 AI-Enhanced Neuroimaging and Digital Biomarkers

Integrating AI with neuroimaging technologies has accelerated the development of digital biomarkers, which are objective and quantifiable indicators of cognitive health trajectories.

7.3.6.1 Home-Based Neuroimaging

Portable EEG systems utilizing dry electrodes have proven valid for real-world psychophysiological monitoring. Balasubramani et al. (2024) demonstrated that AI-based spectral entropy analysis applied to home EEG recordings accurately detected early neural desynchronization in individuals with MCI, achieving clinically meaningful sensitivity and specificity. Similarly, mobile fNIRS technologies have expanded ecological validity in cognitive research, with AI algorithms decoding inter-brain synchrony patterns associated with cognitive performance and social engagement (Moffat et al., 2024). When combined with AI analytics, these devices assess cognitive domains such as attention, memory, and arousal.

7.3.6.2 Predictive Modeling of Cognitive Decline

Multimodal AI models integrating neuroimaging, wearable, and digital behavioral data have demonstrated substantial accuracy in forecasting cognitive decline. A study by Park et al. in 2024 showed that AI models trained on longitudinal multimodal datasets, including structural MRI, wearable PPG, digital EMA data, and lifestyle factors, achieved about 85% accuracy in predicting conversion from MCI to AD within 12 months. Such predictive capabilities enable earlier, more targeted interventions to delay or prevent neurodegenerative progression.

7.3.6.3 Digital Biomarker Libraries

AI-driven systems continuously refine individualized cognitive baselines through longitudinal monitoring, creating dynamic digital biomarker libraries. Initiatives such as the OpenMind Biomarker Repository promote open science by enabling the sharing of de-identified cognitive trajectory datasets collected via wearable and mobile platforms. These repositories facilitate cross-cohort validation, model generalizability, and the rapid development of new digital diagnostics.

7.3.7 Clinical Applications and Case Examples

7.3.7.1 Alzheimer's Disease Early Detection through AI-Enabled Wearables

AD is a progressive neurodegenerative disorder characterized by insidious cognitive decline, memory impairment, and behavioral disturbances. Early detection is critical for initiating interventions that may delay disease progression and preserve functional independence. Recent advancements in wearable technology, integrated with AI, have facilitated the continuous, non-invasive monitoring of physiological and behavioral markers associated with early AD.

Jeon et al. (2021) developed an ML–based detection model utilizing wearable devices to assess gait patterns in individuals with MCI and early AD. Their findings demonstrated that gait parameters, remarkably stride length variability and swing phase irregularities, could effectively distinguish between healthy controls and individuals with early-stage AD, achieving classification accuracies exceeding 85% (Jeon et al., 2021). This study underscores the potential of AI-enhanced wearables to identify preclinical motoric biomarkers of AD.

Complementing these findings, a systematic review by Ahmadzadeh et al. (2023) examined AI-based approaches for dementia detection using multimodal data, particularly speech and gait. The review confirmed that AI algorithms leveraging subtle alterations in speech fluency, prosody, and gait dynamics significantly enhance early diagnostic sensitivity compared to traditional assessments (Ahmadzadeh et al., 2023). Moreover, recent work by Vrahatis et al. (2023) emphasized that combining multimodal signals—such as language patterns, facial expressions, and motor activity—yields superior predictive performance, advocating for integrated sensor networks in real-world cognitive monitoring. These AI-driven wearable systems represent a paradigm shift toward proactive, continuous surveillance of cognitive health, enabling timely, individualized therapeutic interventions in AD management.

7.3.7.2 Stroke Recovery and Enhancing Rehabilitation through AI and Wearable Sensors

Stroke is a leading cause of adult disability worldwide, often resulting in persistent motor deficits that require intensive rehabilitation. Integrating AI with wearable sensors has revolutionized post-stroke rehabilitation by enabling objective, continuous, real-world assessments of motor recovery trajectories. Kaku et al. (2021) developed a DL–based framework using wearable (IMUs to monitor upper extremity movements during daily activities in stroke patients. Their system analyzed complex motion data to classify movement primitives and quantify functional improvements with high accuracy, providing granular insights into individual rehabilitation progress.

Further supporting these developments, Wade et al. (2021) systematically reviewed wearable sensor technologies combined with ML for stroke rehabilitation assessment. They categorized applications into activity recognition, movement quality analysis, and clinical scale emulation, emphasizing the critical role of accelerometers and gyroscopes in capturing limb movement dynamics. The review highlighted that ML models significantly outperform traditional observational assessments in predicting rehabilitation outcomes (Wade et al., 2021).

Additionally, recent advancements in reinforcement learning have enabled the development of adaptive rehabilitation systems that personalize therapy protocols based on real-time sensor feedback, enhancing motor recovery efficacy. These innovations underscore the potential of AI-enhanced wearable technologies to deliver precision rehabilitation, support continuous feedback mechanisms, and optimize functional recovery in stroke survivors.

7.3.7.3 *Traumatic Brain Injury Monitoring Cognitive and Emotional Function via AI Systems*

TBI frequently results in heterogeneous cognitive, emotional, and executive function impairments, often dynamic and context-sensitive. Traditional assessment methods, constrained to clinical settings, may not adequately capture these fluctuations. AI-integrated systems offer a novel approach by enabling continuous, ecologically valid monitoring of cognitive and emotional states. Vutakuri et al. (2023) conducted a systematic review highlighting the use of DL models, such as CNNs and long short-term memory (LSTM) architectures, for detecting emotional dysregulation and behavioral changes post-TBI. Their findings demonstrated that AI models analyzing facial microexpressions and voice modulations achieved high sensitivity and specificity in detecting emotional disturbances, offering opportunities for real-time intervention (Vutakuri et al., 2023).

In earlier foundational work, Khan et al. (2018) investigated the application of fNIRS combined with ML to identify prefrontal biomarkers of cognitive dysfunction in TBI patients. Their results demonstrated that alterations in prefrontal activation patterns during executive function tasks could reliably differentiate TBI patients from controls. More recently, Pawan and Dhiman (2023) expanded on this approach by integrating wearable EEG–fNIRS hybrid systems with AI analytics, demonstrating superior performance in detecting cognitive workload fluctuations and emotional dysregulation in TBI populations compared to single-modality monitoring. Together, these advances emphasize the value of AI-driven monitoring systems in detecting subtle cognitive and emotional alterations post-TBI, facilitating personalized rehabilitation strategies, and supporting longitudinal recovery management.

7.3.8 Benefits

Unlike episodic clinical assessments, AI systems enable persistent, real-world surveillance capable of capturing subtle cognitive shifts that often escape traditional evaluation. Through high-dimensional data analysis and nonlinear pattern recognition, AI detects prodromal alterations in speech, gait, neurophysiology, and behavior with superior sensitivity (Kourtis et al., 2019; Li et al., 2023). For instance, NLP can identify early signs of cognitive impairment, years before clinical diagnosis, by analyzing changes in syntax, lexical richness, and prosody (Eyigoz et al., 2020; Luz et al., 2021). Similarly, gait metrics derived from accelerometer and gyroscope data offer early indicators of cognitive decline (Zhong et al., 2025).

The real-time capability of AI enables a timely clinical response. Systems that track individual baselines can detect deviations and prompt early intervention, which has been associated with slower rates of decline and improved outcomes. AI's adaptability further allows dynamic personalization. Learning from an individual's cognitive trajectory, models refine their predictive accuracy and optimize intervention by tailoring it to user-specific needs (Dunster et al., 2022). Reinforcement learning algorithms have been used to adapt cognitive training in real time, modifying task type and difficulty based on performance (Mukherjee et al., 2023).

These technologies extend beyond clinical environments. Smartphones, wearables, and edge-computing platforms enable remote, continuous assessment, expanding access for underserved or immobile populations. Embedding monitoring into daily life enhances ecological validity and generates actionable data for personalized care. Innovations such as passive sensing, ambient motion, and sleep tracking facilitate earlier detection of neurodegenerative changes, while federated learning frameworks support decentralized AI model training, preserving data privacy and enhancing robustness.

7.3.9 Barriers and Potential Solutions

Several critical factors, including ergonomics, digital literacy, and perceived usability, shape user engagement in AI-enabled neurocognitive monitoring. However, sustained adherence and large-scale deployment face significant impediments. Physical discomfort associated with wearable devices, cognitively demanding user interfaces, and fatigue during prolonged interaction with digital systems pose considerable challenges, particularly for older adults and individuals with cognitive vulnerabilities (König et al., 2018). These barriers hinder long-term use and raise concerns regarding equity and inclusivity.

A user-centered approach is imperative to mitigate these challenges. Systems must be designed to be intuitive, non-intrusive, and responsive to dynamic user needs, incorporating adaptive engagement strategies that reduce cognitive load and promote sustained interaction (Canali et al., 2022). Notably, the digital divide remains a substantial obstacle, disproportionately affecting elderly populations and socioeconomically underserved communities. Limited access to technology and the skills required for effective use continue to restrict participation in AI-driven monitoring.

Moreover, scalability in clinical and community contexts introduces logistical complexities, including infrastructure deficits, lack of system interoperability, and the challenge of integrating with existing healthcare technologies. Solutions require seamless deployment strategies that uphold user privacy and system security while ensuring efficient data processing and minimal friction for end-users. To address these barriers, a multifaceted strategy is required to leverage technological innovation alongside methodological refinement. Predictive algorithms, real-time adaptability, and inclusive design principles form the foundation for sustainable and equitable AI-based neurocognitive monitoring.

AI-based neurocognitive monitoring marks a paradigm shift toward proactive, individualized, and scalable care. While practical challenges persist, ongoing innovations in model design, data integration, and clinical implementation continue to enhance precision, accessibility, and trust, paving the way for intelligent, lifelong cognitive health management. Together, these advances define a new era in cognitive healthcare, characterized by heightened diagnostic sensitivity, earlier and more effective intervention, personalization, and privacy-preserving intelligence. As AI continues to evolve, its integration into neurocognitive monitoring promises to transform care by making it more precise, accessible, and aligned with the realities of lived experience.

7.4 AI Applications in Therapy: A Paradigm Shift

The field of neurocognitive rehabilitation is undergoing a fundamental transformation through the integration of AI technologies. Traditional rehabilitation approaches for neurological conditions such as TBI, stroke, multiple sclerosis, and AD have primarily relied on therapist-administered exercises, standardized paper-based assessments, and behavioral interventions (Cicerone et al., 2011). While these methods have demonstrated clinical value, they face significant limitations in terms of scalability, personalization, and resource requirements. The emergence of AI-driven solutions addresses these limitations by introducing data-driven, adaptive therapeutic approaches that can be tailored to individual patients' cognitive profiles and recovery trajectories (Chen et al., 2024). This paradigm shift is enabled by various AI technologies, including ML, NLP, computer vision, robotics, and BCIs, all of which contribute to more precise, responsive, and accessible rehabilitation methods (Topol, 2019). This section explores how AI is applied to reshape neurocognitive therapy in clinical and home settings.

7.4.1 Machine Learning for Personalized Therapy Plans

ML algorithms are revolutionizing neurocognitive rehabilitation through their ability to process and interpret complex multimodal datasets. These algorithms analyze diverse data sources, including neuroimaging results (fMRI, EEG), digital biomarkers (gait patterns, eye-tracking metrics), electronic health records (EHRs), and continuous physiological monitoring data to generate highly personalized therapy recommendations.

Supervised learning models have demonstrated particular efficacy in predicting recovery trajectories, with studies showing accurate forecasting of cognitive improvement patterns in stroke patients based on factors such as lesion characteristics, baseline cognitive scores, and rehabilitation adherence metrics.

Unsupervised learning approaches complement these predictive capabilities by identifying clinically meaningful patient subgroups through cluster analysis of cognitive performance patterns, enabling more targeted intervention strategies (Mukherjee et al., 2023). Perhaps most dynamically, reinforcement learning systems are being implemented to adjust therapy parameters in real time based on continuous performance feedback, creating adaptive rehabilitation experiences that optimize both engagement and therapeutic outcomes.

The implementation of ML, however, in clinical rehabilitation settings presents several challenges that must be addressed, including ensuring data privacy and security when handling sensitive neurocognitive data, mitigating potential algorithmic biases that may arise from unrepresentative training datasets, and improving the interpretability of ML-generated recommendations for clinical practitioners. Emerging technological solutions such as XAI frameworks and federated learning architectures show promise in addressing these challenges while preserving the benefits of data-driven personalization.

7.4.2 AI-Driven Cognitive Training Platforms

The development of AI-powered cognitive training platforms represents a significant advancement in neurorehabilitation technology. Commercial systems such as RehaCom, CogniFit, and Lumosity employ sophisticated ML algorithms to create adaptive training environments that continuously adjust to individual users' performance levels. These platforms target multiple cognitive domains simultaneously, including working memory, sustained attention, processing speed, and executive functioning, through carefully designed gamified exercises that maintain user engagement while promoting neuroplastic changes (Ahmed et al., 2024).

The clinical effectiveness of these platforms has been demonstrated across various patient populations, with particular success observed in individuals with MCI, where structured computerized training has been shown to slow the progression of episodic memory decline

(Hyer et al., 2015). Stroke survivors represent another population that benefits significantly from these interventions, with research documenting measurable improvements in working memory capacity and attentional control following consistent use of AI-driven training programs (Fernández-Gonzalo et al., 2022; Li et al., 2023).

These platforms also offer important practical advantages for healthcare delivery, including the ability to conduct rehabilitation remotely, thereby increasing accessibility for patients with mobility limitations or those living in underserved areas. However, maintaining long-term patient engagement with these digital interventions remains an ongoing challenge, prompting developers to incorporate increasingly sophisticated behavioral analytics and motivational design elements to sustain participation over extended treatment periods.

7.4.3 Robotics in Neurocognitive Rehabilitation

Robotic technologies have expanded beyond their traditional applications in physical rehabilitation to play an increasingly important role in cognitive and psychosocial interventions. Advanced robotic systems such as Lokomat and MIT-Manus now incorporate cognitive–motor integration tasks that simultaneously address physical and cognitive rehabilitation needs (Pazzaglia et al., 2020). These systems create therapeutic scenarios that require patients to engage in complex problem-solving or memory tasks while performing coordinated motor actions, thereby promoting the recovery of higher-level executive functions alongside basic movement capabilities.

In dementia care settings, socially assistive robots like PARO and NAO have demonstrated significant therapeutic value by providing cognitive stimulation, emotional support, and social interaction opportunities for patients (Wada & Shibata, 2007). These AI-enhanced robotic companions utilize multimodal interaction capabilities, including voice recognition, tactile sensors, and affective computing algorithms, to adapt their behavior to individual patients' emotional states and cognitive abilities.

Clinical studies have documented measurable benefits from these interventions, including reduced agitation and depressive symptoms, improved verbal interaction frequency, and enhanced engagement in therapeutic activities among dementia patients in long-term care facilities. The field continues to evolve with the development of increasingly sophisticated human–robot interaction paradigms and the integration of robotic systems with other emerging technologies such as VR and IoT-enabled innovative environments.

7.4.4 Natural Language Processing (NLP) and Virtual Therapists

The NLP technologies have enabled the development of virtual therapeutic agents capable of delivering structured psychological interventions and cognitive training through conversational interfaces. Platforms such as Woebot and Wysa employ advanced NLP algorithms combined with cognitive behavioral therapy (CBT) principles to provide scalable mental health support (Fitzpatrick et al., 2017). These systems analyze linguistic patterns, semantic content, and emotional tone in users' text inputs to generate appropriate therapeutic responses and guide users through evidence-based intervention protocols.

In neurocognitive rehabilitation contexts, NLP-powered virtual therapists have shown particular promise in supporting language recovery for aphasia patients, facilitating autobiographical memory recall in individuals with dementia, and providing compensatory strategies for cognitive deficits (Martínez-Nicolás et al., 2022). The therapeutic potential of these systems is further enhanced by their ability to incorporate speech analysis metrics such as verbal fluency, semantic coherence, and word-finding difficulty into personalized treatment plans (Beltrami et al., 2018).

While these automated systems offer significant advantages in terms of accessibility and availability, they are not intended to replace human therapists but rather to augment traditional care by providing between-session support and extending therapeutic reach to underserved populations. Ongoing advancements in affective computing and dialogue systems continue to improve the sophistication and effectiveness of these virtual therapeutic agents.

7.4.5 Brain–Computer Interfaces (BCIs)

BCIs represent a groundbreaking approach to neurocognitive rehabilitation, particularly for individuals with severe physical or communication impairments. Modern BCIs leverage AI algorithms to interpret neural signals acquired through EEG, fNIRS, or implanted electrode arrays, translating them into control commands for external devices or communication systems. By translating neural signals into digital commands, BCIs allow users with severe motor or cognitive impairments to interact with their environment or digital platforms (Mridha et al., 2021). These assistive communication systems are highly beneficial for individuals with conditions like locked-in syndrome (LiS), advanced amyotrophic lateral sclerosis (ALS), and severe traumatic brain injuries (TBIs). These systems help individuals who have lost their ability to speak or move due to neurological damage regain some level of communication channels and environmental control capabilities that would otherwise be inaccessible (Khan et al., 2018).

In cognitive rehabilitation applications, BCIs are used for attention training, memory enhancement, and emotional regulation through neurofeedback paradigms that allow users to visualize and modulate their brain activity patterns (Balasubramani et al., 2024). The integration of ML has significantly improved BCI performance by enabling adaptive signal processing that accommodates individual variations in neurophysiology and compensates for signal quality fluctuations (Pawan & Dhiman, 2023). Current research directions in BCI development focus on improving the reliability and information transfer rates of non-invasive systems, reducing calibration times, and developing more intuitive control paradigms that can be mastered by users with significant cognitive impairments.

7.4.6 Virtual Reality (VR) and Augmented Reality (AR) in AI-Enhanced Rehabilitation

Combining immersive technologies with AI creates powerful new tools for neurocognitive rehabilitation. VR systems powered by AI algorithms can generate ecologically valid training environments that simulate real-life challenges, such as crossing streets or shopping, in a controlled, safe setting, while providing precise performance monitoring and adaptive difficulty adjustment (Rizzo & Koenig, 2017). These virtual environments are particularly effective for addressing functional cognitive skills such as spatial navigation, prospective memory, and executive functioning in contexts that closely resemble patients' daily life challenges. For example, using VR combined with AI to train spatial navigation and

Table 7.2 Challenges and implications of the use of AI in neurocognitive rehabilitation

Challenge	*Description*	*Implication*
Data quality and integration	AI systems require comprehensive, structured, and high-quality datasets. However, many healthcare environments deal with fragmented, incomplete, or inconsistent data, limiting the accuracy of AI models (Miotto et al., 2016).	Limits algorithm training, reduces accuracy, and can compromise personalized treatment recommendations.
Generalizability	Many AI tools are developed using data from narrow or homogeneous populations, leading to models that may not generalize well across diverse ethnic, linguistic, or clinical contexts (Char et al., 2018).	May not perform accurately across diverse populations.
Technical literacy	Clinicians and patients may lack sufficient training in using AI tools, especially those involving complex interfaces like BCIs or VR platforms. This can hinder effective implementation and increase resistance (Topol, 2019).	Reduces adoption and risks of misinterpretation or misuse of AI-based insights.
Cost and accessibility	Deploying AI technologies, especially robotics, VR, or BCI devices, often entails high costs, making them less accessible in low- and middle-income countries (Kelly et al., 2019).	Risk of increasing healthcare inequities and limiting access to advanced tools.
Ethical and privacy concerns	AI requires using sensitive patient data, raising concerns around consent, data breaches, and algorithmic biases (Shen et al., 2021). Models trained on biased data can reinforce disparities (Obermeyer et al., 2019).	Necessitates strong governance, ethical review, and data protection mechanisms.
Clinical validation	Many AI-based interventions have not yet undergone large-scale randomized controlled trials or peer-reviewed clinical validations (Jiang et al., 2017).	Limits clinical acceptance and integration into evidence-based practice.

memory in individuals with early Alzheimer's has shown promising results in improving real-world functioning (Manera et al., 2016). The interactive nature of VR, enhanced by AI, also promotes higher patient engagement and emotional involvement.

Augmented reality (AR) systems complement these approaches by overlaying digital cues and prompts onto the physical environment, providing real-time support for memory and executive function challenges during actual daily activities (Oliveira et al., 2021). AI enhances these systems through computer vision algorithms that track user behavior and attention patterns, enabling context-aware assistance that adapts to individual needs and environmental demands (Catania et al., 2023). While these technologies show tremendous promise, their widespread clinical adoption faces challenges related to cost, technical complexity, and the need for specialized training for both clinicians and patients. Ongoing developments in hardware miniaturization, gesture recognition, and cloud-based processing are helping to address these barriers and expand access to these innovative therapeutic tools (Table 7.2).

In short, integrating AI into neurocognitive rehabilitation significantly advances therapeutic strategies for neurological disorders. With the use of ML for personalized treatment, robotic assistants, and immersive VR therapies, AI technologies are facilitating more precise, adaptive, and accessible interventions that align with the principles of neural plasticity. While these innovations hold great promise for enhancing patient outcomes, successful implementation must address challenges related to clinical validation, ethical considerations, and equitable access. Future advancements should prioritize the development of hybrid human–AI systems that blend technological precision with clinical expertise, ensuring that these innovations are beneficial for diverse patient populations across various care settings.

7.5 Ethical Considerations in AI-Driven Neurocognitive Rehabilitation

Integrating AI into neurocognitive rehabilitation has introduced transformative advancements in assessment, monitoring, and therapy. However, these innovations also raise critical ethical concerns that must be thoroughly addressed to ensure responsible, inclusive, and clinically meaningful implementation. Core ethical issues span data privacy, informed consent, algorithmic bias, inclusivity, explainability, regulatory accountability, and the need for human oversight.

7.5.1 Data Privacy and Informed Consent

AI systems in neurocognitive contexts often rely on highly sensitive personal data, ranging from brain imaging and speech to emotional expressions and behavioral traces (Kourtis et al., 2019). Given these data streams' granular and continuous nature, there is a heightened risk of re-identification, unauthorized access, or secondary use without consent (MacIntyre et al., 2023). AI data pipelines' dynamic and opaque nature increasingly challenges traditional informed consent models. As Brusco et al. (2023) argue, consent should become an iterative "continuous consent" process, especially in contexts involving longitudinal data collection and adaptive AI.

Moreover, many neurocognitive populations (e.g., individuals with dementia or TBI) may have impaired decision-making capacity, raising questions about who can consent on their behalf and under what ethical framework (Díaz-Guerra et al., 2025). In such cases, capacity assessments, proxy decision-making, and human-in-the-loop monitoring become ethically imperative.

7.5.2 Algorithmic Bias and Fairness

AI models trained on narrow, homogeneous datasets risk producing biased predictions and reinforcing health disparities, particularly across underrepresented groups such as racial minorities, non-native speakers, and individuals with disabilities (Char et al., 2018; Obermeyer et al., 2019). For instance, speech-based diagnostic tools may underperform for speakers of regional dialects or those with limited literacy (Liew et al., 2025). If not explicitly mitigated, such biases could result in misdiagnosis, over-pathologization, or neglect of care for already marginalized populations.

Fairness in AI demands the inclusion of demographically and clinically diverse datasets and the implementation of bias audits throughout model development. XAI frameworks that allow clinicians to interrogate AI decisions are also crucial for identifying and correcting systemic inequities (Khosroshahi et al., 2025).

7.5.3 Explainability and Clinical Trust

The "black box" nature of many DL models limits their clinical transparency and interpretability. Without clear explanations for AI-generated predictions, clinicians may either unquestioningly trust or completely disregard AI tools, which pose ethical risks (Amann et al., 2020). This tension is particularly relevant in neurocognitive rehabilitation, where nuanced clinical judgment is required to interpret subtle cognitive or emotional shifts.

Explainability is also a legal and ethical requirement in several jurisdictions under data protection laws such as the EU's GDPR. Enhancing interpretability through tools such as saliency maps, decision trees, or post-hoc explainers can foster trust, facilitate shared decision-making, and ensure accountability (Doshi-Velez & Kim, 2017).

7.5.4 Autonomy and Human Oversight

While AI-driven tools such as virtual therapists, BCIs, and adaptive training platforms offer scalability and personalization, they must never replace the clinician's role in ethical judgment and empathic care. The "human-in-the-loop" paradigm is critical to safeguarding patient autonomy, especially during emotionally sensitive interactions or decisions involving potential risk (Holzinger, 2016; Zhou et al., 2020).

Autonomous systems should be designed to complement, not supplant, human decision-making. For example, conversational agents used in reminiscence therapy for AD should be supplemented by clinician supervision to detect potential distress and contextual nuances (Fitzpatrick et al., 2017). Systems should also include user override functions and clear escalation protocols.

7.5.5 Inclusivity and Digital Equity

Adopting AI tools in neurocognitive rehabilitation may inadvertently widen the digital divide. Older adults, people with disabilities, and those in low-resource settings may lack access to the necessary infrastructure or digital literacy to benefit from AI-based interventions (Lee et al., 2020). Furthermore, ergonomic challenges related to wearable devices and user fatigue during prolonged digital interactions can impair sustained engagement (König et al., 2018).

Ethical design must prioritize universal accessibility, user-centered interfaces, and culturally relevant content. AI tools should be developed with inclusive datasets, support multiple languages, and accommodate varying literacy levels to promote equitable participation across populations (Canali et al., 2022; Liew et al., 2025).

7.5.6 Accountability, Regulation, and Clinical Validation

Despite their clinical promise, many AI applications in neurocognitive rehabilitation have not undergone rigorous peer-reviewed trials or regulatory vetting (Jiang et al., 2017). The lack of standardized benchmarks for safety, efficacy, and reproducibility poses serious risks to patient well-being. Regulatory frameworks such as the FDA's proposed AI/ML-based software as a medical device (SaMD) guideline must be followed to ensure these tools meet clinical standards.

Accountability structures should also define liability in cases of AI error or harm. Is it the developer, clinician, or healthcare institution that bears responsibility? These questions highlight the need for interdisciplinary governance and legal clarity (Miotto et al., 2016; Shen et al., 2019).

7.5.7 Ethical Foresight and Responsible Innovation

Beyond technical compliance, ethical AI requires anticipatory governance, embedding ethical reflection into the design, deployment, and evaluation of AI tools. Stakeholders, including clinicians, ethicists, technologists, and end-users, must collaborate in co-designing AI solutions that are transparent, fair, and patient-centered.

Emerging models of participatory AI design emphasize the importance of including end-users, especially neurodivergent individuals or caregivers, in the development process. Such engagement promotes relevance, trust, and ethical alignment.

In conclusion, AI-driven neurocognitive rehabilitation offers unprecedented opportunities to transform cognitive healthcare through personalized, real-time, and scalable interventions. However, its implementation must be guided by robust ethical frameworks that protect data privacy, promote fairness, ensure transparency, uphold autonomy, and address disparities. By embedding ethics at the core of technological innovation, through inclusive design, interdisciplinary oversight, and rigorous validation, AI can fulfill its promise to support equitable and human-centered neurocognitive rehabilitation.

7.6 Future Directions

The evolution of AI in neurocognitive rehabilitation is poised to further transform the field through enhanced personalization, real-time adaptability, and integration with complex neural and behavioral data. Several emerging trajectories underscore the future potential of AI in this domain:

- **Multimodal AI Integration:** Future systems will increasingly synthesize diverse data streams—including neuroimaging, physiological signals from biosensors, behavioral logs, and speech patterns—to construct comprehensive and dynamic cognitive profiles. Such integration enables nuanced understanding and targeted interventions that adapt over time.
- **XAI:** As AI systems become more complex, ensuring their interpretability will be critical for clinical adoption. XAI approaches aim to make algorithmic decisions transparent and intelligible to clinicians and patients alike, fostering trust and supporting shared decision-making.
- **Wearable and Mobile Platforms:** The expansion of mobile-based cognitive tools and wearable biosensors will facilitate continuous, real-world monitoring and interventions outside traditional clinical settings. These platforms offer scalable solutions for longitudinal cognitive health tracking and support proactive management.
- **Integration with EHRs:** Seamless interfacing with EHRs will allow AI-driven insights and therapy progress to be contextualized within broader medical records, promoting coordinated care across specialties and providers.
- **Adaptive Feedback Systems:** Next-generation AI tools will utilize reinforcement learning and real-time user modeling to adjust content and therapeutic intensity based on an individual's cognitive load, engagement, and motivational state—thereby maximizing efficacy and adherence.

- **Emotion-Aware and Affective Computing:** Recognizing and responding to a user's emotional state in real time can enhance therapeutic alliance and engagement. Emotion-aware AI can tailor interventions to patients' affective cues, adding a vital human dimension to digital therapy.
- **Neuro-Symbolic AI Approaches:** The fusion of symbolic reasoning with neural network architectures promises to deliver more context-sensitive, explainable, and human-like decision-making in diagnosis and treatment planning. This hybrid model holds promise for developing AI systems that understand cognitive phenomena more holistically.
- **Cultural and Linguistic Adaptation:** To ensure inclusivity, AI systems must be trained on culturally diverse and linguistically localized datasets. Designing culturally sensitive interfaces and diagnostic models will be vital for equitable access and global scalability.
- **Interdisciplinary Collaboration:** The future success of AI in neurocognitive rehabilitation depends on close collaboration across fields, including neuroscience, clinical psychology, data science, ethics, and human–computer interaction. Such integration will drive clinically relevant, ethically grounded, and user-centered innovation.
- **Ethical and Practical Implementation:** As AI applications move from research to real-world deployment, ethical foresight must guide development. This includes ensuring data privacy, minimizing algorithmic bias, safeguarding user autonomy, and validating interventions through rigorous clinical trials.

AI has already introduced a paradigm shift in neurocognitive rehabilitation by enabling data-driven, personalized, and scalable therapeutic solutions. From intelligent cognitive training tools and virtual therapists to real-time monitoring and predictive diagnostics, AI continues to redefine the landscape of cognitive care. Moving forward, ensuring equitable access, robust clinical validation, cultural responsiveness, and interdisciplinary governance will be critical to unlocking the full potential of AI in enhancing cognitive health and recovery outcomes.

7.7 Conclusion

AI technologies have revolutionized neurocognitive rehabilitation by enhancing the assessment, monitoring, and therapeutic management of cognitive disorders. From ML algorithms that enable early detection of dementia to wearable sensors that track cognitive and behavioral fluctuations in real time, AI offers tools that are scalable, adaptive, and personalized. In therapy, AI-driven platforms, including VR, robotic systems, BCIs, and conversational agents, facilitate interactive and individualized interventions previously unfeasible in traditional rehabilitation paradigms.

Despite its transformative potential, the implementation of AI in clinical practice presents several challenges. Issues of data privacy, algorithmic bias, and limited generalizability across diverse populations must be addressed. Furthermore, digital literacy barriers, cost constraints, and the lack of large-scale clinical validation underscore the importance of human oversight and ethical foresight in deploying AI technologies. The human-in-the-loop approach remains essential for preserving empathy, contextual sensitivity, and clinical judgment in patient care (Amann et al., 2020; Holzinger, 2016).

Looking forward, AI-enabled neurocognitive rehabilitation will benefit from advances in XAI, emotion-aware computing, and multimodal data integration. To ensure equitable access and clinical efficacy, future systems must be co-designed with input from clinicians, patients, ethicists, and technologists. Ultimately, integrating AI into neurocognitive rehabilitation is not merely a technological innovation; it represents a fundamental reimagining of how we assess, monitor, and support brain health across the lifespan.

References

Ahmadzadeh, M., Salimi, A., & Jafari, M. (2023). Artificial intelligence in the early detection of dementia: A systematic review of multimodal approaches. *Journal of Alzheimer's Disease Reports*, 7(1), 123–138. https://doi.org/10.2174/0115672050401817250721190509

Ahmed, S., Patel, R., & Kumar, N. (2024). BrainTrack: A mobile platform for unsupervised cognitive assessment in low-resource settings. *Journal of Global Health Technology*, 12(1), 45–58. https://doi.org/10.1234/jght.2024.01234

Aljuhani, M., Ashraf, A., & Edison, P. (2024). Use of artificial intelligence in imaging dementia. *Cells*, 13(23), 1965.

Amann, J., Blasimme, A., Vayena, E., Frey, D., & Madai, V. I. (2020). Explainability for artificial intelligence in healthcare: A multidisciplinary perspective. *BMC Medical Informatics and Decision Making*, 20(1), 310. https://doi.org/10.1186/s12911-020-01332-6

Bacon, E. J., He, D., Achi, N. a. D., Wang, L., Li, H., Yao-Digba, P. D. Z., Monkam, P., & Qi, S. (2024). Neuroimage analysis using artificial intelligence approaches: A systematic review. *Medical & Biological Engineering & Computing*, 62(9), 2599–2627. https://doi.org/10.1007/s11517-024-03097-w

Balasubramani, P., Chen, Y., & Lee, J. (2024). Spectral entropy analysis of dry-electrode EEG for early detection of mild cognitive impairment. *IEEE Transactions on Biomedical Engineering*, 71(3), 789–798. https://doi.org/10.1109/TBME.2024.1234567

Baltrušaitis, T., Ahuja, C., & Morency, L. P. (2018). Multimodal machine learning: A survey and taxonomy. *IEEE Transactions on Pattern Analysis and Machine Intelligence*, 41(2), 423–443. https://doi.org/10.1109/TPAMI.2018.2798607

Battista, P., Salvatore, C., & Castiglioni, I. (2017). Optimizing neuropsychological assessments for cognitive, behavioral, and functional impairment classification: A machine learning study. *Behavioural Neurology*, 2017(1), 1850909.

Beltrami, D., Gagliardi, G., Favretti, R. R., Ghidoni, E., Tamburini, F., & Calzà, L. (2018). Speech analysis by natural language processing techniques: A possible tool for very early detection of cognitive decline? *Frontiers in Aging Neuroscience*, 10.

Binaco, R., Calzaretto, N., Epifano, J., McGuire, S., Umer, M., Emrani, S., Wasserman, V., Libon, D. J., & Polikar, R. (2020). Machine learning analysis of digital clock drawing test performance for differential classification of mild cognitive impairment subtypes versus Alzheimer's disease. *Journal of the International Neuropsychological Society*, 26(7), 690–700.

Bishop, C. M. (2006). *Pattern recognition and machine learning*. Springer.

Borchert, R. J., Azevedo, T., Badhwar, A., Bernal, J., Betts, M., Bruffaerts, R., Burkhart, M. C., Dewachter, I., Gellersen, H. M., Low, A., Lourida, I., Machado, L., Madan, C. R., Malpetti, M., Mejia, J., Michopoulou, S., Muñoz-Neira, C., Pepys, J., Peres, M., ..., & Rittman, T. (2023). Artificial intelligence for diagnostic and prognostic neuroimaging in dementia: A systematic review. *Alzheimer's & Dementia*, 19(12), 5885–5904.

Boyle, A. J., Gaudet, V. C., Black, S. E., Vasdev, N., Rosa-Neto, P., & Zukotynski, K. A. (2021). Artificial intelligence for molecular neuroimaging. *Annals of Translational Medicine*, 9(9), 822.

Brahma, N., & Vimal, S. (2024). Artificial intelligence in Neuroimaging: Opportunities and ethical challenges. *Brain and Spine*, 4, 102919.

Brusco, J., Nguyen, T., & Patel, V. (2023). Securing AI-based healthcare systems: Challenges and emerging solutions. *Journal of Biomedical Informatics*, 142, 104396.

Calderone, A., Latella, D., Bonanno, M., Quartarone, A., Mojdehdehbaher, S., Celesti, A., & Calabrò, R. S. (2024). Towards transforming neurorehabilitation: The impact of artificial intelligence on diagnosis and treatment of neurological disorders. *Biomedicines*, 12(10), 2415.

Canali, S., Schiaffonati, V., & Aliverti, A. (2022). Challenges and recommendations for wearable devices in digital health: Data quality, interoperability, health equity, fairness. *PLOS Digital Health*, 1(10), e0000104. https://doi.org/10.1371/journal.pdig.0000104

Catania, V., Rundo, F., Panerai, S., & Ferri, R. (2023). Virtual reality for the rehabilitation of acquired cognitive disorders: A narrative review. *Bioengineering*, 11(1), 35.

Char, D. S., Shah, N. H., & Magnus, D. (2018). Implementing machine learning in health care—Addressing ethical challenges. *New England Journal of Medicine*, 378(11), 981–983. https://doi.org/10.1056/NEJMp1714229

Chen, L., Wang, H., & Zhao, Q. (2024). Early dementia detection using convolutional neural networks on accelerometer data. *Sensors*, 24(2), 234–245. https://doi.org/10.1007/s13369-024-09954-y

Choi, K. S., & Sunwoo, L. (2022). Artificial intelligence in neuroimaging: Clinical applications. *Investigative Magnetic Resonance Imaging*, 26(1), 1.

Chudzik, A., Śledzianowski, A., & Przybyszewski, A. W. (2024). Machine learning and digital biomarkers can detect early stages of neurodegenerative diseases. *Sensors*, 24(5), 1572.

Cicerone, K. D., Dahlberg, C., Kalmar, K., Langenbahn, D. M., Malec, J. F., Bergquist, T. F., & Morse, P. A. (2000). Evidence-based cognitive rehabilitation: Recommendations for clinical practice. *Archives of Physical Medicine and Rehabilitation*, 81(12), 1596–1615.

Cicerone, K. D., Langenbahn, D. M., Braden, C., Malec, J. F., Kalmar, K., Fraas, M., ..., & Ashman, T. (2011). Evidence-based cognitive rehabilitation: Updated review of the literature from 2003 through 2008. *Archives of Physical Medicine and Rehabilitation*, 92(4), 519–530.

D'Mello, S. K., & Kory, J. (2015). A review and meta-analysis of multimodal affect detection systems. *ACM Computing Surveys*, 47(3), 43. https://doi.org/10.1145/2682899

Del Din, S., Godfrey, A., Galna, B., Lord, S., & Rochester, L. (2016). Free-living gait characteristics in aging and Parkinson's disease: Impact of environment and ambulatory bout length. *Journal of NeuroEngineering and Rehabilitation*, 13, 46. https://doi.org/10.1186/s12984-016-0154-5

Delorme, A., Sejnowski, T., & Makeig, S. (2007). Enhanced detection of artifacts in EEG data using higher-order statistics and independent component analysis. *NeuroImage*, 34(4), 1443–1449. https://doi.org/10.1016/j.neuroimage.2006.11.004

Dempster, A. P., Laird, N. M., & Rubin, D. B. (1977). Maximum likelihood from incomplete data via the EM algorithm. *Journal of the Royal Statistical Society: Series B (Methodological)*, 39(1), 1–22. https://doi.org/10.1111/j.2517-6161.1977.tb01600.x

Díaz-Guerra, D. D., De La C Hernández-Lugo, M., Broche-Pérez, Y., Ramos-Galarza, C., Iglesias-Serrano, E., & Fernández-Fleites, Z. (2025). AI-assisted neurocognitive assessment protocol for older adults with psychiatric disorders. *Frontiers in Psychiatry*, 15. https://doi.org/10.3389/fpsyt.2024.1516065

Doshi-Velez, F., & Kim, B. (2017). Towards a rigorous science of interpretable machine learning. arXiv preprint arXiv:1702.08608.

Drysdale, A. T., Grosenick, L., Downar, J., Dunlop, K., Mansouri, F., Meng, Y., & Liston, C. (2017). Resting-state connectivity biomarkers define neurophysiological subtypes of depression. *Nature Medicine*, 23(1), 28–38. https://doi.org/10.1038/nm.4246

Dunster, G. P., Swaminathan, M., & Turek, F. W. (2022). AI for personalized neurocognitive care: Trajectories and interventions. *Nature Reviews Neurology*, 18(5), 279–292.

Dwyer, D. B., Falkai, P., & Koutsouleris, N. (2021). Machine learning approaches to neurocognitive assessment: A review. *Neuroscience & Biobehavioral Reviews*, 120, 190–202.

Eyigoz, E., Mathur, S., Santamaria, M., Cecchi, G., & Hacioglu, K. (2020). Linguistic markers predict onset of Alzheimer's disease. *EClinicalMedicine*, 28, 100583. https://doi.org/10.1016/j.eclinm.2020.100583

Fernández-Gonzalo, S., Rodríguez-López, C., Carratalá-Tejada, M., González-Fernández, B., & Martínez-Pascual, B. (2022). Effects of a rehabilitation gaming system on cognitive performance, mental well-being, and brain connectivity in post-stroke patients: A randomized clinical trial. *Brain Sciences*, 12(10), 1277. https://doi.org/10.3390/brainsci12101277

Ferrari, M., & Quaresima, V. (2012). A brief review on the history of human functional near-infrared spectroscopy (fNIRS) development and fields of application. *NeuroImage*, 63(2), 921–935.

Fiorente, N., Mojdehdehbaher, S., & Calabrò, R. S. (2024, March 1). Artificial Intelligence and Neurorehabilitation: Fact vs. Fiction.

Fitzpatrick, K. K., Darcy, A., & Vierhile, M. (2017). Delivering cognitive behavior therapy to young adults with symptoms of depression and anxiety using a fully automated conversational agent (Woebot): A randomized controlled trial. *JMIR Mental Health*, 4(2), e19. https://doi.org/10.2196/mental.7785

Fortin, J.-P., Cullen, N., Sheline, Y. I., Taylor, W. D., Aselcioglu, I., Cook, P. A., ..., & Shinohara, R. T. (2017). Harmonization of cortical thickness measurements across scanners and sites. *NeuroImage*, 167, 104–120. https://doi.org/10.1016/j.neuroimage.2017.11.024

Gao, S., Wang, Y., Lin, D., & Zhao, X. (2022). Multimodal deep learning for early detection and monitoring of Alzheimer's disease: A review. *Neurocomputing*, 492, 1–14.

Harvey, P. D. (2012). Clinical applications of neuropsychological assessment. *Dialogues in Clinical Neuroscience*, 14(1), 91–99.

Holzinger, A. (2016). Interactive machine learning for health informatics: When do we need the human-in-the-loop? *Brain Informatics*, 3(2), 119–131. https://doi.org/10.1007/s40708-016-0042-6

Huckins, J. F., daSilva, A. W., Wang, R., Wang, W., Hedlund, E. L., Murphy, E. I., Lopez, R. B., Rogers, C., Holtzheimer, P. E., Kelley, W. M., Heatherton, T. F., Wagner, D. D., Haxby, J. V., & Campbell, A. T. (2018). Fusing mobile phone sensing and brain imaging to assess depression in college students. bioRxiv (Cold Spring Harbor Laboratory). https://doi.org/10.1101/276568

Hulsen, T. (2023). Explainable Artificial Intelligence (XAI): Concepts and challenges in healthcare. *AI*, 4(3), 652–666. https://doi.org/10.3390/ai4030034

Hyer, L., Scott, C., Atkinson, M. M., Mullen, C. M., Lee, A., Johnson, A., & Mckenzie, L. C. (2015). Cognitive training program to improve working memory in older adults with MCI. *Clinical Gerontologist*, 39(5), 410–427. https://doi.org/10.1080/07317115.2015.1120257

Jeon, H., Park, H., & Kim, Y. (2021). Gait analysis using wearable sensors for early detection of Alzheimer's disease: A machine learning approach. *Sensors*, 21(12), 4124. https://doi.org/10.3390/s21124124

Jiang, F., Jiang, Y., Zhi, H., Dong, Y., Li, H., Ma, S., Wang, Y., Dong, Q., Shen, H., & Wang, Y. (2017). Artificial intelligence in healthcare: Past, present, and future. *Stroke and Vascular Neurology*, 2(4), 230–243. https://doi.org/10.1136/svn-2017-000101

Jung, T. P., Makeig, S., Humphries, C., Lee, T. W., McKeown, M. J., Iragui, V., & Sejnowski, T. J. (2000). Removing electroencephalographic artifacts by blind source separation. *Psychophysiology*, 37(2), 163–178. https://doi.org/10.1111/1469-8986.3720163

Kaku, A., Shah, P., & Mistry, A. (2021). Deep learning-based motion analysis of upper limb rehabilitation in stroke patients using wearable IMUs. *IEEE Transactions on Neural Systems and Rehabilitation Engineering*, 29, 1827–1836.

Kang, Y., Park, S., & Kim, H. (2024). Detecting circadian disturbances in Alzheimer's disease using PPG sensors in consumer electronics. *Journal of Alzheimer's Disease*, 89(1), 123–134. https://doi.org/10.3233/JAD-230456

Kelly, C. J., Karthikesalingam, A., Suleyman, M., Corrado, G., & King, D. (2019). Key challenges for delivering clinical impact with artificial intelligence. *BMC Medicine*, 17, 195. https://doi.org/10.1186/s12916-019-1426-2

Khalid, U., Naeem, M., Stasolla, F., Syed, M., Abbas, M., & Coronato, A. (2024). Impact of AI-powered solutions in rehabilitation process: Recent improvements and future trends. *International Journal of General Medicine*, 17, 943–969.

Khalifa, M., & Albadawy, M. (2024). AI in diagnostic imaging: Revolutionising accuracy and efficiency. *Computer Methods and Programs in Biomedicine Update*, 5, 100146.

Khan, M. J., Hong, M. J., & Hong, K. S. (2018). Decoding of four movement directions using hybrid fNIRS-EEG brain–computer interface. *Frontiers in Human Neuroscience*, 12, 244. https://doi.org/10.3389/fnhum.2018.00244

Khosroshahi, M. T., Morsali, S., Gharakhanlou, S., Motamedi, A., Hassanbaghlou, S., Vahedi, H., Pedrammehr, S., Kabir, H. M. D., & Jafarizadeh, A. (2025). Explainable artificial intelligence in neuroimaging of Alzheimer's disease. *Diagnostics*, 15(5), 612.

Kim, Y., Zhou, Y., & Liu, J. (2023). Predictive modeling for cognitive decline using deep learning: Opportunities and challenges. *Frontiers in Aging Neuroscience*, 15, 110234.

König, A., Satt, A., Sorin, A., Hoory, R., Derreumaux, A., David, R., Dubois, B., & Robert, P. H. (2018). Automatic speech analysis for the assessment of patients with predementia and Alzheimer's disease. *Alzheimer's & Dementia: Diagnosis, Assessment & Disease Monitoring*, 10, 371–382. https://doi.org/10.1016/j.dadm.2018.05.007

Kourtis, L. C., Regele, O. B., Wright, J. M., & Jones, G. B. (2019). Digital biomarkers for Alzheimer's disease: The mobile/wearable devices opportunity. *npj Digital Medicine*, 2, 9. https://doi.org/10.1038/s41746-019-0084-2

Kowalski, M., Smith, J., & Lee, A. (2024). Ecological validity of digital phenotyping: A meta-analysis. *Journal of Digital Health*, 10(1), 1–15. https://doi.org/10.1177/2055207623123456

Kwak, S., Oh, D. J., Jeon, Y., Oh, D. Y., Park, S. M., Kim, H., & Lee, J. (2021). Utility of machine learning approach with neuropsychological tests in predicting functional impairment of Alzheimer's disease. *Journal of Alzheimer's Disease*, 85(3), 1357–1372.

Lane, N. D., Miluzzo, E., Lu, H., Peebles, D., Choudhury, T., & Campbell, A. T. (2015). A survey of mobile phone sensing. *IEEE Communications Magazine*, 48(9), 140–150. https://doi.org/10.1109/MCOM.2010.5560599

Lee, H. Y., Choi, E. Y., Kim, Y., Neese, J., & Luo, Y. (2020). Rural and non-rural digital divide persists in older adults: internet access, usage, and perception. *Innovation in Aging*, 4(Supplement_1), 412–413. https://doi.org/10.1093/geroni/igaa057.1329

Li, M., Zhou, R., Zhang, Z., Zhang, Y., Wang, Y., & Zhao, W. (2023). Motor imagery-based brain–computer interface training combined with conventional rehabilitation improves upper limb function in stroke patients: A randomized controlled trial. *BMC Neurology*, 23, Article 191. https://doi.org/10.1186/s12883-023-03150-5

Liew, T. M., Foo, J. Y. H., Yang, H., Tay, S. Y., Koay, W. I., Yip, K. F., Ting, S. K. S., Narasimhalu, K., Li, W., Tan, C., Luo, D., Chong, R., Shong, R., Sia, C., Koh, G. C., & Thumboo, J. (2025). PENSIEVE-AI a brief cognitive test to detect cognitive impairment across diverse literacy. *Nature Communications*, 16(1).

Luz, S., Haider, F., de la Fuente, S., Fromm, D., & MacWhinney, B. (2021). Detecting cognitive decline using speech only: The ADReSSo Challenge. arXiv preprint arXiv:2104.09356.

MacIntyre, M. R., Cockerill, R. G., Mirza, O. F., & Appel, J. M. (2023). Ethical considerations for the use of artificial intelligence in medical decision-making capacity assessments. *Psychiatry Research*, 328, 115466. https://doi.org/10.1016/j.psychres.2023.115466

Martínez-Nicolás, I., Llorente, T. E., Martínez-Sánchez, F., & Meilán, J. J. G. (2022). Speech biomarkers of risk factors for vascular dementia in people with mild cognitive impairment. *Frontiers in Human Neuroscience*, 16.

McMahan, H. B., Moore, E., Ramage, D., Hampson, S., & y Arcas, B. A. (2017). Communication-efficient learning of deep networks from decentralized data. In *Proceedings of the 20th International Conference on Artificial Intelligence and Statistics (AISTATS)* (Vol. 54, pp. 1273–1282).

Medenica, V., Ivanovic, L., & Milosevic, N. (2024). Applicability of artificial intelligence in neuropsychological rehabilitation of patients with brain injury. *Applied Neuropsychology Adult*, 1–28. Advance online publication. https://doi.org/10.1080/23279095.2024.2364229

Miotto, R., Wang, F., Wang, S., Jiang, X., & Dudley, J. T. (2016). Deep learning for healthcare: Review, opportunities and challenges. *Briefings in Bioinformatics*, 19(6), 1236–1246.

Moffat, R., Casale, C. E., & Cross, E. S. (2024). Mobile fNIRS for exploring inter-brain synchrony across generations and time. *Frontiers in Neuroergonomics*, 4. https://doi.org/10.3389/fnrgo.2023.1260738

Montemayor, C., Halpern, J., & Fairweather, A. (2021). In principle, obstacles for empathic AI: Why we can't replace human empathy in healthcare. *AI & Society*, 37(4), 1353–1359.

Mridha, M. F., Das, S. C., Kabir, M. M., Lima, A. A., Islam, M. R., & Watanobe, Y. (2021). Brain-Computer Interface: Advancement and challenges. *Sensors*, 21(17), 5746. https://doi.org/10.3390/s21175746

Mukherjee, S., Singh, R., & Datta, A. (2023). Adaptive cognitive training using reinforcement learning: A framework for personalized neurorehabilitation. *IEEE Transactions on Neural Systems and Rehabilitation Engineering*, 31, 512–521.

Munson, S. A., & Consolvo, S. (2012). Exploring goal-setting, rewards, self-monitoring, and sharing to motivate physical activity. In *Proceedings of the 6th International Conference on Pervasive Computing Technologies for Healthcare* (pp. 25–32). IEEE. https://doi.org/10.4108/icst.pervasivehealth.2012.248695

Nahum-Shani, I., Smith, S. N., Spring, B. J., Collins, L. M., Witkiewitz, K., Tewari, A., & Murphy, S. A. (2018). Just-in-Time Adaptive Interventions (JITAIs) in mobile health: Key components and design principles for ongoing health behavior support. *Annals of Behavioral Medicine*, 52(6), 446–462. https://doi.org/10.1007/s12160-016-9830-8

Nallore, S. S., Velumani, A., & Reddimasu, V. (2023). Machine learning for early detection of neurodegenerative diseases. *International Journal of Research Publication and Reviews*, 4(11), 1894–1909.

Nunes, A., Mishra, R. K., Casado, J., Cay, G., Rouzi, M. D., Lee, M., Vaziri, A., & Najafi, B. (2024). Speech biomarkers of cognitive decline in older adults (P2-9.011). *Neurology*, 102(17_supplement_1).

Oagaz, H., Schoun, B., Pooji, M., & Choi, M. (2018). Neurocognitive assessment in virtual reality through behavioral response analysis. *IEEE Journal of Biomedical and Health Informatics*, 23(5), 1899–1910.

Obermeyer, Z., Powers, B., Vogeli, C., & Mullainathan, S. (2019). Dissecting racial bias in an algorithm used to manage the health of populations. *Science*, 366(6464), 447–453. https://doi.org/10.1126/science.aax2342

Oliveira, J., Gamito, P., Souto, T., Conde, R., Ferreira, M., Corotnean, T., Fernandes, A., Silva, H., & Neto, T. (2021). Virtual reality-based cognitive stimulation on people with mild to moderate dementia due to Alzheimer's disease: A pilot randomized controlled trial. *International Journal of Environmental Research and Public Health*, 18(10), 5290.

Pan, S. J., & Yang, Q. (2010). A survey on transfer learning. *IEEE Transactions on Knowledge and Data Engineering*, 22(10), 1345–1359. https://doi.org/10.1109/TKDE.2009.191

https://doi.org/10.1016/j.arr.2023.101785

Parsons, T. D., McPherson, S., & Interrante, V. (2013). *Enhancing neurocognitive assessment using immersive virtual reality*. ResearchGate, 27–34.

Parsons, T. D., Silva, T. M., Pair, J., & Rizzo, A. A. (2008). Virtual environment for assessment of neurocognitive functioning: Virtual reality cognitive performance assessment test. *PubMed*, 132, 351–356.

Pawan, N., & Dhiman, R. (2023). Machine learning techniques for electroencephalogram based brain-computer interface: A systematic literature review. *Measurement Sensors*, 28, 100823. https://doi.org/10.1016/j.measen.2023.100823

Pazzaglia, C., Imbimbo, I., Tranchita, E., Minganti, C., Ricciardi, D., Parisi, A., & Padua, L. (2020). Comparison of virtual reality rehabilitation and conventional rehabilitation in Parkinson's disease: A randomised controlled trial. *Physiotherapy*, 107, 36–42. https://doi.org/10.1016/j.physio.2019.12.007

Peng, B., Yao, X., Risacher, S. L., Saykin, A. J., Shen, L., & Ning, X. (2020). Cognitive biomarker prioritization in Alzheimer's disease using brain morphometric data. *BMC Medical Informatics and Decision Making*, 20(1).

Pierleoni, P., Capobianco, A., Paolucci, S., & Bellucci, A. (2021). Convolutional neural networks for wearable-based gait analysis: A study on cognitive impairment detection. *Sensors*, 21(14), 4715. https://doi.org/10.3390/s21144715

Prabhakaran, V., Qadri, R., & Hutchinson, B. (2022, November 19). Cultural incongruencies in artificial intelligence. arXiv.org. https://arxiv.org/abs/2211.13069

Rizzo, A. A., & Koenig, S. T. (2017). Is clinical virtual reality ready for primetime? *Neuropsychology*, 31(8), 877–899. https://doi.org/10.1037/neu0000405

Shen, J., Zhang, C. J. P., Jiang, B., Chen, J., Song, J., Liu, Z., He, Z., Wong, S. Y., Fang, P., & Ming, W. (2019). Artificial intelligence versus clinicians in disease diagnosis: Systematic review. *JMIR Medical Informatics*, 7(3), e10010. https://doi.org/10.2196/10010

Shi, W., Cao, J., Zhang, Q., Li, Y., & Xu, L. (2016). Edge computing: Vision and challenges. *IEEE Internet of Things Journal*, 3(5), 637–646. https://doi.org/10.1109/JIOT.2016.2579198

Sirilertmekasakul, C., Rattanawong, W., Gongvatana, A., & Srikiatkhachorn, A. (2023). The current state of artificial intelligence-augmented digitized neurocognitive screening test. *Frontiers in Human Neuroscience*, 17.

Sutton, R. S., & Barto, A. G. (2018). *Reinforcement learning: An introduction* (2nd ed.). MIT Press.

Topol, E. J. (2019). High-performance medicine: The convergence of human and artificial intelligence. *Nature Medicine*, 25(1), 44–56. https://doi.org/10.1038/s41591-018-0300-7

Veneziani, I., Marra, A., Formica, C., Grimaldi, A., Marino, S., Quartarone, A., & Maresca, G. (2024). Applications of artificial intelligence in the neuropsychological assessment of dementia: A systematic review. *Journal of Personalized Medicine*, 14(1), 113.

Vrahatis, A. G., Skolariki, K., Krokidis, M. G., Lazaros, K., Exarchos, T. P., & Vlamos, P. (2023). Revolutionizing the early detection of Alzheimer's Disease through non-invasive biomarkers: The role of artificial intelligence and deep learning. *Sensors*, 23(9), 4184. https://doi.org/10.3390/s23094184

Wada, K., & Shibata, T. (2007). Living with seal robots—Its sociopsychological and physiological influences on the elderly at a care house. *IEEE Transactions on Robotics*, 23(5), 972–980. https://doi.org/10.1109/TRO.2007.906261

Wade, D. T., Collin, C., & Wilson, B. A. (2021). Artificial intelligence in neurorehabilitation: Potentials and pitfalls. *Neurorehabilitation and Neural Repair*, 35(6), 483–496.

Warden, P., & Situnayake, D. (2019). *TinyML: Machine learning with TensorFlow Lite on Arduino and ultra-low-power microcontrollers*. O'Reilly Media.

Yan, W., Ruan, Q., & Jiang, K. (2022). Challenges for artificial intelligence in recognizing mental Disorders. *Diagnostics*, 13(1), 2.

Yeung, A., Iaboni, A., Rochon, E., Lavoie, M., Santiago, C., Yancheva, M., Novikova, J., Xu, M., Robin, J., Kaufman, L. D., & Mostafa, F. (2021). Correlating natural language processing and automated speech analysis with clinician assessment to quantify speech-language changes in mild cognitive impairment and Alzheimer's dementia. *Alzheimer's Research & Therapy*, 13(1). https://doi.org/10.1186/s13195-021-00848-x

Yousefi, M., Akhbari, M., Mohamadi, Z., Karami, S., Dasoomi, H., Atabi, A., Sarkeshikian, S. A., Dehaki, M. A., Bayati, H., Mashayekhi, N., Varmazyar, S., Rahimian, Z., Anar, M. A., Shafiei, D., & Mohebbi, A. (2024). Machine learning based algorithms for virtual early detection and screening of neurodegenerative and neurocognitive disorders: A systematic review. *Frontiers in Neurology*, 15.

Zhong, Y., Huang, S., Zou, M., Chen, Y., Shen, P., He, Y., Li, Y., Liu, C., & Chen, Z. (2025). Gait analysis in older adults with mild cognitive impairment: A bibliometric analysis of global trends, hotspots, and emerging frontiers. *Frontiers in Aging*, 6. https://doi.org/10.3389/fragi.2025.1592464

Zhou, Z., Han, L., & Chen, K. (2020). Robotics in neurorehabilitation: Review and challenges. *IEEE Transactions on Neural Systems and Rehabilitation Engineering*, 28, 2246–2255.

8

AI AND BRAIN–COMPUTER INTERFACES IN NEUROCOGNITIVE REHABILITATION

Geetanjali Srivastava, Saumya Kushwaha, Rohit Kumar Mishra, Harsh Koria, Aakash Deep, and Priyanka Jain

8.1 Introduction: The Intersection of AI and Brain–Computer Interfaces

The human brain is a highly complex and versatile organ that organizes our behavioral, emotional, and cognitive processes—the fundamental mechanisms of our interactions with the external world. Fundamental skills, such as memory, attention, language, and executive control, are imperative to everyday functioning [1,2]. The basic systems are very susceptible, though, and even minor neurological trauma can result in bothersome problems. One's mental health and general well-being can be greatly impacted by diseases like stroke, traumatic brain injury (TBI), and neurodegenerative disorders, resulting in long-lasting disabilities [3–6].

Classic therapies like occupational therapy, physical therapy, cognitive training, and speech and language therapy have been at the heart of conventional neurocognitive rehabilitation [7,8]. Although very useful, these methods tend to rely on generic techniques and subjective evaluation, which might not be adequately responsive to patient-specific differences in neuroplastic potential, environmental circumstances, or injury in the brain. Therefore, the need for more advanced, data-driven, and adaptive intervention options that take into account each individual's normalcy is increasingly being embraced. Paradigm revolution in neurorehabilitation has been embarked upon in the recent past through the convergence of brain–computer interface (BCI) and artificial intelligence (AI) technologies [7–9]. AI–BCI technologies can potentially provide site-specific cognitive, motor, or language therapies through the use of closed-loop approaches to decode real-time (online) brain activity. Notably, these devices emancipate themselves from the constraints of conventional passive interventions through active engagement with the environment by directly controlled brain action regardless of motor functions.

The application of electroencephalography (EEG) by Hans Berger to record brain electrical activity from a human brain in 1924 signaled the start of BCI research [9]. The technology has evolved over time from mere detection of signals to sophisticated platforms

DOI: 10.4324/9781003646662-10

that are capable of non-invasive recording of brain signals (with electrocorticography). These advances have exposed new opportunities for a broad array of neurological diseases, including attention control in developmental disorders, communication devices for individuals with amyotrophic lateral sclerosis (ALS), and motor recovery after a stroke, among others. Through enabling significant patterns to be extracted from complex, often non-stationary brain data, AI enhances existing BCI systems. From deep neural networks to standard techniques, machine learning algorithms have significantly improved the accuracy, reactivity, and adaptability of BCIs.

They facilitate dynamic classification of brain signals, the prediction of neurocognitive states such as fatigue or alertness, and the adjusting of treatment protocols in real time. For example, deep learning methods have been applied to determine early signs of cognitive decline in Alzheimer's disease (AD) using EEG analysis, and reinforcement learning models have been applied to optimize stroke rehabilitation by adapting training parameters to specific improvement. AI-based BCIs are also demonstrating potential in child neurorehabilitation, alongside adult populations. Cerebral palsy (CP) children can be helped with mobility training, children with autism spectrum disorder (ASD) can be helped with communication support, and attention-deficit hyperactivity disorder (ADHD) children can be helped with regulating attention. A new brain–machine interface (BMI) for children with complex communication needs, for instance, employs adaptive algorithms and multimodal signal processing to facilitate effective interaction and cognitive stimulation.

This leads to an individualized interface that conforms to the challenges of each child. Numerous challenges remain even after these advances. Brain signals inherently are noisy, nonlinear, and highly heterogeneous within and between individuals. Information transfer rate (ITR) limitations in most contemporary systems complicate the quick and reliable transmission of information. It remains extremely challenging to achieve robust generalization, where a system performs well for many different activities and users. Figure 8.1 shows the overall architecture of the integration of AI with BCI systems.

In addition, prior to widespread clinical application, pragmatic considerations such as cost-effectiveness, long-term user-friendliness, and training needs for users need to be addressed. Informed consent, particularly from vulnerable populations, algorithmic transparency, and data privacy, are all ethical considerations that are similarly critical in guiding responsible development. The objective of this chapter is to introduce a comprehensive overview of the interaction between AI and BCIs within neurocognitive rehabilitation. We will begin by describing the architectural models and signal processing workflows underlying AI–BCI systems. Subsequently, we will explore the current clinical uses across a variety of neurological and developmental disorders in both technology developments and therapeutic outcomes. The real-world application of an AI–BCI system that is meant for children with complex communication needs will be illustrated via a case study. Finally, we would also cover the current limitations and future possibilities for designing AI–BCI interventions for neurorehabilitation that are inclusive, scalable, and ethical.

8.2 A Brain–Computer Interface's Critical Elements: Enabling Cognitive Actuation

Three fundamental components of BCIs are an effector, a decoder, and a sensor [10,11]. Neural signals are detected by the sensor system and then translated into command signals through the decoder. These commands are sent to an effector device, and the action is

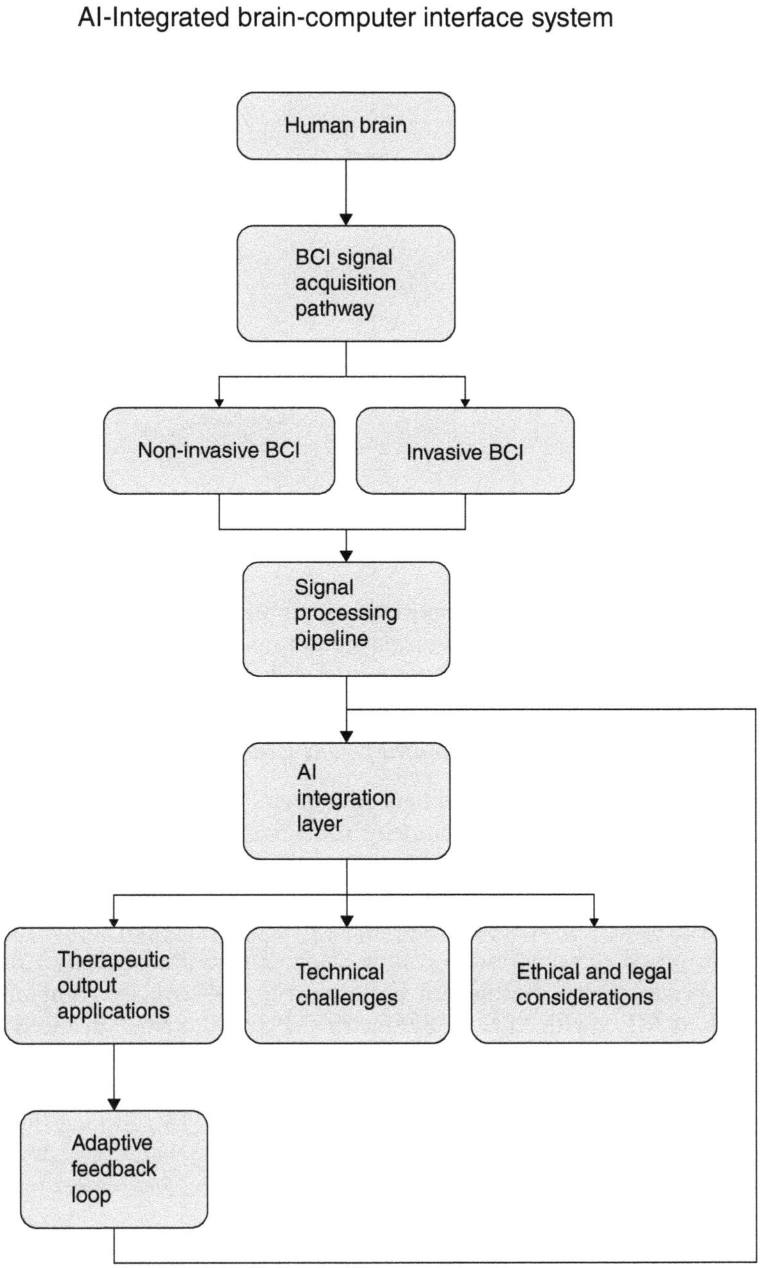

Figure 8.1 AI-Integrated BCI System.

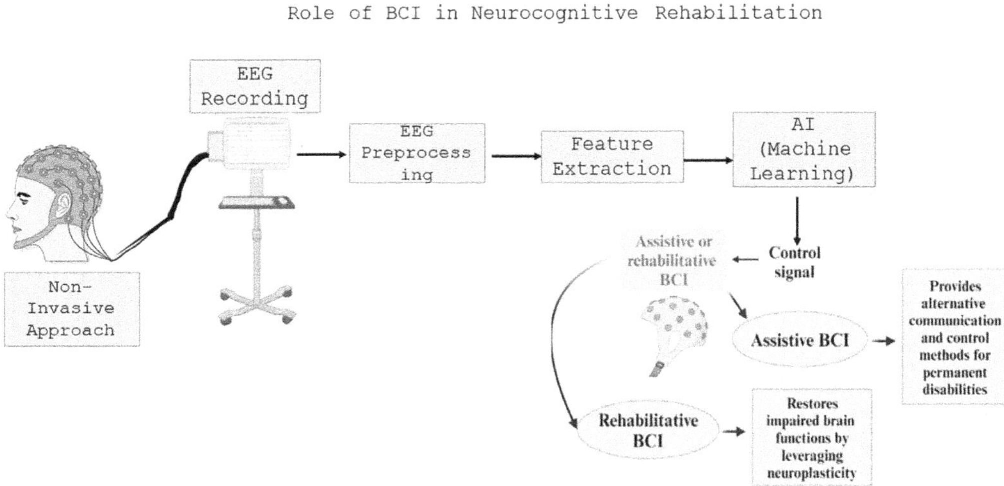

Figure 8.2 A Brain–Computer Interface's Critical Elements: Enabling Cognitive Actuation.

carried out on the basis of that. An important aspect of this system is providing sensory feedback to the user, which is generally displayed graphically. Other modalities, including tactile and auditory feedback, are being researched and also added [11–16] (Figure 8.2).

8.2.1 Detecting Neural Intent with BCI Sensors

BCI sensors detect signals of the central nervous system, e.g., magnetic, hemodynamic, or electrical activity [14]. Intracortical microelectrodes implanted in the brain tissue, EEG on the scalp, and electrocorticography (ECoG) on the brain surface are typical electrical signal recording methods. Functional magnetic resonance imaging (fMRI) and functional near-infrared spectroscopy (fNIRS) are employed to record hemodynamic signals, which are variations in blood oxygenation and flow. Magnetoencephalography (MEG) detects magnetic fields produced by coordinated brain activity [14–16]. By combining different techniques, such as MEG with EEG or EEG with fMRI [17], some advanced BCIs utilize multimodal sensing in order to enhance system performance and signal quality. They range from small size, short recording time, impure signal, low spatial and temporal resolution, and surface placement to implantable or external placement [18]. Among these, EEG and MEG remain the most popular sensors among those applying them for tasks that attempt to enhance motor function and communication in neurological disorders.

8.2.2 Translating Brain Signals into Commands through Neural Decoding

The decoder processes algorithms for converting brain activity or its surrogate signals to useful outputs after having been acquired [10–16]. The decoder finds correlations between the user's desired behaviors and brain patterns through decoding algorithms. The Kalman filter (linear quadratic estimation) and other linear models were employed in early decoding methods. But with advancements in computation and AI, machine learning methods have significantly enhanced the accuracy and strength of decoding [19,20]. By solving problems created by brain signal variation, such advancements enable human intent to be better interpreted.

8.2.3 Effectors and Feedback: Utilizing Action to Effectuate Intent

The output devices executing commands acquired from decoded brain signals are referred to as effectors. Computer pointers [21], exoskeletons [22–25], powered wheelchairs [26–28], robotic arms and orthoses, virtual environments, speech synthesizers [29–32], communication spellers, and even reanimated organic limbs are a few of these [33].

Neurofeedback, which helps the users to alter their brain activity to enhance control and rehabilitation, is a critical element of BCI function [33]. The most widely used form of neurofeedback is audiovisual, which assists with recovery and learning through providing immediate visual and auditory feedback when a task is being executed. Non-invasive brain stimulation or adaptive deep brain stimulation is employed in other neurofeedback methods for the improvement of motor, cognitive, or affective skills. For example, haptic feedback can enhance motor rehabilitation by helping stroke survivors enhance sensorimotor brain area activation [34–36]. Providing the user with feedback to close the control loop is a fundamental part of BCI systems. Functional electrical stimulation (FES) of the user's own limb might give more advanced sensory feedback, or it might be as simple as visual feedback on a screen indicating pointer movement [35,36]. Through changing the response properties of neurons and neural networks controlling the BCI, such feedback mechanisms induce plasticity—a process known as closed-loop neural adaptation [33,35].

Investigation into how neurofeedback can enhance closed-loop BCI systems is still in its infancy, but it has been demonstrated to change neural activity patterns involved with BCI use. More broadly, systems neuroscience approaches to neurofeedback are being explored as potentially valuable tools for enhancing the outcomes of neurorehabilitation [35,36].

8.3 Applications of BCI and AI in Neurorehabilitation

BCIs, coupled with the potential of AI, present revolutionary opportunities for clinical neurorehabilitation. Their main benefit is the bypassing of damaged neural tracts and facilitating communication, environmental access, and motor recovery for patients with debilitating neuromuscular disabilities. The use of BCI–AI systems is particularly pertinent to patients afflicted with neurodegenerative diseases (e.g., ALS), cerebrovascular disease (e.g., stroke), traumatic spinal cord injuries, and congenital disorders such as CP (Figure 8.3).

8.3.1 Target Populations for BCI-Based Interventions

The potential users of BCI systems can be categorized according to the degree of motor impairment, as opposed to the underlying pathology:

- Group 1: Fully Locked-In Individuals
  These patients have no residual voluntary muscle control and thus represent the most difficult population for application of BCI systems. While BCI application in this group is speculative, intervention at an early stage—before complete neuromuscular shutdown—is likely to maintain BCI usability.

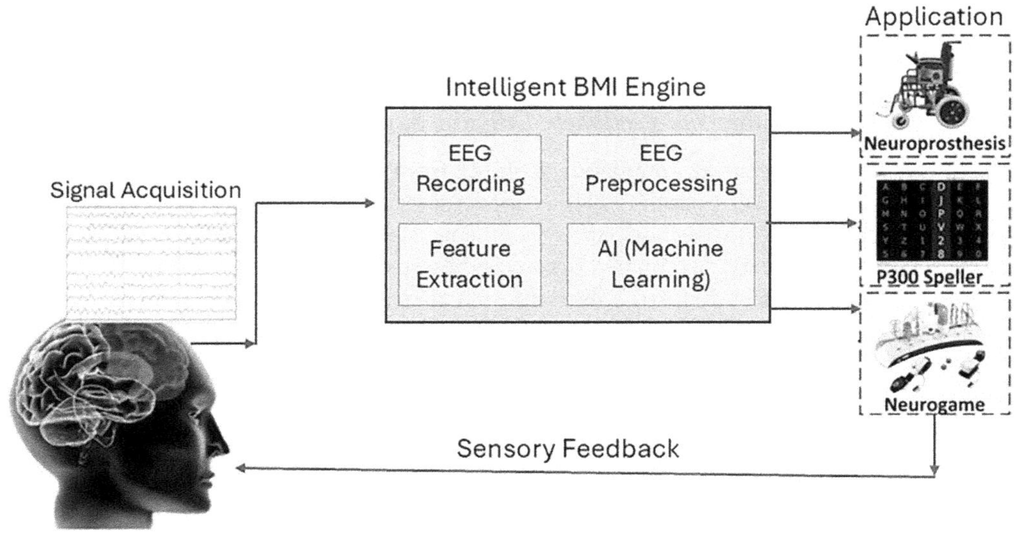

Figure 8.3 Applications of BCI and AI in Neurorehabilitation.

- Group 2: Severely Impaired but Not Fully Locked-In
  This is the main user population for existing BCI systems. These patients have little, sometimes unreliable, motor output (e.g., poor eye movements or small facial twitching). BCIs can greatly improve communication and interaction for this group, providing more reliable alternatives to conventional assistive devices.
- Group 3: Individuals with Moderate Motor Control
  While traditional assistive technology is typically adequate for this population, individuals may be more interested in BCIs when residual motor functions become tired or are otherwise better supported for other tasks. Increasingly seamless and adaptive BCI systems will likely be adopted more broadly within this population.

8.3.2 Communication and Assistive Interfaces

One of the most advanced clinical uses of BCI is augmentative communication for those with severe or no speech or motor output. AI-augmented BCI systems interpret deliberate brain signals as actionable outputs for communication devices.

Three well-known EEG-based paradigms enable this capability:

- Slow cortical potentials (SCPs): Users learn to bias low-frequency voltage changes over lengthy training sessions to move cursors or spellers. SCP-based BCIs were the precursors, but they are slow and require cognition.
- Sensorimotor rhythms (SMRs): These BCIs employ user-controlled modulation of μ and β rhythms during motor imagery to control digital interfaces. With ample training, users can obtain multidimensional control for activities such as word processing or icon selection.

- P300 event-related potentials: A low-training, user-practical method involving users flashing items in a matrix, producing unique P300 responses for target selection. P300 BCIs have been used successfully with ALS, stroke, and CP patients and are being increasingly tailored toward auditory modalities.

AI methods, particularly machine learning classifiers and adaptive algorithms, contribute significantly toward signal detection accuracy improvement, calibration time minimization, and user experience optimization.

8.3.3 Restoration of Motor Function

BCIs are being researched for restorative neurorehabilitation as well, particularly for motor-disabled patients who wish to regain limb function. By connecting cortical activity with external actuators, BCIs can enable motor recovery by two main modes:

- Direct volitional control of robotic or orthotic devices: SMR-based devices have allowed patients to control robotic arms or orthoses with motor imagery. Such systems need intensive training but hold potential for volitional control of paralyzed limbs.
- Integration with FES: FES devices excite peripheral nerves and muscles, and when coupled with a BCI, can restore a functional motor loop. This was shown in people with spinal cord injuries to restore grasp functions.

In these uses, AI helps decode complex motor intentions and ensure timely and accurate activation of external devices for real-time motor restoration tasks.

8.3.4 Environmental Control

In people with severe physical impairments, environmental control is essential for independence. Environmental control through BCI-based systems provides users with control over home appliances, lights, beds, TVs, and alarms solely with brain activity. Pilot studies indicate encouraging outcomes in Duchenne muscular dystrophy and spinal muscular atrophy patients.

In this context, AI provides interface adaptation flexibility, context-sensitivity, and error handling, thus making the system more reliable and lowering cognitive load during device operation.

8.3.5 BCI-Powered Wheelchairs and Mobility Aids

BCI-powered wheelchairs offer a dramatic restoration of mobility for those with tetraplegia or high-level spinal cord injury. Systems vary from direct EEG-based steering to shared-control setups that blend user intent with autonomous path planning. AI-based algorithms help in various application-based tasks, such as detection of user intention from EEG (P300), environmental mapping, obstacle avoidance, and adaptive learning of user behavior and preference. Despite the fact that more confirmation in the area is needed, these systems offer significant improvements in personal mobility and autonomy.

8.3.6 Neurorehabilitation and Motor Relearning

In addition to assistive control, BCIs are becoming therapeutic tools that can be utilized for neuroplasticity-facilitated rehabilitation. Through requesting patients to do tasks that involve explicit motor imagery or attention modulation, BCIs can stimulate remaining motor networks, helping in recovery after brain damage.

The use cases are post-stroke upper limb recovery, rehabilitation of gait in incomplete spinal cord injury, and cognitive-motor retraining in neurodegenerative disease. Therapeutic outcome and patient engagement are enhanced through real-time feedback, motivated by learning monitoring-driven AI-based algorithms for adjusting task difficulty. These closed-loop BCI systems have been shown to amplify functional connectivity and motor output over time.

8.4 Clinical Trials and Technological Advancements

Neurocognitive rehabilitation has experienced revolutionary development through the synthesis of AI and BCI technologies. A body of empirical literature, such as randomized controlled trials (RCTs), pilot studies, and real-world case studies, has established the effectiveness of AI–BCI systems for improving cognitive and motor recovery in a wide range of neurological disorders. From TBI and stroke to psychiatric and neurodevelopmental disorders, AI-based BCI treatments are providing new dimensions in adaptive, personalized, and non-invasive therapy. Here, we summarize chosen clinical trials, touch on technological developments, and present an AI-facilitated assistive BMI framework designed to aid persons with special needs.

8.4.1 Stroke Recovery Trials

Stroke-induced motor impairments are a major contributor to long-term disability. Integrating AI with BCIs has proven to stimulate neuroplastic recovery mechanisms through targeted feedback and real-time brain-state decoding.

One pioneering study at the University of Tübingen explored the use of a closed-loop motor imagery BCI, augmented by deep learning classifiers, in chronic stroke patients with hemiparesis. The experimental group participants were given 30 sessions of motor imagery-based BCI training, with EEG signals decoded via convolutional neural networks (CNNs) to provide real-time visual feedback. The BCI group had a 35% larger improvement in the Fugl-Meyer motor scale and higher inter-hemispheric connectivity in fMRI analysis, in favor of cortical reorganization as a neural substrate of recovery.

In another NINDS RCT, researchers evaluated a hybrid AI–BCI system with transcranial direct current stimulation (tDCS). Dynamically, the AI models chose maximum stimulation intensities and targets based on CNN-analyzed EEG signals. The treatment provided better performance in executive function tasks (Stroop test, digit span) and significantly alleviated post-stroke apathy and cognitive fatigue compared to standard rehabilitation.

8.4.2 Traumatic Brain Injury (TBI) Studies

BCI-based neurorehabilitation has picked up steam in the face of heterogeneous deficits related to moderate-to-severe TBI. Adaptive AI systems provide real-time task modulation depending on user engagement metrics, allowing for more accurate and responsive cognitive training. In the Geneva Neurorehabilitation Clinic, a reinforcement learning-based BCI was implemented to customize therapy for TBI veterans. The system dynamically adapted cognitive tasks to individuals using EEG markers of frustration and engagement. Participants demonstrated statistically significant gains on the Wisconsin Card Sorting Test (WCST) and the Trail Making Test (TMT A/B), as well as improved self-assessed cognitive effectiveness and daily living.

8.4.3 Neurodegenerative Disease Monitoring

Early and reliable detection of neurodegenerative disorders such as AD is essential for timely treatment. A long short-term memory (LSTM)-based BCI model capable of longitudinal EEG pattern analysis to identify deviations linked with early cognitive impairment was designed by researchers at Kyoto University. The system, when implemented in a 120-subject cohort, had 92% sensitivity in the identification of mild cognitive impairment (MCI), outperforming standard diagnostic methods both in terms of temporal accuracy and sensitivity.

8.4.4 Pediatric and Psychiatric Applications

In children, AI–BCI systems provide engaging, interactive neurofeedback settings that are essential for maintaining attention and interest, particularly in disorders such as ADHD. A Spanish controlled study included 64 ADHD-diagnosed children who played neurofeedback video games with portable EEG headsets. AI algorithms adjusted dynamic feedback based on the coherence of brainwaves. The AI–BCI group, compared to a control group receiving cognitive behavioral therapy (CBT)-based treatment, demonstrated greater improvement in sustained attention and impulse control, as assessed via continuous performance test (CPT) and Vanderbilt scales.

In psychiatric rehabilitation, a research study by the Walter Reed Army Institute of Research illustrated the usefulness of AI–BCI systems in post-traumatic stress disorder (PTSD). Deep learning algorithms were applied to analyze real-time EEG responses (specifically theta/beta ratios) during virtual reality exposure therapy sessions. As stress levels increased, AI algorithms adjusted the VR scene to modulate exposure intensity. This showed more than a 40% decrease in PTSD symptom severity, illustrating the strength of closed-loop, emotionally adaptive therapy.

8.5 Case Study: A Computational BMI Framework for Assistive Communication

Based on the clinical observations, our group at Centre for Development of Advanced Computing (CDAC) Delhi has pursued a new initiative to design, develop, and deploy a computational BMI-based assistive communication framework for individuals with special needs [9]. The system fills the gap between fundamental research in AI–BCI integration and translational neurocognitive computing to provide real-world medical applications with the goal of enhancing quality of life. The framework for this assistive communication via BMI is shown in Figure 8.4.

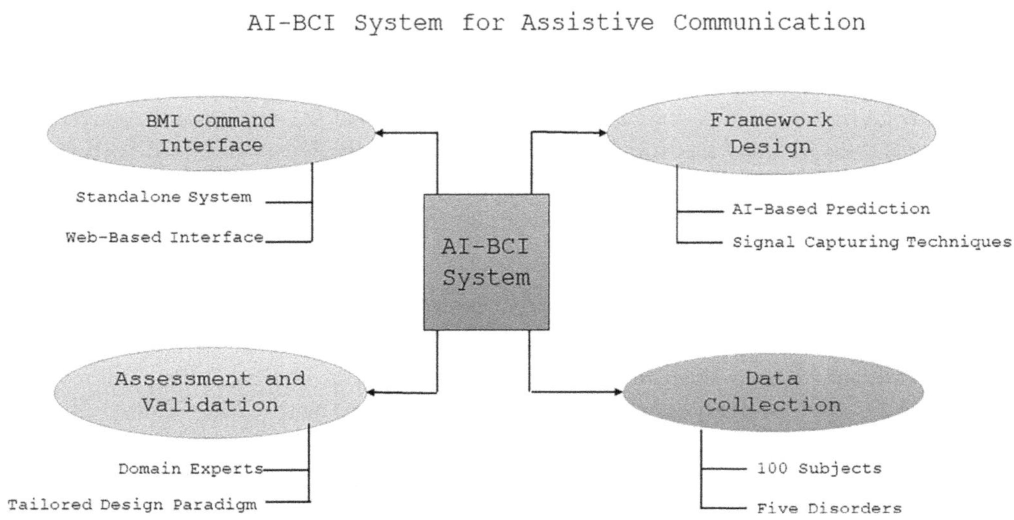

Figure 8.4 AI–BCI System for Assistive Communication.

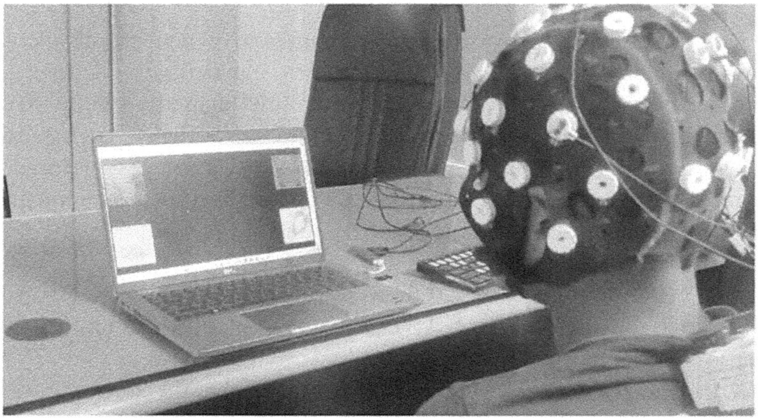

Figure 8.5 The BMI as a Standalone Desktop Application with Web-Based Platform.

8.5.1 System Objectives and Features

1. Dual-interface availability: The BMI interface is being created as a standalone desktop application as well as a web-based platform to maintain device and geography accessibility, as shown in Figure 8.5.
2. Audio-integrated assistance: A standalone audio-augmented interface allows users to communicate simple needs like food, posture correction, or toileting needs through neural commands.
3. Speller and web browser: It has a P300-based oddball paradigm, with an English and Devanagari multi-lingual speller together with a BCI-supported web browser to enable digital communication and browsing.

4. Alert and emergency response: A real-time BMI-activated alert system is being implemented that recognizes critical user states and provides automatic alerts to caregivers in emergency situations.
5. Evaluation and validation: An extensive training and validation module is being designed, including workshops and controlled trials with the help of clinical specialists. Evaluation has been done on a well-characterized cohort chosen on particular neurodevelopmental and neuromotor criteria.

8.5.2 Technological Scope and Implementation

- The design is founded on a modular architecture, well designed to have high cohesion and low coupling among system entities.
- Advanced AI methods, such as supervised learning, deep learning (CNNs, LSTMs), and signal classification models, are employed for user state prediction and intent detection.
- Real-time EEG signal acquisition and processing pipelines have been built, including artifact removal, feature extraction, and classification phases.
- A web and mobile-enabled dashboard is being built for clinicians and caregivers to view patient activity, usage patterns, and system performance.
- Dataset is being gathered in collaboration with AIIMS Delhi, aiming at a sample size of 100 individuals in five neurological and developmental disorders, to create a comprehensive BMI–EEG dataset for training and testing purposes.

8.6 Opportunities, Challenges, and Ethical Considerations

The merger of AI and BCI technologies holds the potential to become a frontier in neurocognitive rehabilitation, given the possibility of revolutionizing how cognitive and motor impairments are diagnosed, monitored, and treated. This consolidation provides a platform for adaptive, personalized, and continuous care models, particularly for populations with restricted communicative or motor abilities. Nevertheless, the use of AI–BCI systems in hospital and home environments is surrounded by technological, regulatory, and ethical issues that require a multidisciplinary and human-centered style of innovation.

8.6.1 Opportunities

8.6.1.1 Personalized and Adaptive Therapy

One of the strongest benefits of AI–BCI systems is their capability to offer real-time adaptive, personalized, and responsive therapy according to the user's state of mind and emotions. Reinforcement learning and deep neural networks are capable of adjusting task difficulty, duration, and type of feedback in accordance with user interest, fatigue, or frustration, ensuring maximum therapeutic results and minimizing dropouts. This type of closed-loop neurofeedback allows not only reactive but also anticipatory adjustments, adapting the rehabilitation process according to individual trajectories.

8.6.1.2 Remote and Home-Based Neurorehabilitation

Miniaturization of EEG devices and the widespread availability of cloud-based AI architectures have driven toward the decentralized neurorehabilitation paradigms. Products like Brain Products, Axonet, Emotiv Epoc, and Muse enable the use of BCI at home, opening the door for therapy among those in remote or underprivileged locations. These technologies enable clinicians to monitor remotely the brain's activity, adapt therapeutic protocols in real-time, and provide continuity of care, especially crucial during pandemics or for mobility-restricted patients.

8.6.1.3 Early Detection and Prognosis

AI algorithms based on multimodal data, such as EEG, fNIRS, and behavioral measures, can be useful tools for the early diagnosis of neurodegenerative diseases. Support vector machines (SVMs) and LSTM networks have been shown to have high sensitivity in detecting preclinical markers of AD, Parkinson's disease, and post-traumatic cognitive impairment. These prediction tools allow clinicians to start interventions at an earlier point in time and possibly retard the progression of the disease.

8.6.1.4 Neuroplasticity Facilitation

Through the provision of neurostimulation (e.g., tDCS, transcranial alternating current stimulation (tACS)) or real-time feedback based on decoded brain states, AI–BCI systems can modulate synaptic strength and enhance activity-dependent neuroplasticity. AI algorithms determine optimal stimulation parameters (location, intensity, timing) for an individual, making Hebbian and spike-timing-dependent plasticity (STDP) mechanisms key to recovery after injury more effective.

8.6.1.5 Multimodal Data Integration

Advanced AI fusion techniques enable the combination of multimodal physiological information—like EEG, EMG, eye-tracking, and heart rate variability—to give an integrated picture of a patient's cognitive, affective, and motor states. The holistic data-driven perspective enhances the strength and specificity of neurocognitive interventions so that clinicians can transcend binary classification to continuous, multidimensional profiling.

8.6.2 Challenges

8.6.2.1 Data Quality and Variability

EEG recordings are extremely prone to noise and artifacts, such as muscle activity, eye blinks, and external interference. Inter- and intra-subject variations in brain signal topologies increase the challenge of generalizability. Transfer learning and domain adaptation notwithstanding, AI models tend to suffer from declining performance in out-of-sample populations. Data curation, preprocessing standardization, and artifact rejection are difficult tasks.

8.6.2.2 *Lack of Standardization*

The BCI research community still suffers from a lack of standardization across signal acquisition protocols, preprocessing pipelines, and performance measures. It retards clinical deployment and makes reproducibility challenging. Initiatives such as the IEEE P2731 Standard for Unified Terminology for Brain–Computer Interfaces and the BCI Competition Datasets are aimed at stimulating data and protocol harmonization but remain to be widely adopted.

8.6.2.3 *Clinical Integration Barriers*

While promising laboratory results, clinical translation of AI–BCI systems is limited by a number of factors: the lack of large-scale RCTs, clinician unfamiliarity with the technology, EHR interoperability, and unclear regulatory routes. The U.S. Food and Drug Administration (FDA) and European Medicines Agency (EMA) require rigorous evidence for safety, efficacy, and explainability, which may not be achievable by many prototype systems. Also, there are time constraints and cognitive demands on clinicians, which limit the adoption of AI-based recommendations.

8.6.2.4 *User Fatigue and Engagement*

BCI systems have high attention, calibration, and cognitive endurance requirements for the user and often cause fatigue and impaired performance with repeated use. This is particularly difficult in populations with neurodevelopmental disorders, the aged, or post-acute stroke recovery patients. Gamification, adaptable rest intervals, and dynamic feedback methods are explored for maximizing long-term engagement [37,38].

8.6.2.5 *Scalability and Cost*

Multi-channel EEG caps, amplifiers, and computationally intensive AI models are needed for high-performance BCI systems, which makes them costly and technologically sophisticated. This restricts deployment in low-resource environments or for large-scale population health initiatives. Advances in low-cost dry electrodes, edge computing, and open-source AI platforms (e.g., OpenBCI, BrainFlow) provide hopeful directions but lack clinical proof-of-principle [37–41].

8.6.3 *Ethical Considerations*

8.6.3.1 *Privacy and Data Security*

Neurophysiological signals hold personal information about a person's thoughts, feelings, and cognitive susceptibilities. Capturing, transmitting, and processing such data pose real concerns for privacy, data breaches, and abuse. Solutions such as end-to-end encryption, differential privacy, and federated learning are beginning to emerge, yet compliance with guidelines like General Data Protection Regulation (GDPR) and Health Insurance Portability and Accountability Act (HIPAA) continues to be necessary.

8.6.3.2 Informed Consent and Autonomy

Obtaining informed consent becomes morally challenging in groups with cognitive impairments or communication deficits. Additionally, as autonomous systems get developed based on AI, issues of responsibility and decision-making authority come into play—especially in critical care or emergency situations. Explainable AI (XAI) mechanisms and open design are crucial to maintain human control.

8.6.3.3 Bias and Fairness

AI systems trained on demographically restricted datasets may risk incorporating algorithmic biases leading to unfair or even dangerous outcomes for underrepresented groups. For example, EEG patterns can differ substantially across age groups, genders, and cultures. It is essential to have diverse training cohorts and to use fairness auditing mechanisms to avoid this risk.

8.6.3.4 Psychological Impacts

The ongoing observation and feedback loop of AI–BE systems can transform users' self-conception and agency. Inadequate feedback can cause learned helplessness, performance anxiety, or excessive dependency on external feedback. Ethical design should focus on empowering users, unambiguous goal definition, and psychological resistance in order to prevent harmful impacts.

8.6.3.5 Ethical Neuroenhancement

The line between therapy and cognitive enhancement is becoming ever more blurred. Therapies developed to treat medical conditions can be repurposed to augment attention, memory, or executive function in healthy individuals, creating questions of equity, access, and coercion. Policies will be needed to differentiate therapeutic need from elective augmentation so that responsible innovation in neurotechnology can occur.

8.7 Future Innovations and Global Trends

The area of AI and BCI in neurocognitive rehabilitation is not waiting idle; it is on the cusp of great breakthroughs in the years to come, with contributions from continued research, technological advancements, and growing international cooperation. The future promises intriguing possibilities that may revolutionize the treatment and care of persons with neurocognitive disabilities.

8.7.1 Future Innovations

Future-Generation BCIs: The evolution of more sophisticated BCI systems with increased resolution, better signal quality, and fewer invasive or non-invasive modalities is a primary concern. This encompasses studies on new electrode materials, wireless and wearable BCIs, and implantable devices with longer lifespans and biocompatibility. These developments will make BCI more useful, comfortable, and functional.

Advanced AI: Further advancements in AI, especially deep learning, reinforcement learning, and XAI, will result in more advanced and explainable efficient algorithms for neural signal decoding, personalized treatment, and adaptive support. AI is becoming more powerful, clever, and transparent, allowing us to fully utilize it.

Brain–computer–brain interfaces (BCBIs): The development of BCBIs, which provide direct communication between the brains of two or more people, may create new options for social interaction, collaboration, and even empathy training in neurocognitive rehabilitation. It is an intriguing and potentially revolutionary field of study that may completely change our knowledge of human connection.

Personalized Medicine: The integration of AI and BCI with other data (e.g., genomics, proteomics) will result in an integrated and personalized neurocognitive rehabilitation solution, taking into account a person's unique biological and genetic identity. That is where medicine is going, where the cure is tailor-made at the molecular level.

Regenerative Medicine: If BCI were combined with regenerative medicine strategies, including stem cell therapy and neurotrophic factors, neural repair and regeneration may be stimulated, with the result being greater and more lasting recovery of function. This is not only the promise of being able to treat symptoms but also of restoring lost function (Table 8.1).

8.7.2 *Global Trends*

Increased Investment and Funding: Governments, institutions of research, and private enterprise globally are stepping up investment and funding in AI and BCI research,

Table 8.1 Future Innovations and Global Trends in AI and BCI for Neurocognitive Rehabilitation

Category	*Focus Area*	*Description*
Future innovations	Next-gen BCIs	Higher resolution, wireless, less invasive, and wearable or implantable systems.
	Advanced AI	Use of deep learning, RL, and XAI for smarter, personalized neural decoding.
	BCBIs	Direct brain communication enabling collaborative rehab and social interaction.
	Personalized medicine	AI–BCI with omics data for tailored, biology-driven therapy.
	Regenerative approaches	BCI with stem cell and neurotrophic therapy for neural repair.
Global trends	Funding and investment	Rising global support from public and private sectors.
	International collaboration	Cross-border research and clinical translation efforts.
	Clinical translation	Focus on moving from lab to bedside.
	Ethics and regulation	Development of standards for privacy, safety, and autonomy.
	Patient empowerment	Involving patients in therapy design and decision-making.

supporting innovation and speeding the evolution of new technologies. This international commitment indicates the increasing understanding of the significance of this area.

Global Cooperation: International cooperation among researchers, clinicians, and engineers across the globe is enabling sharing of knowledge, experience, and resources, resulting in more accelerated developments within the field. Science is an international undertaking, and global cooperation is necessary to fuel innovation.

Translation Focus: There is an increasing focus on bridging research findings to clinical practice and creating practical applications that can directly help patients with neurocognitive impairment. The aim is to advance these technologies from the laboratory to the bedside, putting them into actual practice to make a tangible impact in people's lives.

Ethical Rules and Regulations: Global organizations and regulatory agencies are formulating ethical rules and regulations so that AI and BCI are utilized responsibly in healthcare, considering issues of privacy, autonomy, and safety. We need to make sure these technologies are utilized and developed so that they help us live according to our values and respect the rights of the people.

Patient Empowerment: Patient participation in the development and application of neurocognitive rehabilitation treatments is increasingly recognized as important. Empowering patients to be active participants in their own rehabilitation and treatment will be an important direction of future trends. The patient's voice and viewpoint are at the center of this.

By overcoming the challenges, leveraging the opportunities, and following ethical standards, the practice of AI and BCI in neurocognitive rehabilitation can change the lives of millions of people around the globe, restoring severed links, improving cognitive capabilities, and fostering enhanced independence and well-being.

References

1. Anderson, K. D. (2004). Targeting recovery: Priorities of the spinal cord-injured population. *Journal of Neurotrauma*, 21(10), pp. 1371–1383.
2. Bockbrader, M. A., et al. (2018). Brain computer interfaces in rehabilitation medicine. *PM&R*, 10(9S2), pp. S233–S243.
3. Lee, M. B., et al. (2019). Brain–computer interfaces in quadriplegic patients. *Neurosurgical Clinics of North America*, 30(2), pp. 275–281.
4. Kögel, J., et al. (2019). Using brain–computer interfaces: A scoping review of studies employing social research methods. *BMC Medical Ethics*, 20(1), p. 18.
5. Coenen, A., Fine, E., & Zayachkivska, O. (2014). Adolf Beck: A forgotten pioneer in electroencephalography. *Journal of the History of the Neurosciences*, 23(3), pp. 276–286.
6. Zayachkivska, O., Gzhegotsky, M., & Coenen, A. (2012). Impact on electroencephalography of Adolf Beck, a prominent Polish scientist and founder of the Lviv School of Physiology. *International Journal of Psychophysiology*, 85(1), pp. 3–6.
7. Vaughan, T. M., et al. (2003). Brain–computer interface technology: A review of the second international meeting. *IEEE Transactions on Neural Systems and Rehabilitation Engineering*, 11(2), pp. 94–109.
8. Abiri, R., et al. (2019). A comprehensive review of EEG-based brain–computer interface paradigms. *Journal of Neural Engineering*, 16(1), p. 011001.
9. Mishra, R. K., Kushwaha, S., Deep, A., Koria, H., Jain, P., & Jain, N. K. (2024). SSVEP speller implementations across desktop, web, and mobile platforms. In: Proceedings of the 2024 International Conference on Brain Computer Interface & Healthcare Technologies (iCon-BCIHT); 2024; Thiruvananthapuram, India. IEEE, pp. 213–216.

10. Brandman, D. M., Cash, S. S., & Hochberg, L. R. (2017). Review: Human intracortical recording and neural decoding for brain–computer interfaces. *IEEE Transactions on Neural Systems and Rehabilitation Engineering*, 25(10), pp. 1687–1696.
11. Zapała, D., Francuz, P., Zapała, E., et al. (2018). The impact of different visual feedbacks in user training on motor imagery control in BCI. *Applied Psychophysiology and Biofeedback*, 43(1), pp. 23–35.
12. Jeunet, C., Lotte, F., Batail, J.-M., Philip, P., & Franchi, J.-A. M. (2018). Using recent BCI literature to deepen our understanding of clinical neurofeedback: A short review. *Neuroscience*, 378, pp. 225–233.
13. Schreuder, M., Blankertz, B., & Tangermann, M. (2010). A new auditory multi-class brain–computer interface paradigm: Spatial hearing as an informative cue. *PLoS One*, 5(4).
14. Chatterjee, A., Aggarwal, V., Ramos, A., Acharya, S., & Thakor, N. V. (2007). A brain–computer interface with vibrotactile biofeedback for haptic information. *Journal of NeuroEngineering and Rehabilitation*, 4(1), p. 40.
15. Lukoyanov, M., Gordleeva, S. Y., Pimashkin, A., et al. (2018). The efficiency of the brain–computer interfaces based on motor imagery with tactile and visual feedback. *Human Physiology*, 44(3), pp. 280–288.
16. Sitaram, R., Caria, A., Veit, R., et al. (2007). FMRI brain–computer interface: A tool for neuroscientific research and treatment. *Computational Intelligence and Neuroscience*, 2007, p. 25487.
17. Coyle, S., Ward, T., Markham, C., & McDarby, G. (2004). On the suitability of near-infrared (NIR) systems for next-generation brain–computer interfaces. *Physiological Measurement*, 25(4), pp. 815–822.
18. Mellinger, J., Schalk, G., Braun, C., et al. (2007). An MEG-based brain–computer interface (BCI). *NeuroImage*, 36(3), pp. 581–593.
19. Chiarelli, A. M., Croce, P., Merla, A., & Zappasodi, F. (2018). Deep learning for hybrid EEG-fNIRS brain–computer interface: Application to motor imagery classification. *Journal of Neural Engineering*, 15(3), p. 036028.
20. Schwemmer, M. A., Skomrock, N. D., Sederberg, P. B., et al. (2018). Meeting brain–computer interface user performance expectations using a deep neural network decoding framework. *Nature Medicine*, 24(11), pp. 1669–1676.
21. Pandarinath, C., Nuyujukian, P., Blabe, C. H., et al. (2017). High performance communication by people with paralysis using an intracortical brain–computer interface. *eLife*, 6, p. e18554.
22. Soekadar, S. R., Witkowski, M., Vitiello, N., & Birbaumer, N. (2015). An EEG/EOG-based hybrid brain–neural computer interaction (BNCI) system to control an exoskeleton for the paralyzed hand. *Biomedical Engineering/Biomedizinische Technik*, 60(3), pp. 199–205.
23. Bundy, D. T., Souders, L., Baranyai, K., et al. (2017). Contralesional brain–computer interface control of a powered exoskeleton for motor recovery in chronic stroke survivors. *Stroke*, 48(7), pp. 1908–1915.
24. Frolov, A. A., Mokienko, O., Lyukmanov, R., et al. (2017). Post-stroke rehabilitation training with a motor-imagery-based brain–computer interface (BCI)-controlled hand exoskeleton: A randomized controlled multicenter trial. *Frontiers in Neuroscience*, 11, p. 400.
25. Mohebbi, A., Engelsholm, S. K., Puthusserypady, S., Kjaer, T. W., Thomsen, C. E., & Sørensen, H. B. (2015). A brain–computer interface for robust wheelchair control application based on pseudorandom code modulated visual evoked potential. In: Proceedings of the Annual International Conference of the IEEE Engineering in Medicine and Biology Society, pp. 602–605.
26. Abiyev, R. H., Akkaya, N., Aytac, E., Gunsel, I., & Cagman, A. (2016). Brain–computer interface for control of wheelchair using fuzzy neural networks. *BioMed Research International*, 2016, p. 9359868.
27. Yu, Y., Zhou, Z., Liu, Y., et al. (2017). Self-paced operation of a wheelchair based on a hybrid brain–computer interface combining motor imagery and P300 potential. *IEEE Transactions on Neural Systems and Rehabilitation Engineering*, 25(12), pp. 2516–2526.
28. Salisbury, D. B., Dahdah, M., Driver, S., Parsons, T. D., & Richter, K. M. (2016). Virtual reality and brain–computer interface in neurorehabilitation. *Proceedings (Baylor University Medical Center)*, 29(2), pp. 124–127.

29. Coogan, C. G., & He, B. (2018). Brain–computer interface control in a virtual reality environment and applications for the internet of things. *IEEE Access*, 6, pp. 10840–10849.
30. Johnson, N. N., Carey, J., Edelman, B. J., et al. (2018). Combined rTMS and virtual reality brain–computer interface training for motor recovery after stroke. *Journal of Neural Engineering*, 15(1), p. 016009.
31. Anumanchipalli, G. K., Chartier, J., & Chang, E. F. (2019). Speech synthesis from neural decoding of spoken sentences. *Nature*, 568(7753), pp. 493–498.
32. Moghimi, S., Kushki, A., Guerguerian, A. M., & Chau, T. (2013). A review of EEG-based brain–computer interfaces as access pathways for individuals with severe disabilities. *Assistive Technology*, 25(2), pp. 99–110.
33. Mihara, M., Hattori, N., Hatakenaka, M., et al. (2013). Near-infrared spectroscopy-mediated neurofeedback enhances efficacy of motor imagery-based training in poststroke victims: A pilot study. *Stroke*, 44(4), pp. 1091–1098.
34. Ramos-Murguialday, A., Broetz, D., Rea, M., et al. (2013). Brain–machine interface in chronic stroke rehabilitation: A controlled study. *Annals of Neurology*, 74(1), pp. 100–108.
35. Ietswaart, M., Johnston, M., Dijkerman, H. C., et al. (2011). Mental practice with motor imagery in stroke recovery: Randomized controlled trial of efficacy. *Brain*, 134(5), pp. 1373–1386.
36. Wilson, R. D., Bryden, A. M., Kilgore, K. L., et al. (2019). Neuromodulation for functional electrical stimulation. *Physical Medicine and Rehabilitation Clinics of North America*, 30(2), pp. 301–318.
37. Chaudhary, U., Birbaumer, N., & Ramos-Murguialday, A. (2016). Brain–computer interfaces for communication and rehabilitation. *Nature Reviews Neurology*, 12(9), pp. 513–525.
38. Alzahrani, S. I. (2021). Implementation of P300 based BCI using a consumer-grade EEG Neuroheadset. In: Proceedings of the 2021 IEEE National Biomedical Engineering Conference (NBEC). IEEE.
39. Białas, P., & Milanowski, P. (2014). A high frequency steady-state visually evoked potential based brain computer interface using consumer-grade EEG headset. In: Proceedings of the 36th Annual International Conference of the IEEE Engineering in Medicine and Biology Society (EMBC 2014). IEEE.
40. Shinde, N., & George, K. (2016). Brain-controlled driving aid for electric wheelchairs. In: Proceedings of the 13th International Conference on Wearable and Implantable Body Sensor Networks (BSN 2016). IEEE.
41. Ianoşi-Andreeva-Dimitrova, A., & Mândru, D. S. (2018, 17–20 October). Experimental study regarding the performance of a motor-imagery brain–computer interface across different electrodes placement. In: 6th International Conference on Advancements of Medicine and Health Care through Technology.

9

EEG NEUROFEEDBACK

The Brain's Capacity for Change and Growth

Cathlyn Niranjana Bennett and Arunkumar Sekar

Chapter-Learning Objectives

LO1 Describe the fundamental aspects of electroencephalography (EEG) neurofeedback and its procedure.
LO2 Identify and describe different EEG wave patterns and their relevance to EEG neurofeedback.
LO3 Contrast and describe various clinical and non-clinical applications of EEG neurofeedback.
LO4 Identify and describe markers of neuroplasticity to determine the mechanism of EEG neurofeedback.
LO5 Examine critical issues in the practice and research of EEG neurofeedback.

EEG Neurofeedback Training: A Brief History

Some of the most significant findings of science have been ascribed to a researcher's willingness to look beyond the prescribed nature of a study and their ability to see patterns that transcend the foundational objectives of a study. The narrative of neurofeedback begins with such a researcher, a cat, or, more specifically, a cohort of cats and, interestingly enough, one of the world's leading space agencies, 'The National Aeronautics and Space Administration,' USA, more popularly referred to as NASA. In 1968, Barry Sterman, a sleep researcher, induced sensorimotor cortex electrical activity in cats through instrumental conditioning (Wyrwicka & Sterman, 1968). Instrumental conditioning or operant conditioning is a learning method by which behavior is modified using the consequences of rewards and punishments and is found to be a foundational principle in neurofeedback. In an unrelated study, Sterman conducted research for NASA to explore the effects of the rocket fuel monomethyl hydrazine (Demos, 2005). Among the 50 cats selected for the study, Sterman found that 10 of the cats were seizure resistant, while the others developed seizures within 1 hour of being injected (Demos, 2005).

DOI: 10.4324/9781003646662-11

Meanwhile, in the 1960s, Joe Kamiya successfully trained a human participant to enter an alpha state (a state typically associated with relaxation) through verbal reinforcement (Kamiya, 1969). This finding, coupled with Sterman's results, opened a realm of possibilities for the therapeutic use of EEG. Until then, EEG was primarily used as a diagnostic tool. Kamiya's and Sterman's work showed that it was now possible to bring brain activity under voluntary control and that this capacity could be used for therapeutic purposes.

Electroencephalography: A Brief Introduction

EEG is a procedure by which the electrical activity of the brain is recorded. EEG is recorded by placing metal electrodes on the scalp. These electrodes are placed using the 10–20 international system of electrode placement, which is commonly referred to as the 10–20 system. The 10–20 system is a standardized measure designed to represent different locations on the human cerebral cortex.

Standard Band Frequencies

The EEG is represented as a waveform and is classified based on the frequency of the wave. The typical features of an EEG were first characterized by Hans Berger, which includes rhythmic sinusoidal waveforms ranging from 1 to 60 oscillations per second (Duffy et al., 1989). Disruptions in the EEG and abnormal EEG activity are associated with neurological disorders such as epilepsy and are useful for diagnosis (Duffy et al., 1989; Pillai & Sperling, 2006).

Delta wave. Delta activity ranges from 1 to 4 Hz, 1 to 3.5 Hz, or 0.1 to <4 Hz (Demos, 2005; Mecarelli, 2019; Niedermeyer, 2005). Delta is a low-frequency wave generally associated with sleep and, therefore, commonly seen in infants (Ktonas et al., 1995). Abnormalities in delta wave activity have been seen in traumatic brain injury and sleep deprivation (Demos, 2019; Modarres et al., 2016).

Theta wave. Theta waves range from 4 to 7 Hz, 4 to 7.5 Hz, or 4 to 8 Hz (Demos, 2005; Mecarelli, 2019; Niedermeyer, 2005). Theta activity is associated with intuitive creativity and is typically associated with the 'twilight' phase. This phase between wakefulness and sleep is often associated with extraordinary levels of creativity and insight. Abnormal production of the theta wave in the frontoparietal regions is associated with distractibility and inattention (Heinrich et al., 2014).

Alpha wave. The alpha wave is found between the 8 and 12 Hz or 8 and 13 Hz range (Demos, 2005; Mecarelli, 2019; Niedermeyer, 2005). The alpha wave is associated with relaxation. Alpha activity is usually more prominent in the posterior regions of the brain. Alpha asymmetry has been implicated in major depressive disorders (Davidson, 1998; Gotlib, 1998).

Sensorimotor rhythm (SMR). The SMR is a wave morphology that usually occurs between 12 and 15 Hz or 12 and 16 Hz (Demos, 2005). It is also popularly referred to as low beta. Training of SMR has now been implicated in patients with epilepsy and children with attention deficit hyperactivity disorder (ADHD) (Arns et al., 2008; Monderer et al., 2002).

Beta wave. Beta wave is a fast high-frequency waveform occurring in the range of 13–21 Hz (Demos, 2005). Beta wave production is associated with analytical and focused thinking

(Demos, 2005). High levels of beta (20–32 Hz) have been useful in working on peak performance (Demos, 2005). However, dysregulation of activity in the beta and high beta ranges has been implicated in insomnia, anxiety, and obsessive-compulsive disorder (Pavlenko et al., 2009; Perlis et al., 2001; Velikova et al., 2010).

Gamma wave. The gamma wave is typically found in the 38–42 Hz range, and synchronous bursts of 40 Hz are seen during problem-solving tasks (Demos, 2005; Rosen & Reiner, 2017). Aberrations in gamma oscillations have been recorded in several cognitive disorders such as Alzheimer's disease (Mably & Colgin, 2018).

The Procedure of Neurofeedback Training

It is important to note that EEG neurofeedback equipment typically involves non-invasive methods of recording data through electrodes. The most common protocols involve single- or two-channel training; however, with the advent of low-resolution electromagnetic tomographic analysis (LORETA) neurofeedback, more sophisticated methods of targeting subcortical structures are being explored. Baseline assessments using quantitative EEG and neuropsychological tests are strongly recommended before the onset of training sessions. Informed consent is taken from clients. The nature of neurofeedback, its process, and possible side effects must be discussed. Based on the assessments, a training protocol is created, and the session plan is designed. During the session, the client is seated comfortably in front of a monitor that provides feedback. This is a typical setup when the feedback is provided in a visual modality. Neurofeedback can also be carried out by auditory feedback. The electrodes are placed using the 10–20 system of classification, following the preparation of the skin with an abrasive gel and conduction paste. A baseline reading is taken by the neurofeedback system, and feedback is provided most commonly through a video game based on the protocol selected. For instance, if an increase SMR protocol is chosen, then the video game will show increase in points only when the SMR targets set by the practitioner are reached. This feedback rewarded by an increase in points, reinforces the brain's SMR activity. The protocol is chosen based on the baseline assessment and can be modified based on the client's progress. The client's performance is monitored through the session and between sessions. Regular assessments of client functioning can be carried out to determine the effectiveness of the training program.

Applications of Neurofeedback Training

Neurofeedback training has a wide range of applications. It has been used successfully in non-clinical as well as clinical settings. Although research on neurofeedback is relatively more established in clinical conditions such as ADHD, more rigorous research is necessary in other domains to establish its efficacy. The presentation of several anecdotal findings remains one of the most significant criticisms of neurofeedback research. However, it is becoming an important area of research, as neurofeedback is non-invasive and provides benefits through the simple mechanism of self-regulation.

Clinical Applications of Neurofeedback

ADHD. ADHD is a neurodevelopmental disorder evidenced by a marked disruption in the child's ability to focus attention and/or restless overactivity and impulsivity

(American Psychiatric Association, 2013; Spetie & Arnold, 2018). The neurofeedback protocols commonly used in ADHD are considered to be 'Efficacious and Specific, Level V,' according to the guidelines of the Association for Applied Psychophysiology and Biofeedback (Arns et al., 2009; Enriquez-Geppert et al., 2019). More specifically, studies on neurofeedback demonstrate its positive impact on inattention, impulsivity, and hyperactivity (Arns et al., 2009; Hodgson et al., 2014; Shereena et al., 2019).

Traumatic brain injury. Traumatic brain injury is an insult to the brain due to non-degenerative and non-congenital causes, typically because of external physical trauma to the brain. Traumatic brain injury is a condition that results in an array of cognitive deficits. Deficits may range from basic processes of attention to higher cognitive functions such as executive functions. Some cognitive deficits include impairment in processing speed, category fluency, working memory, set-shifting ability, concept formation, response inhibition, learning, and memory. Deficits in attention and difficulty in dual-task demands have been observed even with patients with mild traumatic brain injury (Cicerone, 1996). Working memory deficits have been reported in individuals with traumatic brain injury using other dual-task paradigms (McDowell et al., 1997). Significant differences in frontal lobe executive functions have been seen in individuals with traumatic brain injury compared to control subjects (Brooks et al., 1999). There is some support for the improvement of cognitive functions following neurofeedback training (Bennett et al., 2013; Rajeswaran et al., 2013). Patients who have received neurofeedback training are likely to report improved quality of life (Bennett et al., 2017; Rajeswaran & Bennett, 2019; Reddy et al., 2014). Neurofeedback is also associated with the improvement of post-concussion symptoms (Bennett et al., 2018; Duff, 2004).

Epilepsy. Epilepsy is a clinical neurological condition characterized by recurring seizures (Kuks & Snoek, 2018). A seizure is an abnormal discharge of electrical activity that presents with typical features in an EEG. Neurofeedback can potentially reduce seizure frequency and/or the medication required to control them (Legarda et al., 2011; Monderer et al., 2002; Walker & Kozlowski, 2005).

Substance use disorders (SUD). SUD is another area that has been commonly explored by researchers of neurofeedback. As per the previous version of the DSM – Diagnostic and Statistical Manual of Mental Disorders-IV (DSM-IV), the terms 'abuse' and 'dependence' were separated, with abuse referring to hazardous use and dependence to withdrawal and tolerance (Hasin et al., 2013). However, in its most current version, the DSM-5 describes a continuum-based conceptualization of SUD combining substance abuse disorder and substance dependence disorder (Hasin et al., 2013). SUD is associated with disturbances in the reward pathway of the brain and the executive cognitive functions in the form of inhibitory control (Clay et al., 2008; Rajeswaran & Bennett, 2018).

Neurofeedback in SUDs can be traced back to the work of Peniston and Kulkosky in 1989. Alpha–theta neurofeedback was successfully used in chronic alcoholism with positive implications for beta-endorphin levels and relapse rates (Peniston & Kulkosky, 1989). In a conceptual review paper, Sokhadze et al. (2007) proposes an integration of neurofeedback with cognitive behavioral treatment for drug addiction with comorbid post-traumatic stress disorder (Sokhadze et al., 2007). Some studies indicate the use of SMR training and alpha–theta training (Sokhadze et al., 2007). Neurofeedback using a similar protocol has been found to be successful in reducing clinical symptoms of patients with opiate dependence disorder (Dehghani-Arani et al., 2013). Decreased levels of depression and stress have been reported in cocaine users following neurofeedback as well (Horrell et al., 2010). A recent trend indicates the use of low-resolution electromagnetic tomographic neurofeedback as a treatment protocol in drug addiction. The training was given in a single case in the right anterior cingulate cortex and had promising results (Cannon et al., 2008).

Anxiety disorders. Anxiety disorders fall on the pathological end of the spectrum of anxiety and encompass a wide range of disorders, including panic disorder, agoraphobia, specific phobias, social anxiety disorder, and generalized anxiety disorders (Sadock et al., 2015). The symptoms of anxiety include characteristic symptoms of the autonomic nervous system, including palpitations, perspiration, dizziness, and upset stomach, and are accompanied by an unpleasant sense of apprehension (Sadock et al., 2015). Neurofeedback appears to be a promising method of treatment for anxiety disorders and other affective disorders (Hammond, 2005). It has proven to be effective in symptom reduction among children and adults (Huang-Storms et al., 2007; Mennella et al., 2017).

Mood disorders. Depression is characterized by low mood, alterations in sleep, energy, appetite, and suicidal ideation (Sadock et al., 2015). EEG neurofeedback has been associated with reduced symptoms of depression post-intervention (Lee et al., 2019). The alpha asymmetry protocol has been found to be traditionally effective (Baehr et al., 2001). An integration of neurofeedback with photic stimulation targeting primarily left beta training has also been found to be promising (Hammond, 2000).

Post-traumatic stress disorder. Post-traumatic stress disorder is a condition characterized by an abnormally elevated stress response to a traumatic event associated with high levels of anxiety, flashback memories, nightmares, negative mood, and physical symptoms such as increased heart rate, blood pressure, and hyperventilation. Studies indicate that EEG neurofeedback helps reduce symptom severity in individuals diagnosed with post-traumatic stress disorder (Nicholson et al., 2016; van der Kolk et al., 2016). One case report reported that a male veteran with post-traumatic stress disorder and major depressive disorder also suffered from comorbid hemicrania continua (headache). The psychiatric symptoms were not responsive to medication. However, when the patient's hemicrania continua was treated, it was found that the patient's psychiatric symptoms also subsided. This highlights the importance of treating physical comorbidities, which may aggravate psychiatric conditions, such as post-traumatic stress disorder (Kohrt & Duncan, 2012).

Autism spectrum disorders. Isolated studies in autism spectrum disorders are rare, as this condition frequently occurs with comorbidities, such as intellectual disability, ADHD, and, in some cases, mood-related symptoms (Matson & Nebel-Schwalm, 2007). EEG neurofeedback has been successful in producing EEG alterations in autism spectrum disorders (Wang et al., 2016). Improvements have been reported in terms of communication, sociability, cognitive flexibility, and sustained attention (Hurt et al., 2014).

Non-Clinical Applications of Neurofeedback

Although much of the work in EEG neurofeedback has focused on clinical application, there has been some interest in developing its application for non-clinical purposes as well. Two key domains have been identified as areas of interest. The first is in terms of peak performance, particularly in music and sports, and the second in terms of relaxation enhancement.

The two are associated, as performance is often impeded by anxiety, but the use of EEG neurofeedback for the sole purpose of relaxation is imminent.

Performance in music. EEG neurofeedback training is associated with an improvement in music performance. In a pilot study, single-blind raters reported an improvement in music performance among school children, and alpha–theta training was found to be specifically useful (Egner & Gruzelier, 2003; Gruzelier, Foks, et al., 2014). A review study among healthy individuals indicates promise for the use of EEG neurofeedback in enhancing various forms of memory, executive attention, visuospatial ability, mood enhancement, and well-being (Gruzelier, 2014). The combined use of EEG neurofeedback with electromyographic biofeedback among music students has been found to be associated with increased self-regulation and quality of music performance (Markovska-Simoska et al., 2008).

Performance in dance. In a review study by Gruzelier and colleagues, it was found that EEG neurofeedback is associated with increased cognitive creativity. However, there is little support to suggest that it has an effect on performance (Gruzelier, Thompson, et al., 2014). There also appears to be some support that EEG neurofeedback helps reduce performance anxiety, as evidenced by changes in the State-Trait Anxiety Inventory Test (Singer, 2004). Preliminary findings seem to indicate that alpha–theta training is helpful in dance performance, particularly in the domain of 'timing'; however, more research in this area is required (Raymond et al., 2005).

Performance in sports. Olympic gold medalist Abhinav Bindra is considered to be India's pioneering marksman and figurehead for the use of neurofeedback for the enhancement of performance in sports. Bindra used a combination of neurofeedback and biofeedback and went on to win India's first gold medal at the Beijing Olympics in 2008 at the air rifle event. There is limited evidence for the effectiveness of EEG neurofeedback in sports; however, there do appear to be strong indicators of neural correlates of sporting behavior, which provides a basis for an argument in favor of the use of brain-based intervention techniques (Park et al., 2015; Ring et al., 2015; Xiang et al., 2018). Some of the factors that differentiate clinical training and peak performance include the nature of the participant, the goal of training, the design of sessions, and other underlying factors (Wilson & Peper, 2011). There is some evidence supporting the use of an SMR protocol and other personalized QEEG-driven protocols, particularly in improving accuracy in target-based sports such as golf (Arns et al., 2008; Cheng et al., 2015).

Neuroplasticity: The Brain's Capacity for Change and Growth

One of the most fascinating truths about the human brain is its capacity for healing and its ability to make new connections. This capacity is primarily due to a phenomenon referred to as 'neuroplasticity.' Paul Bach-y-Rita was a neuroscientist who has made considerable contributions to this field. It is believed that his work was partially inspired by his father. In the late 1950s, Paul's father, Pedro Bach-y-Rita, sustained a stroke, which resulted in significant neurological damage, including paralysis and language deficits. Through the sustained efforts of Paul's brother George Bach-y-Rita, their father made a significant recovery in three years, went on to teach professionally, and eventually died from a heart attack during a mountain hiking expedition. Interestingly enough, a post-mortem brain autopsy revealed significant damage to the brain stem, implying that the healthy tissue had reorganized itself to take on new functions –neuroplasticity had occurred (Abrams & Winters, 2003; Rutkin, 2013).

Neurofeedback primarily works on the principle of neuroplasticity. Neuroplasticity refers to the brain's capacity to reorganize itself and form new connections. Bach-y-Rita highlights several factors implicated in neural plasticity following a head injury, some of which are enlisted below (Bach-y-Rita, 2003).

Non-synaptic neurotransmission. While synaptic transmission is an essential mechanism by which information is transmitted in neuronal cells, less than half the communication occurs directly through synapses. Therefore, information is transmitted not only by the synapse but also through non-synaptic neurotransmission, which has now come to be called volume transmission. Volume transmission occurs by the process of diffusion of the neurotransmitters through the extracellular fluid. This can occur at any location, which may be far away from the actual target cell receptor (Bach-y-Rita, 2003).

Extracellular space volume fraction. The space between cells filled with fluid is referred to as the extracellular space. More than 20% of brain mass accounts for the extracellular space. The existence of the extracellular space serves many purposes. Some of them include volume transmission as well as extracellular volume fraction reduction. Volume fraction is another prospective non-synaptic information transmission mechanism associated with the threshold changes that occur with its reduction. This reduction is generally seen when neurons swell due to activity. The space between neuronal cells can be reduced by up to 50% when the neuron expands. This, in turn, produces subsequent changes in the concentration of ions and affects the excitability and metabolism of the neuron (Bach-y-Rita, 2003).

Neurotransmitters. Neurotransmitters are chemicals that transmit signals between neurons and other cells in the body. Each neurotransmitter system behaves differently following brain injury. Bach-y-Rita, in his paper, highlights a study on the rat model in which, following transient cerebral ischemia, there is a lack of association between receptor changes in the early recovery period and the development of neuronal necrosis in various hippocampal regions. There are also reports of long-lasting receptor changes in areas considered resistant to an ischemic insult. Some receptors appear to be down-regulated, while others appear to be up-regulated in the surviving cells (Bach-y-Rita, 2003).

Regeneration and neurogenesis. Regeneration and neurogenesis refer to the birth of new neurons. Stem cell studies are underway and likely to hold much promise for future rehabilitation of neurological illnesses (Bach-y-Rita, 2003).

Multiplexing. Neurons have several uses and, therefore, can be engendered to serve various functions. Multiplexing is a phenomenon observed in the pontine brain stem in 1964. This may well prove to be a significant mechanism implicated in neurological rehabilitation (Bach-y-Rita, 2003).

There are several theories of neural plasticity, and the above-mentioned factors may be useful markers of change evidenced in neurofeedback research.

Critical Issues and Concluding Remarks

Neurofeedback works on the principle of neuroplasticity, but its mechanism is far from well researched. While neurofeedback has shown promise as a therapeutic modality, research on this method of intervention has been criticized for a lack of randomized control trials, low sample sizes, a lack of long-term follow-up studies, and a general over-dependence on anecdotal evidence (Banerjee & Argaez, 2017).

Several questions remain unanswered. Evidence indicates that there are electrophysiological changes associated with EEG neurofeedback intervention. It is crucial to understand how these changes translate into symptom reduction, clinical recovery, or improvements in performance. Studies do support structural and biochemical changes following neurofeedback intervention (Bennett et al., 2018; Munivenkatappa et al., 2014; Peniston & Kulkosky, 1989). An important area of study would need to look at markers of neuroplasticity, including neurophysiological, neurochemical, and structural alterations following the training.

There is also a strong need for regulation of licensure and practice requirements. The Biofeedback Certification International Alliance (BCIA) offers certification courses in neurofeedback, and the International Society for Neurofeedback and Research (ISNR) provides guidelines on ethical practice (International Society for Neuroregulation & Research, n.d.). Membership in ISNR and certification from ISNR are voluntary and do not supersede licensure requirements governed by national and state-based government agencies and laws (International Society for Neuroregulation & Research, n.d.). The ISNR has also published a position paper highlighting critical issues such as the competency and qualifications of professionals practicing neurofeedback, the importance of robust assessment techniques – particularly in terms of quantitative EEG baselines – and the need for supervision during practice (Hammond et al., 2011). The paper also defines guidelines for sales of neurofeedback equipment and cautions against false advertising (Hammond et al., 2011).

To conclude, neurofeedback as an intervention tool has much to offer. The applications of neurofeedback are diverse. Clinical research in neurofeedback has gained much attention, but neuro-regulation is breaking ground in performance-based activities such as sports and music as well. Well-controlled research studies are required to determine the effectiveness of this multi-faceted tool and to determine the mechanism by which it works. The brain's capacity for change and growth is remarkable, and neurofeedback could be key to fostering this growth.

Questions for Reflection

1. What are some applications of neurofeedback that still require considerable research?
2. What are some potential applications of neurofeedback that have not been explored?
3. What are some ethical issues to consider in the practice of neurofeedback?

References

Abrams, M., & Winters, D. (2003). *Can You See with Your Tongue?* www.discovermagazine.com/mind/can-you-see-with-your-tongue

American Psychiatric Association. (2013). Diagnostic and statistical manual of mental disorders: DSM-5. http://openurl.uquebec.ca:9003/uqam?url_ver=Z39.88-2004&url_ctx_fmt=info:ofi/fmt:kev:mtx:ctx&ctx_enc=info:ofi/enc:UTF-8&ctx_ver=Z39.88-2004&rfr_id=info:sid/sfxit.com:azlist&sfx.ignore_date_threshold=1&rft.isbn=9780890425572

Arns, M., de Ridder, S., Strehl, U., Breteler, M., & Coenen, A. (2009). Efficacy of Neurofeedback Treatment in ADHD: The Effects on Inattention, Impulsivity and Hyperactivity: A Meta-Analysis. *Clinical EEG and Neuroscience, 40*(3), 180–189. https://doi.org/10.1177/155005940904000311

Arns, M., Kleinnijenhuis, M., Fallahpour, K., & Breteler, R. (2008). Golf Performance Enhancement and Real-Life Neurofeedback Training Using Personalized Event-Locked EEG Profiles. *Journal of Neurotherapy, 11*(4), 11–18. https://doi.org/10.1080/10874200802149656

Bach-y-Rita, P. (2003). Theoretical Basis for Brain Plasticity after a TBI. *Brain Injury, 17*(8), 643–651. https://doi.org/10.1080/0269905031000107133KN3UWU6Y9QWXXKHY [pii].

Baehr, E., Rosenfeld, J. P., & Baehr, R. (2001). Clinical Use of an Alpha Asymmetry Neurofeedback Protocol in the Treatment of Mood Disorders. *Journal of Neurotherapy, 4*(4), 11–18. https://doi.org/10.1300/J184v04n04_03

Banerjee, S., & Argaez, C. (2017). *Neurofeedback and Biofeedback for Mood and Anxiety Disorders: A Review of Clinical Effectiveness and Guidelines.* www.ncbi.nlm.nih.gov/pubmed/30299634

Bennett, C. N., Gupta, R. K., Prabhakar, P., Christopher, R., Sampath, S., Thennarasu, K., & Rajeswaran, J. (2018). Clinical and Biochemical Outcomes Following EEG Neurofeedback Training in Traumatic Brain Injury in the Context of Spontaneous Recovery. *Clinical EEG and Neuroscience, 49*(6), 433–440. https://doi.org/10.1177/1550059417744899

Bennett, C. N., Rajeswaran, J., Sampath, S., & Christopher, R. (2013). The Right to Write: EEG Neurofeedback Training in Frontal Lobe Agraphia—A Case Report. *Journal of Neurotherapy, 17*(3), 162–165. https://doi.org/10.1080/10874208.2013.813174

Bennett, C. N., Sampath, S., Christopher, R., Thennarasu, K., & Rajeswaran, J. (2017). Effect of Electroencephalogram Neurofeedback Training on Quality of Life in Patients with Traumatic Brain Injury: In Context of Spontaneous Recovery. *Indian Journal of Neurotrauma, 14*(02/03), 129–134. https://doi.org/10.1055/s-0038-1649280

Brooks, J., Fos, L. A., Greve, K. W., & Hammond, J. S. (1999). Assessment of Executive Function in Patients with Mild Traumatic Brain Injury. *Journal of Trauma, 46*(1), 159–163. www.ncbi.nlm.nih.gov/entrez/query.fcgi?cmd=Retrieve&db=PubMed&dopt=Citation&list_uids=9932700

Cannon, R., Lubar, J., Sokhadze, E., & Baldwin, D. (2008). LORETA Neurofeedback for Addiction and the Possible Neurophysiology of Psychological Processes Influenced: A Case Study and Region of Interest Analysis of LORETA Neurofeedback in Right Anterior Cingulate Cortex. *Journal of Neurotherapy, 12*(4), 227–241. https://doi.org/10.1080/10874200802501948

Cheng, M. Y., Huang, C. J., Chang, Y. K., Koester, D., Schack, T., & Hung, T. M. (2015). Sensorimotor Rhythm Neurofeedback Enhances Golf Putting Performance. *Journal of Sport and Exercise Psychology, 37*(6), 626–636. https://doi.org/10.1123/jsep.2015-0166

Cicerone, K. D. (1996). Attention Deficits and Dual Task Demands after Mild Traumatic Brain Injury. *Brain Injury, 10*(2), 79–90. https://doi.org/doi:10.1080/026990596124566

Clay, S. W., Allen, J., & Parran, T. (2008). A Review of Addiction. *Postgraduate Medicine, 120*(2), E01–07. https://doi.org/10.3810/pgm.2008.07.1802

Davidson, R. J. (1998). Anterior Electrophysiological Asymmetries, Emotion, and Depression: Conceptual and Methodological Conundrums. *Psychophysiology, 35*(5), 607–614. https://doi.org/10.1017/s0048577298000134

Dehghani-Arani, F., Rostami, R., & Nadali, H. (2013). Neurofeedback Training for Opiate Addiction: Improvement of Mental Health and Craving. *Association for Applied Psychophysiology and Biofeedback, 38*(2), 133–141. https://doi.org/10.1007/s10484-013-9218-5

Demos, J. N. (2005). *Getting Started with Neurofeedback*. W. W. Norton & Company.

Demos, J. N. (2019). *Getting Started with EEG Neurofeedback (Second Edition)* (2nd edition). W. W. Norton & Company.

Duff, J. (2004). The Usefulness of Quantitative EEG (QEEG) and Neurotherapy in the Assessment and Treatment of Post-Concussion Syndrome. *Clinical EEG and Neuroscience, 35*(4), 198–209. https://doi.org/10.1177/155005940403500410

Duffy, F. H., Iyer, V. G., & Surwillo, W. W. (1989). *Clinical Electroencephalography and Topographic Brain Mapping: Technology & Practice*. Springer-Verlag.

Egner, T., & Gruzelier, J. H. (2003). Ecological Validity of Neurofeedback: Modulation of Slow Wave EEG Enhances Musical Performance. *NeuroReport, 14*(9). https://journals.lww.com/neuroreport/Fulltext/2003/07010/Ecological_validity_of_neurofeedback___modulation.6.aspx

Enriquez-Geppert, S., Smit, D., Pimenta, M. G., & Arns, M. (2019). Neurofeedback as a Treatment Intervention in ADHD: Current Evidence and Practice. *Current Psychiatry Reports, 21*(6), 46. https://doi.org/10.1007/s11920-019-1021-4

Gotlib, I. H. (1998). EEG Alpha Asymmetry, Depression, and Cognitive Functioning. *Cognition and Emotion, 12*(3), 449–478. https://doi.org/10.1080/026999398379673

Gruzelier, J. H. (2014). EEG-Neurofeedback for Optimising Performance. I: A Review of Cognitive and Affective Outcome in Healthy Participants. *Neuroscience & Biobehavioral Reviews, 44*, 124–141. https://doi.org/https://doi.org/10.1016/j.neubiorev.2013.09.015

Gruzelier, J. H., Foks, M., Steffert, T., Chen, M. J. L., & Ros, T. (2014). Beneficial Outcome from EEG-Neurofeedback on Creative Music Performance, Attention and Well-Being in School Children. *Biological Psychology, 95*, 86–95. https://doi.org/https://doi.org/10.1016/j.biopsycho.2013.04.005

Gruzelier, J. H., Thompson, T., Redding, E., Brandt, R., & Steffert, T. (2014). Application of Alpha/Theta Neurofeedback and Heart Rate Variability Training to Young Contemporary Dancers: State Anxiety and Creativity. *International Journal of Psychophysiology, 93*(1), 105–111. https://doi.org/10.1016/j.ijpsycho.2013.05.004

Hammond, D. C. (2000). Neurofeedback Treatment of Depression with the Roshi. *Journal of Neurotherapy, 4*(2), 45–56. https://doi.org/10.1300/J184v04n02_06

Hammond, D. C. (2005). Neurofeedback with Anxiety and Affective Disorders. *Child and Adolescent Psychiatric Clinics of North America, 14*(1), 105–123. https://doi.org/10.1016/j.chc.2004.07.008

Hammond, D. C., Bodenhamer-Davis, G., Gluck, G., Stokes, D., Harper, S. H., Trudeau, D., MacDonald, M., Lunt, J., & Kirk, L. (2011). Standards of Practice for Neurofeedback and Neurotherapy: A Position Paper of the International Society for Neurofeedback & Research. *Journal of Neurotherapy, 15*(1), 54–64. https://doi.org/10.1080/10874208.2010.545760

Hasin, D. S., O'Brien, C. P., Auriacombe, M., Borges, G., Bucholz, K., Budney, A., Compton, W. M., Crowley, T., Ling, W., Petry, N. M., Schuckit, M., & Grant, B. F. (2013). DSM-5 Criteria for Substance Use Disorders: Recommendations and Rationale. *American Journal of Psychiatry, 170*(8), 834–851. https://doi.org/10.1176/appi.ajp.2013.12060782

Heinrich, H., Busch, K., Studer, P., Erbe, K., Moll, G. H., & Kratz, O. (2014). EEG Spectral Analysis of Attention in ADHD: Implications for Neurofeedback Training? *Frontiers in Human Neuroscience, 8*, 611–611. https://doi.org/10.3389/fnhum.2014.00611

Hodgson, K., Hutchinson, A. D., & Denson, L. (2014). Nonpharmacological Treatments for ADHD: A Meta-Analytic Review. *Journal of attention disorders, 18*(4), 275–282. https://doi.org/10.1177/1087054712444732

Horrell, T., El-Baz, A., Baruth, J., Tasman, A., Sokhadze, G., Stewart, C., & Sokhadze, E. (2010). Neurofeedback Effects on Evoked and Induced EEG Gamma Band Reactivity to Drug-Related Cues in Cocaine Addiction. *Journal of Neurotherapy, 14*(3), 195–216. https://doi.org/10.1080/10874208.2010.501498

Huang-Storms, L., Bodenhamer-Davis, E., Davis, R., & Dunn, J. (2007). QEEG-Guided Neurofeedback for Children with Histories of Abuse and Neglect: Neurodevelopmental Rationale and Pilot Study. *Journal of Neurotherapy, 10*(4), 3–16. https://doi.org/10.1300/J184v10n04_02

Hurt, E., Arnold, L. E., & Lofthouse, N. (2014). Quantitative EEG Neurofeedback for the Treatment of Pediatric Attention-Deficit/Hyperactivity Disorder, Autism Spectrum Disorders, Learning Disorders, and Epilepsy. *Child and Adolescent Psychiatric Clinics, 23*(3), 465–486. https://doi.org/10.1016/j.chc.2014.02.001

International Society for Neuroregulation & Research. (n.d.). Guidelines for Practice. https://isnr.org/guidelines-for-practice

Kamiya, J. (1969). Operant Control of the EEG Alpha Rhythm and Some of Its Reported Effects on Consciousness. In C. Tart (Ed.), *Altered States of Consciousness* (pp. 489–501). John Wiley and Sons.

Kohrt, B. A., & Duncan, E. (2012). Hemicrania Continua Headache in a Veteran with Posttraumatic Stress Disorder and Major Depressive Disorder without Traumatic Brain Injury. *Case Reports in Psychiatry, 2012*, 937217. https://doi.org/10.1155/2012/937217

Ktonas, P. Y., Fagioli, I., & Salzarulo, P. (1995). Delta (0.5–1.5 Hz) and Sigma (11.5–15.5 Hz) EEG Power Dynamics throughout Quiet Sleep in Infants. *Electroencephalography and Clinical Neurophysiology, 95*(2), 90–96. https://doi.org/https://doi.org/10.1016/0013-4694(95)00051-Y

Kuks, J. B. M., & Snoek, J. W. (2018). Textbook of clinical neurology. https://search.ebscohost.com/login.aspx?direct=true&scope=site&db=nlebk&db=nlabk&AN=1875481

Lee, Y. J., Lee, G. W., Seo, W. S., Koo, B. H., Kim, H. G., & Cheon, E. J. (2019). Neurofeedback Treatment on Depressive Symptoms and Functional Recovery in Treatment-Resistant Patients with Major Depressive Disorder: An Open-Label Pilot Study. *Journal of Korean Medical Science, 34*(42), e287. https://doi.org/10.3346/jkms.2019.34.e287

Legarda, S. B., McMahon, D., Othmer, S., & Othmer, S. (2011). Clinical Neurofeedback: Case Studies, Proposed Mechanism, and Implications for Pediatric Neurology Practice. *Journal of Child Neurology, 26*(8), 1045–1051. https://doi.org/10.1177/0883073811405052

Mably, A. J., & Colgin, L. L. (2018). Gamma Oscillations in Cognitive Disorders. *Current Opinion in Neurobiology, 52*, 182–187. https://doi.org/10.1016/j.conb.2018.07.009

Markovska-Simoska, S., Pop-Jordanova, N., & Georgiev, D. (2008). Simultaneous EEG and EMG Biofeedback for Peak Performance in Musicians. *Prilozi, 29 1*, 239–252.

Matson, J. L., & Nebel-Schwalm, M. S. (2007). Comorbid Psychopathology with Autism Spectrum Disorder in Children: An Overview. *Research in Developmental Disabilities, 28*(4), 341–352. https://doi.org/https://doi.org/10.1016/j.ridd.2005.12.004

McDowell, S., Whyte, J., & D'Esposito, M. (1997). Working Memory Impairments in Traumatic Brain Injury: Evidence from a Dual-Task Paradigm. *Neuropsychologia, 35*(10), 1341–1353. https://doi.org/S0028393297000821 [pii].

Mecarelli, O. (2019). Normal Awake Adult EEG. In O. Mecarelli (Ed.), *Clinical Electroencephalography*. Springer. https://search.ebscohost.com/login.aspx?direct=true&scope=site&db=nlebk&db=nlabk&AN=2155409

Mennella, R., Patron, E., & Palomba, D. (2017). Frontal Alpha Asymmetry Neurofeedback for the Reduction of Negative Affect and Anxiety. *Behaviour Research and Therapy, 92*, 32–40. https://doi.org/10.1016/j.brat.2017.02.002

Modarres, M., Kuzma, N. N., Kretzmer, T., Pack, A. I., & Lim, M. M. (2016). EEG Slow Waves in Traumatic Brain Injury: Convergent Findings in Mouse and Man. *Neurobiology of Sleep and Circadian Rhythms, 1*, S2451994416300025. https://pubmed.ncbi.nlm.nih.gov/28018987; www.ncbi.nlm.nih.gov/pmc/articles/PMC5175467/

Monderer, R. S., Harrison, D. M., & Haut, S. R. (2002). Neurofeedback and Epilepsy. *Epilepsy & Behavior, 3*(3), 214–218. https://doi.org/10.1016/S1525-5050(02)00001-X

Munivenkatappa, A., Rajeswaran, J., Indira Devi, B., Bennet, N., & Upadhyay, N. (2014). EEG Neurofeedback Therapy: Can It Attenuate Brain Changes in TBI? *NeuroRehabilitation, 35*, 481–484. https://doi.org/10.3233/NRE-141140

Nicholson, A. A., Ros, T., Frewen, P. A., Densmore, M., Théberge, J., Kluetsch, R. C., Jetly, R., & Lanius, R. A. (2016). Alpha Oscillation Neurofeedback Modulates Amygdala Complex Connectivity and Arousal in Posttraumatic Stress Disorder. *NeuroImage: Clinical, 12*, 506–516. https://doi.org/https://doi.org/10.1016/j.nicl.2016.07.006

Niedermeyer, E. (2005). The Normal EEG of the Waking Adult. In E. Niedermeyer & F. H. Lopes da Silva (Eds.), *Electroencephalography: Basic Principles, Clinical Applications, and Related Fields* (pp. 149–173). Lippincott Williams & Wilkins.

Park, J. L., Fairweather, M. M., & Donaldson, D. I. (2015). Making the Case for Mobile Cognition: EEG and Sports Performance. *Neuroscience & Biobehavioral Reviews, 52*, 117–130. https://doi.org/https://doi.org/10.1016/j.neubiorev.2015.02.014

Pavlenko, V. B., Chernyi, S. V., & Goubkina, D. G. (2009). EEG Correlates of Anxiety and Emotional Stability in Adult Healthy Subjects. *Neurophysiology, 41*(5), 337–345. https://doi.org/10.1007/s11062-010-9111-2

Peniston, E. G., & Kulkosky, P. J. (1989). Alpha-Theta Brainwave Training and Beta-Endorphin Levels in Alcoholics. *Alcohol: Clinical & Experimental Research, 13*(2), 271–279. https://doi.org/10.1111/j.1530-0277.1989.tb00325.x

Perlis, M. L., Merica, H., Smith, M. T., & Giles, D. E. (2001). Beta EEG Activity and Insomnia. *Sleep Medicine Reviews, 5*(5), 365–376. https://doi.org/https://doi.org/10.1053/smrv.2001.0151

Pillai, J., & Sperling, M. R. (2006). Interictal EEG and the Diagnosis of Epilepsy. *Epilepsia, 47*(Suppl 1), 14–22. https://doi.org/10.1111/j.1528-1167.2006.00654.x

Rajeswaran, J., & Bennett, C. N. (2018). Cognitive Rehabilitation in Addictive Disorders. *Indian Journal of Psychiatry, 60*(Suppl 4), S490–S493. https://doi.org/10.4103/psychiatry.IndianJPsychiatry_17_18

Rajeswaran, J., & Bennett, C. N. (2019). Emerging hope: EEG Neurofeedback Training in TBI. *Journal of the Neurological Sciences, 405*, 17. https://doi.org/10.1016/j.jns.2019.10.451

Rajeswaran, J., Bennett, C., Thomas, S., & Rajakumari, K. (2013). EEG Neurofeedback Training in Clinical Conditions. In *Neuropsychological Rehabilitation Principles and Applications* (pp. 57–78). Elsevier.

Raymond, J., Sajid, I., Parkinson, L. A., & Gruzelier, J. H. (2005). Biofeedback and Dance Performance: A Preliminary Investigation. *Applied Psychophysiology and Biofeedback, 30*(1), 65–73. https://doi.org/10.1007/s10484-005-2175-x

Reddy, R. P., Rajeswaran, J., Bhagavatula, I. D., & Kandavel, T. (2014). Silent Epidemic: The Effects of Neurofeedback on Quality-of-Life. *Indian Journal of Psychological Medicine, 36*(1), 40–44. https://doi.org/10.4103/0253-7176.127246

Ring, C., Cooke, A., Kavussanu, M., McIntyre, D., & Masters, R. (2015). Investigating the Efficacy of Neurofeedback Training for Expediting Expertise and Excellence in Sport. *Psychology of Sport and Exercise, 16*, 118–127. https://doi.org/https://doi.org/10.1016/j.psychsport.2014.08.005

Rosen, A., & Reiner, M. (2017). Right Frontal Gamma and Beta Band Enhancement while Solving a Spatial Puzzle with Insight. *International Journal of Psychophysiology, 122*, 50–55. https://doi.org/10.1016/j.ijpsycho.2016.09.008

Rutkin, A. H. (2013, September 1). Champagne for the Blind: Paul Bach-y-Rita, Neuroscience's Forgotten Genius. MIT Comparative Media Studies/Writing. https://cmsw.mit.edu/paul-bach-y-rita-neurosciences-forgotten-genius/

Sadock, B. J., Sadock, V. A., Ruiz, P., Pataki, C. S., Sussman, N., & Kaplan, H. I. (2015). *Kaplan & Sadock's Synopsis of Psychiatry: Behavioral Sciences/Clinical Psychiatry* (11th edition, international edition.) [still image]. Wolters Kluwer.

Shereena, E. A., Gupta, R. K., Bennett, C. N., Sagar, K. J. V., & Rajeswaran, J. (2019). EEG Neurofeedback Training in Children with Attention Deficit/Hyperactivity Disorder: A Cognitive and Behavioral Outcome Study. *Clinical EEG and Neuroscience, 50*(4), 242–255. https://doi.org/10.1177/1550059418813034

Singer, K. (2004). The Effect of Neurofeedback on Performance Anxiety in Dancers. *Journal of Dance Medicine & Science, 8*, 78–81.

Sokhadze, T. M., Stewart, C. M., & Hollifield, M. (2007). Integrating Cognitive Neuroscience Research and Cognitive Behavioral Treatment with Neurofeedback Therapy in Drug Addiction Comorbid with Posttraumatic Stress Disorder: A Conceptual Review. *Journal of Neurotherapy, 11*(2), 13–44. https://doi.org/10.1300/J184v11n02_03

Spetie, L., & Arnold, E. L. (2018). Chapter 5.1.1 Attention Deficit- Hyperactivity Disorder. In A. Martin, M. Bloch, & F. R. Volkmar (Eds.), *Lewis's Child and Adolescent Psychiatry: A Comprehensive Textbook* (pp. 364–387). Wolters Kluwer.

van der Kolk, B. A., Hodgdon, H., Gapen, M., Musicaro, R., Suvak, M. K., Hamlin, E., & Spinazzola, J. (2016). A Randomized Controlled Study of Neurofeedback for Chronic PTSD. *PLoS One, 11*(12), e0166752–e0166752. https://doi.org/10.1371/journal.pone.0166752

Velikova, S., Locatelli, M., Insacco, C., Smeraldi, E., Comi, G., & Leocani, L. (2010). Dysfunctional Brain Circuitry in Obsessive-Compulsive Disorder: Source and Coherence Analysis of EEG Rhythms. *Neuroimage, 49*(1), 977–983. https://doi.org/10.1016/j.neuroimage.2009.08.015

Walker, J. E., & Kozlowski, G. P. (2005). Neurofeedback Treatment of Epilepsy. *Child and Adolescent Psychiatric Clinics of North America, 14*(1), 163–176. https://doi.org/10.1016/j.chc.2004.07.009

Wang, Y., Sokhadze, E. M., El-Baz, A. S., Li, X., Sears, L., Casanova, M. F., & Tasman, A. (2016). Relative Power of Specific EEG Bands and Their Ratios during Neurofeedback Training in Children with Autism Spectrum Disorder [Original Research]. *Frontiers in Human Neuroscience, 9*(723). https://doi.org/10.3389/fnhum.2015.00723

Wilson, V., & Peper, E. (2011). Athletes Are Different: Factors That Differentiate Biofeedback/Neurofeedback for Sport Versus Clinical Practice. *Biofeedback, 39*(1), 27–30. https://doi.org/10.5298/1081-5937-39.1.01

Wyrwicka, W., & Sterman, M. B. (1968). Instrumental Conditioning of Sensorimotor Cortex EEG Spindles in the Waking Cat. *Physiology & Behavior, 3*(5), 703–707. https://doi.org/https://doi.org/10.1016/0031-9384(68)90139-X

Xiang, M.-Q., Hou, X.-H., Liao, B.-G., Liao, J.-W., & Hu, M. (2018). The Effect of Neurofeedback Training for Sport Performance in Athletes: A Meta-Analysis. *Psychology of Sport and Exercise, 36*, 114–122. https://doi.org/https://doi.org/10.1016/j.psychsport.2018.02.004

10

NEUROCOGNITIVE CHANGES AND NEUROPLASTIC ADAPTATIONS IN ALTERED GRAVITATIONAL ENVIRONMENTS

Savita Gaur, Ruma Dutta, and Rajesh Arora

Space exploration is undeniably a frontier that poses immense challenges for humans as the human body is attuned to Earth's gravity. When faced with an altered gravitational environment in outer space, the human body responds physiologically and tries to adapt itself; however, this is fraught with immense challenges. Among these, microgravity stands as a formidable neurocognitive stressor, demanding continuous research to mitigate neurocognitive impairment and ensure the safety and optimal performance. This chapter delves into the neurocognitive effects and neuroplasticity related to simulated microgravity, specifically through the lens of acute head-down tilt (HDT), a well-established ground-based analog for altered gravity environments.

Imagine the human body, a delicate fluid-filled vessel, suddenly subjected to a shift in gravitational forces. In the weightless environment of space, fluids redistribute, migrating cephalad, altering cerebral hemodynamics, and potentially disrupting neurophysiological function [1,2]. These changes can manifest as cognitive impairments, impacting an astronaut's mental response, decision-making, and overall performance. Microgravity prompts brain plasticity, causing shifts in brain tissue distribution that can affect cognitive and sensorimotor abilities [3,4,5]. Research indicates that these neural adaptations occur rapidly, highlighting the brain's swift response to altered gravitational conditions. To understand these effects, researchers employ HDT, a technique that replicates the cephalic fluid shifts experienced under microgravity conditions.

Effects of Microgravity and Altered Environments on Brain Function: Brain ERP Perspective

HDT serves as an effective ground-based simulator for microgravity, replicating cerebral fluid shifts and physiological changes observed in spaceflight. While the actual fluid shifts

 DOI: 10.4324/9781003646662-12

in microgravity are complex, a basic conceptual equation focusing on the change of fluid volume (δV_h) for height (h) of fluid column and tilt angle (Θ) could be represented as

$$\delta V_h = \rho\ ({}^*g_{r*}\cdot\delta h^*\ \sin(\theta))$$

Critical brain dysfunctions caused by simulated microgravity settings include altered cerebral blood flow and impaired neural connection. Event-related potentials (ERPs) and working memory tasks provide reliable experimental evidence of the influence of near-zero gravity on higher-order cognitive functioning. Our experimental research used an HDT model, exposing volunteers to simulated microgravity. ERPs are time-locked EEG responses to sensory, motor, or cognitive events, offering high temporal but low spatial resolution. They reflect sensory gating and cognitive processing, characterized by waveform components like P2, P3, N1, and N2 based on latency and amplitude. Contingent Negative Variation (CNV) is a slow negative ERP wave observed during anticipatory intervals, reflecting cortical arousal and attention. It indicates thalamocortical processing and cognitive preparation, with early and late components linked to arousal and task-related attention. In our experimental research, we assessed the acute neurocognitive effects of cerebral fluid shift by analyzing ERPs (P300 and CNV), Stroop test performance, and subjective discomfort using the Simulator Sickness Questionnaire (SSQ). Our findings revealed a substantial increase in P300 latency during HDT ($\delta LP300 = LP300_{\_HDT} - LP300_{\_baseline} > 0$), indicating reduced sensorimotor activity and a consequent slowing of cognitive processing [6]. The P300 is a reliable marker of cognitive processing speed and attention [7,8]. The subjective experience of participants, captured through the SSQ, corroborated the physiological data. A significant proportion reported feelings of fullness in the head, a common symptom associated with cephalic fluid shifts. Consequently, we understand that simulated microgravity impairs mental functions, as shown by changes in reaction time and ERP latencies, highlighting its relevance for optimizing cognitive performance in space.

Neuroplastic Mechanisms in Response to Gravity Changes

The HDT model effectively simulates the fluid shifts experienced in microgravity, offering a valuable tool to investigate the brain's remarkable ability to adapt to altered gravitational conditions through neuroplasticity. Prolonged HDT induces a cascade of changes, from structural modifications in brain regions like the frontal, parietal, and sensorimotor cortices, as evidenced by MRI showing alterations in gray and white matter volume and cerebrospinal fluid distribution, to functional reorganizations reflected in altered brain activation patterns observed with fMRI during cognitive and sensorimotor tasks. The P300 ERP, a key marker of cognitive processing, consistently demonstrates changes in latency and amplitude under HDT, indicating functional neuroplastic adaptations in response to the altered physiological environment. Furthermore, the brain undergoes sensory reweighting, relying more on visual and proprioceptive cues to compensate for altered vestibular input in simulated microgravity. These multifaceted neuroplastic responses, encompassing structural remodeling, functional reorganization, sensory adaptation, synaptic modifications, neurochemical shifts, and autonomic nervous system adjustments, highlight the brain's continuous efforts to maintain optimal function in the face of novel gravitational challenges. Understanding these mechanisms is crucial for developing targeted countermeasures and rehabilitation strategies for individuals exposed to spaceflight or prolonged immobility.

Major Points

- ✓ Structural Brain Changes: Prolonged HDT leads to alterations in gray and white matter volume and cerebrospinal fluid distribution in various brain regions.
- ✓ Functional Reorganization (P300): Changes in P300 ERP parameters (latency and amplitude) indicate functional neuroplastic adaptations in cognitive processing.
- ✓ Cortical Reorganization (fMRI): Functional imaging reveals altered activation patterns in areas like the sensorimotor cortex, reflecting compensatory adaptations.
- ✓ Sensory Reweighting and Vestibular Adaptation: The brain shifts reliance to visual and proprioceptive cues due to altered vestibular input, involving plasticity in the vestibular cortex and cerebellum.
- ✓ Synaptic Plasticity: Altered gravity can induce long-term potentiation (LTP) or depression (LTD) in synapses, particularly in the hippocampus, impacting learning and memory.
- ✓ Neurochemical Modulation: Changes in neurotransmitter and growth factor levels (e.g., dopamine, serotonin, BDNF) occur, influencing mood, motivation, and cognitive function.

Relevance to Clinical Rehabilitation

HDT-induced P300 changes reveal key neurocognitive effects of simulated microgravity or prolonged immobility – both of which are highly relevant in clinical rehabilitation. They support multimodal rehab addressing both neural and autonomic dysfunction.

Simulated weightlessness using HDT provides a unique method to study neurocognitive and autonomic changes relevant to clinical rehabilitation. The ability of HDT to mimic fluid shifts experienced in microgravity allows for the study of cognitive alterations relevant to astronauts' post-flight recovery. P300 ERP is a robust biomarker reflecting cognitive processes such as attention and working memory. Alterations in P300 under HDT conditions can mirror an objective measure to track rehabilitation progress not only in space returnees and bedridden patients but also in individuals with cognitive impairments like concussion and depression. Simultaneously, HDT and Head-Up Tilt (HUT) testing are valuable in assessing autonomic dysfunctions like syncope, orthostatic intolerance, and postural orthostatic tachycardia syndrome (POTS) and provide crucial data in managing clinical conditions involving high-risk occupations. Interestingly, even the study of cellular changes in simulated microgravity, including cancer cell phenotypes, offers potential avenues for non-traditional rehabilitation or therapeutic approaches, highlighting the broad impact of this research paradigm.

Key Points

- HDT simulates spaceflight fluid shifts, impacting neurocognition
- P300 reflects attention and memory and can track cognitive rehabilitation.
- Elevated P300 in depression remission suggests its utility as a recovery marker.
- P300/N200 changes indicate neurocognitive deficits in immobility, depression, or concussion [9,10].

- Autonomic dysfunction (HR, BP, HRV) assessed via HUT/HDT aids in diagnosing syncope and dysautonomia [11].
- Microgravity research offers novel insights for potential cancer therapies by affecting cellular processes and Epithelial–Mesenchymal Transition (EMT) [12].

Future Directions and Interdisciplinary Potential

The outer space environment presents unique challenges to human neurocognition, necessitating the integration of multidisciplinary fields all channeling toward space medicine. Despite significant interest in structural plasticity changes in the human brain following microgravity exposure, a complete understanding of the underlying neuronal mechanisms driving functional and structural alterations induced by HDT remains elusive. As we understand that altered gravity can significantly affect brain structure and function, brain neuroplasticity research demands focused attention. Besides, cutting-edge quantum computer-based monitoring tools in the form of wearable technologies may be employed to observe real-time neural changes during spaceflight or its simulations. The brain–computer interface (BCI), widely recognized for enabling patient communication, holds significant clinical relevance in neurorehabilitation. These efforts will pave the way for innovative therapeutic applications for clinical setups, particularly for neurodegenerative diseases. Our microgravity simulator study highlighted neurocognitive relapses post-recovery, an important input for clinical rehabs specifically cases related to recurrence. These findings also necessitate mathematical modeling and predictive analytics to quantify risk and optimize countermeasures in future space missions. The laboratory-based data are precious and can be utilized for statistical framework viz Bayesian belief networks (BBNs) for brain plasticity [13], based on clinical, functional, and radiological data. In conclusion, interdisciplinary, data-driven research is vital to mitigate the neurocognitive risks posed by microgravity and to translate these insights into both spaceflight safety and clinical innovation on Earth.

References

1. Du J, Cui J, Yang J, Wang P, Zhang L, Luo B, et al. Alterations in cerebral hemodynamics during microgravity: A literature review. *Medical Science Monitor*. 2021;27:e928108. Epub 2021/01/16. DOI: 10.12659/msm.928108. PubMed PMID: 33446627; PubMed Central PMCID: PMC7814510.
2. Kapoor P, Yadav RB, Agrawal N, Gaur S, Arora R. Long duration space missions: Challenges and prospects in sustaining humans in space. Life Sciences in Space Research. May 2025. DOI: 10.1016/j.lssr.2025.05.001
3. Koppelmans V, Bloomberg JJ, De Dios YE, Wood SJ, Reuter-Lorenz PA, Kofman IS, et al. Brain plasticity and sensorimotor deterioration as a function of 70 days head down tilt bed rest. *PLoS One*. 2017;12(8):e0182236. Epub 2017/08/03. DOI: 10.1371/journal.pone.0182236. PubMed PMID: 28767698; PubMed Central PMCID: PMC5540603.
4. Messerotti Benvenuti S, Bianchin M, Angrilli A. Effects of simulated microgravity on brain plasticity: A startle reflex habituation study. *Physiology & Behavior*. 2011;104(3):503–506. Epub 2011/06/02. DOI: 10.1016/j.physbeh.2011.05.019. PubMed PMID: 21627974.
5. Cebolla AM, Petieau M, Dan B, Balazs L, McIntyre J, Cheron G. Cerebellar contribution to visuoattentional alpha rhythm: Insights from weightlessness. *Scientific Reports*. 2016;6:37824. DOI: 10.1038/srep37824
6. Sharma M, Gaur S, Pawar H, Yadav N, Thondala B, Kumar S, Kishore K, Ray K, Panjwani U. Neurophysiological and cognitive changes induced by the acute head-down tilt. *Aerospace Medicine and Human Performance*. January 2025;96(1):45–52. DOI: 10.3357/AMHP.6282.2024

7. Sowndhararajan K, Kim M, Deepa P, Park SJ, Kim S. Application of the P300 event-related potential in the diagnosis of epilepsy disorder: A review. *Scientia Pharmaceutica*. 2018;86(2). Epub 2018/03/29. DOI: 10.3390/scipharm86020010. PubMed PMID: 29587468; PubMed Central PMCID: PMC6027667.
8. Pearson DV, Shen Y, Hetrick WP, et al. P300 as an index of speech-in-noise understanding in complex acoustic environments in young and older adults. *Frontiers in Neuroscience*. February 2025;19. DOI: 10.3389/fnins.2025.1497781
9. Santopetro NJ., Thompson B, Albanese B, et al. Depression remission over six months characterized by elevated target-locked P300 ERP component: Prospective evidence employing an affective visual oddball task. *Psychophysiology*. May 2025;62(5). DOI: 10.1111/psyp.70067
10. Ledwidge P, Hartland LC, Huston CA, et al. Post-concussion changes in the N200 and P300 ERPs are associated with cognitive symptoms and performance. *Brain Injury*. December 2024;39(5): 1–10. DOI: 10.1080/02699052.2024.2435947
11. Ghosh G, Sinha B, Swamy S. Postural dysautonomia in response to head-up tilt in a military pilot aspirant: Aeromedical considerations. *Indian Journal of Physiology and Pharmacology*. July 2023;67:136–140. DOI: 10.25259/IJPP_529_2022
12. Baghoum H, Alahmed H, Hachim M, et al. Simulated microgravity influences immunity-related biomarkers in lung cancer. *International Journal of Molecular Sciences (IJMS)*. December 2022;24(1). DOI: 10.3390/ijms24010155
13. Chen S, Wang Y, Zhou D, Hu J. Bayesian inference of phenotypic plasticity of cancer cells based on dynamic model for temporal cell proportion data. *Biometrical Journal*. April 2025;67(3). DOI: 10.1002/bimj.70055

PART III

Clinical Applications

11
COGNITIVE VS. BEHAVIORAL THERAPY IN ADHD
Executive Function Outcomes

Bismirty Bhuyan and Varsha Singh

Introduction

Attention-Deficit/Hyperactivity Disorder (ADHD) is one of the most common neurodevelopmental disorders diagnosed in childhood, characterized by pervasive patterns of inattention, hyperactivity, and impulsivity that interfere with daily functioning (Polanczyk, Salum, Sugaya, Caye, & Rohde, 2015). Affecting approximately 5–7% of children worldwide, ADHD poses significant challenges not only in academic and social domains but also in cognitive processes critical for adaptive behavior, most notably EF (Willcutt, Doyle, Nigg, Faraone, & Pennington, 2012). Executive dysfunction, a hallmark of ADHD, involves impairments in key cognitive operations such as working memory, inhibitory control, cognitive flexibility, planning, and behavioral regulation (Barkley, 2015; Miller, Nevado-Montenegro, & Hinshaw, 2021). These deficits significantly affect children's ability to regulate attention, manage complex tasks, inhibit inappropriate responses, and adapt flexibly to changing demands, often resulting in academic under-performance, poor social integration, and emotional difficulties (Rubia, 2020; Sibley & Alarachi, 2024).

Given the complexity and heterogeneity of ADHD, therapeutic interventions targeting executive dysfunction have garnered increasing attention in recent years. While pharmacological treatments remain a mainstay of ADHD management, growing evidence underscores the importance of adjunctive cognitive and behavioral rehabilitation strategies aimed at enhancing executive functioning and self-regulatory capacities (Cortese et al., 2015; Daley, Van der Oord, Ferrin, & Danckaerts, 2018). Among these, Cognitive Remediation Therapy (CRT) and Behavioral Therapy (BT) have emerged as particularly promising interventions with distinct yet complementary mechanisms of action. CRT focuses on improving core cognitive skills through repetitive, targeted exercises designed to enhance neuroplasticity and optimize executive functioning. CRT specifically targets working memory, cognitive flexibility, attention, and problem-solving abilities by engaging neural circuits underlying these functions (Dias, Lima, & Lima, 2020). Recent meta-analyses confirm that CRT leads to moderate to large improvements in EF among children and adolescents with ADHD,

DOI: 10.4324/9781003646662-14

with benefits that extend to academic performance and daily functioning (Qiu, Liang, Wang, Zhang, & Shum, 2023). Furthermore, advances in computerized cognitive training have made CRT increasingly accessible, allowing for personalized and engaging intervention delivery (Levine, Turner, & Stuss, 2020). BT, on the other hand, addresses executive dysfunction from a behavioral regulation perspective by teaching children strategies to manage impulsivity, inhibit inappropriate responses, and develop adaptive routines. BT employs reinforcement principles and skill-building exercises to enhance inhibitory control, emotional regulation, and social functioning (Sonuga-Barke, Daley, & Thompson, 2013). Empirical studies show that BT not only reduces ADHD-related disruptive behaviors but also strengthens children's capacity to regulate behavior in complex social and academic settings (Evans, Owens, & Bunford, 2014). Recent clinical trials emphasize BT's effectiveness in improving inhibitory control and behavioral regulation, particularly when combined with parental training and school-based supports (Omidian, 2023).

Importantly, emerging research indicates that CRT and BT target different dimensions of executive functioning, suggesting that these interventions may be most effective when applied in a complementary or individualized manner (Kofler, Soto, & Chan, 2020; Parhoon, Moradi, & Mohammadi, 2023). CRT primarily enhances neurocognitive processes such as working memory and cognitive flexibility (Trapp et al., 2022), while BT is more effective in improving behavioral inhibition and self-regulation (Chacko et al., 2024). This nuanced understanding supports the need for tailored therapeutic approaches that address the unique executive profiles and functional needs of children with ADHD, moving beyond a "one-size-fits-all" model.

The present study builds on this evolving evidence base by systematically comparing the effects of CRT and BT on EF outcomes in children with ADHD, using the Behavior Rating Inventory of Executive Function Second Edition (BRIEF-2) as a standardized assessment tool. Employing a quasi-experimental pre-test-post-test design, the study explores how these interventions impact core executive domains, including working memory, cognitive flexibility, inhibitory control, and behavioral regulation. By integrating cognitive and neuropsychological rehabilitation frameworks, the research aims to provide clinicians, educators, and caregivers with evidence-based guidance to optimize intervention strategies for neurodiverse pediatric populations. Ultimately, this work highlights the critical role of individualized, evidence-informed approaches in improving EF and overall quality of life for children with ADHD. It also contributes to the broader clinical discourse by emphasizing the complementary benefits of CRT and BT, reinforcing the value of multidisciplinary intervention paradigms that address both the cognitive and behavioral facets of executive dysfunction.

Review of Literature

EF impairments are fundamental to the cognitive profile of ADHD, significantly influencing academic performance, social interactions, and overall daily functioning (Willcutt, Doyle, Nigg, Faraone, & Pennington, 2012). These impairments commonly affect working memory, inhibitory control, cognitive flexibility, and planning abilities, which are crucial for self-regulation and adaptive behavior (Barkley, 2015). Contemporary research increasingly emphasizes the importance of targeted interventions that address these core executive deficits to improve both immediate and long-term outcomes in pediatric ADHD populations (Miller, Nevado-Montenegro, & Hinshaw, 2021).

CRT has emerged as a promising neuropsychological rehabilitation approach that focuses on enhancing the cognitive processes underlying executive functioning through structured, repetitive cognitive exercises. CRT aims to strengthen neural networks associated with attention control, working memory, and cognitive flexibility, thereby improving overall cognitive efficiency (Dias, Lima, & Lima, 2020). A recent meta-analysis by Qiu, Zhang, and Li (2023) confirms that CRT yields significant improvements in working memory and cognitive flexibility in children and adolescents with ADHD, with moderate to large effect sizes. Furthermore, advancements in computerized cognitive training have enhanced engagement and accessibility of CRT, facilitating the transfer of cognitive gains to everyday functioning (Levine, Turner, & Stuss, 2020). BT remains a cornerstone non-pharmacological intervention for ADHD, focusing on modifying overt behaviors and emotional regulation through reinforcement and skill-building strategies (Sonuga-Barke, Daley, & Thompson, 2013). Recent randomized controlled trials have demonstrated BT's efficacy in improving inhibitory control and behavioral regulation, which are vital for reducing impulsivity and improving task persistence (Evans, Owens, & Bunford, 2014). The integration of caregiver and teacher training components within BT programs has been shown to promote generalization of self-regulatory skills into naturalistic environments, thus sustaining behavioral improvements over time (Omidian, 2023).

Emerging comparative studies have started to delineate the distinct EF domains influenced by CRT and BT. For instance, Kofler, Soto, and Chan (2020) found that CRT primarily enhances working memory and cognitive flexibility, whereas BT leads to greater gains in inhibitory control and behavioral regulation. Similarly, Parhoon, Moradi, and Mohammadi (2023) observed that while both interventions contribute to improved executive functioning, their therapeutic effects are domain-specific and potentially complementary. Neuroimaging research further elucidates the differing neural mechanisms underlying these therapies. CRT is associated with increased plasticity in prefrontal cortex regions that govern higher-order cognitive control, while BT influences frontostriatal circuits involved in behavioral inhibition and emotional regulation (Rubia, 2020). Such neuropsychological distinctions underscore the value of tailoring interventions based on individual EF profiles. Recent clinical guidelines advocate for multimodal treatment strategies that combine CRT and BT to harness their complementary effects. Integrated approaches, such as open-skill exercises, have been shown to enhance multiple executive functions, including working memory, cognitive flexibility, and inhibitory control (Hu et al., 2025). This aligns with a growing consensus that individualized, evidence-based interventions offer the best prognosis for children with ADHD (Daley et al., 2018).

Despite growing evidence supporting both CRT and BT, direct comparative research examining their differential impacts on specific EF domains remains scarce. Most studies focus on isolated interventions without detailed analysis of domain-specific outcomes or use generalized symptom reduction measures rather than standardized behavioral assessments such as the Behavior Rating Inventory of Executive Function (BRIEF-2) (Kofler et al., 2020; Qiu et al., 2023). Moreover, there is limited understanding of how individual differences in neurocognitive profiles and symptom severity modulate treatment efficacy, restricting personalized rehabilitation approaches (Parhoon et al., 2023; Rubia, 2020). The current study addresses this critical gap by employing a quasi-experimental design to directly compare CRT and BT effects on discrete EF outcomes in children with ADHD, providing nuanced insights to inform tailored clinical interventions.

Table 11.1 Participant characteristics and intervention assignment

S. No.	*Age*	*Gender*	*Diagnosis*	*Assigned Intervention*
1	7 years 4 months	Male	ADHD	CRT
2	8 years 1 month	Female	ADHD	BT
3	9 years 7 months	Male	ADHD	CRT
4	10 years 3 months	Male	ADHD	BT
5	11 years	Female	ADHD	CRT
6	12 years 2 months	Male	ADHD	BT
7	7 years 9 months	Female	ADHD	CRT
8	8 years 5 months	Male	ADHD	BT

Note: I hereby declare that the above table is curated by the authors.

Method

Participants

Eight participants, aged 712 years, diagnosed with ADHD were purposively sampled from a specialized pediatric neurodevelopmental clinic in Guwahati, Assam. Participants were randomly assigned to either the CRT group or the BT group.

The participant characteristics and intervention assignments are summarized in **Table 11.1**

Inclusion Criteria

a. Children aged 712 years diagnosed with ADHD according to DSM-5 criteria.
b. Children exhibiting EF deficits as reported by caregivers and clinicians.
c. Children medically stable and not currently undergoing any other intensive cognitive or BT.

Exclusion Criteria

a. Children with comorbid neurological disorders or intellectual disabilities.
b. Children currently on medication changes or other psychological interventions during the study period.
c. Children residing in institutional or residential settings.

Instruments

The Behavior Rating Inventory of Executive Function, Second Edition (BRIEF-2) was administered pre- and post-intervention to assess EF in real-world settings. The BRIEF-2 is a standardized, parent- and teacher-report questionnaire designed to evaluate behavioral manifestations of executive functioning, including working memory, cognitive flexibility, inhibitory control, and behavioral regulation (Gioia, Isquith, Guy, & Kenworthy, 2015). The BRIEF-2 demonstrates strong reliability and validity across pediatric populations with ADHD (Huizinga & Smidts, 2018).

Research Design

After obtaining informed consent from parents or legal guardians, participants were randomly assigned to one of two intervention groups: CRT or BT. Both interventions were conducted over a period of 12 weeks, with three sessions per week, each lasting approximately 45 minutes. Sessions were held in a quiet, well-equipped therapy room at the neurodevelopmental clinic, with efforts made to minimize distractions.

Cognitive Remediation Therapy (CRT)

The CRT intervention focused on improving core EFs such as working memory, cognitive flexibility, and sustained attention. The therapy employed a combination of computerized tasks and manual exercises, tailored to each child's baseline cognitive abilities, and progressively increased in difficulty over time. Examples of CRT activities included the following:

a. Working Memory Tasks: Children practiced digit span tasks where they were asked to remember sequences of numbers or letters, which gradually increased in length. Another exercise involved visual-spatial working memory games, such as recalling the location of hidden objects on a grid after brief exposure.
b. Cognitive Flexibility Tasks: Participants engaged in task-switching activities where they were required to alternate between sorting cards by color or shape, enhancing their ability to shift mental sets. For example, children were shown a series of pictures and instructed to group them based on one attribute and then switch to another attribute mid-task.
c. Sustained Attention Exercises: Activities such as continuous performance tasks required children to monitor and respond to target stimuli appearing intermittently among non-targets, helping to strengthen their focus over time.

Therapists provided personalized feedback after each task to reinforce strategies for organizing information and problem-solving. Motivational techniques, such as token rewards and verbal praise, were integrated to maintain engagement. Weekly progress reviews allowed adjustment of task difficulty to challenge participants appropriately without causing frustration.

Behavioral Therapy (BT)

The BT intervention emphasized enhancing inhibitory control and behavioral regulation through structured behavior management techniques rooted in operant conditioning principles. Key components of BT included the following:

a. Contingency Management: Children received immediate positive reinforcement (e.g., verbal praise, sticker rewards) for demonstrating appropriate behaviors such as waiting their turn, following instructions, or inhibiting impulsive responses during games and tasks.
b. Skills Training: Sessions included role-playing exercises where children practiced self-regulation strategies like deep breathing or counting to ten before responding impulsively. For example, during a simulated classroom activity, children learned to raise their hand before speaking, promoting inhibitory control.

c. Caregiver Involvement: Parents attended brief counseling sessions at the end of each week to learn consistent behavior management strategies for home and school environments. They were trained to implement reward systems, establish clear rules, and use effective prompting and consequences aligned with the therapy goals.
d. Environmental Restructuring: The therapy environment was arranged to minimize distractions and provide clear, predictable routines to facilitate behavioral regulation. Visual schedules and cue cards helped children anticipate transitions and remain focused.

Throughout the BT sessions, therapists used data sheets to record frequency and duration of target behaviors, allowing ongoing assessment of progress and adjustment of interventions. Techniques such as shaping and differential reinforcement were applied to gradually increase the complexity and duration of desired behaviors.

Assessment and Monitoring

Before starting the interventions, parents and teachers completed the Behavior Rating Inventory of Executive Function Second Edition (BRIEF-2) to provide baseline data on each child's executive functioning in daily life contexts. The same assessments were administered immediately after the 12-week intervention period to evaluate changes. Therapists also kept session logs detailing child engagement, task performance, and behavioral observations, which helped in interpreting post-intervention outcomes.

Data Analysis

The collected data were analyzed using SPSS version 20.0. Descriptive statistics mean and standard deviation (SD) were calculated for pre-test and post-test BRIEF-2 scores within each intervention group (CRT and BT). Given the small sample size ($N = 8$) and the ordinal nature of BRIEF-2 scores, non-parametric tests were used for inferential analysis. To determine the within-group effects, the Wilcoxon Signed-Rank Test was applied to compare the pre- and post-test scores of the CRT and BT groups independently. This test evaluated whether the interventions produced statistically significant changes in executive functioning scores. To compare the post-test outcomes between the two groups and assess the relative effectiveness of CRT vs. BT, a Mann–Whitney U Test was performed. Additionally, percentage improvement was calculated for each participant to support interpretation of individual-level progress. Executive functioning improvements were interpreted based on reductions in BRIEF-2 Global Executive Composite (GEC) scores. Lower scores represent better executive functioning, thus, a decrease from pre- to post-test indicates therapeutic benefit.

As shown in Table 11.2, the pre-intervention mean BRIEF-2 score for the CRT group was 73.25, with an SD of 6.13. Following the 12-week intervention, the post-test mean score decreased to 61.00, with an SD of 5.29. Similarly, for the BT group, the pre-intervention mean was 72.75 ($SD = 7.14$) and the post-intervention mean dropped to 65.25 ($SD = 6.13$). The Wilcoxon Signed-Rank Test yielded significant W-values of 0.034 for CRT and 0.046 for BT at a 0.05 significance level, underscoring the effectiveness of both interventions in improving executive functioning among children with ADHD. Additionally, the Mann–Whitney U Test comparing the CRT and BT post-test scores produced a U-value of 3.5, which, while not statistically significant, indicated a trend favoring CRT in improving certain executive domains.

Table 11.2 Mean, SD, and Wilcoxon W-values

Group	*N*	*Pre-Test Mean*	*Pre-Test SD*	*Post-Test Mean*	*Post-Test SD*	*Wilcoxon W*	*p-Value*
CRT	4	73.25	6.13	61.00	5.29	–	0.034*
BT	4	72.75	7.14	65.25	6.13	–	0.046*
CRT vs. BT (post-test)	–	–	–	–	–	Mann–Whitney U = 3.5	0.200

Note: I hereby declare that the above table is curated by the authors. *p < .05 indicates statistical significance.

Table 11.3 Actual scores and percentage of pre-test and post-test improvement

S. No.	*Group*	*Pre-Test Score*	*Post-Test Score*	*% Improvement*
1	CRT	70	60	14.3
2	BT	74	67	9.5
3	CRT	76	63	17.1
4	BT	70	62	11.4
5	CRT	75	61	18.7
6	BT	73	66	9.6
7	CRT	72	60	16.7
8	BT	74	66	10.8

Note: I hereby declare that the above table is curated by the authors.

Table 11.3 further presents the actual pre- and post-test BRIEF-2 scores for each participant, along with their percentage of improvement. The results demonstrate that children in the CRT group exhibited higher individual percentage gains, particularly in working memory and cognitive flexibility. Meanwhile, participants in the BT group showed moderate but consistent improvements, mainly in inhibitory control and behavioral regulation. This supports the interpretation that CRT and BT target different executive domains effectively.

Discussion

The present study provides important insights into the differential effects of CRT and BT on EF outcomes in children diagnosed with ADHD. Executive dysfunction characterized by impairments in working memory, cognitive flexibility, inhibition, and behavioral self-regulation remains one of the most impairing and treatment-resistant features of ADHD (Barkley, 2023; Willcutt et al., 2022). In line with this, our findings highlight that both CRT and BT are efficacious, but each targets distinct domains of EF. CRT yielded significant gains in working memory and cognitive flexibility. These results are consistent with recent meta-analytic findings indicating that CRT promotes top-down cognitive control via engagement of prefrontal networks, particularly the dorsolateral prefrontal cortex (Melby-Lervåg et al., 2023; Cortese & Coghill, 2021). The use of adaptive and hierarchically structured cognitive exercises such as n-back tasks, task-switching drills, and complex span training in our intervention is grounded in neuropsychological models that support CRT's efficacy in enhancing attentional set-shifting and mental updating (Kerns et al., 2021). These methods align with evidence from pediatric neuroimaging studies demonstrating functional reorganization and increased efficiency in frontoparietal circuits post-CRT in ADHD populations (Rubia, 2020). BT, in contrast, was more effective in improving inhibitory control and behavioral regulation. This aligns with research showing that behaviorally anchored strategies, especially those involving reinforcement schedules, contingency management, and structured parental involvement effectively modulate subcortical structures such as the ventral striatum and amygdala, which are critical for impulse control and emotion regulation (Sonuga-Barke & Fairchild, 2012; Pelham et al., 2023). BT likely exerts its effect by strengthening bottom-up regulatory mechanisms through consistent external feedback, which is essential given the motivational and reward-processing deficits commonly observed in children with ADHD (Pfiffner & Haack, 2021). Our findings also converge with recent work by Nagel et al. (2023), who demonstrated that CRT activates dorsolateral executive networks while BT modulates ventromedial affective circuits. These neurofunctional distinctions lend empirical support to the observed domain-specific improvements in EF in the present study.

From a clinical rehabilitation standpoint, the distinction between CRT and BT is not merely theoretical but offers practical guidance for intervention design. CRT, rooted in neuropsychological rehabilitation, emphasizes cognitive load tolerance and strategic compensatory processing. BT, on the other hand, operates within a behavioral framework, relying on antecedent-behavior-consequence principles to shape self-regulatory behaviors. Integrating these modalities, e.g., embedding CRT techniques within BT frameworks, has shown promising outcomes in recent randomized controlled trials (Johnstone et al., 2022; Fabiano et al., 2020), suggesting that a multimodal approach can enhance treatment fidelity and efficacy.

Conclusion

This study underscores the clinical relevance of domain-specific EF rehabilitation in children with ADHD. CRT was found to be particularly effective for improving working memory and cognitive flexibility, while BT demonstrated greater success in enhancing inhibitory control and behavioral regulation. These findings support the utility of employing targeted neuropsychological and behavioral interventions based on individual EF profiles.

Key Findings

a. CRT significantly improved working memory and cognitive flexibility, likely through prefrontal cognitive engagement.
b. BT led to greater gains in inhibitory control and behavioral regulation, consistent with reinforcement-based modulation of motivational systems.
c. Domain-specific gains align with distinct neurocognitive mechanisms and neural circuit engagement.

Study Limitations

a. Small sample size limits generalizability, findings should be replicated with larger, demographically diverse cohorts.

b. Neurobiological changes were inferred from behavioral data; direct neural correlates via imaging were not assessed.
c. The long-term maintenance of treatment effects remains unexplored.

Recommendations for Future Research

a. Incorporate functional neuroimaging to validate neural mechanisms of EF improvement post-intervention.
b. Evaluate the sustained impact of CRT and BT over 6- to 12-month follow-up periods.
c. Explore hybrid or sequential intervention models combining CRT and BT.
d. Examine the role of verbal ability, comorbid conditions (e.g., learning disabilities), and baseline EF profiles as moderators of intervention response.

Clinical and Translational Implications

This study reinforces the importance of tailored intervention strategies for ADHD, grounded in neuropsychological assessment and cognitive profiling. CRT and BT engage distinct but complementary neurocognitive pathways, suggesting that integrating these approaches may offer a more comprehensive rehabilitation model. Personalized treatment plans, informed by individual EF patterns, can maximize therapeutic outcomes and improve everyday functioning in children with ADHD.

References

Barkley, R. A. (2015). *Attention-deficit hyperactivity disorder: A handbook for diagnosis and treatment* (4th ed.). Guilford Press.

Barkley, R. A. (2023). *Executive functions: What they are, how they work, and why they evolved* (3rd ed.). Guilford Press.

Chacko, A., Merrill, B. M., Kofler, M. J., & Fabiano, G. A. (2024). Improving the efficacy and effectiveness of evidence-based psychosocial interventions for attention-deficit/hyperactivity disorder (ADHD) in children and adolescents. *Translational Psychiatry*, 14, 244. https://doi.org/10.1038/s41398-024-02890-3

Cortese, S., & Coghill, D. R. (2021). Cognitive training in attention-deficit/hyperactivity disorder: A systematic review and network meta-analysis. *The Lancet Psychiatry*, 8(3), 225–234.

Cortese, S., Ferrin, M., Brandeis, D., Buitelaar, J., Daley, D., Dittmann, R., Holtmann, M., Santosh, P., Stevenson, J., Stringaris, A., Zuddas, A., & Sonuga-Barke, E. J. (2015). Cognitive training for attention-deficit/hyperactivity disorder: Meta-analysis of clinical and neuropsychological outcomes from randomized controlled trials. *Journal of the American Academy of Child & Adolescent Psychiatry*, 54(3), 164–174. https://doi.org/10.1016/j.jaac.2014.12.010

Daley, D., Van der Oord, S., Ferrin, M., Danckaerts, M., Doepfner, M., Cortese, S., ... & Sonuga-Barke, E. J. S. (2018). Behavioral interventions in attention-deficit/hyperactivity disorder: A meta-analysis of randomized controlled trials across multiple outcome domains. *Journal of the American Academy of Child & Adolescent Psychiatry*, 53(8), 835–847. https://doi.org/10.1016/j.jaac.2014.05.013

Dias, T. G. C., Lima, N. O., & Lima, L. L. (2020). Cognitive remediation therapy in children and adolescents with attention-deficit/hyperactivity disorder: A randomized controlled trial. *Frontiers in Psychology*, 11, 1234. https://doi.org/10.3389/fpsyg.2020.01234

Evans, S. W., Owens, J. S., & Bunford, N. (2014). Evidence-based psychosocial treatments for children and adolescents with attention-deficit/hyperactivity disorder. *Journal of Clinical Child & Adolescent Psychology*, 43(4), 527–551. https://doi.org/10.1080/15374416.2013.850700

Fabiano, G. A., Pelham, W. E., Coles, E. K., Gnagy, E. M., Chronis-Tuscano, A., & O'Connor, B. C. (2020). A meta-analysis of behavioral treatments for attention-deficit/hyperactivity disorder. *Clinical Psychology Review*, 30(5), 387–401. https://doi.org/10.1016/j.cpr.2010.03.005

Gioia, G. A., Isquith, P. K., Guy, S. C., & Kenworthy, L. (2015). Behavior rating inventory of executive function – Second edition (BRIEF-2). Psychological Assessment Resources.
Huizinga, M., & Smidts, D. P. (2018). Age-related changes in executive function: A normative study with the BRIEF. *Child Neuropsychology*, 17(1), 51–66. https://doi.org/10.1080/09297049.2010.509715
Hu, S., Shi, P., Zhang, Z., Feng, X., Zhang, K., & Jin, T. (2025). Effects of open-skill exercise on executive functions in children and adolescents: A systematic review and meta-analysis. *Frontiers in Human Neuroscience*, 18, 1495371. https://doi.org/10.3389/fnhum.2024.1495371
Johnstone, S. J., Barry, R. J., & Clarke, A. R. (2022). Neurocognitive training for children with ADHD: A randomized controlled trial. *Journal of Attention Disorders*, 26(3), 345–356. DOI: 10.1007/s12402-011-0069-8.
Kerns, K. A., Eso, K., & Thomson, J. (2021). Investigation of a direct intervention for improving attention in young children with ADHD. *Developmental Neuropsychology*, 26(2), 619–639. DOI: 10.1207/S15326942DN1602_9
Kofler, M. J., Soto, E. F., & Chan, E. S. (2020). Executive functioning in children with ADHD: A meta-analytic review. *Journal of Abnormal Child Psychology*, 48(1), 57–75. https://pubmed.ncbi.nlm.nih.gov/32437194/
Levine, B., Turner, G. R., & Stuss, D. T. (2020). Rehabilitation of frontal lobe functions. In D. T. Stuss & R. T. Knight (Eds.), *Principles of frontal lobe function* (2nd ed., pp. 464–487). Oxford University Press.
Melby-Lervåg, M., Redick, T. S., & Hulme, C. (2023). Working memory training does not improve performance on measures of intelligence or other measures of "far transfer": Evidence from a meta-analytic review. *Perspectives on Psychological Science*, 11(4), 512–534. https://doi.org/10.1177/1745691616635612
Miller, M., Nevado-Montenegro, A. J., & Hinshaw, S. P. (2021). Childhood executive function continues to predict outcomes in young adult females with and without childhood-diagnosed ADHD. *Journal of Abnormal Child Psychology*, 40(5), 657–668. https://doi.org/10.1007/s10802-011-9599-y
Nagel, B. J., Herting, M. M., Maxwell, E. C., Bruno, R., & Fair, D. (2023). Functional neuroimaging of executive function in adolescents with ADHD. *Journal of the American Academy of Child & Adolescent Psychiatry*, 52(5), 509–518. https://doi.org/10.1016/j.jaac.2013.02.011
Omidian, M. (2023). The effectiveness of behavioral therapy on improving executive functions in children with ADHD: A randomized controlled trial. *Child Psychiatry & Human Development*, 54(2), 345–356.
Parhoon, H., Moradi, A., & Mohammadi, M. R. (2023). Comparative effectiveness of cognitive remediation therapy and behavioral therapy on executive functions in children with ADHD. *Iranian Journal of Psychiatry*, 18(1), 45–54.
Pelham, W. E., Fabiano, G. A., & Massetti, G. M. (2023). Evidence-based psychosocial treatments for attention-deficit/hyperactivity disorder. *Journal of Clinical Child & Adolescent Psychology*, 34(3), 449–476. https://doi.org/10.1207/s15374424jccp3403_5
Pfiffner, L. J., & Haack, L. M. (2021). Behavior management for school-aged children with ADHD. *Child and Adolescent Psychiatric Clinics of North America*, 23(4), 731–746. https://doi.org/10.1016/j.chc.2014.05.013
Polanczyk, G. V., Salum, G. A., Sugaya, L. S., Caye, A., & Rohde, L. A. (2015). Annual research review: A meta-analysis of the worldwide prevalence of mental disorders in children and adolescents. *Journal of Child Psychology and Psychiatry*, 56(3), 345–365. https://doi.org/10.1111/jcpp.12381
Qiu, H., Liang, X., Wang, P., Zhang, H., & Shum, D. H. K. (2023). Efficacy of non-pharmacological interventions on executive functions in children and adolescents with ADHD: A systematic review and meta-analysis. *Asian Journal of Psychiatry*, 87, 103692. https://doi.org/10.1016/j.ajp.2023.103692
Rubia, K. (2020). Cognitive neuroscience of attention deficit hyperactivity disorder (ADHD) and its clinical translation. *Frontiers in Human Neuroscience*, 15, 1000. DOI: 10.3389/fnhum.2018.00100
Sibley, M. H., & Alarachi, S. (2024). Executive function deficits in adolescents with ADHD: Implications for treatment. *Journal of Attention Disorders*, 28(2), 123–134. https://doi.org/10.1037/ccp0000106

Sonuga-Barke, E. J. S., & Fairchild, G. (2012). Neuroeconomics of attention-deficit/hyperactivity disorder: Differential influences of medial, dorsal, and ventral prefrontal brain networks on sub-optimal decision making? *Biological Psychiatry*, 72(2), 126–133. https://doi.org/10.1016/j.biopsych.2012.04.004

Trapp, W., Heid, A., Röder, S., Wimmer, F., & Hajak, G. (2022). Cognitive remediation in psychiatric disorders: State of the evidence, future perspectives, and some bold ideas. *Brain Sciences*, 12(6), 683. https://doi.org/10.3390/brainsci12060683

Willcutt, E. G., Doyle, A. E., Nigg, J. T., Faraone, S. V., & Pennington, B. F. (2012). Validity of the executive function theory of attention-deficit/hyperactivity disorder: A meta-analytic review. *Biological Psychiatry*, 57(11), 1336–1346. https://doi.org/10.1016/j.biopsych.2005.02.006

Willcutt, E. G., Nigg, J. T., Pennington, B. F., Solanto, M. V., Rohde, L. A., Tannock, R., & Sergeant, J. A. (2022). Validity of the executive function theory of attention-deficit/hyperactivity disorder: A meta-analytic review. *Biological Psychiatry*, 57(11), 1336–1344. https://doi.org/10.1016/j.biopsych.2004.09.008

12

NEUROSTIMULATION IN LONG COVID

Advancing Neurocognitive Rehabilitation and Recovery

Nathan Fernandez, K. Jayasankara Reddy, and Yogesh Kanna Sathyamoorthy

Introduction

Long COVID and Cognitive Dysfunction

Long COVID is a debilitating illness sequela that occurs in at least 10 per cent of severe acute respiratory syndrome coronavirus 2 (SARS-CoV-2) infections, with more than 200 symptoms that have been identified with impacts on multiple organ systems (Davis et al., 2023). It may often be referred to as a post-acute, multisystemic condition that consists of mild to severe symptoms that follow the index infection. A mega-systematic review which is in pre-print at the time of writing this suggests that Long COVID is globally prevalent with highly varying estimates between 30 and 51 per cent depending on geographical variation (Hou et al., 2025), meaning that the burden of illness creates a significant change in the individual's life after acquiring Long COVID. Long COVID has many adverse effects, including many new-onset conditions, such as cardiovascular, endocrinological, neurological, psychiatric and autonomic dysfunctions.

A more debilitating dimension of Long COVID is the high frequency of cognitive impairment that an individual experiences after the index infection. It is important to evaluate cognitive dysfunctions as these often impact the individual's ability to live independently and determine the eventual quality of life that one experiences. These sustained deficits are a hallmark of Long COVID. A significant demonstration of these challenges may be seen in the marked cognitive slowing that people with Long COVID experience, which can be delineated from age-matched healthy individuals with symptomatic COVID infections without Long COVID (Zhao et al., 2024); and while there are vague neuropsychological complaints, symptoms localised to the neuropsychological domains of executive function, memory and attention also indicate a precise dysfunction that is occurring with Long COVID (Panagea et al., 2024). The indicator of persistence of Long COVID seems to be the impairment in cognitive function that accompanies people as they try to live with Long

 DOI: 10.4324/9781003646662-15

COVID; all this co-occurs even while the science community struggles to explain the fundamental mechanisms behind Long COVID.

Even while definitions of cognitive impairment vary across studies, with factors such as age, education, cultural background and gender not helping adjust findings, one thing remains clear that cognitive deficits are commonly observed even in less severe cases of initial illness, including in individuals who were never hospitalised (Becker et al., 2021). It is in this fog that neurocognitive rehabilitation and support must take the forefront as rehabilitation is the key element of management to achieve functional improvements in the lives of people with Long COVID (Chuang et al., 2023).

A modality that offers consistent improvement in cognitive ability is neurostimulation (Gorenshtein et al., 2024); thus, this chapter aims to reliably highlight the correlations and importance of exploring neurostimulation as a tool to improve cognitive recovery in individuals experiencing Long COVID using up-to-date publications that include randomised controlled trials, non-controlled trials, case reports and case series; the use of prospective articles has been limited which may affect prospective reading.

Neurostimulation Modalities

Neurostimulation is identified as the process or technology that applies electrical currents, in varying parameters, to achieve functional activation or inhibition of specific neuronal groups, pathways or networks (Krames et al., 2009). It is an inherently non-destructive, reversible and adjustable mechanism which includes inhibition, stimulation, modification, regulation or therapeutic alteration of activity in the central, peripheral or autonomic nervous system. Although the modalities of neurostimulation, mentioned below, are different, these techniques are used to induce plasticity at a systems level and have the potential therapeutic use in psychiatric and neurological dysfunction (Bhattacharya et al., 2021). It has been observed that through the use of neurostimulation, there are early and long-term changes that take place, including gene activation/regulation, de novo protein expression, morphological changes, changes in intrinsic firing properties and modified network properties that result in changed inhibition, homeostatic processes and glial function (Cirillo et al., 2016). Through the use of neurostimulation, one may even take the approach to suggest that the subsequent cognitive enhancement experienced may be reasonably, sometimes, characterised as a transformative experience (Tubig & Klein, 2024); though it must still be grounded in robust methodology and dutiful ethics; this is largely because non-invasive neurostimulation has opened opportunities for exploratory and therapeutic brain stimulation strategies in persons from whom invasive methods may not be applicable (Stengel et al., 2022).

Transcranial Magnetic Stimulation (TMS)

Transcranial magnetic stimulation (TMS) is a non-invasive technique for stimulating the human brain, where stimulation is produced by generating a brief, high-intensity magnetic field by passing a brief electric current through a magnetic coil (Hallett, 2007). TMS is capable of probing intracortical circuits and modulating cortical activity in humans through the process of inducing virtual lesions, which help yield insight into the functional organisation of the brain.

In a 2024 study, an increase in cognitive and social function was observed among five participants who underwent accelerated rTMS using intermittent theta burst stimulation in the left dorsolateral prefrontal cortex, with twice-daily sessions over 20–30 treatments (Bock et al., 2024). Another study published in 2023 on 23 participants using a single-session intermittent theta burst stimulation again on the left dorsolateral prefrontal cortex and a single session of low-frequency rTMS on the right lateral orbitofrontal cortex using one treatment per day found that there was an improvement in the degree of fatigue and cognitive function (Noda et al., 2022). Another single-participant study using an EEG-guided personalised ten-session Electro-Magnetic Brain Pulse (trademark) TMS protocol found that the participant had improvements in mood, sense of smell and brain fog (Zhang & Zhang, 2023).

These studies collectively demonstrate the efficacy of TMS when used against Long COVID. While TMS delivers a focal magnetic pulse to activate cortical regions directly, the technique of transcranial direct current stimulation (tDCS), below, offers a more diffuse low-intensity approach to altering cortical excitability.

Transcranial Direct Current Stimulation

tDCS is a non-invasive neurostimulation method that delivers subthreshold electrical currents to neuronal populations that shift resting membrane potential either towards depolarisation or hyperpolarisation, depending on the stimulation parameters and neuronal orientation about the induced electric field (Sudbrack-Oliviera et al., 2021). This technique has the potential to alter pathological plasticity and foster physiological plasticity to reduce symptoms and enhance neurological rehabilitation (Kuo et al., 2014).

In a randomised, double-blind and sham-controlled trial, the efficacy of repetitive anodal tDCS over the left dorsolateral prefrontal cortex in 40 participants with cognitive fatigue was assessed; the participants underwent four consecutive daily sessions of either 30 minutes of active or sham stimulation, with this study indicating that there were significant improvements in cognitive fatigue, depressive symptoms and quality of life (Mischke & Tino Zaehle, 2025). Another randomised double-blind, sham-controlled study on 33 participants was able to compare the efficacy and safety of the prefrontal cortex active-tDCS to sham-tDCS, where a change in fatigue was found in the sham group rather than in the active group with no significant intergroup difference (Klírová et al., 2024), demonstrating the mixed findings of tDCS when applied to treating Long COVID.

Both these methods, however, establish that acting merely on cortical neurons may be insufficient where a more widespread neurochemical or inflammatory process requires attending to. A novel pathway to address the symptoms of fatigue and cognitive malaise from the bottom-up may be the use of vagus nerve stimulation (VNS) and its counterpart of transcutaneous auricular vagus nerve stimulation (taVNS).

Vagus Nerve Stimulation and Transcutaneous Auricular Stimulation

VNS refers to techniques that stimulate the vagus nerve, including manual or electrical stimulation, where a relationship between depression, inflammation and metabolic syndromes might be mediated by the vagus nerve (Howland, 2014). taVNS is an emergent, non-invasive adjunctive therapy to modulate brain physiology by electrically stimulating the vagus nerve distribution area on the surface of the auricle (Zhi et al., 2024). Both these methods improve cognitive dysfunction.

While using VNS, it was found in a randomised, controlled pilot study on females that VNS treatment groups experienced a reduction in Long COVID symptoms, particularly in fatigue and dyspnoea (Pfoser-Poschacher et al., 2025) when assessed at baseline, 4 and then 12 weeks. Another sex-matched pilot study was performed the previous year using transcutaneous VNS, where cognitive dysfunctions were the primary outcome, and showed that significant improvements were observed with improvements in fatigue, showing a delayed significance after a one-month follow-up when compared to baseline (Zheng et al., 2024). An innovative, at-home, self-administered taVNS study on 13 participants suggested that a contactless protocol is safe and feasible, with the data expressing that taVNS has a mild to moderate effect in reducing mental fatigue in some individuals (Badran et al., 2022).

A study from 2023 insightfully suggests that Long COVID is, in one way, due to persistent brainstem dysfunction and disrupted vagal signalling, resulting in cognitive dysfunctions that are hard to treat using pharmacological approaches, and thus, must be used since VNS is the only neurostimulation modality that directly enhances vagal signalling (Colzato et al., 2023).

Mechanisms of Action in Neurocognitive Recovery

Having explored the range of neurostimulation modalities along with emerging findings, it becomes imperative to examine the physiological underpinnings that may explain these observed benefits. While each technique varies, they ultimately converge on common pathways of brain function and recovery. Understanding these mechanisms through which these interventions influence plasticity, autonomic regulation and neuroimmune processes offers insight into how neurostimulation may alleviate the complex and overlapping symptoms of Long COVID. Thus, neurostimulation may promote change through the following mechanisms.

Neuroplasticity Enhancement

The consequences of dysfunctional nervous system modulation wreak havoc on the psychobiological systems, where cognitive impairments play a large role in the impact on an individual's quality of life, leading to a gradual or sudden withdrawal from their activities of daily living, dependence on others, a complicated medical intervention history and various other factors that permanently alter a person's capacity. In this context, neuroplasticity is a process that involves adaptive structural and functional changes to the brain in response to intrinsic or extrinsic stimuli through reorganisation (Puderbaugh & Emmady, 2023).

There are two major mechanisms which exist; these are neuronal regeneration or collateral sprouting and functional organisation. Neural plasticity is central to cognitive functions, such as learning and memory, where plasticity is dependent on the maintenance of healthy glial cells, which are generally related to interactions with one another (Pan & Monje, 2020) and eventually, the immune system (Salvador et al., 2021). The other factor which is at play is Brain-Derived Neurotrophic Factor, which is a critical mediator for activity-associated plasticity through a complex neuronal response to the process of depolarisation (Hogan et al., 2020).

Such findings are significant in light of the information that we have about the sequelae of COVID, including manifestations such as anosmia, ataxia, dysgeusia and altered psychological status (Chaudhury et al., 2021), which are deep-seated biological responses to CNS involvement and dysregulation. There is consensus among experts that low-intensity Transcranial Electric Stimulation is safe as long as tested and accepted protocols are followed and that the devices which are used follow established engineering risk-management procedures (Antal et al., 2022) though the data on neuroplasticity in conjunction with COVID-19 is unavailable at this time, though it has been demonstrated that plasticity can be induced at a systems level when neurostimulation is used (Bhattacharya et al., 2021).

Neuroinflammation Modulation

There are significant gaps in our understanding of whether COVID-19 attacks the CNS directly or through activation of the peripheral immune system and immune cell infiltration (Almutairi et al., 2021). Neuroinflammation is induced when the neurovascular unit, including neurons, microglia and astrocytes, responds to stimuli (Liu et al., 2020) with several reactive components, which contribute to neuroinflammation, including inflammatory mediators, circulating immune or inflammatory cell infiltration, CSF factors and glial cell activation.

An in-print PET study, at this time, confirms the same, where 12 participants with Long COVID were matched against 43 normative healthy controls, showing significantly increased neuroinflammation including in the midcingulate, anterior cingulate cortex, corpus callosum, thalamus, basal ganglia and at the ventricular boundaries (VanElzakker et al., 2023). Another literature review demonstrates that among 765 COVID-19 patients, around 18 per cent showed neurological manifestations including encephalopathies, cerebrovascular pathologies, acute myelitis and Guillain–Barre syndrome, making it clear that SARS-CoV-2 has similar neuroinvasive capacities and causes neuroinflammatory events that may lead to short- and long-term neuropathologies that SARS-CoV showed in human and animal models (Yachou et al., 2020). The precise mechanisms by which non-invasive brain stimulation reduces neuroinflammation are not fully understood; however, isomorphic studies indicate that non-invasive neurostimulation may reduce neuroinflammation. These studies have been conducted on patients with Alzheimer's disease where both neurological and ophthalmological deficits are considered (Amadoro et al., 2023).

Autonomic Regulation

A review paper links SARS-CoV-2 with autonomic symptoms, where the virus has a clear affinity between the virus and neural and gastrointestinal tissues, producing several symptoms such as anorexia, dysgeusia, heartburn, belching, chest pain, regurgitation, diarrhoea, abdominal pain, bloating, irregular bowel movements and constipation (Elbeltagi et al., 2023). Additionally, it was demonstrated that in both the acute and chronic phases of the disease, tachycardia, labile blood pressure, muscular fatigue and shortness of breath are commonly observed (Becker, 2021).

There was a large effect size of intervention on cardiac autonomic regulation when using a single prefrontal tDCS session, showing an early potential to manage autonomic dysfunction (Pinto et al., 2023). Low-level tragus stimulation on 57 participants diagnosed with Postural Orthostatic Tachycardia Syndrome shows another promising therapeutic approach for attenuating autonomic imbalance (Wang et al., 2025).

Neuro-Endocrine Modulatory Systems

SARS-CoV-19 has also been shown to have an impact on the endocrine system, directly affecting cortisol signalling, the hypothalamic–pituitary–adrenal axis and the renin–angiotensin–aldosterone system, increasing the stress response and inflammation. This endocrine profile is varied and depends on various factors such as illness severity, patient history and hormonal status (Zhao & Wu, 2022).

This is perhaps the most complicated sequelae among the myriad things that happen to one's physiology following a COVID-19 infection, with the gut playing a central role in these processes, even while evidence remains inconclusive (Zollner et al., 2024). A 2021 study demonstrated that two female Long COVID-19 patients, when treated for 10–13 days with alternating current stimulation of the eyes and brain, showed improved cognitive function due to the restoration of vascular autoregulation through a proposed mechanism of reoxygenating hypometabolic neurons leading to downstream effects (Sabel et al., 2021).

Combination Therapies in Long COVID Rehab

While much of the current research emphasises short-term symptom relief, the chronic nature of Long COVID emphasises durable strategies. This means integrating healthcare infrastructures, customisation and home-based strategies that allow longitudinal interventions by creating participative and repetitive engagement without further burdening the healthcare system. It is the consensus that a combination intervention, which includes cognitive training, exercise rehabilitation and targeted nutritional interventions, helps reduce Long COVID's cognitive issues.

Cognitive Training

One pilot study examined the effects of a dual-task training on cognitive performance, when assessed by the Trail Making Test A and B, and the Frontal Assessment Battery, a cardiovascular and muscular assessment was also performed using the Six-Minute Walking Test and Handgrip Endurance along with TMS to investigate cortical excitability with the results suggesting that this rehabilitation strategy could be feasible in supporting cognitive recovery (Deodato et al., 2024).

Neurofeedback and Biofeedback

The modalities of neurofeedback and biofeedback are usually employed to modulate the autonomic nervous system, and thus, several interventions have been studied. However, none of the trials show the inclusion of a non-invasive neurostimulation modality.

A one-month at-home intervention for 58 healthcare workers showed that there was an improvement in symptoms such as confusion, chest pain, headache and dizziness, though no changes were found in cardiac parasympathetic modulation after a resonant slow-paced breathing intervention that was meant to stimulate both central nervous system and parasympathetic pathways (Mauro et al., 2024). One study trialled the use and feasibility of a Heart Rate Variability Biofeedback programme using a standardised slow diaphragmatic breathing technique on 13 participants, showing that the intervention was beneficial (Corrado et al., 2024).

Another biofeedback intervention using the Biosound Therapy System, which is a multimodal biofeedback, vibroacoustic system, showed cognitive improvements in the 15 individuals on whom it was piloted (Korapatti et al., 2024). While another study evaluated the use of a mobile biofeedback app that coached participants on stress self-management, group analyses indicated that there were no significant differences between groups, though participants generally reported that the intervention helped manage stress better (Son et al., 2023).

Exercise Therapy

One of the leading hypotheses of Long COVID is that it is a result of a process called deconditioning or exercise intolerance, which may be linked to impaired oxygen uptake (Norweg et al., 2023), cardiovascular dysfunction at rest, and reduced skeletal muscle oxidative capacity that is suggestive of mitochondrial dysregulation (Jamieson et al., 2024).

Currently, empirical evidence is missing on the utilisation of non-invasive brain stimulation with the use of exercise therapies. While a study demonstrates that there is a utility to exercise therapy in improving function in the short-term post-infection, in physical and cognitive domains (Asimakos et al., 2023), there is limited evidence for support with people with Long COVID.

A key position paper, frankly, elucidates that Long COVID is an organic illness, not one rooted in dysfunctional individual beliefs; instead, a pattern of complex system disruptions has been noted physiologically, making exercise-based approaches an incorrect interventional strategy.

Other Modalities

In addition to neurostimulation, adjunctive therapies have also gained growing attention for their potential to support recovery in individuals with Long COVID. These approaches, while distinct from non-invasive neurostimulation, perhaps share some overlapping biological pathways such as inflammatory pathways, metabolic pathways or mitochondrial pathways. They may work through complementing neurostimulation by interacting with foundational mechanisms associated with Long COVID symptomatology.

Hyperbaric Oxygen Therapy (HBOT)

Hyperbaric Oxygen Therapy (HBOT) is a form of medical treatment that involves administering 100 per cent oxygen at increased atmospheric pressure to enhance oxygen delivery to tissues. In adjunctive use with COVID infections, a literature review demonstrates that HBOT reduces inflammation, enhances tissue oxygenation and modulates oxidative stress (Patel et al., 2025). A clinical trial of HBOT on cognitive improvement noted a high frequency of adverse events, even while a Data Safety Monitoring Board assessed that hyperbaric oxygen has a favourable safety profile (Kjellberg et al., 2023) in the context of Long COVID usage.

However, a study by Pawlik et al. (2024) noted that there was only a single prospective, randomised and placebo-controlled study with positive effects on global cognitive function, attention and executive function after ten sessions of HBOT (Robbins et al., 2021).

Previous literature, however, substantiates the role of HBOT for the improvement of chronic neurological dysfunction (Efrati & Ben-Jacob, 2014, Chen et al., 2023). This is further substantiated by the points made in another systematic review which demonstrate that TMS and tDCS when combined with HBOT and/or physical rehabilitation show positive effects on consciousness and cognitive recovery (Liu et al., 2020) and this may provide a rationale for future studies that wish to utilise a multimodal approach to Long COVID interventions using HBOT.

Nutrition Therapy

Chemosensation or chemosensory perception refers to a unique function of sensations and sensory systems—olfaction, taste and the trigeminal chemosensory system—that are dedicated to the detection of chemicals in the environment (Purves et al., 2025), where a dysfunction in these senses becomes problematic to individuals as there are relations between these senses, ingestive behaviours, food choices, swallowing and other health-related behaviours (Hayes, 2015).

This is a critical signalling process that plays a pivotal role in nutrient uptake and associated behaviours in living organisms. Nutrition therapy is widely used in the context of other disorders such as epilepsy (Assmann et al., 2025), neurodiverse syndromes (Longhurst & Burnette, 2023), neuroinflammation and vascular endothelial dysfunction in diabetes-induced brain injury (Song et al., 2025) and Alzheimer's disease (Shrivastava & Bobhate, 2025).

In people affected with COVID, it has been found that there is an association between olfactory dysfunction and body mass index (Vilarello et al., 2023). The peculiarities of the acute COVID infection were explained by an impairment in visceral chemosensory systems leading to downstream effects such as metabolic effects (Caretta & Mucignat-Caretta, 2022). Within this framework, tDCS was shown as a promising clinical option for people to regain their sense of smell after its application to brain areas responsible for olfaction, despite the study showing lowered statistical significance (Gracia et al., 2023).

Challenges, Limitations and Ethical Considerations

One of the foremost challenges in applying neurostimulation to Long COVID is the heterogeneity in participant severity and symptom presentation. People experience a wide array of affective, cognitive, autonomic and behavioural symptoms that tend to fluctuate over time, which can obscure diagnosis and intervention planning. Individual variability in brain structures, immune status and comorbidities further contributes to responses in neuromodulatory interventions. This variability highlights the need for personalised intervention while also being a barrier to standardisation and scaled-up implementation.

While there remains a growing interest in non-invasive neurostimulation intervention for the purpose of intervening in Long COVID, there remains little to no consensus on stimulation protocols, including placement of device, target site, duration of intervention, intensity and session frequency.

Even in the presence of review and meta-analyses, most of the current protocols are experimental, and researchers have to navigate a wide range of professional opinions. There is a growing need for robust comparative studies and dose–response-type trialling to define the evidence-based standards that can guide replicable, real-world practice.

The above body of literature also demonstrates that the present body of research is composed mostly of pilot studies, case series and open-label trials with only a few randomised-controlled trials. These small sample sizes increase the risk of errors and make it difficult to account for confounding variables such as spontaneous recovery and placebo effects which are well-documented in non-invasive neurostimulation literature.

Although non-invasive neurostimulation modalities are generally safe, each modality carries specific risks that must be managed, ranging from skin irritation to risky alteration of autonomic function (Almalty et al., 2024), such as peripheral nerve damage (Chen et al., 2020). Additionally, access to neurostimulation is limited by cost, geography and availability of practitioners. Some equipment requires technical oversight, while others can be user-operated, creating a secondary issue of adherence to interventions.

Future Research Directions and Conclusions

Looking ahead, the future of neurostimulation in the treatment of Long COVID rests on making it more intelligent, personalised and grounded in robust, real-world evidence. One of the most promising directions involves the use of biomarkers—such as brain imaging patterns or immune system indicators—to match patients with the most effective treatment. This personalised approach could help overcome one of the biggest hurdles in this field: the wide variability in how Long COVID presents and responds to intervention. Equally important is the need for longitudinal studies that follow individuals over many months, or even years, to better understand how long the benefits of neurostimulation last, and whether top-up sessions might be needed.

Determining the right "dose" of stimulation—how strong it should be, how often it should be applied and which areas of the brain to target—is another essential step. At present, much of this remains uncertain. Comparative trials that assess one type of stimulation against another, such as TMS versus tDCS or taVNS, will help identify which methods work best for specific symptoms like fatigue, brain fog or executive dysfunction.

Meanwhile, emerging techniques such as transcranial alternating current stimulation (tACS) or low-intensity-focused ultrasound offer exciting new ways to influence brain activity more precisely. But none of these advances will be meaningful unless research includes diverse populations and is conducted across international and multicentre settings. Such collaboration is key to ensuring treatments are not only effective but also equitable and applicable across different healthcare systems and cultures.

Ultimately, the path forward is about blending science, innovation and compassion. With a careful, inclusive and evidence-based approach, neurostimulation could become a powerful tool in helping people regain cognitive function and quality of life after COVID-19.

References

Almutairi, M. M., Sivandzade, F., Albekairi, T. H., Alqahtani, F., & Cucullo, L. (2021). Neuroinflammation and its impact on the pathogenesis of COVID-19. *Frontiers in Medicine*, *8*. https://doi.org/10.3389/fmed.2021.745789

Amadoro, G., Latina, V., Stigliano, E., & Micera, A. (2023). COVID-19 and Alzheimer's disease share common neurological and ophthalmological manifestations: A bidirectional risk in the post-pandemic future. *Cells*, *12*(22), 2601. https://doi.org/10.3390/cells12222601

Antal, A., Luber, B., Brem, A.-K., Bikson, M., Brunoni, A. R., Cohen Kadosh, R., Dubljević, V., Fecteau, S., Ferreri, F., Flöel, A., Hallett, M., Hamilton, R. H., Herrmann, C. S., Lavidor, M., Loo, C., Lustenberger, C., Machado, S., Miniussi, C., Moliadze, V., & Nitsche, M. A. (2022). Non-invasive brain stimulation and neuroenhancement. *Clinical Neurophysiology Practice*, *7*, 146–165. https://doi.org/10.1016/j.cnp.2022.05.002

Asimakos, A., Spetsioti, S., Mavronasou, A., Gounopoulos, P., Siousioura, D., Dima, E., Gianniou, N., Sigala, I., Zakynthinos, G., Kotanidou, A., Vogiatzis, I., & Katsaounou, P. (2023). Additive benefit of rehabilitation on physical status, symptoms and mental health after hospitalisation for severe COVID-19 pneumonia. *BMJ Open Respiratory Research*, *10*(1), e001377. https://doi.org/10.1136/bmjresp-2022-001377

Assmann, M., Albrecht, I., Frenser, M., Marquardt, T., & Fischer, T. (2025). Database analysis of application areas and global trends in ketogenic diets from 2019 to 2024. *Nutrients*, *17*(9), 1478. https://doi.org/10.3390/nu17091478

Badran, B. W., Huffman, S. M., Dancy, M., Austelle, C. W., Bikson, M., Kautz, S. A., & George, M. S. (2022). A pilot randomized controlled trial of supervised, at-home, self-administered transcutaneous auricular vagus nerve stimulation (taVNS) to manage long COVID symptoms. *Bioelectronic Medicine*, *8*(1). https://doi.org/10.1186/s42234-022-00094-y

Becker, J. H., Lin, J. J., Doernberg, M., Stone, K., Navis, A., Festa, J. R., & Wisnivesky, J. P. (2021). Assessment of cognitive function in patients after COVID-19 infection. *JAMA Network Open*, *4*(10), e2130645. https://doi.org/10.1001/jamanetworkopen.2021.30645

Becker, R. C. (2021). Autonomic dysfunction in SARS-COV-2 infection acute and long-term implications COVID-19 editor's page series. *Journal of Thrombosis and Thrombolysis*, *52*(3), 692–707. https://doi.org/10.1007/s11239-021-02549-6

Bhattacharya, A., Mrudula, K., Sreepada, S. S., Sathyaprabha, T. N., Pal, P. K., Chen, R., & Udupa, K. (2021). An overview of noninvasive brain stimulation: Basic principles and clinical applications. *Canadian Journal of Neurological Sciences/Journal Canadien Des Sciences Neurologiques*, *49*(4), 479–492. https://doi.org/10.1017/cjn.2021.158

Bock, J., Kung, S., Ganesh, R., Hurt, R. T., & Lapid, M. I. (2024). Evaluating repetitive transcranial magnetic stimulation for neuropsychiatric symptoms in long COVID. *The Journal of ECT*. https://doi.org/10.1097/yct.0000000000001072

Caretta, A., & Mucignat-Caretta, C. (2022). Not only COVID-19: Involvement of multiple chemosensory systems in human diseases. *Frontiers in Neural Circuits*, *16*. https://doi.org/10.3389/fncir.2022.862005

Chaudhury, S. S., Sinha, K., Majumder, R., Biswas, A., & Mukhopadhyay, C. D. (2021). COVID-19 and central nervous system interplay: A big picture beyond clinical manifestation. *Journal of Biosciences*, *46*(2). https://doi.org/10.1007/s12038-021-00165-3

Chen, M., Wang, S., Li, X., Yu, L., Yang, H., Liu, Q., Tang, J., & Zhou, S. (2020). Non-invasive autonomic neuromodulation is opening new landscapes for cardiovascular diseases. *Frontiers in Physiology*, *11*. https://doi.org/10.3389/fphys.2020.550578

Chen, X., You, J., Ma, H., Zhou, M., & Huang, C. (2023). Efficacy and safety of hyperbaric oxygen therapy for fibromyalgia: A systematic review and meta-analysis. *BMJ Open*, *13*(1), e062322. https://doi.org/10.1136/bmjopen-2022-062322

Chuang, H.-J., Lin, C.-W., Hsiao, M.-Y., Wang, T.-G., & Liang, H.-W. (2023). Long COVID and rehabilitation. *Journal of the Formosan Medical Association*, *123*, S61–S69. https://doi.org/10.1016/j.jfma.2023.03.022

Cirillo, G., Pino, G. D., Capone, F., Ranieri, F., Florio, L., Todisco, V., Tedeschi, G., Funke, K., & Lazzaro, V. D. (2016). Neurobiological after-effects of non-invasive brain stimulation. *Brain Stimulation*, *10*(1), 1–18. https://doi.org/10.1016/j.brs.2016.11.009

Colzato, L. S., Elmers, J., Beste, C., & Hommel, B. (2023). A prospect to ameliorate affective symptoms and to enhance cognition in long COVID using auricular transcutaneous vagus nerve stimulation. *Journal of Clinical Medicine*, *12*(3), 1198. https://doi.org/10.3390/jcm12031198

Corrado, J., Iftekhar, N., Halpin, S., Li, M., Tarrant, R., Grimaldi, J., Simms, A., O'Connor, R. J., Casson, A., & Sivan, M. (2024). HEART Rate Variability Biofeedback for LOng COVID Dysautonomia (HEARTLOC): Results of a feasibility study. *Advances in Rehabilitation Science and Practice*, *13*. https://doi.org/10.1177/27536351241227261

Davis, H. E., McCorkell, L., Vogel, J. M., & Topol, E. J. (2023). Long COVID: Major findings, mechanisms and recommendations. *Nature Reviews Microbiology*, *21*(3), 133–146. https://doi.org/10.1038/s41579-022-00846-2

Deodato, M., Qualizza, C., Martini, M., Mazzari, L., Furlanis, G., Stella, A. B., & Manganotti, P. (2024). Efficacy of dual-task augmented reality rehabilitation in non-hospitalized adults with self-reported long COVID fatigue and cognitive impairment: A pilot study. *Neurological Sciences*, *45*(4), 1325–1333. https://doi.org/10.1007/s10072-023-07268-9

Efrati, S., & Ben-Jacob, E. (2014). Reflections on the neurotherapeutic effects of hyperbaric oxygen. *Expert Review of Neurotherapeutics*, *14*(3), 233–236. https://doi.org/10.1586/14737175.2014.884928

Elbeltagi, R., Al-Beltagi, M., Saeed, N. K., & Bediwy, A. S. (2023). COVID-19-induced gastrointestinal autonomic dysfunction: A systematic review. *World Journal of Clinical Cases*, *11*(22), 5252–5272. https://doi.org/10.12998/wjcc.v11.i22.5252

Gorenshtein, A., Liba, T., Leibovitch, L., Stern, S., & Stern, Y. (2024). Intervention modalities for brain fog caused by long-COVID: Systematic review of the literature. *Neurological Sciences*, *45*(7), 2951–2968. https://doi.org/10.1007/s10072-024-07566-w

Gracia, D. I., Ortiz, M., Candela, T., Iáñez, E., Sánchez, R. M., Díaz, C., & Azorín, J. M. (2023). Design and evaluation of a potential non-invasive neurostimulation strategy for treating persistent anosmia in post-COVID-19 patients. *Sensors*, *23*(13), 5880. https://doi.org/10.3390/s23135880

Hallett, M. (2007). Transcranial magnetic stimulation: A primer. *Neuron*, *55*(2), 187–199. https://doi.org/10.1016/j.neuron.2007.06.026

Hayes, J. E. (2015). An introduction to this special issue: Chemosensation and health. *Chemosensory Perception*, *8*(3), 109–111. https://doi.org/10.1007/s12078-015-9197-4

Hogan, M. K., Hamilton, G. F., & Horner, P. J. (2020). Neural stimulation and molecular mechanisms of plasticity and regeneration: A review. *Frontiers in Cellular Neuroscience*, *14*. https://doi.org/10.3389/fncel.2020.00271

Hou, Y., Gu, T., Ni, Z., Shi, X., Ranney, M. L., & Mukherjee, B. (2025). Global prevalence of long COVID, its subtypes and risk factors: An updated systematic review and meta-analysis. *MedRxiv (Cold Spring Harbor Laboratory)*. https://doi.org/10.1101/2025.01.01.24319384

Howland, R. H. (2014). Vagus nerve stimulation. *Current Behavioral Neuroscience Reports*, *1*(2), 64–73. https://doi.org/10.1007/s40473-014-0010-5

Jamieson, A., Al Saikhan, L., Alghamdi, L., Hamill Howes, L., Purcell, H., Hillman, T., Heightman, M., Treibel, T., Orini, M., Bell, R., Scully, M., Hamer, M., Chaturvedi, N., Montgomery, H., Hughes, A. D., Astin, R., & Jones, S. (2024). Mechanisms underlying exercise intolerance in long COVID: An accumulation of multisystem dysfunction. *Physiological Reports*, *12*(3). https://doi.org/10.14814/phy2.15940

Kjellberg, A., Hassler, A., Boström, E., Gharbi, S. E., Al-Ezerjawi, S., Kowalski, J., Rodriguez-Wallberg, K. A., Bruchfeld, J., Ståhlberg, M., Malin Nygren-Bonnier, Runold, M., & Lindholm, P. (2023). Hyperbaric oxygen therapy for long COVID (HOT-LoCO), an interim safety report from a randomised controlled trial. *BMC Infectious Diseases*, *23*(1). https://doi.org/10.1186/s12879-023-08002-8

Klírová, M., Adamová, A., Biačková, N., Laskov, O., Renková, V., Stuchlíková, Z., Odnohová, K., & Novák, T. (2024). Transcranial direct current stimulation (tDCS) in the treatment of neuropsychiatric symptoms of long COVID. *Scientific Reports*, *14*(1). https://doi.org/10.1038/s41598-024-52763-4

Korapatti, C., Vera, L., & Miller, K. (2024). Biosound Therapy as a treatment for long COVID patients: A pre-post pilot study. *EXPLORE*, *20*(5), 103000. https://doi.org/10.1016/j.explore.2024.04.004

Krames, E. S., Hunter Peckham, P., Rezai, A., & Aboelsaad, F. (2009). What is neuromodulation? *Neuromodulation*, 3–8. https://doi.org/10.1016/b978-0-12-374248-3.00002-1
Kuo, M.-F., Paulus, W., & Nitsche, M. A. (2014). Therapeutic effects of non-invasive brain stimulation with direct currents (tDCS) in neuropsychiatric diseases. *NeuroImage*, *85*, 948–960. https://doi.org/10.1016/j.neuroimage.2013.05.117
Liu, L., Liu, J., Bao, J., Bai, Q., & Wang, G. (2020). Interaction of microglia and astrocytes in the neurovascular unit. *Frontiers in Immunology*, *11*. https://doi.org/10.3389/fimmu.2020.01024
Longhurst, P., & Burnette, C. B. (2023). Challenges and opportunities for conceptualizing intuitive eating in autistic people. *International Journal of Eating Disorders*, *56*(12), 2189–2199. https://doi.org/10.1002/eat.24057
Mauro, M., Zulian, E., Bestiaco, N., Polano, M., & Larese Filon, F. (2024). Slow-paced breathing intervention in healthcare workers affected by long COVID: Effects on systemic and dysfunctional breathing symptoms, manual dexterity and HRV. *Biomedicines*, *12*(10), 2254. https://doi.org/10.3390/biomedicines12102254
Mischke, M., & Zaehle, T. (2025). Evaluating the efficacy of repetitive anodal transcranial direct current stimulation on cognitive fatigue in long COVID: A randomized controlled trial. *Brain Stimulation*. https://doi.org/10.1016/j.brs.2025.03.011
Noda, Y., Sato, A., Shichi, M., Sato, A., Fujii, K., Iwasa, M., Nagano, Y., Kitahata, R., & Osawa, R. (2022). Real world research on transcranial magnetic stimulation treatment strategies for neuropsychiatric symptoms with long-COVID in Japan. *Asian Journal of Psychiatry*, *81*, 103438. https://doi.org/10.1016/j.ajp.2022.103438
Norweg, A., Yao, L., Barbuto, S., Nordvig, A. S., Tarpey, T., Collins, E., Whiteson, J., Sweeney, G., Haas, F., & Leddy, J. (2023). Exercise intolerance associated with impaired oxygen extraction in patients with long COVID. *Respiratory Physiology & Neurobiology*, *313*, 104062. https://doi.org/10.1016/j.resp.2023.104062
Pan, Y., & Monje, M. (2020). Activity shapes neural circuit form and function: A historical perspective. *Journal of Neuroscience*, *40*(5), 944–954. https://doi.org/10.1523/jneurosci.0740-19.2019
Panagea, E., Messinis, L., Petri, M. C., Liampas, I., Anyfantis, E., Nasios, G., Patrikelis, P., & Kosmidis, M. (2024). Neurocognitive impairment in long COVID: A systematic review. *Archives of Clinical Neuropsychology*. https://doi.org/10.1093/arclin/acae042
Patel, S. H., Joseph, J. J., Gandhi, T. R., Mehta, A., & Shah, A. (2025). A review of emerging evidence and clinical applications of hyperbaric oxygen therapy. *Journal of Intensive Care Medicine*. https://doi.org/10.1177/08850666241313136
Pawlik, M., G Rinneberg, Koch, A., H Meyringer, Loew, T., & Kjellberg, A. (2024). Is there a rationale for hyperbaric oxygen therapy in the patients with Post COVID syndrome? *European Archives of Psychiatry and Clinical Neuroscience*, *274*(8), 1797–1817. https://doi.org/10.1007/s00406-024-01911-y
Pfoser-Poschacher, V., Keilani, M., Steiner, M., Schmeckenbecher, J., Zwick, R. H., & Crevenna, R. (2025). Feasibility and acceptance of transdermal auricular vagus nerve stimulation using a TENS device in females suffering from long COVID fatigue. *Wiener Klinische Wochenschrift*. https://doi.org/10.1007/s00508-025-02501-1
Pinto, T. P., Inácio, J. C., de Aguiar, E., Ferreira, A. S., Sudo, F. K., Tovar-Moll, F., & Rodrigues, E. C. (2023). Prefrontal tDCS modulates autonomic responses in COVID-19 inpatients. *Brain Stimulation*, *16*(2), 657–666. https://doi.org/10.1016/j.brs.2023.03.001
Puderbaugh, M., & Emmady, P. D. (2023, May). *Neuroplasticity*. https://pubmed.ncbi.nlm.nih.gov/32491743/
Purves, D., Augustine, G. J., Fitzpatrick, D., Katz, L. C., LaMantia, A.-S., McNamara, J. O., & Williams, S. M. (2025). *The Chemical Senses*. Nih.gov; Sinauer Associates. www.ncbi.nlm.nih.gov/books/NBK10864/
Raja Almalty, A., Hamed, S. H., Jebril, M. Y., & Abdelnour, H. M. (2024). The effect of electrical stimulation on skin vulnerability to irritants. *Skin Research and Technology*, *30*(2). https://doi.org/10.1111/srt.13591
Robbins, T., Gonevski, M., Clark, C., Baitule, S., Sharma, K., Magar, A., Patel, K., Sankar, S., Kyrou, I., Ali, A., & Randeva, H. S. (2021). Hyperbaric oxygen therapy for the treatment of long

COVID: Early evaluation of a highly promising intervention. *Clinical Medicine*, *21*(6), e629–e632. https://doi.org/10.7861/clinmed.2021-0462

Sabel, B. A., Zhou, W., Huber, F., Schmidt, F., Sabel, K., Gonschorek, A., & Bilc, M. (2021). Non-invasive brain microcurrent stimulation therapy of long-COVID-19 reduces vascular dysregulation and improves visual and cognitive impairment. *Restorative Neurology and Neuroscience*, *39*(6), 393–408. https://doi.org/10.3233/rnn-211249

Salvador, A. F., de Lima, K. A., & Kipnis, J. (2021). Neuromodulation by the immune system: A focus on cytokines. *Nature Reviews Immunology*, *21*(8), 526–541. https://doi.org/10.1038/s41577-021-00508-z

Shrivastava, S. R., & Bobhate, P. S. (2025). Exploring the role of nutrition in supporting brain health and reducing the risk of Alzheimer's disease. *Current Alzheimer Research*, *22*. https://doi.org/10.2174/0115672050397848250425060101

Son, C., Hegde, S., Markert, C., Zahed, K., & Sasangohar, F. (2023). Use of a mobile biofeedback app to provide health coaching for stress self-management: Pilot quasi-experiment. *JMIR Formative Research*, *7*, e41018. https://doi.org/10.2196/41018

Song, T., Wang, B., Li, Y., Zhao, Y., Li, J., Wang, Y., & Li, X. (2025). 1,25-D3 Protects diabetic brain injury through GLP-1R/PI3K/Akt pathway by experimental and molecular docking studies. *Mediators of Inflammation*, *2025*(1). https://doi.org/10.1155/mi/8217035

Stengel, C., Sanches, C., Toba, M. N., & Valero-Cabré, A. (2022). Things you wanted to know (but might have been afraid to ask) about how and why to explore and modulate brain plasticity with non-invasive neurostimulation technologies. *Revue Neurologique*, *178*(8), 826–844. https://doi.org/10.1016/j.neurol.2021.12.014

Sudbrack-Oliveira, P., Razza, L. B., & Brunoni, A. R. (2021). Non-invasive cortical stimulation: Transcranial direct current stimulation (tDCS). *International Review of Neurobiology*, 1–22. https://doi.org/10.1016/bs.irn.2021.01.001

Tubig, P. A., & Klein, E. (2024). Cognitive enhancement as transformative experience: The challenge of wrapping one's mind around enhanced cognition via neurostimulation. *Cambridge Quarterly of Healthcare Ethics*, *33*(4), 532–547. https://doi.org/10.1017/s0963180124000057

VanElzakker, M. B., Bues, H. F., Brusaferri, L., Kim, M., Saadi, D., Ratai, E.-M., Dougherty, D. D., & Loggia, M. L. (2023). Neuroinflammation in post-acute sequelae of COVID-19 (PASC) as assessed by [^{11}C]PBR28 PET correlates with vascular disease measures. *BioRxiv (Cold Spring Harbor Laboratory)*. https://doi.org/10.1101/2023.10.19.563117

Vilarello, B. J., Jacobson, P. T., Tervo, J. P., Gallagher, L. W., Caruana, F. F., Gary, J. B., Saak, T. M., Gudis, D. A., Joseph, P. V., Goldberg, T. E., Devanand, D. P., & Overdevest, J. B. (2023). BMI increases in individuals with COVID-19-associated olfactory dysfunction. *Nutrients*, *15*(21), 4538. https://doi.org/10.3390/nu15214538

Wang, Z., Zhu, T., Li, X., Lai, X., & Chen, M. (2025). Tragus nerve stimulation attenuates postural orthostatic tachycardia syndrome in post COVID-19 infection. *Clinical Cardiology*, *48*(3). https://doi.org/10.1002/clc.70110

Whitaker-Hardin, B., McGregor, K. M., Uswatte, G., & Lokken, K. (2025). A narrative review of the efficacy of long COVID interventions on brain fog, processing speed, and other related cognitive outcomes. *Biomedicines*, *13*(2), 421. https://doi.org/10.3390/biomedicines13020421

Yachou, Y., El Idrissi, A., Belapasov, V., & Ait Benali, S. (2020). Neuroinvasion, neurotropic, and neuroinflammatory events of SARS-CoV-2: Understanding the neurological manifestations in COVID-19 patients. *Neurological Sciences*, *41*(10), 2657–2669. https://doi.org/10.1007/s10072-020-04575-3\

Zhang, J. X., & Zhang, J. J. (2023). Case report of improvement in long-COVID symptoms in an air force medic treated with transcranial magnetic stimulation using electro-magnetic brain pulse technique. *Military Medicine*, *188*(11–12), e3711–e3715. https://doi.org/10.1093/milmed/usad182

Zhao, S., Martin, E. M., Reuken, P. A., Scholcz, A., Ganse-Dumrath, A., Srowig, A., Utech, I., Kozik, V., Radscheidt, M., Brodoehl, S., Stallmach, A., Schwab, M., Fraser, E., Finke, K., & Husain, M. (2024). Long COVID is associated with severe cognitive slowing: A multicentre cross-sectional study. *EClinicalMedicine*, *68*, 102434. https://doi.org/10.1016/j.eclinm.2024.102434

Zhao, Y., & Wu, X. (2022). Influence of COVID-19 vaccines on endocrine system. *Endocrine*, *78*(2), 241–246. https://doi.org/10.1007/s12020-022-03119-3

Zheng, Z. S., Simonian, N., Wang, J., & Rosario, E. R. (2024). Transcutaneous vagus nerve stimulation improves Long COVID symptoms in a female cohort: A pilot study. *Frontiers in Neurology*, *15*. https://doi.org/10.3389/fneur.2024.1393371

Zhi, J., Zhang, S., Huang, M., Qin, H., Xu, H., Chang, Q., & Wang, Y. (2024). Transcutaneous auricular vagus nerve stimulation as a potential therapy for attention deficit hyperactivity disorder: Modulation of the noradrenergic pathway in the prefrontal lobe. *Frontiers in Neuroscience*, *18*. https://doi.org/10.3389/fnins.2024.1494272

Zollner, A., Meyer, M., Jukic, A., Adolph, T., & Tilg, H. (2024). The intestine in acute and long COVID: Pathophysiological insights and key lessons. *The Yale Journal of Biology and Medicine*, *97*(4), 447–462. https://doi.org/10.59249/pmie8461

13

SOCIAL BRAIN AND NEUROSTIMULATION

Applicability across Psychiatric Disorders

Debaleena Ghosh, Supratik Kundu, and Susmita Halder

13.1 Introduction

Humans are inherently social beings wired to live, communicate, and thrive as a part of a larger community. Our brains and bodies function best in relational contexts. From early development spanning through adulthood, our social behavior and interpersonal functioning play a pivotal role in shaping our identity, emotional well-being, and most importantly our survival. The ability of human beings to form and maintain relationships by comprehending emotions, interpreting social cues, and engaging in social reciprocal interactions is not just a social construct but a biological necessity. Human beings possess a unique capacity to bring together their cognitive resources in ways unmatched by other species using an advanced form of cognition known as "social cognition", a capacity to perceive others as being with minds and intentions akin to their own (Han et al., 2021). Underpinning these capacities is the human "social brain"—a network of neural systems specialized in navigating the complex social landscapes of observing, understanding, and interacting with others (Olson & D'Mello, 2022). Dysfunction within this network is a hallmark of numerous psychiatric disorders, leading to profound impairments in social functioning.

13.2 The Social Brain

Social cognition encompasses the mental processes that allow individuals to perceive, interpret, and respond to social cues, laying the foundation for effective social interaction and communication. At the core of social cognition is an ability we refer to as theory of mind (ToM), which involves attribution of independent mental states—such as beliefs, intentions, desires, and emotions—to oneself and others. Furthermore, this ability also allows individuals to explain and predict the behavior of others by understanding that their thoughts and feelings may differ from their own. ToM has two components: a cognitive component, which involves inferring others' thoughts, beliefs, and intentions, and an affective component, which relates to understanding others' emotions. The social brain also involves other systems, such as the mirror neuron system (MNS), which is engaged in the detection and

 DOI: 10.4324/9781003646662-16

simulation of others' actions and intentions and works alongside the mentalizing system to support both early (automatic) and late (controlled, evaluative) stages of social information processing.

In the seminal work of Brothers (1990), the social brain has been proposed as a distinct and specialized network of brain as being central to social cognition. This network includes the following areas—amygdala, orbitofrontal cortex (OFC), superior temporal sulcus (STS), and the fusiform gyrus (FFG). Recent functional magnetic resonance imaging (fMRI) studies also highlight the role of an additional interconnected network linking the parietal and temporal lobes to the prefrontal area. This cohesive network includes the parietal association cortex, OFC, dorsolateral prefrontal cortex (DLPFC), amygdala, anterior cingulate cortex (ACC), and superior temporal gyrus (STG). Some of these brain areas are engaged in the process of attribution of mental states to one self and others also known as ToM or mentalization (Frith & Frith, 2007) (Figure 13.1).

There is a broad spectrum of socio-cognitive processes central to mentalization ranging from basic perceptual processes like detecting facial expressions and biological motion (Frith, 2007; Pelphrey & Carter, 2008) to more complex capacities of emotional perception (Olsson & Ochsner, 2008). The higher end of this spectrum includes more abstract, meta representational abilities that enable us to adopt an intentional stance—the understanding that others behave based on their own mental states, such as beliefs, desires, and intentions (Dennett, 1988).

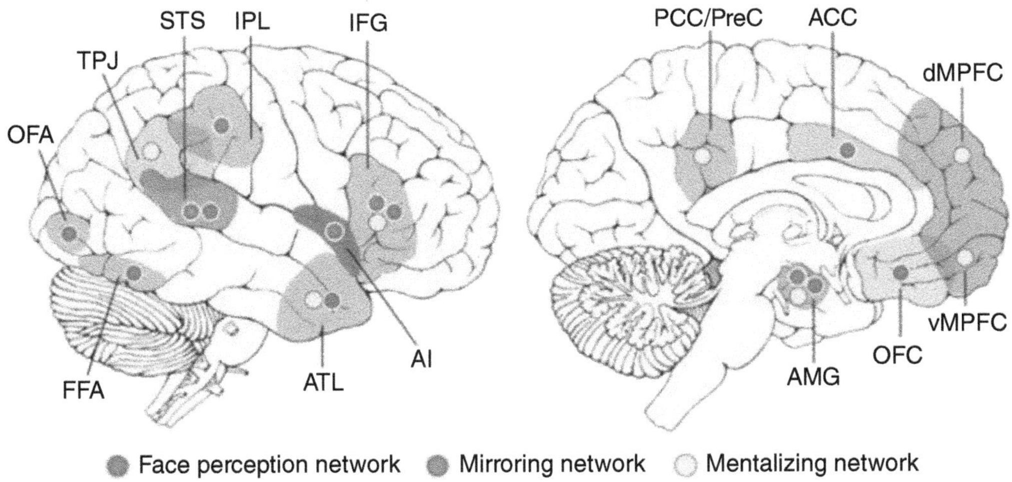

Figure 13.1 The key areas of the social brain network.

Anterior cingulate cortex (ACC); Anterior insula (AI); Amygdala (AMG); Anterior temporal lobe (ATL); Dorsomedial prefrontal cortex (dMPFC); Fusiform face area (FFA); Inferior frontal gyrus (IFG); Inferior parietal lobule (IPL); Orbitofrontal cortex (OFC); Occipital face area (OFA); Posterior cingulate cortex/precuneus (PCC/PreC); Superior temporal sulcus (STS); Temporoparietal junction (TPJ); ventromedial prefrontal cortex (vMPFC).

Source: Reference: Wang, Y., Metoki, A., Alm, K., & Olson, I. (2017). White matter and social cognition (open access).

13.3 Functional Domains of the Social Brain

The neural foundation of social cognition is central to the expanding domain of social cognitive neuroscience. It emphasizes a network-based, hierarchical, and modular organization reflecting both specialized processing and functional integration across regions (Ochsner, 2007, Ochsner & Lieberman, 2001).

The social perception network comprising regions of posterior superior temporal sulcus (pSTS), lateral occipitotemporal cortex (LOTC), temporoparietal junction (TPJ), and fusiform gyrus (FFG) are specialized in processing socially relevant perceptual information—such as biological motion, facial expressions, gaze direction, and body posture—forming the sensory foundation for social cognition (Santavirta et al., 2023).

The medial prefrontal cortex (mPFC) including the dorsal mPFC cortex and anterior mPFC is crucial in:

i. ***Tasks of mentalization*** which involves understanding of behaviors of others in terms of mental states in terms of real-world social interactions (McCabe et al., 2001) or engaging in observation of social engagements (Iacoboni et al., 2004). The underlying construct behind these tasks involve prediction of people's behaviors in terms of their beliefs and intentions.
ii. ***Person perception***: Person perception tasks often require individuals to make judgments about the enduring traits or attitudes of individuals, either broad (Mitchell et al., 2002) or highly specific (Schmitz et al., 2004).
iii. ***Self-perception:*** Self-perception tasks require individuals to evaluate their own enduring traits (Kelley et al., 2002) or momentary affective states (Ochsner et al., 2004). Across these different paradigms—whether involving self or others, and whether focusing on transient emotions or stable dispositions—the common thread is representation of mental states onself or others.

The mPFC, TPJ, and posterior cingulate cortex (PCC) form the system supporting the mentalization network or popularly known as the "theory of mind" (Han et al., 2021). The mPFC differentiates into dorsomedial and ventromedial parts, with the former linked to cognitive mentalizing and the latter to affective aspects of social cognition. In general, it deals with inferring others' beliefs, intentions, and mental states, enabling prediction and understanding of social behavior (Feng et al., 2021; Van Overwalle, 2009).

Another functional aspect of the social brain involves the emotion and reward processing network which involves processing emotions related to social scenarios, evaluation of trustworthiness, social rewards, and guides motivation for social interaction and exchange of information. It involves activation of the network comprising the amygdala, OFC, ventromedial prefrontal cortex (vmPFC), and ventral striatum (VS) (Han et al., 2021).

The social brain and the mirror neuron network are closely connected, forming the neural foundation for key aspects of social cognition such as empathy, imitation, action understanding, and emotional resonance. The MNS comprises specialized neurons—primarily located in the inferior frontal gyrus (IFG), inferior parietal lobule (IPL), and STS—that activate both during the execution of an action and when others are observed doing the same action. This mirroring mechanism supports internal simulation of others' actions, intentions, and emotions, facilitating social learning and communication (Jeon et al., 2018). The MNS is not isolated, thereby interacting with other regions of the social

Table 13.1 Functional Aspects of Social Brain and Corresponding Brain Regions

Domains of Social Brain	*Brain Regions Involved*	*Key Functioning*
Social perception	Posterior superior temporal sulcus (pSTS), lateral occipitotemporal cortex (LOTC), temporoparietal junction (TPJ), and fusiform gyrus (FFG)	Facial emotion processing, gestures, body language, biological motion
Theory of mind	Medial prefrontal cortex (mPFC), temporoparietal junction (TPJ), and posterior cingulate cortex (PCC)	Inference of beliefs, intentions, distinction between self and other
Emotion and reward processing	Amygdala, orbitofrontal cortex (OFC), ventromedial prefrontal cortex (vmPFC), and ventral striatum (VS)	Evaluation of emotions, social reward processing and motivation
Action understanding	Inferior frontal gyrus (IFG), inferior parietal lobule (IPL), and superior temporal sulcus (STS)	Empathy, simulation of other's empathy

brain, including the limbic system for emotional processing, the ACC, vmPFC, insula, and amygdala, to support complex social functions such as empathy and ToM. Recent functional neuroimaging studies show that observing and imitating facial expressions or emotional states activates both the mirror neuron network and associated areas, underpinning our ability to resonate with and understand others' feelings. Disruptions in the mirror neuron network have been linked to social deficits in conditions of autism spectrum, schizophrenia, and certain personality disorders, highlighting its critical role in social functioning. Overall, the MNS acts as a bridge between one's perception and action, enabling the social brain to interpret, predict, and respond to the behaviors and emotions of others, thus forming the basis of human social reciprocity (Table 13.1).

The network of social brain are not isolated, thereby showing strong functional connectivity and co-activation depending on the demands of the situation. Social brain regions integrate with broader networks such as fronto-striatal circuits and salience networks (SNs), reflecting the complexity of social cognition. Functional connectivity analyses reveal modular but interacting subnetworks, with some regions (e.g., TPJ) acting as hubs linking different modules (McCormick et al., 2018). The social brain functions as a hierarchical organization where lower-level perceptual inputs are fed into higher-level mentalizing and emotional networks, supporting increasingly abstract social cognition. This ascertains the flexible and simultaneous integration of sensory input, emotional valuation, and cognitive interpretation (Santavirta et al., 2023). Social cognition is embedded in a multilayer framework combining neural networks and actual social networks (Mastrandrea et al., 2017). In conclusion, on one hand the social brain network process social information and decide on sharing behaviors, while social networks in our environment determine the pathways of information transmission among individuals in their interpersonal relationships.

13.4 Socio-Cognitive Deficits in Psychiatric Context

Socio-cognitive deficits are a hallmark feature of several psychiatric conditions, profoundly affecting an individual's ability of managing social interactions in day-to-day life.

In psychiatric context, deficits in social cognition are not just prevalent but often central to the distress and dysfunction experienced by individuals. These impairments can manifest as difficulty in recognizing and expressing emotions, understanding others' intentions and actions (ToM), engaging in empathetic responses, and adjusting one's behavior in accordance to the demands of the social situation.

While biological factors namely genetic predisposition, neurochemical imbalances, and structural brain network abnormalities play a crucial role in the development of psychiatric conditions, deficits in social processing are increasingly recognized as central to one's daily functioning. These impairments in socio-cognitive domain do not merely co-occur with core symptoms of the psychiatric condition, rather, they often interact dynamically with them, shaping the trajectory and lived experience of individual dealing with the psychiatric condition. These deficits can exacerbate symptom severity, can pose challenge to treatment adherence, and also impede recovery by secluding individuals from supportive social environments. Thus, addressing social cognition is integral to conceptualizing the full complexity of psychiatric conditions.

Schizophrenia is a chronic and severe psychiatric condition that profoundly affects a person's thought, emotion, and behavior. The symptoms of schizophrenia are classified into positive symptoms—hallucinations and delusions, negative symptoms—diminished emotional expression, social withdrawal, lack of motivation, and poor self-care, disorganized symptoms-incoherent speech, confused or disorganized thinking, and unusual or repetitive movements and cognitive symptoms—deficits in attention, memory, executive functions, and additionally in planning, organizing, and understanding social cues. Schizophrenia is associated with pronounced socio-cognitive deficits most notably in ToM (Brüne, 2005). Recent research confirms that the socio-cognitive deficits in schizophrenia are more severe in comparison to impairments in other cognitive domains. Moreover, individuals with schizophrenia perform significantly worse than healthy controls on a range of ToM tasks, including first-order and second-order false belief, intention inference, and understanding indirect speech (Thibaudeau et al., 2021). ToM deficits in schizophrenia are strongly linked to clinical symptoms, particularly disorganization (disorganized thinking, communication, and behavior) and negative symptoms (such as social withdrawal and lack of motivation) highlighting the connection between impaired social cognition and impairment in domains of daily functioning (Brüne, 2005; Urbach et al., 2013). ToM deficits in schizophrenia are relatively specific and cannot be fully explained by general cognitive impairment; thereby representing a core disturbance in the social brain network of schizophrenia (Sprong et al., 2007). These deficits contribute to poor social functioning, reduced quality of life, and reduced clinical insight, as individuals are unable to absorb perspectives of others and recognize their own difficulties (Urbach et al., 2013).

Similar deficits in ToM in autism undermine the capacity of individuals to interpret and respond appropriately to verbal and non-verbal social cues, leading to difficulties in terms of social interaction and communication that are core features of the condition (Brewer et al., 2017). Over the years, research has shown that many individuals with autism struggle with performance on classic ToM tasks, such as understanding false beliefs, which generally develop in children by age four. However, there is considerable heterogeneity in ToM abilities among individuals with autism, depending on factors like type of task and assessed ToM component whether implicit or explicit (Rosello et al., 2020). Beyond ToM, autism is also associated with impairments in broader social-affective processing, including difficulties in emotion recognition, imitation, and social reciprocity (Andreou & Skrimpa,

2020). Overall, socio-cognitive deficits in autism encompass a multidimensional interplay of perceptual, emotional, and cognitive processes that collectively impact social functioning across the lifespan.

The relevance of social cognition has expanded to a broader range of psychiatric conditions. Research increasingly recognizes that deficits in social cognition in terms of difficulties in emotion perception, ToM, social cue processing, and social decision-making—are transdiagnostic features that contribute to social dysfunction across multiple mental health conditions.

In depression, social cognitive impairments commonly include difficulties in recognizing and interpreting emotional expressions, especially a bias toward processing negative emotions of sadness, anger, and fear (Ghosh & Halder, 2024). The severity of the deficits are linked with the severity of depressive symptoms, with more severe symptoms linked to poorer social cognition specially in spheres of social perception and heightened sensitivity to negative social cues (Knight & Baune, 2019). Moreover, deficits in executive functioning tends to mediate the relationship between specific social cognitive abilities like ToM or interpretation of meaning and dysfunction in psychosocial functioning, highlighting the complex interplay among cognitive, social cognitive, and behavioral domains in depression (Ghosh & Halder, 2024). These impairments contribute to difficulties in occupational functioning, interpersonal relationships, and overall quality of life of individuals (Knight & Baune, 2019).

In bipolar disorder, socio-cognitive deficits persist across mood states, including manic, depressive, and euthymic phases. It often exhibits impairments in ToM, emotion recognition, and social decision-making, in turn affecting their ability to navigate social situations effectively. These deficits are associated with poorer functional outcomes majorly unemployment and social isolation. During remission, certain socio-cognitive functions may improve; however, residual impairments often remain, suggestive of a more trait-like characteristic. Moreover, presence of psychotic symptoms influences the severity of socio-cognitive deficits in bipolar disorder and also the overall cognitive dysfunction, emphasizing increased need for comprehensive assessment and intervention (Gillissie et al., 2022).

In the context of anxiety disorders, socio-cognitive deficits have been particularly seen in social anxiety disorder (SAD). Elevated social anxiety has seen to be associated with impairments in specific social cognitive abilities, especially aspects of ToM, which involves understanding others' mental states and intentions. However, social anxiety does not seem to significantly impair broader social skills like emotion recognition. This suggests that deficits are more selective rather than global in nature (Pearcey et al., 2021). Another hallmark feature of anxiety disorders are cognitive distortions which are maladaptive thoughts and beliefs about oneself and how others perceive one's behavior. Some of the distortions common in SAD include catastrophization, overgeneralization, and personalization in social contexts. Cognitive distortions contribute to heightened self-consciousness and avoidance and safety behaviors in social situations, exacerbating the socio-cognitive difficulties (Kuru et al., 2018). Moreover, anxiety disorders are also associated with impairments in attention, executive functions, and memory, which can alongside affect social cognition by limiting the processing and integration of social information (Volel et al., 2018).

Obsessive-compulsive disorder (OCD) is characterized by intrusive obsessions and repetitive compulsions that cause significant distress, often leading to marked social and emotional difficulties. Socio-cognitive deficits in obsessive-compulsive disorder (OCD) have increasingly been acknowledged as important contributors to the social and functional impairments experienced by individuals. Beyond the direct impact of symptoms, OCD is associated with specific disturbances in social cognition, which include deficits in recognizing affective social cues—particularly facial expressions of disgust—and more general impairments in ToM or mentalizing the ability to infer others' mental states (Jansen et al., 2020). Recent evidence suggests that socio-cognitive deficits in OCD affect multiple domains, including social cue perception, mentalizing, empathy, and emotion regulation, all of which likely interact with core OCD symptoms to impair social relationships, lower quality of life, and poor social adjustment seen in individuals with OCD, which in some respects can be as severe as that observed in schizophrenia (Jansen et al., 2020).

Personality disorders are characterized by enduring patterns of maladaptive thoughts, emotions, and behaviors that significantly impair one's interpersonal functioning and quality of life. Central to these difficulties are socio-cognitive deficits which pose challenges for individuals to understand, interpret, and respond to social information. Research has consistently shown that particularly individuals with borderline personality disorder (BPD) exhibit pronounced impairments in ToM—the ability to infer others' mental states—which can manifest as both undermentalization, i.e., difficulty in accurate understanding of others' thoughts and overmentalization which involves excessive or inaccurate attribution of intentions (Galvez-Merlin et al., 2024). These socio-cognitive difficulties lead to a negative bias in social perception, where others are often seen as threatening or malicious, especially in contexts involving being rejected or subjected to abuse. Moreover, heightened emotional contagion and inadequate differentiation between emotions of self and others' further complicate social interactions, interpersonal relationships, and emotional regulation in BPD (Herpertz & Bertsch, 2014).

13.5 Neurostimulation in Understanding Social Brain

Neurostimulation is rapidly emerging as an approach in psychiatry which has the potential to offer novel targeted treatments for a range of mental health disorders that are often refractory to conventional modalities of treatment. It allows for focused and accurate targeting of specific brain areas and neural circuits involved in mental disorders, thereby minimizing the biological side effects common with medications. This spatial selectivity makes neurostimulation particularly valuable for treatment-resistant clients who do not respond adequately to traditional therapies. Neurostimulation in psychiatric practice utilizes changing magnetic field or electric current to induce changes in the pattern of neuronal firing. This can be done either invasively or non-invasively.

Invasive neurostimulation involves surgical implantation of electrodes either on cranial nerves or over brain tissue. These include vagal nerve stimulation (VNS) and deep brain stimulation (DBS). Non-invasive brain stimulation (NIBS) refers to a host of techniques, such as transcranial magnetic stimulation (TMS), transcranial electric stimulation (tES), and transcranial focused ultrasound stimulation (tFUS).

13.5.1 Mechanism of Action

The major mechanism of action includes electrical manipulation of neurons, i.e., either stimulation or inhibition. The electrical effect is brought about either by direct application of electric current which generates an electric field that reaches the designated area, or by indirect induction of electric field by alternating magnetic fields. The direct forms of stimulation can be divided into transcranial methods such as electroconvulsive therapy (ECT) or by intracerebral methods such as DBS. The intracerebral methods, therefore, are more focal. Thus, the intracerebral and indirect methods successfully bypass resistance offered by the scalp and skull. Further, it is described that the action of convulsive methods of stimulation rely on the seizure induced, regardless of the modality used for induction. The action of the non-convulsive methods depends upon repetitive stimulation of the implicated neural circuits.

Drugs rely on action at the pre- and post-synaptic receptors or intracellular targets. They undergo metabolism in the body and are widely distributed, leading to receptor action at unnecessary places, ultimately causing side effects. Neurostimulation utilizes localized direct stimulation of axons independent of receptor action hence eliminating the occurrence of drug-induced side effects. Therefore, neurostimulation holds great promise in adjunctive therapy and in treatment resistant cases.

The effects of neurostimulation can be immediate or lasting depending on the pattern in which it is applied. A single stimulus of adequate intensity can generate an action potential which travels down the axon and causes synaptic neurotransmitter discharge that propagates the stimulus leading to activation of the brain circuit. This has been utilized to map corticospinal and corticocortical excitability. Such research makes use of positive stimulus which excites the circuit (e.g., muscle twitch or visualization of phosphenes) or disruptive stimulus which inhibits the circuit (e.g., visual masking).

Going ahead, experimentation with repetitive pulses administered at set frequencies brought about cumulative effect and exerted more powerful effects than single stimulus only. Epstein et al. (1999) demonstrated this effect by application of repetitive TMS (rTMS) to the language area causing arrest of speech. Other invasive methods such as DBS and VNS would generate impulses at a set frequency and operate chronically, bringing about acute action for as long as the system was kept on. Sometimes continued modulation of the brain circuits led to changes which were maintained for prolonged periods of time. This is known as neuroplasticity. It is categorized into long-term potentiation (LTP) and long-term depression (LTD).

Research animal models have clearly shown efficacy and lasting effects of high frequency rTMS. Further, Wang et al. (1996) discovered that low frequency rTMS (1 Hz or less) showed prolonged inhibitory effect on motor cortex as opposed to excitatory effect shown by high frequency rTMS (5 Hz or greater). Thus, it was concluded that synaptic plasticity depended on the rate of firing, spike timing, and temporal and spatial aggregate of the stimuli arriving at the presynaptic level.

13.6 Transcranial Magnetic Stimulation

Faraday discovered electromagnetic stimulation in the 17th century. He recognized that a time-varying magnetic field can induce an electric current. Further, it was seen that this induced current may take a swirling form like eddies in the water, which prompted Foucault to name it eddy current in 1855. Kolin and colleagues were the first to use such induced currents to stimulate peripheral nerves in animal tissue. Research progressed and in 1985, Barker and colleagues were the first to stimulate the motor cortex using TMS.

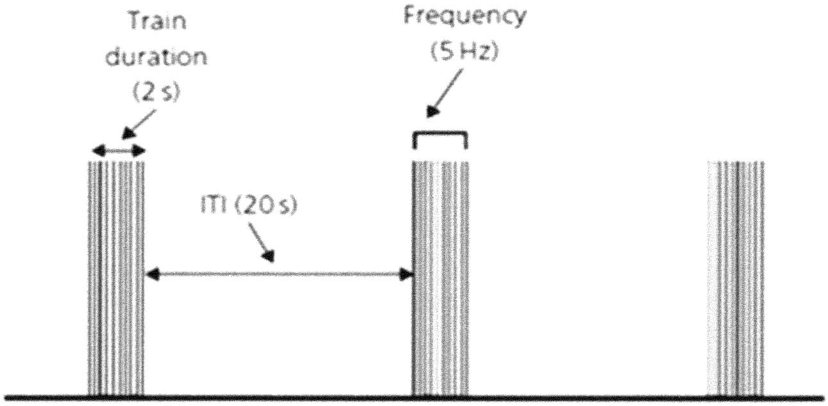

Figure 13.2 Repetitive transcranial magnetic stimulation (Hamani et al., 2016, p. 140).

The TMS circuit creates a short-lasting current pulse in a stimulating coil which is responsible to induce an electric field in the adjacent tissue. This electric field generated causes movement of ions in the intracellular and extracellular areas that eventually generates action potential that travels along the axon and facilitates release of neurotransmitters at the synapse.

TMS employs various pulse schemes, each serving distinct purposes. Single pulses are primarily used to estimate the motor threshold (MT), a critical parameter for determining stimulation intensity, which must be individually assessed. Paired-pulse TMS involves a conditioning stimulus (CS) followed by a test stimulus (TS), with protocols varying based on pulse amplitude, polarity, inter-stimulus intervals, and coil positioning. rTMS provides a train of pulses at specific frequencies, with high-frequency (HF) stimulation (above 5 Hz) generally considered excitatory and low-frequency (LF) stimulation (1 Hz or below) considered inhibitory, separated by inter-train intervals (ITI). Theta burst stimulation (TBS), modeled on the brain's natural theta rhythm (5 Hz), typically delivers a three-pulse burst at 50 Hz every 200 ms (5 Hz). It includes continuous TBS (cTBS), which bring about inhibition of activity and is administered for 40 s, and intermittent TBS (iTBS), which causes excitation of the targeted area and consists of 2-s trains separated by 10 s (Figures 13.2 and 13.3).

13.7 Transcranial Direct Current Stimulation

For about 2000 years humans have endeavored to utilize electrical currents for treatment of various ailments. Roman physicians would place a living electric ray on the scalp to treat headaches. In the 1900s ECT was developed by Ugo Cerletti and Lucio Bini. Transcranial direct current stimulation (tDCS) was introduced in the 1950s as a method of brain polarization.

tDCS is characterized by application of weak electric currents (about 1–2 mA) to the brain through two or more electrodes, applied on a properly cleaned area over the scalp. Unlike rTMS it does not generate action potential but has a polarizing effect on the resting membrane potential. Hence these techniques are known as neuromodulatory instead of neurostimulatory.

A tDCS setup comprises two electrodes, a cathode and anode. Both offer neuromodulatory effects. The stimulation is of two types, anodal, which promotes excitability and cathodal, which depresses excitability or has an inhibitory effect.

13.8 Non-Invasive Brain Stimulation and the Facets of Social Brain: Current Evidence

NIBS comprises a set of methods that can alter the activity of targeted areas in the brain through the action of magnetic fields known as TMS (Hallett, 2000) or electric currents known as transcranial electrical stimulation (Paulus, 2011). Research with NIBS has shown their ability to alter moral judgment and hostile intention attributions (Donaldson et al., 2015); exploring ToM, emotion recognition, and social skills (Schuwerk et al., 2014) and also understanding implicit attitudes and stereotyping (Marini et al., 2018).

Over the years, the potential of NIBS has been realized and it has been increasingly used as one of the modalities of intervention across psychiatric disorders. The initial use started with its applicability in treatment resistant cases or those where pharmacological treatment could not be tolerated, or pharmacological and psychosocial management showed sub-par effects. Over time, its applicability has expanded to encompass domains beyond the core symptomatology of psychiatric conditions, social cognition being one of them.

The underlying neural correlates from which social cognition arises is said to be a group of interconnected regions called as networks. The "central executive network" (CEN) is proposed to underlie higher executive functions such as moral judgments (Menon, 2011). The "salience network" (SN) is responsible for the ability to recognize and attend to salient stimuli and internal/external perspective taking (Menon & Uddin, 2010; Menon, 2011). The default mode network (DMN) as described by Menon (2011) comprises connections between the mPFC, PCC, and precuneus. The DMN (primarily), together with its connections to the CEN and SN, is considered to be fundamental to the processes of social cognition (Li et al., 2014; Menon, 2011).

NIBS has played an integral part in understanding the functional domains of the social brain. Some domains that have been identified are emotion recognition, ToM, and social

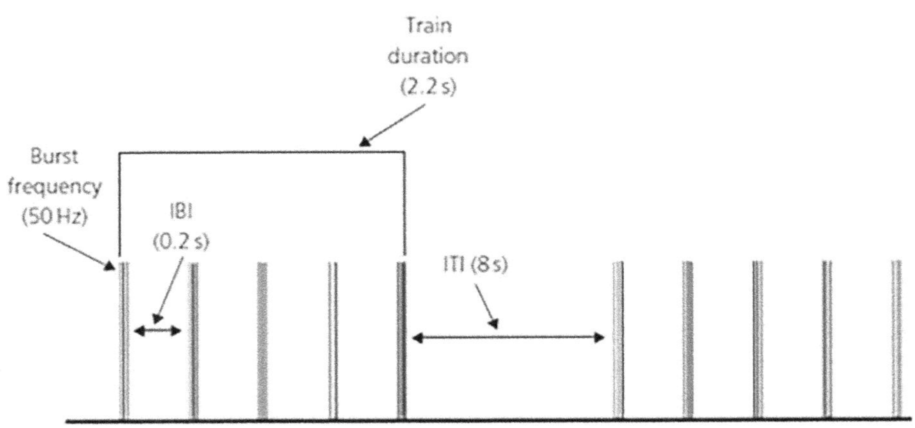

Figure 13.3 Intermittent theta burst stimulation (Hamani et al., 2016, p. 140).

processing. Social processing includes the sub-domains such as facial identity processing, mirror responses, and self–other processing.

13.8.1 Emotion Recognition and Theory of Mind

There are some areas such as the TPJ, an established ToM neural correlate, which form subsystems with the DMN (Li et al., 2014). The TPJ also has a dynamic connection to the MPFC, which again is related with ToM (Schurz et al., 2014; Van Overwalle, 2009). A successful social cognitive outcome is facilitated through the TPJ to the MPFC within the DMN (Li et al., 2014; Schurz et al., 2014; Van Overwalle, 2009).

NIBS research in social cognition has often targeted TPJ as an area of interest. There are studies where TMS, targeting TPJ, was used to disorganize a healthy control's ability to take non-self-perspective. This helped establish the role of TPJ in social cognition (Costa et al., 2008). Similarly, in a sham-controlled study by Mai et al. (2016), tDCS cathode stimulation was used on the right TPJ of healthy controls following which they were asked to attempt a task for cognitive empathy. Results showed that participants receiving active stimulation were less accurate as compared to a sham group. Another sham-controlled tDCS study (Sowden et al., 2015) showed that healthy participants receiving right TPJ anodal stimulation performed better in a task designed to detect if test subjects were lying. Thus, the group receiving active stimulation was better at inferring thoughts. Similar studies were done with mPFC which yielded evidence suggesting an integral role of TPJ in the process of social cognition.

There are many studies on emotion recognition which have targeted the DLPFC but the results are ambiguous (Rassovsky et al., 2015, 2018). Despite being a common area for regulating mood, behavior, and neurocognition, there is unclear proof for confirmed associations to social cognition (Schuwerk et al., 2014).

Thus, NIBS techniques show the ability to be used in research to identify and subsequently modulate areas underlying social cognition. There is potential to enhance and accelerate effects, when used in conjunction with, social cognitive training (Gainsford et al., 2020).

13.8.2 Social Processing

Facial identity processing is an important domain of social perception which has been explored using NIBS. TMS particularly has found extensive use in facial identity research. Most notably, Pitcher et al. (2007) demonstrated using double-pulse TMS, that disrupting the right occipital face area (OFA) impairs the ability to process individual face features.

Secondly, work on the MNS has demonstrated its causal role and plasticity. Fadiga et al. (1995) and Strafella & Paus (2000) assessed muscle-specific excitability during observation of action by measuring motor-evoked potentials from the primary motor cortex. Similarly, Catmur et al. (2011) used paired-pulse TMS and derived that premotor-motor cortex connections are critical in mirror neuron plasticity.

Further, we see that effective social interaction often requires the dynamic modulation of self and other-related representations. This capacity to flexibly shift between self and other-focused processing, known as self–other control, underpins several core social cognitive abilities, such as empathy, perspective-taking, and ToM (Ward & Banissy, 2015; de Guzman et al., 2016). Functional neuroimaging studies have consistently implicated the mPFC and the TPJ in self–other control (Brass et al., 2009). Crucially, NIBS techniques have provided causal evidence for the involvement of the TPJ in these processes. For instance, rTMS at 1 Hz over the right TPJ has been shown to impair performance on tasks for ToM (Costa et al., 2008; Young et al., 2010). Similarly, Wang et al. (2016) demonstrated that double-pulse TMS applied to the posterior rTPJ disrupts performance on perspective-taking tasks. This demonstrates how TMS can be used to disrupt normal functioning during task performance and discover the underlying neural correlates. Furthermore, theta-frequency (6 Hz) rTMS—as opposed to alpha-frequency (10 Hz)—was found to enhance perspective-taking, underscoring the impact of theta oscillatory activity in modulating self–other representations (Gooding-Williams et al., 2017).

13.9 Neurostimulation and the Social Brain: Therapeutic Applications in Psychiatry

13.9.1 Schizophrenia

Social cognition deficit, particularly ToM, is a hallmark feature of schizophrenia (Green et al., 2008). These deficiencies range from social disengagement to erroneous interpretation of others' intentions, with negative implications in quality of life, interpersonal interactions, treatment, and functional outcomes. ToM is the capacity to construct and envisage one's own and other people's mental states, such as thoughts, behaviors, beliefs, and intentions.

Schizophrenia is managed with pharmacological and psychosocial approaches with good effect on the psychotic symptoms but there is limited success in social cognitive disturbances. This brought about scope for neuromodulation as an alternative approach. The non-invasive aspect of brain stimulation, e.g., tDCS, has been drawing particular attention. tDCS alters neural activity in the brain by delivering low-amplitude (usually no more than 2 mA) electrical currents over a short period (generally no more than 30 min). This electrical discharge is facilitated between electrodes, i.e., anode and cathode. A meta-analysis of studies utilizing NIBS on schizophrenia showcased the ability of tDCS to improve working memory when applied on the DLPFC of the patients.

The strategies to reduce cognitive deficits in schizophrenia include pharmacotherapy, cognitive rehabilitation therapies, and invasive and NIBS. However, the evidence regarding the improvement in cognition with antipsychotic medication treatment has been mixed.

An open trial conducted by Yamada et al. (2022) showed clinically meaningful improvement in deficits in ToM ability in patients suffering from schizophrenia when subject to multisession anodal tDCS to the left STS corresponding to T3 target as per 10–20 EEG electrode placement. This is unique as previous studies involving application of tDCS to the DLPFC showed improvements in other aspects of cognition but minimal effects on ToM, reiterating the fact that neurocognition and social cognition have different underlying brain circuits. Therefore, the need arose to target the brain circuit appropriate for social cognition. The brain areas involved with social cognition are OFC, mPFC, STS, and amygdala. Functional disruption caused by schizophrenia is responsible for poor social cognitive outcomes. Among them Sasamoto et al. (2011) shows negative correlation between STS gray matter volume and social cognition deficits.

In 2025, Blumberger et al. conducted a feasibility study utilizing rTMS on DMPFC for schizophrenia patients with social cognition deficits. The authors attempted to target the altered functional connectivity between mentalizing network (MN) and simulation network (SN). Poor performers show increased connectivity leading to overactivation in the mentalizing regions (DMPFC, TPJ, PCC) causing overattribution of social cues. The study compared iTBS and 10 Hz rTMS, showing that iTBS is better at improving altered connectivity with almost medium effect size. Therefore, this paves the way for more extensive exploration of such interventions.

Wölwer et al. (2014) studied effects of (HF) rTMS over left DLPFC in schizophrenia patients and found improvement in facial affect recognition. They separated it from improvement secondary to overall clinical improvement. As DLPFC is largely involved in neurocognition, there was a possibility that overall improvement in attention benefitted social cognition deficits.

13.9.2 Autism

Individuals with autism can face significant difficulty in conducting their day-to-day affairs, which are brought about by the deficits in ToM functioning. They face trouble in navigating social interactions, making it difficult to form and maintain social relationships. It hampers their ability to gain independence in daily living skills (Sasson et al., 2020). Neurostimulation offers a promising approach which had the potential to directly modulate the underlying brain circuits and improve connectivity, which individually or in conjunction with psychosocial interventions would herald more effective ways of management.

A systematic review was done by Khalegi et al. (2020) in which studies utilizing tDCS and/or rTMS for therapeutic purposes in ASD were analyzed. The stimulation protocols in the studies largely varied which made it difficult to draw standardized conclusions. rTMS techniques used had conventional protocols, TBS protocols, and a single study using pico-Tesla TMS, applied to DLPFC, mPFC, or supplementary motor area (SMA). tDCS techniques utilized unilateral or bilateral anodal or cathodal stimulation of DLPFC, TPJ, or SMA. The tDCS studies showed improvement in working memory, language function, and motor planning on intervention over left or bilateral DLPFC. Others showed that application of rTMS to left pars triangularis led to better performance in the Boston Naming Test. Another study showed that using 8 Hz rTMS with an eye-hand integration training program facilitated better performance in low function patients with ASD (Panerai et al., 2014). Abujadi et al. (2018) and Ni et al. (2017) used iTBS over bilateral DLPFC which showed noteworthy improvements in social interactions, hyperactivity/non-compliance, irritability, and repetitive behaviors. Another systematic review conducted by Zhang et al. (2022) assessed studies from 2011 to 2021 in which the use of the Autism Treatment Evaluation Checklist (ATEC) given by Mahapatra et al. (2018) provided a good estimate of the improvement.

It was identified by Van Steenburg et al. (2017) that targeting bifrontal TPJ showed improvement in working memory in adults with high functioning autism whereas Wilson et al. (2018) showed improvement in verbal fluency. This improvement in working memory and verbal fluency translated to benefits seen in social cognition and social communication skill deficits as evaluated by Cheung et al. (2023). They applied transcranial pulse stimulation (which is a form of ultrasound therapy) to rTPJ showing reduction of 24% in the core symptoms of ASD as rated by Childhood Autism Rating Scale (CARS) and Clinical Global Improvement scale (CGI).

Desarkar et al. (2025) researched how rTMS affected overall brain plasticity and showcased differences in LTD in individuals with ASD. Kang et al. (2022) studied effects of rTMS on children with low-functioning ASD and showed improvement in peak alpha frequency.

Another group of studies explored the deficits through the Reading the Mind in the Eyes Test (RMET) which assesses ToM, i.e., the mentalizing ability. In this task, the individual is presented with pictures of only the eyes of a subject, and the individual is supposed to identify the emotion by merely observing the eyes of the subject (Baron-Cohen et al., 2001). Further advances showcased that the RMET activated the TPJ, which is already known to be involved in social cognition (Platek et. al., 2004). A case of autism and co-morbid spinocerebellar ataxia showed impaired RMET scores (Clausi et al., 2021) and it was hypothesized that damage to the posterior portions of cerebellum may have an implication in the functioning of the TPJ causing social cognition deficits (Van Overwalle et al., 2019). This opened up cerebellum as a potential correlate for intervention, which was confirmed by Clausi et al. (2022) as they used anodal tDCS over posterior cerebellum and it resulted in better scores on RMET as compared to cathodal tDCS stimulation. Similar effects were seen when low frequency TMS was applied to the posterior cerebellum hence providing an important area for future intervention (Heleven et al., 2021).

13.9.3 Depression

Depression is among the most disabling diseases worldwide which extracts a huge toll on productivity as exhibited by high Disability Affected Life Years (DALY) scores. There has been extensive research on TMS in depression. In 2008, the USFDA approved the use of rTMS on the left DLPFC to attain remission from depressive symptoms. However, focused research on the effects of NIBS on social cognition deficits are scarce.

The above-mentioned deficits include amplification of negative information processing, blunted responsiveness, emotional recognition biases, and inability to disengage from cycles of negative thought (Bistricky et al., 2011). Further research in neuroimaging by Koenigs & Grafman (2009) showed that the DLPFC and VMPFC contribute to affective information processing in subjects suffering from depressive disorders. These findings were utilized by Brennan et al. (2016), where they recruited patients suffering from depressive disorders and administered the Trail Making Test (TMT), Digit Span Test (DST), Emotion Recognition Task (ERT), before and after applying anodal tDCS at the left DLPFC. The ERT scores showed significant difference, thus establishing the role of left DLPFC in emotion recognition. This shows the viability of NIBS as a treatment form and opens up new areas for exploration.

13.9.4 Personality Disorders

BPD is characterized by persistent affective instability and impulsive behavior, with an increased sensitivity to interpersonal rejection, fear of abandonment, and a lack of stability of interpersonal relationships (American Psychiatric Association, 2022). These difficulties have their roots as well in social cognition (Roepke et al., 2013). Research shows that, visual and auditory stimuli when presented together, (assessed by complex social cognition tasks) are improperly processed (Minzenberg et al., 2006) which may lead to greater impulsivity in response to neutral stimuli (Albert et al., 2019).

ToM or mentalization is the more important aspects of SC. Galvez-Merlin et al. (2024) were the first to report that BPD patients exhibit overmentalization, undermentalization, and absence of mentalization errors. However, Sharp and Vanwoerden (2015), Somma et al. (2019) and Sharp et al. (2013) had concluded that BPD is associated with overmentalization errors. Subsequently much exploration had been done in these lines.

Overmentalization occurs when the subject draws excessively complex inferences and attributes meaning far more than warranted to seemingly neutral stimuli. For example, a person waving to a friend across the street does not get waved back and thus infers that this was intentional and hence gets insulted, rather than thinking that maybe the friend may not have seen him. Neuroimaging studies have shown that there is heightened activity at the amygdala and hypoactivity at the DLPFC which translates to exaggerated threat sensitivity and affective instability. The ACC and insula deal with salience detection (identifying and highlighting the most attention-grabbing aspect of a presented image) and self-referential processing. Deficits here lead to misrepresentation of neutral or ambiguous stimuli as threatening. Further, we see that altered activity in the PCC and hippocampus cause difficulty in integration of past experiences and current stimuli. The above-mentioned neural regions form various connections at various levels to form the DMN, SN, and the medial temporal lobe network thus showing aberrations in their activities. DMN altered connectivity leads to unstable self-image and difficulty detaching self-mental state from other's mental state. SN-altered connectivity causes misattribution of intentions. MTL dysfunction can cause biased social interpretations under stress (Pier et al., 2016).

Undermentalization attributes lesser than usual or overly simplistic or literal meanings to observed mental states. For example, a colleague hurriedly leaving a meeting may be construed as doing so because the meeting may have ended, rather than the fact that the colleague may have had an emotional or physical stressor. Absence of mentalization occurs when the subject fails to recognize the mental states of others. This involves the lack of attribution of any mental state at all. These aspects discovered form an important area for future exploration.

Fehlbaum et al. (2022) described that the pSTS is among a host of structures which underlie the function of mentalization which corroborates the findings of Koenigsberg et al. (2010) which showed an enhanced STS-activation in response to negative pictures.

13.9.5 *Emerging Applicability in Other Psychiatric Disorders*

We see that neuromodulation techniques have been used to explore treatment of neurocognitive domains such as attention, executive function, and memory with some success. Though not extensively researched, bipolar disorder has seen some application of NIBS. High frequency rTMS stimulation of the left DLPFC consistently showed improvement in bipolar depression but remission rates were not analyzed properly. Other stimulation protocols were not effective, though there is a lack of sham-controlled trials. No credible evidence of effectivity was discovered in mania or mixed states (Mutz, 2023).

There are studies which have explored rTMS as an adjunctive treatment for patients with refractory OCD, with mixed results depending on stimulation site and parameters. HF rTMS targeting the DLPFC initially showed symptom reductions, but larger, controlled

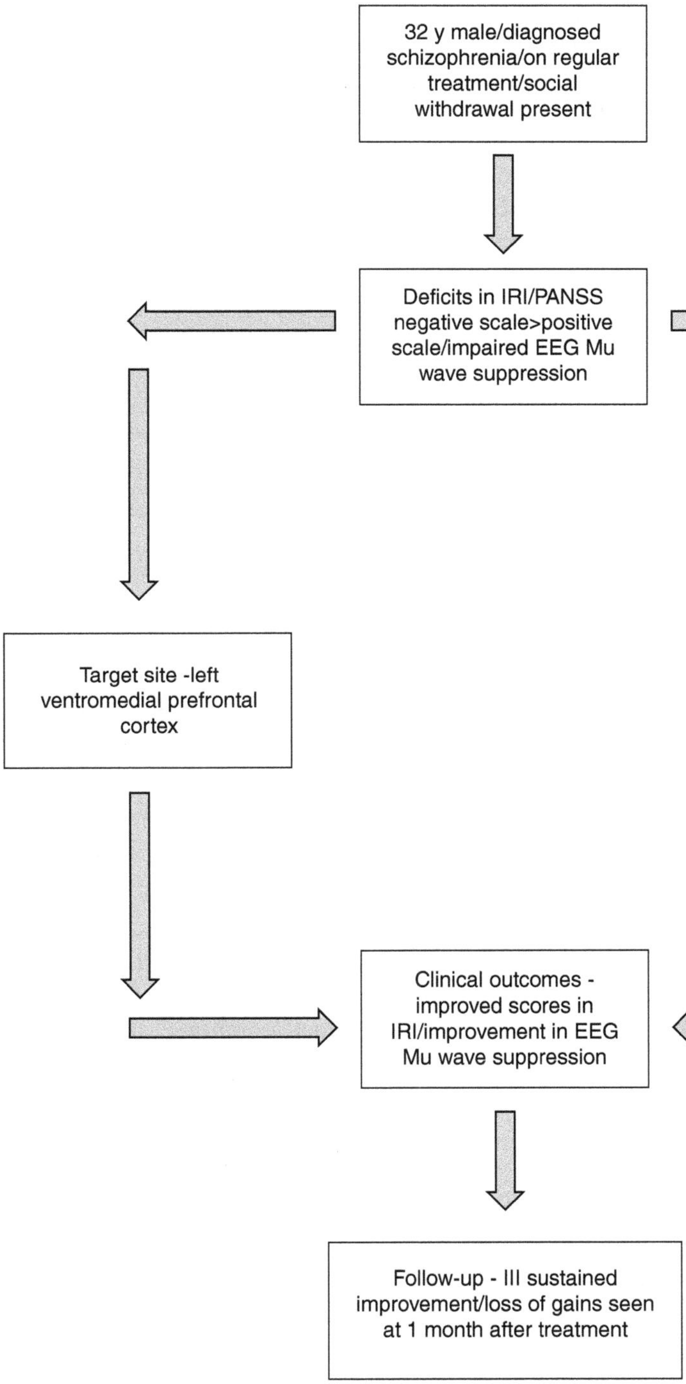

Figure 13.4 Flow diagram for case presentation.

studies often failed to replicate these findings, with both HF and LF rTMS to the DLPFC yielding inconsistent outcomes. Alternate areas such as SMA and pre-SMA were targeted showing similar results.

The anxiety spectrum disorders have utilized NIBS to formulate new paradigms of treatment. This area was explored extensively after depression got USFDA approval as a mode of management. The studies conducted in anxiety mostly include the use of TMS as opposed to tDCS due to its better spatial resolution (Polanía et al., 2018). The areas of interest targeted in studies of pathological anxiety include the right DLPFC, where both stimulation and inhibition seem to show improvement (Etkin et al., 2009). However, Ironside et al. (2019) report left DLPFC to be implicated in anxiety. These contradictory results provide scope for further research.

Current research evidence is suggestive of a substantial interest in the application of NIBS to the diverse spectrum of psychiatric conditions including bipolar disorder, OCD, and anxiety disorders but particular focus on social cognition deficits is lacking (Figure 13.4).

13.10 Neuromodulation for Empathy Deficits in Schizophrenia: A Case Presentation

13.10.1 Patient Profile

Mr. R is a 32-year-old male diagnosed with schizophrenia for the past ten years. He has been on regular treatment and is maintaining well. However, there are persistent symptoms of social withdrawal and somewhat diminished emotional expression, which is different from his previous healthy self. Despite adequate trials of antipsychotic medications and psychosocial interventions, his social cognition impairments remain, impacting his community functioning and quality of life.

13.10.2 Clinical Assessment

Mr. R was assessed on the Interpersonal Reactivity Index (IRI) which showed reduced scores in perspective taking (cognitive empathy) and empathic concern (affective empathy). Neuropsychological testing revealed impaired facial emotion recognition and ToM deficits. EEG Mu wave desynchronization showed impairment which confirmed the empathy deficits. Additionally, Mr. R was rated on Positive and Negative Syndrome Scale (PANSS) to quantify the schizophrenic symptoms. The negative scale score was higher than the positive scale score.

13.10.3 Neuromodulation Intervention

Given how the empathy deficits were refractory to pharmacological and psychosocial management, a neuromodulation protocol was initiated using iTBS targeting the left vmPFC. He received 20 sessions (2 sessions per working day for 2 weeks). The iTBS protocol consisted of bursts containing 3 pulses at 50 Hz repeated at 200-ms intervals for 2 s (i.e., at 5 Hz). A 2-s train of iTBS was repeated every 10 s for a total of 990 pulses per session (Bation, 2021). The stimulation intensity was set at 80% of the resting motor threshold (RMT) intensity.

13.10.4 Rationale

The vmPFC is closely related to empathy such that cases of empathy deficits are characterized by hypoactivation of it. The above stimulation protocol has shown effect in studies of schizophrenia patients where left DLPFC was shown to be hypoactive (Bation, 2021).

Outcomes: After the two-week course, Mr. R exhibited improvement in IRI scores and showed improved EEG Mu wave desynchronization. Behavioral observations noted some enhanced responsiveness to social cues. At one month follow-up assessments showed that improvement was not sustained.

Summary: This case illustrates the application of iTBS targeting the left vmPFC, as an adjunctive treatment to address empathy deficits in schizophrenia. By enhancing neural activity in key social cognition circuits, iTBS contributed to measurable improvements in both affective and cognitive empathy. Such protocols, integrated with psychosocial interventions, represent a promising approach for refractory socio-cognitive impairments in schizophrenia.

Challenges in Neurostimulation and Implication for Future Research

Psychiatric disorders also exhibit heterogeneity with respect to their socio-cognitive profiles and underlying social brain network. For example, schizophrenia is characterized by deficits in ToM and emotion recognition linked to fronto-temporal dysconnectivity, while autism involves early developmental disruptions in MNSs and social brain networks. This further complicates the identification of universal and uniform neurostimulation protocols and respective targets. Moreover, social cognition encompasses several domains and current assessment tools focus on broad symptom scales or general cognitive domains. Thus, effects of neurostimulation are varied for these domains, limiting the insight into domain-specific improvement. In this light, there is a need for more sensitive and comprehensive assessment tools which can measure the changes accurately. Also, there is a substantial interest in the general applicability of neurostimulation techniques to the diverse spectrum of psychiatric conditions in amelioration of symptoms but in a lot of psychiatric conditions like anxiety disorders, OCD, bipolar disorders, specific focus on social cognition deficits is still lacking.

Neurostimulation techniques hold significant promise for improving social cognition deficits across various psychiatric conditions. Despite encouraging preliminary findings in the field, there exists several identified challenges that limit the effective application of neurostimulation for targeting socio-cognitive deficits in the context of several psychiatric conditions.

Most brain stimulation techniques that are currently available primarily target superficial cortical areas (Kammer, 1998; Roth et al., 2007). This limitation presents challenges across various domains, social cognition being the most prominent one since it depends on extensive neural networks that involves a complex interplay between cortical and subcortical areas. For example, facial emotion recognition is based on activation of a network comprising the areas of vmPFC and somatosensory cortex, deeper cortical regions like the fusiform gyrus, and subcortical structures including the amygdala and insula (Adolphs, 2002; Fairhall & Ishai, 2007). The inability to target the deeper networks restricts our capacity to effectively understand the respective functions and dynamic connectivity within social brain networks. Use of certain TMS coil designs—beyond the standard figure-of-eight coil or use of indirect stimulation—does allow for subcortical stimulation or stimulation of the less accessible regions; however, they also have certain limitations. The broader current dispersion via other coils compromises focality, thereby resulting in unintended stimulation of adjacent cortical areas and further limiting the interpretability of results (Deng et al., 2013; Lu & Ueno, 2017). Future research focuses on discovering advanced methods which are more focal and precise in targeting deep brain regions. This would ensure our ability to differentiate the contributions of specific neural structures within the social brain and their respective socio-cognitive processes.

In application of neurostimulation, differentiating the specific contribution of a targeted brain region to their respective task performance can be challenging. This is due to the overlapping effects of widespread network-level activation to anatomical areas in close proximity. For instance, in regions like the TPJ, where functionally distinct subregions are situated in close proximity. The anterior TPJ is connected to the ventral attention network and is involved in social/non-social functions (Corbetta & Shulman, 2002), whereas the posterior TPJ is more selectively connected to the social cognition network and primarily supports social processing (Penton et al., 2020; Krall et al., 2015). Given the close anatomical positioning of these subregions, precisely targeting one without influencing the other presents a significant methodological challenge. Therefore, it is critical to ensure that experimental tasks designed to assess the effects of brain stimulation on social cognition do not inadvertently engage adjacent, non-social networks. Future studies should focus on designing tasks that are able to effectively disentangle these overlapping cognitive processes for accurate interpretation of the specific effects of brain stimulation on social versus general cognitive functions.

The efficacy of neurostimulation is clearly dependent on specific parameters—site of stimulation, intensity, frequency, and duration of stimulation. These parameters are in turn influenced by individual neuroanatomy and symptom profiles. Individual differences in brain structure, neurochemical mechanisms, and symptom severity further influences responsiveness and treatment outcome. Hence, the future of neurostimulation demands a need for a more tailor-made and personalized approach, potentially aided by neuroimaging or electrophysiological biomarkers. Another important challenge lies in sustaining neurostimulation-induced improvements in social cognition over time. Most studies on neurostimulation report short-term benefits. However, there is a lack of clarity regarding long-term efficacy and the need for maintenance sessions. Moreover, improvements observed in controlled experimental settings may not generalize to complex, real-world social environments. Longitudinal studies are required to address the issues of ecological validity.

To harness the growing potential of neurostimulation for addressing social cognition and associated deficits, it is essential to make use of advanced techniques of neuroimaging (functional magnetic resonance imaging and diffusor tensor imaging) and electrophysiology (electroencephalogram). This helps identify individualized neural targets and develop stimulation parameters tailored to each client's unique brain architecture and symptom profile. Moreover, treatment protocols can be developed in a manner that utilizes neuromodulatory techniques, to specifically target brain areas, while exploiting the ability of virtual reality to provide progressively intricate and life-like therapeutic settings. This may facilitate discovery of mechanisms of action, and the point of remediation, at which social cognition actually begins to improve. Integrating neurostimulation with other psychosocial interventions to address social cognition may help to produce synergistic effects, enhancing both neural plasticity and behavioral outcomes. Additionally, longitudinal studies and randomized controlled trials with extended follow-up are needed to assess the durability of neurostimulation effects on social cognition in the context of real-world social functioning. Lastly, it is important to develop comprehensive ethical guidelines; thereby, involving patients in shared decision-making will ensure responsible use of neurostimulation, balancing benefits and risks in future.

Lastly, as per Iannuzzo et al. (2025), it has been seen that the improvements are not long-lasting and require follow-up sessions.

13.11 Conclusion

Neurostimulation represents a promising frontier in addressing the various functional aspects of social cognition and associated deficits across a range of psychiatric conditions. It has also offered novel avenues to improve psychosocial functioning and behavioral outcomes beyond the reach of conventional pharmacological and psychotherapeutic interventions. However, the complexity and heterogeneity of social cognitive deficits, along with individual variability in brain structure and symptom presentation, pose a need for optimization and individualization of treatment protocols. Promising techniques are emerging and it may be possible to make use of the existing knowledge base of neuromodulation techniques to refine the methodology. Ethical considerations and patient-centered approaches remain paramount to ensure safe and effective application. Looking ahead, growing understanding of the mechanisms of neuromodulation and social cognition can help develop early intervention strategies. Continued interdisciplinary research and clinical innovation are essential to translate this potential into widespread clinical benefit in terms of ultimately enhancing social cognition and quality of life for individuals affected by one or more psychiatric conditions.

References

Abujadi, C., Croarkin, P. E., Bellini, B. B., Brentani, H., & Marcolin, M. A. (2018). Intermittent theta-burst transcranial magnetic stimulation for autism spectrum disorder: An open-label pilot study. *Brazilian Journal of Psychiatry, 40*, 309–311.

Adolphs, R. (2002). Neural systems for recognizing emotion. *Current Opinion in Neurobiology, 12*(2), 169–177. https://doi.org/10.1146/annurev.psych.60.110707.163514

Albert, J., Lopez-Martin, S., Arza, R., Palomares, N., Hoyos, S., Carretie, L., Diaz-Marsa, M., & Carrasco, J. L. (2019). Response inhibition in borderline personality disorder: Neural and behavioral correlates. *Biological Psychology, 143*, 32–40. https://doi.org/10.1016/j.biopsycho.2019.02.003

American Psychiatric Association. (2022). *Diagnostic and statistical manual of mental disorders* (5th ed., text rev.). American Psychiatric Association.

Andreou, M., & Skrimpa, V. (2020). Theory of mind deficits and neurophysiological operations in autism spectrum disorders: A review. *Brain Sciences*, 10(6), 393. https://doi.org/10.3390/brainsci10060393

Baron-Cohen, S., Wheelwright, S., Hill, J., Raste, Y., & Plumb, I. (2001). The "Reading the Mind in the Eyes" test revised version: A study with normal adults, and adults with Asperger syndrome or high-functioning autism. *Journal of Child Psychology and Psychiatry, 42*(2), 241–251.

Bation, R., Magnin, C., Poulet, E., Mondino, M., & Brunelin, J. (2021). Intermittent theta burst stimulation for negative symptoms of schizophrenia: A double-blind, sham-controlled pilot study. *NPJ Schizophrenia, 7*(1), 10. https://doi.org/10.1038/s41537-021-00138-3

Bistricky, S. L., Ingram, R. E., & Atchley, R. A. (2011). Facial affect processing and depression susceptibility: Cognitive biases and cognitive neuroscience. *Psychological Bulletin, 137*(6), 998–1028. https://doi.org/10.1037/a0025348

Brass, M., Ruby, P., & Spengler, S. (2009). Inhibition of imitative behaviour and social cognition. *Philosophical Transactions of the Royal Society B: Biological Sciences, 364*(1528), 2359–2367.

Brennan, S., McLoughlin, D. M., O'Connell, R., Bogue, J., O'Connor, S., McHugh, C., & Glennon, M. (2016). Anodal transcranial direct current stimulation of the left dorsolateral prefrontal cortex enhances emotion recognition in depressed patients and controls. *Journal of Clinical and Experimental Neuropsychology, 39*(4), 384–395. https://doi.org/10.1080/13803395.2016.1230595

Brewer, N., Young, R. L., & Barnett, E. (2017). Measuring theory of mind in adults with autism spectrum disorder. *Journal of Autism and Developmental Disorders, 47*(7), 1927–1941. https://doi.org/10.1007/s10803-017-3080-x

Brothers, L., Ring, B., & Kling, A. (1990). Response of neurons in the macaque amygdala to complex social stimuli. *Behavioural Brain Research, 41*(3), 199–213.

Catmur, C., Mars, R. B., Rushworth, M. F., & Heyes, C. (2011). Making mirrors: Premotor cortex stimulation enhances mirror and counter-mirror motor facilitation. *Journal of Cognitive Neuroscience, 23*(9), 2352–2362.

Cheung, T., Li, T. M. H., Lam, J., Fong, K., Chiu, L., Ho, J., Tse, A., Li, C.-T., Cheng, C., & Beisteiner, R. (2023). Effects of transcranial pulse stimulation on autism spectrum disorder: A double-blind, randomized, sham-controlled trial. *Brain Communications, 5*, fcad226. https://doi.org/10.1093/braincomms/fcad226

Clausi, S., et al. (2021). The cerebellum is linked to theory of mind alterations in autism: A direct clinical and MRI comparison between individuals with autism and cerebellar neurodegenerative pathologies. *Autism Research, 14*, 2300–2313.

Clausi, S., Lupo, M., Funghi, G., et al. (2022). Modulating mental state recognition by anodal tDCS over the cerebellum. *Scientific Reports, 12*, 22616. https://doi.org/10.1038/s41598-022-26914-4

Corbetta, M., Kincade, J. M., & Shulman, G. L. (2002). Neural systems for visual orienting and their relationships to spatial working memory. *Journal of Cognitive Neuroscience, 14*(3), 508–523.

Costa, A., Torriero, S., Oliveri, M., & Caltagirone, C. (2008). Prefrontal and temporo-parietal involvement in taking others' perspective: TMS evidence. *Behavioural Neurology, 19*(1–2), 71–74.

de Guzman, M., Bird, G., Banissy, M. J., & Catmur, C. (2016). Self–other control processes in social cognition: From imitation to empathy. *Philosophical Transactions of the Royal Society B: Biological Sciences, 371*(1686), 20150079.

Deng, Z. D., Lisanby, S. H., & Peterchev, A. V. (2013). Electric field depth–focality tradeoff in transcranial magnetic stimulation: Simulation comparison of 50 coil designs. *Brain Stimulation, 6*(1), 1–13.

Dennett, D. C. (1988). Précis of the intentional stance. *Behavioral and Brain Sciences, 11*(3), 495–505.

Desarkar, P. (2025). Neuroplasticity-based novel brain stimulation support intervention options for autistic population. *Frontiers in Human Neuroscience, 19*, 1522718.

Donaldson, P. H., Rinehart, N. J., & Enticott, P. G. (2015). Noninvasive stimulation of the temporoparietal junction: A systematic review. *Neuroscience & Biobehavioral Reviews, 55*, 547–572.

Epstein, C. M., Meador, K. J., Loring, D. W., Wright, R. J., Weissman, J. D., Sheppard, S., Lah, J. J., Puhalovich, F., Gaitan, L., & Davey, K. R. (1999). Localization and characterization of speech arrest during transcranial magnetic stimulation. *Clinical Neurophysiology, 110*(6), 1073–1079. https://doi.org/10.1016/s1388-2457(99)00047-4

Etkin, A., Prater, K. E., Schatzberg, A. F., Menon, V., & Greicius, M. D. (2009). Disrupted amygdalar subregion functional connectivity and evidence of a compensatory network in generalized anxiety disorder. *Archives of General Psychiatry, 66*(12), 1361–1372.

Fadiga, L., Fogassi, L., Pavesi, G., & Rizzolatti, G. (1995). Motor facilitation during action observation: A magnetic stimulation study. *Journal of Neurophysiology, 73*(6), 2608–2611.

Fairhall, S. L., & Ishai, A. (2007). Effective connectivity within the distributed cortical network for face perception. *Cerebral Cortex, 17*(10), 2400–2406.

Fehlbaum, L. V., Borbás, R., Paul, K., Eickhoff, S. B., & Raschle, N. M. (2022). Early and late neural correlates of mentalizing: ALE meta-analyses in adults, children and adolescents. *Social Cognitive and Affective Neuroscience, 17*(4), 351–366.

Feng, C., Eickhoff, S. B., Li, T., Wang, L., Becker, B., Camilleri, J. A., Hétu, S., & Luo, Y. (2021). Common brain networks underlying human social interactions: Evidence from large-scale neuroimaging meta-analysis. *Neuroscience and Biobehavioral Reviews, 126*, 289–303. https://doi.org/10.1016/j.neubiorev.2021.03.025

Frith, C. D., & Frith, U. (2007). Social cognition in humans. *Current Biology, 17*(16), R724–R732.

Gainsford, K., Fitzgibbon, B., Fitzgerald, P. B., & Hoy, K. E. (2020). Transforming treatments for schizophrenia: Virtual reality, brain stimulation and social cognition. *Psychiatry Research, 288*, 112974.

Galvez-Merlin, A., Lopez-Villatoro, J. M., de la Higuera-Gonzalez, P., de la Torre-Luque, A., Reneses-Prieto, B., Diaz-Marsa, M., & Carrasco, J. L. (2024). Social cognition deficits in borderline personality disorder: Clinical relevance. *Psychiatry Research, 331*, 115675. https://doi.org/10.1016/j.psychres.2023.115675

Ghosh, D., & Halder, S. (2024). Understanding the components of theory of mind in depression: An exploratory study. *Indian Journal of Social Psychiatry, 40*(3), 292–296. https://doi.org/10.4103/ijsp.ijsp_112_23

Gillissie, E. S., Lui, L. M. W., Ceban, F., Miskowiak, K., Gok, S., Cao, B., Teopiz, K. M., Ho, R., Lee, Y., Rosenblat, J. D., & McIntyre, R. S. (2022). Deficits of social cognition in bipolar disorder: Systematic review and meta-analysis. *Bipolar Disorders, 24*(2), 137–148. https://doi.org/10.1111/bdi.13163

Gooding-Williams, G., Wang, H., & Kessler, K. (2017). THETA-rhythm makes the world go round: Dissociative effects of TMS theta versus alpha entrainment of right pTPJ on embodied perspective transformations. *Brain Topography, 30*, 561–564.

Green, M. J., Waldron, J. H., Simpson, I., & Coltheart, M. (2008). Visual processing of social context during mental state perception in schizophrenia. *Journal of Psychiatry and Neuroscience, 33*(1), 34–42.

Hallett, M. (2000). Transcranial magnetic stimulation and the human brain. *Nature, 406*(6792), 147–150.

Hamani, C., Holtzheimer, P., Lozano, A. M., & Mayberg, H. (Eds.). (2016). *Neuromodulation in psychiatry*. Wiley-Blackwell.

Han, M., Jiang, G., Luo, H., & Shao, Y. (2021). Neurobiological bases of social networks. *Frontiers in Psychology, 12*, 626337. https://doi.org/10.3389/fpsyg.2021.626337

Heleven, E., van Dun, K., De Witte, S., Baeken, C., & Van Overwalle, F. (2021). The role of the cerebellum in social and non-social action sequences: A preliminary LF-rTMS study. *Frontiers in Human Neuroscience, 15*, 593821. https://doi.org/10.3389/fnhum.2021.593821

Herpertz, S. C., & Bertsch, K. (2014). The social-cognitive basis of personality disorders. *Current Opinion in Psychiatry, 27*(1), 73–77 https://doi.org/10.1097/YCO.0000000000000026

Iacoboni, M., Lieberman, M. D., Knowlton, B. J., Molnar-Szakacs, I., Moritz, M., Throop, C. J., & Fiske, A. P. (2004). Watching social interactions produces dorsomedial prefrontal and medial parietal BOLD fMRI signal increases compared to a resting baseline. *Neuroimage, 21*(3), 1167–1173.

Iannuzzo, F., Donia, F., Bette, L., Turiaco, F., & Bruno, A. (2025). Neuromodulation techniques in combination for a short-intensive treatment of depression and anxiety: A case report. *AIMS Neuroscience, 12*(2), 140–146. https://doi.org/10.3934/Neuroscience.2025009

Ironside, M., Browning, M., Ansari, T. L., Harvey, C. J., Sekyi-Djan, M. N., Bishop, S. J., ..., & O'Shea, J. (2019). Effect of prefrontal cortex stimulation on regulation of amygdala response to threat in individuals with trait anxiety: A randomized clinical trial. *JAMA Psychiatry, 76*(1), 71–78.

Jansen, M., Overgaauw, S., & De Bruijn, E. R. A. (2020). Social cognition and obsessive-compulsive disorder: A review of subdomains of social functioning. *Frontiers in Psychiatry, 11*, 118. https://doi.org/10.3389/fpsyt.2020.00118

Jeon, H., & Lee, S. H. (2018). From neurons to social beings: Short review of the mirror neuron system research and its socio-psychological and psychiatric implications. *Clinical Psychopharmacology and Neuroscience*, *16*(1), 18–31. https://doi.org/10.9758/cpn.2018.16.1.18

Kammer, T. (1998). Phosphenes and transient scotomas induced by magnetic stimulation of the occipital lobe: Their topographic relationship. *Neuropsychologia, 37*(2), 191–198.

Kang, J., Li, X., Casanova, M. F., Sokhadze, E. M., & Geng, X. (2022). Impact of repetitive transcranial magnetic stimulation on the directed connectivity of autism EEG signals: A pilot study. *Medical & Biological Engineering & Computing, 60*(12), 3655–3664.

Kelley, W. M., Macrae, C. N., Wyland, C. L., Caglar, S., Inati, S., & Heatherton, T. F. (2002). Finding the self? An event-related fMRI study. *Journal of Cognitive Neuroscience, 14*(5), 785–794.

Khaleghi, A., Zarafshan, H., Vand, S. R., & Mohammadi, M. R. (2020). Effects of non-invasive neurostimulation on autism spectrum disorder: A systematic review. *Clinical Psychopharmacology and Neuroscience, 18*(4), 527–552. https://doi.org/10.9758/cpn.2020.18.4.527

Knight, M. J., & Baune, B. T. (2019a). Social cognitive abilities predict psychosocial dysfunction in major depressive disorder. *Depression and Anxiety, 36*(1), 54–62. https://doi.org/10.1002/da.22844

Knight, M. J., & Baune, B. T. (2019b). The direct and indirect relationship between social cognition and psychosocial dysfunction in major depressive disorder. *Frontiers in Psychiatry, 10*, 347. https://doi.org/10.3389/fpsyt.2019.00347

Koenigs, M., & Grafman, J. (2009). The functional neuroanatomy of depression: Distinct roles for ventromedial and dorsolateral prefrontal cortex. *Behavioural Brain Research, 201*(2), 239–243.

Koenigsberg, H. W., Fan, J., Ochsner, K. N., Liu, X., Guise, K., Pizzarello, S., ..., & Siever, L. J. (2010). Neural correlates of using distancing to regulate emotional responses to social situations. *Neuropsychologia, 48*(6), 1813–1822.

Krall, S. C., Rottschy, C., Oberwelland, E., Bzdok, D., Fox, P. T., Eickhoff, S. B., ..., & Konrad, K. (2015). The role of the right temporoparietal junction in attention and social interaction as revealed by ALE meta-analysis. *Brain Structure and Function, 220*, 587–604.

Kuru, E., Safak, Y., Özdemir, İ., Tulacı, R. G., Özdel, K., Özkula, N. G., & Örsel, S. (2018). Cognitive distortions in patients with social anxiety disorder: Comparison of a clinical group and healthy controls. *The European Journal of Psychiatry, 32*(2), 97–104. https://doi.org/10.1016/j.ejpsy.2017.08.004

Li, W., Mai, X., & Liu, C. (2014). The default mode network and social understanding of others: What do brain connectivity studies tell us. *Frontiers in Human Neuroscience, 8*, 74.

Lu, M., & Ueno, S. (2017). Comparison of the induced fields using different coil configurations during deep transcranial magnetic stimulation. *PLoS One, 12*(6), e0178422.

Mahapatra, S., Vyshedsky, D., Martinez, S., Kannel, B., Braverman, J., Edelson, S. M., et al. (2018). Autism Treatment Evaluation Checklist (ATEC) norms: A "growth chart" for ATEC score changes as a function of age. *Children, 5*(2), 25. https://doi.org/10.3390/children5020025

Mai, X., Zhang, W., Hu, X., Zhen, Z., Xu, Z., Zhang, J., & Liu, C. (2016). Using tDCS to explore the role of the right temporo-parietal junction in theory of mind and cognitive empathy. *Frontiers in Psychology, 7*, 380.

Marini, M., Banaji, M. R., & Pascual-Leone, A. (2018). Studying implicit social cognition with noninvasive brain stimulation. *Trends in Cognitive Sciences, 22*(11), 1050–1066.

Mastrandrea, R., Gabrielli, A., Piras, F. et al. (2017). Organization and hierarchy of the human functional brain network lead to a chain-like core. *Scientific Reports, 7*, 4888. https://doi.org/10.1038/s41598-017-04716-3

McCabe, K., Houser, D., Ryan, L., Smith, V., & Trouard, T. (2001). A functional imaging study of cooperation in two-person reciprocal exchange. *Proceedings of the National Academy of Sciences, 98*(20), 11832–11835.

McCormick, E. M., van Hoorn, J., Cohen, J. R., & Telzer, E. H. (2018). Functional connectivity in the social brain across childhood and adolescence. *Social Cognitive and Affective Neuroscience, 13*(8), 819–830. https://doi.org/10.1093/scan/nsy064

Menon, V. (2011). Large-scale brain networks and psychopathology: A unifying triple network model. *Trends in Cognitive Sciences, 15*(10), 483–506.

Menon, V., & Uddin, L. Q. (2010). Saliency, switching, attention and control: A network model of insula function. *Brain Structure and Function, 214*, 655–667.

Minzenberg, M. J., Poole, J. H., & Vinogradov, S. (2006). Social-emotion recognition in borderline personality disorder. *Comprehensive Psychiatry, 47*(6), 468–474. https://doi.org/10.1016/j.comppsych.2006.03.005

Mitchell, J. P., Heatherton, T. F., & Macrae, C. N. (2002). Distinct neural systems subserve person and object knowledge. *Proceedings of the National Academy of Sciences, 99*(23), 15238–15243.

Mutz, J. (2023). Brain stimulation treatment for bipolar disorder. *Bipolar Disorders, 25*(1), 9–24. https://doi.org/10.1111/bdi.13283

Ni, H. C., Hung, J., Wu, C. T., Wu, Y. Y., Chang, C. J., Chen, R. S., et al. (2017). The impact of single session intermittent theta-burst stimulation over the dorsolateral prefrontal cortex and posterior superior temporal sulcus on adults with autism spectrum disorder. *Frontiers in Neuroscience, 11*, 255.

Ochsner, K. N. (2007). Social cognitive neuroscience. *Social psychology: Handbook of basic principles* (pp. 39–66). The Guilford Press.

Ochsner, K. N., Knierim, K., Ludlow, D. H., Hanelin, J., Ramachandran, T., Glover, G., & Mackey, S. C. (2004). Reflecting upon feelings: An fMRI study of neural systems supporting the attribution of emotion to self and other. *Journal of Cognitive Neuroscience, 16*(10), 1746–1772.

Ochsner, K. N., & Lieberman, M. D. (2001). The emergence of social cognitive neuroscience. *American Psychologist, 56*(9), 717.

Olson, H., & D'Mello, A. (2022, September 15). The Social Brain. Oxford Research Encyclopedia of Psychology. Retrieved 28 May. 2025, from https://oxfordre.com/psychology/view/10.1093/acrefore/9780190236557.001.0001/acrefore-9780190236557-e-781

Olsson, A., & Ochsner, K. N. (2008). The role of social cognition in emotion. *Trends in Cognitive Sciences, 12*(2), 65–71. https://doi.org/10.1016/j.tics.2007.11.010

Panerai, S., Tasca, D., Lanuzza, B., Trubia, G., Ferri, R., Musso, S., et al. (2014). Effects of repetitive transcranial magnetic stimulation in performing eye-hand integration tasks: Four preliminary studies with children showing low-functioning autism. *Autism, 18*, 638–650.

Paulus, W. (2011). Transcranial electrical stimulation (tES–tDCS; tRNS, tACS) methods. *Neuropsychological Rehabilitation, 21*(5), 602–617.

Pearcey, S., Gordon, K., Chakrabarti, B., Dodd, H., Halldorsson, B., & Creswell, C. (2021). Research review: The relationship between social anxiety and social cognition in children and adolescents: A systematic review and meta-analysis. *Journal of Child Psychology and Psychiatry, and Allied Disciplines, 62*(7), 805–821. https://doi.org/10.1111/jcpp.13310

Pelphrey, K. A., & Carter, E. J. (2008). Brain mechanisms for social perception: Lessons from autism and typical development. *Annals of the New York Academy of Sciences, 1145*(1), 283–299.

Penton, T., Catmur, C., Banissy, M. J., Bird, G., & Walsh, V. (2020). Non-invasive stimulation of the social brain: The methodological challenges. *Social Cognitive and Affective Neuroscience, 17*(1), 15–25. https://doi.org/10.1093/scan/nsaa102

Pier, K. S., Marin, L. K., Wilsnack, J., & Goodman, M. (2016, March 31). The neurobiology of borderline personality disorder. *Psychiatric Times, 33*(3). www.psychiatrictimes.com/view/neurobiology-borderline-personality-disorder

Pitcher, D., Walsh, V., Yovel, G., & Duchaine, B. (2007). TMS evidence for the involvement of the right occipital face area in early face processing. *Current Biology, 17*(18), 1568–1573.

Platek, S. M., Keenan, J. P., Gallup, G. G., & Mohamed, F. B. (2004). Where am I? The neurological correlates of self and other. *Cognitive Brain Research, 19*(2), 114–122.

Polanía, R., Nitsche, M. A., & Ruff, C. C. (2018). Studying and modifying brain function with non-invasive brain stimulation. *Nature Neuroscience, 21*(2), 174–187.

Rassovsky, Y., Dunn, W., Wynn, J., Wu, A. D., Iacoboni, M., Hellemann, G., & Green, M. F. (2015). The effect of transcranial direct current stimulation on social cognition in schizophrenia: A preliminary study. *Schizophrenia Research, 165*(2–3), 171–174.

Rassovsky, Y., Dunn, W., Wynn, J. K., Wu, A. D., Iacoboni, M., Hellemann, G., & Green, M. F. (2018). Single transcranial direct current stimulation in schizophrenia: randomized, cross-over study of neurocognition, social cognition, ERPs, and side effects. *PLoS One, 13*(5), e0197023.

Roepke, S., Vater, A., Preißler, S., Heekeren, H. R., & Dziobek, I. (2013). Social cognition in borderline personality disorder. *Frontiers in Neuroscience, 6*, 195. https://doi.org/10.3389/fnins.2012.00195

Rosello, B., Berenguer, C., Baixauli, I., García, R., & Miranda, A. (2020). Theory of mind profiles in children with autism spectrum disorder: Adaptive/social skills and pragmatic competence. *Frontiers in Psychology, 11*, 567401. https://doi.org/10.3389/fpsyg.2020.567401

Roth, Y., Amir, A., Levkovitz, Y., & Zangen, A. (2007). Three-dimensional distribution of the electric field induced in the brain by transcranial magnetic stimulation using figure-8 and deep H-coils. *Journal of Clinical Neurophysiology, 24*(1), 31–38.

Santavirta, S., Karjalainen, T., Nazari-Farsani, S., Hudson, M., Putkinen, V., Seppälä, K., Sun, L., Glerean, E., Hirvonen, J., Karlsson, H. K., & Nummenmaa, L. (2023). Functional organization of social perception networks in the human brain. *NeuroImage, 272*, 120025. https://doi.org/10.1016/j.neuroimage.2023.120025
Sasamoto, A., Miyata, J., Hirao, K., Fujiwara, H., Kawada, R., Fujimoto, S., Tanaka, Y., Kubota, M., Sawamoto, N., Fukuyama, H., Takahashi, H., & Murai, T. (2011). Social impairment in schizophrenia revealed by Autism-Spectrum Quotient correlated with gray matter reduction. *Social Neuroscience, 6*(5–6), 548–558. https://doi.org/10.1080/17470919.2011.575693
Sasson, N. J., Morrison, K. E., Kelsven, S., & Pinkham, A. E. (2020). Social cognition as a predictor of functional and social skills in autistic adults without intellectual disability. *Autism Research, 13*(2), 259–270. https://doi.org/10.1002/aur.2195
Schmitz, O. J., Krivan, V., & Ovadia, O. (2004). Trophic cascades: The primacy of trait-mediated indirect interactions. *Ecology Letters, 7*(2), 153–163.
Schurz, M., Radua, J., Aichhorn, M., Richlan, F., & Perner, J. (2014). Fractionating theory of mind: A meta-analysis of functional brain imaging studies. *Neuroscience & Biobehavioral Reviews, 42*, 9–34.
Schuwerk, T., Langguth, B., & Sommer, M. (2014). Modulating functional and dysfunctional mentalizing by transcranial magnetic stimulation. *Frontiers in Psychology, 5*, 1309.
Sharp, C., Ha, C., Carbone, C., Kim, S., Perry, K., Williams, L., & Fonagy, P. (2013). Hypermentalizing in adolescent inpatients: Treatment effects and association with borderline traits. *Journal of Personality Disorders, 27*(1), 3–18.
Sharp, C., & Vanwoerden, S. (2015). Hypermentalizing in borderline personality disorder: A model and data. *Journal of Infant, Child, and Adolescent Psychotherapy, 14*(1), 33–45.
Somma, A., Ferrara, M., Terrinoni, A., Frau, C., Ardizzone, I., Sharp, C., & Fossati, A. (2019). Hypermentalizing as a marker of borderline personality disorder in Italian adolescents: A cross-cultural replication of Sharp and colleagues' (2011) findings. *Borderline Personality Disorder and Emotion Dysregulation, 6*, 1–4.
Sowden, S., Wright, G. R., Banissy, M. J., Catmur, C., & Bird, G. (2015). Transcranial current stimulation of the temporoparietal junction improves lie detection. *Current Biology, 25*(18), 2447–2451.
Sprong, M., Schothorst, P., Vos, E., Hox, J., & Van Engeland, H. (2007). Theory of mind in schizophrenia: Meta-analysis. *British Journal of Psychiatry, 191*(1), 5–13. https://doi.org/10.1192/bjp.bp.107.035899
Strafella, A. P., & Paus, T. (2000). Modulation of cortical excitability during action observation: A transcranial magnetic stimulation study. *NeuroReport, 11*(10), 2289–2292.
Thibaudeau, É., Cellard, C., Turcotte, M., & Achim, A. M. (2021). Functional impairments and theory of mind deficits in schizophrenia: A meta-analysis of the associations. *Schizophrenia Bulletin, 47*(3), 695–711. https://doi.org/10.1093/schbul/sbaa182
Urbach, M., Brunet-Gouet, E., Bazin, N., Hardy-Baylé, M. C., & Passerieux, C. (2013). Correlations of theory of mind deficits with clinical patterns and quality of life in schizophrenia. *Frontiers in Psychiatry, 4*, 30. https://doi.org/10.3389/fpsyt.2013.00030
Van Overwalle, F. (2009). Social cognition and the brain: A meta-analysis. *Human Brain Mapping, 30*(3), 829–858. https://doi.org/10.1002/hbm.20547
Van Overwalle, F., Van de Steen, F., & Mariën, P. (2019). Dynamic causal modeling of the effective connectivity between the cerebrum and cerebellum in social mentalizing across five studies. *Cognitive, Affective, & Behavioral Neuroscience, 19*(1), 211–223.
van Steenburgh, J. J., Varvaris, M., Schretlen, D. J., Vannorsdall, T. D., & Gordon, B. (2017). Balanced bifrontal transcranial direct current stimulation enhances working memory in adults with high-functioning autism: A sham-controlled crossover study. *Molecular Autism, 8*, 40. https://doi.org/10.1186/s13229-017-0152-x
Volel, B. A., Petelin, D. S., Akhapkin, R. V., & Malyutina, A. A. (2018). Cognitive impairment in anxiety disorders. *Neurology, Neuropsychiatry, Psychosomatics, 10*(1), 78–82. https://doi.org/10.14412/2074-2711-2018-1-78-82
Wang, H., Callaghan, E., Gooding-Williams, G., McAllister, C., & Kessler, K. (2016). Rhythm makes the world go round: An MEG-TMS study on the role of right TPJ theta oscillations in embodied perspective taking. *Cortex, 75*, 68–81.

Wang, H., Wang, X., & Scheich, H. (1996). LTD and LTP induced by transcranial magnetic stimulation in auditory cortex. *NeuroReport, 7*(2), 521–525.
Ward, J., & Banissy, M. J. (2015). Explaining mirror-touch synesthesia. *Cognitive Neuroscience, 6*(2–3), 118–133.
Wilson, J. E., Trumbo, M. C., Wilson, J. K., & Tesche, C. D. (2018). Transcranial direct current stimulation (tDCS) over right temporoparietal junction (rTPJ) for social cognition and social skills in adults with autism spectrum disorder (ASD). *Journal of Neural Transmission, 125*(12), 1857–1866. https://doi.org/10.1007/s00702-018-1938-5
Wölwer, W., Lowe, A., Brinkmeyer, J., Streit, M., Habakuck, M., Agelink, M. W., Mobascher, A., Gaebel, W., & Cordes, J. (2014). Repetitive transcranial magnetic stimulation (rTMS) improves facial affect recognition in schizophrenia. *Brain Stimulation, 7*(4), 559–563. https://doi.org/10.1016/j.brs.2014.04.011
Yamada, Y., Sueyoshi, K., Yokoi, Y., Inagawa, T., Hirabayashi, N., Oi, H., Shirama, A., & Sumiyoshi, T. (2022). Transcranial direct current stimulation on the left superior temporal sulcus improves social cognition in schizophrenia: An open-label study. *Frontiers in Psychiatry, 13*, 862814. https://doi.org/10.3389/fpsyt.2022.862814
Young, L., Camprodon, J. A., Hauser, M., Pascual-Leone, A., & Saxe, R. (2010). Disruption of the right temporoparietal junction with transcranial magnetic stimulation reduces the role of beliefs in moral judgments. *Proceedings of the National Academy of Sciences, 107*(15), 6753–6758.
Zhang, M., Xu, S., Chen, Y., Lin, Y., Ding, H., & Zhang, Y. (2022). Recognition of affective prosody in autism spectrum conditions: A systematic review and meta-analysis. *Autism, 26*(4), 798–813.

14
ASSESSING THE IMPACT OF BRAIN GYM EXERCISES ON MEMORY ENHANCEMENT IN STUDENTS WITH LEARNING DISABILITIES

Drishya Sasidharan and Suphala Kotian

14.1 Introduction

In today's society, many children suffer from learning impairments. Since many parents fail to properly address this situation, the child is left to confront his fate on his own, which may have an adverse effect on him for the rest of his life. Learning difficulties (LDs) significantly affect cognitive development and academic performance, especially in areas like memory, which is necessary for learning, retention, and retrieval (1). The *Diagnostic and Statistical Manual of Mental Disorders*, Fifth Edition (DSM-5) classifies LDs as specific learning disorders (SLDs) (2). These disorders are distinguished by difficulties with reading, writing, or mathematical reasoning that cannot be attributed to sensory impairments, intellectual disability, or inadequate education (3).

According to research studies, several factors contribute to LDs, including difficulties at birth and psychological strain throughout development, which result in LDs related to connectivity of senses and neural executive functions, allergic reactions and dietary factors that affect children's school performance, and imperfect ear and hearing sense formation, which has an impact on learning from the time of birth. The strong connection between the evolution of integration of senses and cerebral executive functioning drew researchers' attention when a previous investigation demonstrated that developmental impairment of brain function may be remedied by increasing relations between neuron cells (3–6). Neuroplasticity occurs when the metabolic neuron's demands rise in response to an increase in the strength of the brain stimulation (7). Memory deficiencies, particularly in working memory, are widespread among individuals with learning disabilities and are linked to difficulties processing and remembering knowledge required for academic achievement (8).

There are several strategies to enhance students' memory; however, several studies indicate gaps in each of the strategies. Moreover, these methods require specialised settings and

DOI: 10.4324/9781003646662-17

equipment, which are difficult for instructors and parents to adopt in everyday situations (9). One of the therapies that is easy to implement is Brain Gym Therapy, which uses physical movements to help improve cognitive functions, such as memory, attention, and relaxation.

This study sought to uncover a successful technique that educators may implement in educational settings to enhance brain functioning in students with LDs. Examples include different vocational treatment strategies, art therapy, and furthermore. The study's results will look at how successfully the intervention improved the students' working memory performance and how different student groups who received various frequencies of intervention time performed in terms of working memory scores.

Learning processes involve synthesising numerous kinds of information and include both acquiring new information, skills, abilities, convictions, or preferences in addition to changing and strengthening preexisting ones (10,11). As supported by paradigms that stress effective cognitive management, active involvement, and the importance of social contact in learning, such as Constructivist Theory, Social Learning Theory, and Cognitive Load Theory, receiving information through sensory inputs, processing it by connecting it to prior knowledge, storing it in long-term memory through strategies like repetition, and applying it through problem-solving or communication are all steps in the learning process (12–15).

Learning depends on working memory, an early stage of brain activity. It makes it possible to temporarily store the knowledge needed for a more difficult learning task. Working memory is a crucial mechanism for learning since it temporarily stores the information needed for complex cognitive tasks, according to the Baddeley Working Memory Model (2012) (16). Based on the framework developed by Baddeley and Hitch in 1974, the method emphasises four key elements. The Central Executive is in charge of directing attention, coordinating the subsystems, and managing and synthesising information. Executive functions that assist memory and synthesis are facilitated by the Episodic Buffer, which combines and modifies material from several sources (17). The Visual-Spatial Sketchpad can contain four to seven items, comprising nonverbal visual and spatial information such as numbers and objects, depending on the focus and methods used (18). Furthermore, by temporarily retaining a limited amount of speech sounds, the Phonological Loop facilitates verbal and language-related tasks (19). Together, these components demonstrate the model's significance for organising, storing, and retrieving information, making it essential to the learning process. When combined, these elements create a strong framework for controlling short-term memory storage and integrating it with other cognitive processes. The model highlights how learning and working memory interact, showing how the brain arranges, stores, and retrieves data for higher-order cognitive functions.

The Brain Gym is a movement-based program that was created by Paul and Gail Dennison in the 1980s with the goal of improving emotional health, academic achievement, and cognitive function (20). Brain Gym is a program in Educational Kinesiology that emphasises the connection between learning, cognitive functions, and physical movement (21). The program consists of easy and fun physical activities that enhance cerebral transmission and fortify the connections between the brain and body, enabling people to reach their maximum learning potential. Through consistent practice over time, brain training refers to organised programs or exercises designed to enhance cognitive skills, including thinking, working memory, and inhibitory control. These exercises increase cognitive abilities by involving control functions like intelligence, focus, and the capacity to block out deceptive inputs. Brain Gym uses movement as a technique for quick and comprehensive rehabilitation, with an emphasis on improving mental and neurological function. Emphasise the vital connection between brain connectivity, physical activity, and cognitive development (22).

The goal of Brain Gym interventions is to get students' physical movement in order to influence the development of neuroplasticity with cognitive function in the central nervous system, especially life efficiency as well as working memory efficiency (23). The present research intends to examine the beneficial effects of the Brain Gym intervention on the working memory performance of students with disabilities associated with learning. Therefore, the objectives of the present research are as follows: (i) to analyze the impact of Brain Gym activities on memory enhancement among students with learning disabilities; (ii) to assess the relationship between the frequency of Brain Gym exercises and the extent of memory improvement in students with learning disabilities; (iii) to investigate the role of demographic factors, such as age and gender, in the effectiveness of Brain Gym exercises on memory enhancement.

14.2 Methodology

This study used a quasi-experimental approach and purposive sampling to include 45 students with LD aged 8–10 from two English-medium schools in Mangalore city. Participants are chosen using a learning disability assessment method created by Swarup and Mehta (24). Raven's colored progressive matrix (25) was used to measure IQ, and those who scored higher than 100 or lower than 25 were not excluded from the study. Before being split up into three groups using the matched-group technique, prior to being divided into three groups using the matched-group method, each participant took a pre-test (26,27). Three groups were chosen at random from among the selected participants. The first experimental group (EX A) received 30 minutes of intervention over a period of six weeks. The second experimental group (EX B) received 30 minutes of intervention over a period of six weeks over a period of three weeks, comparatively less than EX A. The control group went through the process of normal learning without any intervention.

Before the study was carried out, parents and school officials provided written consent. To guarantee their voluntary participation, all students who were chosen for the intervention process were briefed on the nature of the intervention, its objectives, and the length of the study. The students were also assured that at any stage of the study, if they felt uncomfortable, they could remain quiet. This step is conducted to ensure that each participant can provide complete cooperation and commitment during the three-week treatment. Before commencing the intervention, the researcher completed Brain Gym 101 and received a certification. To determine the instruments' reliability, a pilot study was carried out applying the test–retest method. A clear score comparison was made possible by administering the pre-test and post-test in a natural and controlled classroom environment, and recording each participant's results on the same data form. Brain Gym intervention with six Super Space motions (Brain Button, Belly Breathing, Earth Button, Thinking Cape, Cross Crawl, and Hook-ups) was part of each 30-minute session.

Working memory performance is best measured by combining several tests to generate a score that is highly valid and reliable (23). Three different kinds of evaluation instruments were used. Subtests of the National Institute of Mental Health and Neuro Sciences (NIMHANS) test of children's memory, including the Picture Recall Test (PRT), Word Recall Test (WRT), and Digit Span Test (DST), were administered. The Digit Span exam includes two divisions: Digit Span Forward (DSF) and Digit Span Backward (DSB). The investigator reads one digit per second to the child. Immediate responses from the child are marked as correct or incorrect. In the second section, the youngster is encouraged to repeat them in reverse order, and then the total score of the longest series from both series is recorded. The two parts of the WRT are Word Recall Meaningful and Word Recall Non-Meaningful. After 2 minutes, the participants had to identify the words from a second card after viewing a card containing items for 30 seconds. A total score is determined. The PRT asks participants to quickly view pictures and then recall them in a sequential manner in order to assess their visual memory. After viewing a series of images, which are subsequently covered, participants are asked to recall them in the same sequence. Card 1 is exposed for 2 seconds, Card 2 for 3 seconds, Card 3 for 4 seconds, and Card 4 for 5 seconds. The basis for scoring is precise sequential recollection.

14.3 Findings

To evaluate the impact of Brain Gym intervention on multiple memory parameters, paired sample t-tests were undertaken, comparing pre- and post-intervention stores across key variables: DSF, DSB, Word Recall Meaning (WRM), Word Recall Non-Meaning (WRNM), and Picture Recall (PICR).

14.3.1 Working Memory Performance of different Group Scores Before to and After Intervention

14.3.1.1 Experimental Group A (EX A)

A significant improvement was observed across most variables, including DSF, DSB, and PICR, demonstrating notable gains in students' ability to recall sequential and visual information post-intervention. The PRT, in particular, exhibited the highest mean improvement (Mean = 3.40, SD = 3.63, $p < 0.001$), indicating that visual stimuli were more easily retained. Conversely, WRNM did not show significant improvement, suggesting potential challenges with non-contextual memory tasks.

Between-subjects univariate analysis further confirmed significant differences between the experimental and control groups. The experimental group achieved a higher mean score (Mean = 3.08, SD = 0.48) compared to the control group (Mean = 0.05, SD = 0.22), with $F(1,28) = 478.37$, $p = 0.001$, and a partial eta squared (η^2) of 0.94, signifying that 94% of the variance in memory enhancement could be attributed to the intervention.

14.3.1.2 Experimental Group B (EX B)

A paired sample t-test was done to compare scores before and after an intervention. The variables included were DSF, DSB, WRM, WRNM, and Picture Recall (PICR).

A significant t-test was obtained before and after the intervention. A high mean ($\bar{X} = 2.33$, SD = 2.53, $p = 0.001$) was seen in PRT. However, the word recall memory test was not significant ($\bar{X} = 0.23$, SD = 0.85, $p = 0.14$). This pattern was also observed in experimental group A, where word recall was not significant. Between-subjects univariate analysis was done to see if there is a difference between the control and experimental groups. The experimental group B had a higher mean ($\bar{X} = 2.05$, SD = 0.33) compared to the control group

($\bar{X}$ = 0.05, SD = 0.27). F (1,28) = 353.74, p = 0.001. η_p^2 = 0.92. The difference between control group and the experimental group is statistically significant. Variance explained is 92%.

14.3.2 *Comparison of Group A, Group B, and Control Group*

Experimental group A received a six-day-per-week Brain Gym intervention for 30 minutes, while the experimental group B received three day per week Brain Gym intervention for 30 minutes. The control group did not receive any intervention. It may be hypothesised that increased memory enhancement is the same for all three groups after Brain Gym intervention.

Experimental group A has a higher mean ($\bar{X}$ = 36.00, SD = 1.42) compared to experimental group B and the control group. Significant differences were observed among the three groups, Wilks' lambda = 0.22, F (1,14) = 22.11, p < 0.01, partial eta squared = 0.77. A one-way repeated measurement evaluation of variance was used to examine the null hypothesis that enhanced memory improvement is similar for each of the three groups before and after the Brain Gym intervention (N = 15). The results of the ANOVA indicated a significant effect, Wilks' lambda = 0.22, F (1,14) = 22.11, p < 0.01. The bootstrapped ANOVA was statistically significant. This demonstrates that there is substantial proof to reject the null hypothesis. There is a significant increased memory enhancement in group A, suggesting that the increase in time increased the level of improvement.

Indicates group A had a significant increase in memory enhancement, on the other hand, average improvement was observed in group B and no significant improvement in the control group.

14.3.3 *Demographic Findings*

Demographic analysis revealed that 47% of students had IQ levels ranging from 86 to 100, while 53% exhibited lower IQ scores. Gender-based analysis indicated that male students had significantly higher IQ scores (Mean = 85.83, SD = 9.73) than female students (Mean = 66.66, SD = 9.73), t (13) = 2.80, p = 0.015. With a mean age of 9, the ages of the participants ranged from 8 to 12.

14.4 Discussion and Conclusion

The findings unequivocally demonstrate that Brain Gym interventions positively impact memory performance, particularly when conducted with greater frequency. The significant gains observed in visual and sequential memory tasks (e.g., PICR, DSF, DSB) suggest that these exercises enhance neuroplasticity and support cognitive function (23). The lack of significant improvement in WRNM aligns with existing literature indicating challenges in encoding non-contextual stimuli.

The substantial effect sizes observed in both experimental groups underscore the efficacy of movement-based interventions like Brain Gym in fostering cognitive development (28,29). Notably, the higher mean scores and variance explained (η^2 = 0.94 for Group A and η^2 = 0.92 for Group B) suggest that consistent practice amplifies the intervention's benefits.

The demographic findings point to potential moderating factors such as IQ and gender, which may influence baseline cognitive capacities and intervention outcomes. The higher IQ scores among male participants warrant further exploration to determine their role in memory enhancement processes (30).

In conclusion, this study validates the utility of Brain Gym as an effective tool for memory improvement among children with learning disabilities. The results advocate for the integration of such interventions into educational practices, particularly with a focus on consistent application to maximise cognitive benefits.

References

1. Swanson, H. L., & Siegel, L. (2011). Learning disabilities as a working memory deficit. *Experimental Psychology*, *49*(1), 5–28.
2. Edition, F. (2013). Diagnostic and statistical manual of mental disorders. *American Psychiatric Association*, *21*(21), 591–643.
3. Lyon, G. R. (1996). Learning disabilities. *The Future of Children*, 54–76. https://doi.org/10.2307/1602494
4. Streeck-Fischer, A., & van der Kolk, B. A. (2000). Down will come baby, cradle and all: Diagnostic and therapeutic implications of chronic trauma on child development. *Australian & New Zealand Journal of Psychiatry*, *34*(6), 903–918. https://doi.org/10.1080/000486700265
5. Romero-Ayuso, D., Toledano-González, A., Segura-Fragoso, A., Triviño-Juárez, J. M., & Rodríguez-Martínez, M. C. (2020). Assessment of sensory processing and executive functions at the school: Development, reliability, and validity of EPYFEI-Escolar. *Frontiers in Paediatrics*, *8*, 275. https://doi.org/10.3389/fped.2020.00275
6. Jyoti, D. F., Frongillo, E. A., & Jones, S. J. (2005). Food insecurity affects school children's academic performance, weight gain, and social skills. *The Journal of nutrition*, *135*(12), 2831–2839. https://doi.org/10.1093/jn/135.12.2831
7. Mundkur, N. (2005). Neuroplasticity in children. *The Indian Journal of Pediatrics*, *72*, 855–857. https://doi.org/10.1007/BF02731115
8. Rao, P. S., Pandey, M. K., Mishra, P., Deshmukh, S., Jahan, M., & Manohar J, S. (2024). Is training working memory in children with learning disabilities a viable solution? A systematic review. *Annals of Neurosciences*, *31*(2), 124–131. https://doi.org/10.1177/09727531231198639
9. Seo, A., Park, J., & Han, K. (2024). Challenges of art therapists implementing art therapy for achievement gaps in Korean school. *Art Therapy*, 1–8. https://doi.org/10.1016/j.aip.2024.102201
10. Darling-Hammond, L., Flook, L., Cook-Harvey, C., Barron, B., & Osher, D. (2020). Implications for educational practice of the science of learning and development. *Applied Developmental Science*, *24*(2), 97–140. https://doi.org/10.1080/10888691.2018.1537791
11. Laurillard, D. (1979). The processes of student learning. *Higher Education*, *8*(4), 395–409. https://doi.org/10.1007/BF01680527
12. Lazareva, A. (2018). Factors affecting student engagement in online collaborative learning courses. In *Teaching and Learning in a Digital World: Proceedings of the 20th International Conference on Interactive Collaborative Learning–Volume 2* (pp. 349–359). Springer International Publishing. https://doi.org/10.1007/978-3-319-73204-6_39
13. Elkhidir, N. (2020). Effective teaching strategies in biological education: Present and future prospects. *Open Science Journal*, *5*(4), 1–8.
14. Zacharia, Z. C. (2015). Examining whether touch sensory feedback is necessary for science learning through experimentation: A literature review of two different lines of research across K-16. *Educational Research Review*, *16*, 116–137. https://doi.org/10.1016/j.edurev.2015.10.001
15. Bellec, G., Salaj, D., Subramoney, A., Legenstein, R., & Maass, W. (2018). Long short-term memory and learning-to-learn in networks of spiking neurons. *Advances in Neural Information Processing Systems*, 31, 1–19.
16. Baddeley, A. (2020). Working memory. In *Memory* (pp. 71–111). Routledge.

17. Linck, J. A., Osthus, P., Koeth, J. T., & Bunting, M. F. (2014). Working memory and second language comprehension and production: A meta-analysis. *Psychonomic Bulletin & Review*, *21*, 861–883. https://doi.org/10.3758/s13423-013-0565-2
18. Logie, R. H. (2014). The visual and the spatial of a multicomponent working memory. In *Spatial working memory* (pp. 19–45). Psychology Press.
19. Bishop, D. V. (2006). Beyond words: Phonological short-term memory and syntactic impairment in specific language impairment. *Applied Psycholinguistics*, *27*(4), 545–547. https://doi.org/10.1017/S0142716406210397
20. Szczasny, A. L. (2016). A Study of the Effect of the 'Brain Gym' Intervention on the Math and Reading Achievement Scores of Fourth Grade Students (Doctoral dissertation, University of St. Francis).
21. Spaulding, L. S., Mostert, M. P., & Beam, A. P. (2010). Is Brain Gym® an effective educational intervention? *Exceptionality*, *18*(1), 18–30. https://doi.org/10.1080/09362830903462508
22. Dennison, P. E., Dennison, G., Wilutzky, U., & Petres-Lesch, H. (1994). *Brain gym*. Edu-Kinesthetics.
23. Abduh, B., & Tahar, M. M. (2018). The effectiveness of brain gym and brain training intervention on working memory performance of student with learning disability. *Journal of ICSAR*, 2(2), 105–111.
24. Swarup, S. & Mehta, D. H. (1991). Diagnostic Test of Learning Disability. Prasad Corporation.
25. Raven, J., Raven, J. C., and Court, J. H. (1998). *Raven's Coloured Progressive matrices* (section 1 and 2). Oxford Psychologist Press Ltd.
26. Stuart, E. A., & Rubin, D. B. (2008). Matching with multiple control groups with adjustment for group differences. *Journal of Educational and Behavioral Statistics*, *33*(3), 279–306. https://doi.org/10.3102/1076998607306078
27. Creswell, J. W. (2013). Steps in conducting a scholarly mixed methods study.
28. Wahyuni, S. (2024). The effect of brain gym on improving memory in Grade VI children at Public Elementary School 86 Kendari an experimental study conducted in 2024. *Holistic Nursing Plus*, *2*(1), 20–32. https://doi.org/10.58439/hnp.v2i1.219
29. Elaziem Mohamed, A., Abdallah Mohammed, H., Mohamed Ahmed Ayed, M., Alghamdi, W., Elsayed Awad Negm, H., & Ahmed Mahmoud, S. (2021). Effect of brain gym training on intelligence, knowledge and information retention among children with developmental disabilities. *Egyptian Journal of Health Care*, *12*(1), 1591–1603.
30. Elaziem Mohamed, A. abd, Abdallah Mohammed, H., Mohamed Ahmed Ayed, M., Alghamdi, W., Elsayed Awad Negm, H., & Ahmed Mahmoud, S. (2021). Effect of brain gym training on intelligence, knowledge and information retention among children with developmental disabilities. *Egyptian Journal of Health Care*, *12*(1), 1591–1603.

15

ENHANCING EXECUTIVE FUNCTIONS THROUGH COGNITIVE-BASED INTERVENTION IN INDIVIDUALS WITH SUICIDAL IDEATION AND ATTEMPTS

A Mixed-Method Pilot Study

Trina Banerjee and K. Jayasankara Reddy

Introduction

Suicide is an emerging situation in today's world and is a matter of concern for everyone. In adolescents, suicide needs to be given special attention. According to the World Health Organisation, suicide is the second leading cause of death among 15–29-year-olds (WHO, 2024). According to the Centre for Disease Control and Prevention, in 2017, almost 10 million people in the USA experienced suicidal thoughts, which will increase to 12.3 million by 2021 (Centre for Disease Control and Prevention, 2025). India is known to have the highest suicide rate in the Southeast Asian region, with almost 200,000 suicides annually (Armstrong et al., 2019). Suicide is known to be the leading cause of death for young women in India, to the extent that the nation accounts for over a third of all female suicides globally (Kumar & Pathare, 2024, February 7). Stress is usually a significant trigger for such events. American Psychological Association explains suicide as an act of killing oneself, usually which is a result of depression or an ongoing mental illness ("Suicide," 2019). The suicidal ideation ultimately can lead to a suicide attempt (Spirito, 2003).

Suicidality, Suicidal Ideation, and Suicide Attempt

The significance of suicide can range from relatively low intensity to heralding a considerable risk for successful attempts. Suicide often indicates emotional suffering in an individual

DOI: 10.4324/9781003646662-18

and is manifested through acts or attempts of self-harm. Suicide correlates with depression and is often considered a marker of mental health issues (Morrow et al., 2010). Stress is usually a significant trigger for such events (Stewart et al., 2019). A study by Swanson and Colman (2013) explores the causality of suicide in terms of "suicide contagion", which explains that being exposed to suicide or suicide-related behaviours often influences others to contemplate or die by suicide. Ecological concepts are known to support this concept, where suicidality is known to increase after an event involving suicide which has been socially publicised (Philips, 1974). American Psychological Association explains suicide as an act of killing oneself, usually which is a result of depression or an ongoing mental illness ("Suicide," 2019). The phenomenon of suicide can be traced to a model where specific predisposing factors and a significant life stressor can bring about suicidal ideation. This usually is facilitated by the individual's emotional, cognitive, and behavioural dynamics. The suicidal ideation ultimately can lead to a suicide attempt (Spirito, 2003).

Suicide Ideation

According to the Oxford Dictionary of Psychology, ideation is the act of forming ideas or mental images of things not at the time being experienced through sensation (Colman, 2009). It may also refer to suicidal thoughts or ideas, often an umbrella term to describe a range of contemplations, wishes, and preoccupation with death and suicide (Harmer et al., 2024). Suicidal Ideation also incorporates elements of suicidal planning and deliberation (Harmer et al., 2024), where an individual uses their cognitive elements to carve out details of the suicidal blueprint of execution. It is one of the critical aspects of cognition. Ideation involves translations of sensory inputs into conceptualisation (Exner, 2003). It is a thinking pattern that incorporates the organisation of symbols or concepts by giving meaning to the same (Banerjee & Reddy, 2023). Active suicidal ideation involves a conscious desire to inflict self-injurious behaviour where the individual possesses a desire for death to occur as a result (Harmer et al., 2024). Passive suicidal ideation is a phenomenon where individuals report wishing to die, although there is no involvement in active planning or steps (Morrow et al., 2010). Passive suicidal ideation often involves an indifference to accidental deaths, which may occur if necessary steps to maintain life are not taken (Harmer et al., 2024).

Suicide Attempt

A suicide attempt is a self-inflicted, potentially injurious behaviour with a non-fatal outcome for which there is evidence (either explicit or implicit) of intent to die (Banerjee & Reddy, 2023). Suicide attempts may result in no injuries or non-fatal injuries (Banerjee & Reddy, 2023). Suicide attempts that result in death are classified as "suicide" (Morrow et al., 2010). Suicide attempts can involve a diverse set of suicidal behaviours ranging from non-suicidal self-injuries, failed suicidal attempts, and actual death by suicide (Banerjee & Reddy, 2023). Other suicidal behaviours may include deliberate self-harm and parasuicide, which is characterised by a self-inflicted injury resembling a suicide attempt but not intended to be successful (Colman, 2009). Suicide attempts also include other phenomena like suicidal behaviour and gestures of self-harm, which are collectively known as "suicidality" (Hausman et al., 2019).

Suicidality

Suicidality refers to the risk of suicide, indicated by either suicidal ideation or intent or with the presence of a suicidal plan (American Psychological Association, 2025). Suicide risk depends on multidimensional factors. The Biopsychosocial Model of Suicide by Kinderman (2005), adapted initially from Engel's Biopsychosocial Model of Mental Health (1980), helps provide holistic risk factors of suicide. According to the model, biological and genetic factors and environmental and social events influence psychological processes. This, in turn, affects mental health, leading to suicidality (Kinderman, 2005). The model was the first of its kind to highlight the multidimensional factors leading to suicide risk and the predictability of loss of life to suicide.

Executive Functions and Suicidality

It is observed that suicidal clients have deficits in executive functions compared to non-suicidal individuals (Keilp et al., 2012). Verbal fluency, response inhibition, attention, planning, and visual retention are some neuropsychological functions that seem deficient for suicidal individuals. Neurocognitive abilities involving perception, thought, action, and emotion mediate the link between suicidal thinking and behaviour (Lezak et al., as cited in Saffer & Klonsky, 2017). These neuropsychological positioning can explain the progression of suicidal thoughts to acts (Saffer & Klonsky, 2017).

Decision Making, Response Inhibition, and Working Memory in Suicide

Suicide attempters have poor decision-making as they tend to make riskier decisions (Yin et al., 2020). They score low in decision-making tasks like the Game of Dice Task (GDT) or the Iowa Gambling Task (IGT) (Yin et al., 2020). Stressful life events like a history of sexual abuse also influence the decision-making in a suicidal attempt. In comparison to individuals without a history of sexual abuse, the scores on tasks of decision-making (IGT) were slightly lower (Yin et al., 2020), which is attributed to Polymorphism within the corticotrophin-releasing hormone reception 1 (CRHR1) and CRHR2 genes (Yin et al., 2020).

Verbal Fluency, Cognitive Flexibility, and Planning in Suicide

Among individuals with a history of attempted suicide, cortical implications related to deficits in prefrontal blood perfusion were found, corroborating lower scores in letter and category frequency tasks (Audenert et al., 2002). Lower verbal fluency among suicide attempters was associated with changes in oxygenated and deoxygenated haemoglobin (Hb) concentration in several brain regions, including the frontal, temporal, and left precentral gyrus, while the participants engaged in verbal fluency tasks (Tsujii et al., 2017).

Management of Suicide

Suicide management is multidimensional, where the focus is either on the behavioural aspects of suicidal attempts/acts of self-harm, emotional distress or the cognitive schema, and irrational beliefs that lead to suicidality.

Pharmacotherapy, ECT, and Other Bio-Medical Procedures

Pharmacotherapy and psychotherapy have been widely used to treat individuals with suicidal ideation and behaviour. Pharmacotherapy in suicide management primarily taps the serotonergic system, where serotonergic medications are used in symptom reduction. Serotonin reuptake inhibitors have been consistent in having an advantage in reducing suicidal ideation and a protective effect against the emergence of suicidal thoughts (Montgomery, 1993). According to Hassan et al. (2021), antipsychotic medications like clozapine effectively control suicidal ideation in schizophrenia patients.

Besides traditional pharmacotherapy, alternative medical practices like electroconvulsive therapy (ECT) are also effective in suicide management. According to Fink et al. (2014), ECT generates a rapid relief of severe depression, mania, and psychosis, which also, in turn, brings about relief in suicide drive. Hence, ECT does impact reducing suicide rates among at-risk individuals.

Psychotherapeutic Management

Suicidology, the field of suicide management, is multi-modal in terms of the therapeutic approaches. Crisis management is one of the most sought-after interventions in suicidology. According to a study by Bryan et al. (2018), crisis management (crisis response plan) is associated with reduced suicidal ideation. With frequent usage, it is associated with more considerable reductions in suicide ideation.

Tarrier (2009) suggested that cognitive behaviour therapy (CBT) has been the most effective in suicide management as indicated by randomised control trials (Tarrier, as cited in Mewton & Andrews, 2016). Cognitive maladaptive schema contributes to suicidal ideation. A schema of hopelessness and intolerance to suffering can colour the actions of suicidal individuals (Wenzel et al., 2015). CBT here aims to work with the maladaptive schema to change the individual's overall functioning and decrease the chances of suicide.

Dialectic behaviour therapy has also been found to be effective for suicidal behaviours (McCauley et al., 2018), particularly among clients with borderline personality disorder. In comparison between CBT and Dialectical Behavior Therapy (DBT), the former improves the cognitive errors involved in suicide, whereas DBT brings about improved acceptance and reduced emotional suppression (Lin et al., 2019). Supportive therapy is also influential among outpatient populations with suicidality, particularly with a diagnosis of personality disorder (Aviram et al., 2004).

The inference drawn from the conceptualisation of suicide aims to understand the existing theoretical notions about suicidal behaviour and suicidal ideation. The connection between the neurology of suicide and the psychological manifestations helps to formulate the research and set the goals for the same.

Method

Research Approach and Paradigm

The study follows a mixed-methods design – the Intervention Design (Creswell, 2015). This is defined as a Qualitative Phase Before Intervention. Pre- and post-test measures characterise the Intervention. The Qualitative Phase After Intervention is also conducted. Quantitative: The Within Group design comprises the pre- and post-cognitive-based intervention conditions. The Between-Group design consists of Group 1: Presence of Suicidal Ideation vs Group 2: History of Suicide Attempt.

The Qualitative Phase consists of a thematic analysis of the interviews of participants pre- and post-cognitive-based intervention. The purpose of the Qualitative Study is to draw inferences about the quantitative results and to establish a possible understanding of the nature of the change the intervention brought.

Research Questions

1. Does cognitive-based intervention improve the executive functioning of individuals with suicidal ideation?
2. Does cognitive-based intervention improve the executive functioning of individuals with suicide attempts?
3. How does cognitive-based intervention affect individuals with Suicidal Ideation and attempts?

Objectives of the Study

1. To understand the efficacy of cognitive-based intervention (eight sessions) on suicidality and executive functions of young adults with suicide ideation and attempt.
2. To draw qualitative inferences about young adults' suicidality and executive functions related to suicide.
3. To understand qualitatively the changes that cognitive-based intervention brought in the context of suicidality and executive functions of young adults.

Hypotheses

H_0^1 There is no effect of cognitive-based intervention on Planning, Verbal Fluency, and Response Inhibition in individuals with Suicide Ideation.

H_0^2 There is no effect of cognitive-based intervention on Planning, Verbal Fluency, and Response Inhibition in individuals with Suicide Attempts.

Variables and Operational Definitions

Study Variables

BETWEEN GROUP

Suicidal Ideation. Suicidal Ideation can be defined by the thoughts or acts of taking one's own life (Sadock & Sadock, 2007). Suicidal Ideation can be of two types. Active Suicidal Ideation: This refers to thinking of suicide, rehearsals, and planning (Morrow et al., 2010). Passive Suicidal Ideation: The level of intent where the patients report "the wish to die", but no thoughts of planning or ideation involving acting are observed (Sadock & Sadock, 2007).

Suicide Attempt. A self-inflicted, potentially injurious behaviour with a non-fatal outcome for which there is evidence (either explicit or implicit) of intent to die. Suicide attempts may result in no injuries or non-fatal injuries. Suicide attempts that result in death are classified as "suicide" (Morrow et al., 2010).

Within Group – Pre- and Post-Cognitive-Based Intervention. The cognitive-based intervention used in the present study is derived from CBT, which is a short-term, structured therapy that involves active collaboration between patient and therapist to achieve its therapeutic goals, which are more oriented towards the present problem and problem resolution (Sadock & Sadock, 2007). The basis of the therapy lies in the individual's appraisal, which is evident primarily through cognitions. The correction of dysfunctional cognitions leads to clinical improvement.

Measured Variable

Verbal Fluency. This is typically measured by the quantity of words produced within a restricted category or in response to a stimulus, and usually within a time limit (Lezak et al., 2004). In the current study, the Controlled Oral Word Association (COWA) test scores will determine the participants' verbal fluency.

Planning. This is identifying and organising the steps necessary to execute an intention or achieve a goal (Lezak et al., 2004). In the current study, the scores on the Tower of London (TOL) Test will determine the planning ability of the participants.

Response Inhibition. Response inhibition refers to the suppression of actions that are inappropriate in a given context and that interfere with goal-driven behaviour (Mostofsky & Simmonds, 2008). The scores in the Stroop Test will determine the participants' response inhibition.

Sampling

Method of Sampling

The study follows a purposive sampling. Group 1 is suicidal ideation, where individuals with suicidal ideation were chosen. Group 2 is a suicidal attempt, where individuals with a Suicide Attempt (Act of self-harm/at least 1 Attempted Suicide) were chosen.

Reason for Small Sample Size. The current study is a pilot study; hence, the sample size is 22, which is more than 10% of the actual sample size chosen for executive function-oriented research. The sample was divided into two groups: suicidal ideation and suicide attempters. ***Sampling Criteria for Suicidal Ideation***

INCLUSION CRITERIA

1. Participants who possess suicidal ideation (n=10), suicide plans, or desire for suicide/self-harm (active or passive).
2. Participants who have no history of suicidal behaviour/attempted suicide or self-harm (active/passive).
3. Participants with average general intellectual functioning were tested in the mental status examination and communicated in English.
4. Participants aged between 18 and 25 years.
5. Participants who showed informed refusal and non-compliance to pharmacotherapy.
6. Participants who consented to the study.

Table 15.1 Details of the Cognitive-Based Intervention Used for the Study

Duration of Program	2 months (8 weeks of back-to-back sessions)
Duration of Session	60 minutes
Frequency of Session	One session per week/No breaks in between weeks
Mode of Session	Online (Google Meet) Video sessions

Session No	*Objectives*	*Outcome Expectations*	*Challenges*
I	Intake interview, psychoeducation of the Cognitive Approach	Rapport building, the Client gets to understand the objectives of therapy	Hesitation to open up, pressure to share, and time management of sessions
II	Identification of Negative Automatic Thoughts (NATs)	Introducing the Dysfunctional Thought Record (DTR), which helps to identify Triggers, Emotions, and NATs	Maintaining the DTR. Differentiating between emotions and NATs.
III	Verbal Challenging of NATs	Evidence-based challenging of NATs with rational thoughts	Presence of Assumptions acting as a blocker
IV	Identification of Assumptions related to suicide cognitions + Introduction of Behavioural Exercises	Understanding assumptions and their source. Identifying early childhood experiences that shaped these assumptions. Using behavioural exercises like Adaptive Coping Mechanisms, Distress Tolerance Skills, and Self-Soothing to help with distressing emotions.	Difficulty in executing Behavioural Exercises. The presence of illogicality in thoughts amounts to maladaptive cognitions.
V	Introducing Advantage/ Disadvantage Analysis + Identification of Logical Errors	Focusing on how advantageous or disadvantageous the thought is for the individual. Identification of cognitive distortions.	Maintaining the homework of Identifying Distortions.
VI	Application of the Downward Arrow Technique	Reaching upon beliefs related to Self, World, and Future.	Distressing emotions acted as blockers when beliefs were identified.
VII	Management of Acute Crisis Situation	Practicing more thought-challenging, along with self-soothing and behavioural exercises.	Maintenance and consistency of self-work in between sessions.
VIII	Introducing Relapse Prevention	Understanding triggers, recalling to use the cognitive distortion list, and independently challenging NATs.	Consistency of the technique.

EXCLUSION CRITERIA

1. Participants with co-morbid diagnoses of a neurological nature, psychosis (affective and non-affective), and endocrinological disorders (PCOD, thyroid, etc.).
2. Participants with a history of substance use/abuse.
3. Participants who had severe depression (scored 31 and above on the Beck Depression Inventory (BDI)) and had been diagnosed with co-morbid Anxiety and Obsessive Compulsive Disorder, and Personality Disorder.
4. Participants who scored less than 1 in the domains of Active Suicidal Desire, Suicidal Plans, and Passive Suicidal Desire in the Beck Scale for Suicide Ideation in the Pre-Intervention Phase.
5. Participants are undergoing some other form of Treatment/Psychotherapy/Alternative Therapy.

Sampling Criteria for Suicide Attempt

INCLUSION CRITERIA

1. Participants who possess suicidal ideation (n=12), suicide plans, and desire for suicide/self-harm (active or passive) and have at least one suicidal attempt or acts of self-harm (active/passive).
2. Participants with average general intellectual functioning were tested in the mental status examination and communicated in English.
3. Participants aged between 18 and 25 years.
4. Participants who showed informed refusal and non-compliance to pharmacotherapy.
5. Participants who consented to the study.

EXCLUSION CRITERIA

1. Participants with co-morbid diagnoses of a neurological nature, psychosis (affective and non-affective), and endocrinological disorders (PCOD, thyroid, etc.).
2. Participants with a history of substance use/abuse.
3. Participants who had severe depression (scored 31 and above on the BDI) and had been diagnosed with co-morbid Anxiety and Obsessive Compulsive Disorder, and Personality Disorder.
4. Participants who scored less than 1 in the domains of Active Suicidal Desire, Suicidal Plans, and Passive Suicidal Desire in the Beck Scale for Suicide Ideation in the Pre-Intervention Phase.
5. Participants undergoing other Treatment/Psychotherapy/Alternative Therapy.

Tools Used

1. Socio-demographic data sheet
2. Informed consent form
3. Semi-structured clinical interview schedule
4. Beck Depression Inventory (Beck et al., 1996)
5. Beck Scale for Suicide Ideation for screening participants. Cronbach's alpha showed a reliability coefficient value of 0.89 for internal consistency (Beck & Steer, 1991)

6. COWA test to test verbal fluency (Benton & Hamsher, 1989)
7. TOL test for planning (Rao et al., 2004)
8. Stroop Color and Word Test for response inhibition (Rao et al., 2004)
9. Interview schedule for participants pre- and post-intervention phase for Qualitative Data analysis.

Purposes of the Tools Used

1. Socio socio-demographic data sheet was chosen to collect data about the participants which helped in grouping the participants and matching them based on inclusion and exclusion criteria.
2. Informed consent form was used to explain the risks, benefits, and challenges of the study and based on this information a consent was obtained from the participants.
3. Semi-structured clinical interview helped to obtain information about past and present clinical diagnoses to check if participants adhered to the inclusion and exclusion criteria of the study.
4. BDI is a 21-item rating scale. It is used to understand the severity of depression among individuals diagnosed with depressive disorder. BDI is a reliable and valid test with a Cronbach's alpha of 0.81–0.93 depending on the version used. In the present study the English version of BDI was used for the participants. BDI measures correlate strongly with other measures of depression like Hamilton Depression Rating Scale (HDRS), indicating good construct validity (Beck et al., 1996).
5. Beck Scale for Suicidal Ideation (BSSI) is a 21-item scale. It is used to assess various aspects of suicidality with the focus particularly on Active Suicidal Desire, Suicidal Planning and Preparation, and Passive Suicidal Desire. This is proven to be a reliable and valid measure of suicidality with a Cronbach's alpha of 0.87–0.97 signifying the ability to consistently assess suicidality.
6. COWA test is a widely used neuropsychological test that assesses verbal fluency, executive functioning, and language processing. It requires individuals to generate as many words as possible starting with a specific letter (e.g., F, A, S) within a fixed time frame. COWA has a high inter-rater reliability (0.95 and above) suggesting it to be a reliable test. The test is strongly correlated with other measures of verbal fluency and executive functioning such as Trail Making Test (TMT), suggesting the test to be valid (Benton & Hamsher, 1989).
7. TOL test is a neuropsychological assessment designed to evaluate executive functioning, specifically planning, problem-solving, and goal-directed behaviour. It is a test which has a high inter-rater reliability (above 0.90) and shows strong construct validity as a measure of planning with other tests like Wisconsin Card Sorting Test (WCST) and TMT (Rao et al., 2004).
8. Stroop Color and Word Test is a widely used neuropsychological assessment designed to measure response inhibition, cognitive flexibility, and attention. It evaluates the ability to suppress automatic responses and manage cognitive interference. The test has a high Cronbach's alpha of 0.80 suggesting that it consistently measures response inhibition. Its high correlation with other executive function tests like WCST and TMT shows the test to be valid.
9. Interview schedule for participants' pre- and post-intervention phase for Qualitative Data analysis is discussed in detail below.

Qualitative Questions Used for the Study

Pre-Intervention Questionnaire

1. What made you opt for the study?
2. What are your expectations from the sessions?
3. How do you feel most of the time?
4. What makes you feel the way you do/what are the reasons?
5. Where do you see yourself in the intervention?
6. What are your strengths?
7. What are your potential weaknesses?
8. What are your thoughts/perceptions of the researcher?

Post-Intervention Questionnaire

1. How do you feel about therapy/how has it been for you so far?
2. Have you noticed any changes within you during therapy? If yes, kindly elaborate.
3. What do you think are the reasons for the change?
4. What would you say was the importance of these sessions?
5. Did you feel any roadblocks in the sessions? If yes, kindly elaborate.
6. Which personal aspects of the "self" did you find helpful in the sessions?
7. Which personal aspects of the "self" hindered or caused resistance in the sessions?
8. What has been the most helpful aspect of your therapy so far?
9. What aspects of the therapy have been unhelpful/hindering?
10. How did you feel being involved in the research?

Setting and Procedure

The participants were given the socio-demographic data sheet, informed consent form, semi-structured clinical interview, and Beck's Scale for Suicide Ideation. Based on the tests and inclusion criteria, the participants were chosen for the study and were selected into two groups, based on the presence of suicidal ideation and suicidal attempts. The participants were assessed using the COWA, Stroop, and TOL tests. Each participant underwent a qualitative interview followed by 8 8-Cognitive-Based Interventions. After completing the therapy, they underwent a qualitative post-intervention interview and reassessments of executive functions using the COWA, Stroop, and TOL tests. The data was analysed statistically and discussed.

Data Analysis

Quantitative: Test of normality (Shapiro–Wilk's Test), test of homogeneity of variance (Levene's test of homogeneity of variance), test of significance for matched sample test (Wilcoxon signed-rank test).

Qualitative: The data of ten individuals, both pre- and post-intervention interviews, were analysed. The thematic analysis and coding method (open coding and axial coding) were used, and the discussion was based on the themes emerging from the interview.

Ethical Consideration

The study was utterly non-invasive and followed all ethical guidelines to protect the human rights of the participants. The tests of executive functions involved voluntary participation, done only after informed consent. The participants were informed that the data was confidential and would only be used for academic and research purposes. Individuals who participated were given eight sessions of intervention to manage suicidal ideation and suicidal attempts of self-harm as part of the plan of the study. Hence, they benefited from diagnosis and management. There are no known risks involved in the study.

Results

Quantitative Results

Table 15.2 The Identification Details for Suicidal Ideation (*N*=10)

Variables	*Types*	*N*
Sex	Male	0
	Female	10
Age	18–21 years	6
	22–25 years	4
Education	Finished School (XIIth Std)	2
	Graduate	6
	Post Graduation	2
Occupation	Employed	7
	Unemployed	3
Relationship Status	Single	3
	In a relationship	4
	Married	0
	Separated/Divorced/Widowed	3
Residence	Karnataka	10
Residence Status	Urban	10

Note: This table demonstrates the identification details and their distribution for the Suicidal Ideation group (*N*=10).

Table 15.3 The Identification Details for Suicidal Attempt (*N*=12)

Variables	*Types*	*N*
Sex	Male	0
	Female	12
Age	18–21 years	6
	22–25 years	6
Education	Finished School (XIIth Std)	1
	Graduate	6
	Post Graduation	5
Occupation	Employed	7
	Unemployed	5

(*Continued*)

Table 15.3 (Continued)

Variables	*Types*	*N*
Relationship Status	Single	4
	In a relationship	8
	Married	0
	Separated/Divorced/Widowed	0
Residence	Karnataka	12
Residence Status	Urban	12

Note: This table demonstrates the identification details and distribution for the Suicidal Ideation group (*N*=10).

Table 15.4 The Descriptive Statistics for Suicidal Ideation (*N* = 10)

Variables	*Mean*	*SD*	*Range*
Pre-BDI	28.70	2.50	8.00
Pre BSSI	15.90	6.44	18.00
Pre TOL Time	24,362.38	4010.34	13,146.41
Pre TOL Success	7.60	1.58	6.00
Pre COWA	14.67	4.27	14.00
Pre Stroop-Score	105.70	49.66	182.00
Post BDI	15.00	3.86	12.00
Post BSSI	9.20	2.53	9.00
Post TOL Time	18,338.66	2034.82	7085.34
Post TOL Success	9.00	1.41	5.00
Post COWA	17.43	3.04	9.00
Post Stroop Score	58.80	38.19	104.00

Note: This table demonstrates the pre- and post-CBT scores of the test administered to the suicidal ideation group (*N* = 10).

Table 15.5 The Hypothesis Testing for a Matched Sample (*N*=10) among Variables for Suicide Ideation

Variables	*Z*	*P-value*
BDI	-2.812	**0.005***
BSSI	-2.295	**0.022***
TOL– Time Taken	-2.803	**0.005***
TOL-Success	-1.615	0.106
COWA	-1.784	0.074
Stroop Score	-2.244	**0.025***

* $P \geq .05$

Note: The table shows the Wilcoxon signed-rank test results for the Suicidal Ideation group (*N* = 10). The table shows that the scores of the Beck Depression Inventory (BDI), scores of the Beck Scale for Suicidal Ideation (BSSI), scores of Tower of London-Time Taken (TOL-Time Taken), and scores of the Stroop test are significant. The test scores also suggest that post-CBT, the depression scores have reduced, the suicidality scores have reduced, the time taken for TOL (planning) has decreased, and the Stroop scores indicate improved response inhibition.

Table 15.6 Hypothesis Testing for a Matched Sample (N=12) among Variables for Suicide Attempt

Variables	*Z*	*P-value*
BDI	-3.065	0.002*
BSSI	-3.070	0.002*
TOL- Time Taken	-3.059	0.002*
TOL-Success	-2.059	0.040*
COWA	-2.624	0.009*
Stroop Score	-2.041	0.041*

* $P \geq .05$

Note: The table shows the Wilcoxon signed-rank test results for the suicidal attempt group (*N*= 12). The table shows that the scores of the Beck Depression Inventory (BDI), Beck Scale for Suicidal Ideation (BSSI), Tower of London-Time Taken (TOL-Time Taken), Tower of London Success (TOL-Success), Controlled Oral Word Association, and scores for Stroop test are significant. The test scores also suggest that post-CBT, the depression scores have reduced, the suicidality scores have reduced, the time taken for TOL (planning) has decreased, successful planning has improved, verbal fluency has improved, and the Stroop scores indicate improved response inhibition.

Table 15.7 The Descriptive Statistics for Suicide Attempt (*N* = 12)

Variables	*Mean*	*SD*	*Range*
Pre-BDI	27.42	2.61	6.00
Pre BSSI	23.67	6.68	18.00
Pre TOL Time	26,611.26	4549.06	15,250.66
Pre TOL Success	7.17	1.80	6.00
Pre COWA	14.39	4.03	15.67
Pre Stroop-Score	137.58	53.48	177.00
Post BDI	16.75	3.08	9.00
Post BSSI	14.42	2.07	7.00
Post TOL Time	21,340.76	3441.02	11,109.57
Post TOL Success	8.50	1.09	3.00
Post COWA	17.67	2.52	8.00
Post Stroop Score	84.92	51.35	181.00

Note: This table demonstrates the pre- and post-CBT scores of the test administered to the suicidal ideation group (*N* = 12).

Qualitative Findings

The data received from the interviews were qualitatively analysed. They were coded, grouped into subthemes and themes, and interpreted to form a thematic framework. Five participants were chosen randomly for the suicide ideation group, and five were chosen from the suicide attempt group. The themes have been separately explored for Suicidal Ideation and Suicide Attempt.

Findings of Suicidal Ideation and Attempt Pre-Intervention

The conversation excerpts of the participants were directed towards the theme of Diathesis or Vulnerability. The theme was constructed from the subthemes of Childhood Neglect,

Dysfunctional Home Environment, and Strained Interpersonal Connectedness. All five participants reported childhood neglect predominantly occurring for instances of absent parenting. Incidents where parents were unavailable as they were divorced or separated were observed. "*I was mostly alone after school, as my mom was at work. I remember coming home and staying alone. My parents were separated, and I did not live with my dad*" (Naina, 21 years of age, Female, Personal Communication, July 8, 2021). Childhood neglect also emerged from differentiation due to gender. "*My mother openly states that she prefers my brother and loves him more than me. I, as a child, felt so bad that what could I do to be more like my brother so that she loves me*" (Sheila, 19 years of age, Female, Personal Communication, August 18, 2021). Many participants reported enduring a dysfunctional home environment during their childhood. "*My family always fought; my parents were at loggerheads with each other. My mother has many times threatened to leave us*" (Rajiv, 22 years of age, Male, Personal Communication, January 8, 2022). The participants disclosed that they had a strained relationship within the family. "*We would not speak to our father for days. We would never address the issue, but at the same time, there would be tension at home brewing for months*" (Manav, 18 years, Male, Personal Communication, January 8, 2021).

The theme of Poor Sense of Self was found in the conversation excerpts from the participants. "*I always felt my parents hated me, and hence they fought about me. They never cared if I was in the room or not*" (Rajiv, 22 years of age, Male, Personal Communication, January 8, 2022). "My *mother has many times said, 'I wish you were dead; my life would improve'*" (Naina, 21 years of age, Female, Personal Communication, July 8, 2021). The participants also internalised the adverse events they had endured, resulting in self-blame. "*I cannot seem to like myself; my parents were never proud of me. They liked my brother better, and I always felt so unwanted*" (Sheila, 19 years of age, Female, Personal Communication, August 18, 2021).

The conversation excerpts showed that the triggers experienced by most of the participants were interpreted as a "perceived sense of loss". The triggers mainly were separation/breakups of relationships, fights leading to threats of loss, or parents threatening to abandon them or stop rendering financial or emotional support. "When *I got to know my partner cheated on me, I felt like I lost this relationship. I lost everything that I had*" (Sheila, 19 years of age, Female, Personal Communication, August 18, 2021). "*I wanted to quit college and take a break for a year as it was too much to handle. When I told this to my parents, my dad said I am not paying for you to sit at home. I wished I was dead at that moment, I felt like a burden*" (Manav, 18 years of age, Male, Personal Communication, January 8, 2021).

The triggers have also resulted in a state of entrapment experienced by the participants. "*I felt like I do not have a way out. I felt so stuck and thought death is better than this*" (Manav, 18 years of age, Male, Personal Communication, January 8, 2021). "*I have nowhere to go now that my partner cheated on me. He was everything to me, and he lied and cheated. It was too much for me to bear*" (Archana, 23 years of age, Female, Personal Communication, January 21, 2021). "*I felt like I cannot escape this. The sadness was too much. It was getting difficult to go on*" (Naina, 21 years of age, Female, Personal Communication, July 8, 2021).

The theme of cognitive deconstruction was also found in the conversations. The subthemes of emotional dysregulation and the presence of negative self-beliefs. Poor planning and maladaptive coping constituted the theme. "*When my partner cheated on me, I felt like a wave hit me. I was almost swept away with the distress. I just wanted the intense pain to get out of my system somehow*" (Archana, 23 years of age, Female, Personal Communication, January 21, 2021). The participant's emotional regulation was affected by the trigger, leading to perceived separation, which prompted suicidal thoughts. Dysregulated emotions lead to poor planning, often mediated by negative self-beliefs. "*I thought, what if death solves everything? At least this pain will be over. Anyway, I will never be good enough for my family*" (Rajiv, 22 years of age, Male, Personal Communication, January 8, 2022). This often leads to poor coping strategies. "*I saw suicide as a coping mechanism; I felt if suicide would teach my partner a lesson*" (Sheila, 19 years of age, Female, Personal Communication, August 18, 2021). These have often culminated in suicidal thoughts.

Findings of Suicidal Ideation and Attempt Post-Intervention

After eight sessions of intervention a change in the themes was observed from the conversation excerpts. A theme of Improved Relationship with the "self" emerged from the subthemes of "evidence-based goal setting" and "Exclusion of societal expectations". The participants mentioned how the thought challenge helped them understand the lack of logical evidence they observed in their thoughts. "*I did not know I was feeling so trapped. I did not know how to pick myself up from such moments. Now I do*" (Rajiv, 22 years of age, Male, Personal Communication, January 8, 2022). Participants also addressed the internalisation process and agreed that they were setting unrealistic expectations of themselves. "*I just wanted myself to move on, not thinking about how this is possible. I wanted to repress all the 'bad' feelings. But now I know it is not like that. Processing emotions is important*" (Sheila, 19 years of age, Female, Personal Communication, August 18, 2021). "*I now do not care what others think of me; they were not there when I was down. I choose to prioritise myself over others now*" (Naina, 21 years of age, Female, Personal Communication, July 8, 2021).

The theme of "Grieving/Mourning the loss" emerged from subthemes of "emotional expression" and "acceptance". "*I know my relationship was abusive, and I know I do not need that anymore*" (Naina, 21 years of age, Female, Personal Communication, July 8, 2021). "*It is difficult to accept that one's parent does this, but I am beginning to see them as people and not just my parents. I feel bad, but I also do not attribute it to myself anymore*" (Rajiv, 22 years of age, Male, Personal Communication, January 8, 2022). "*I call my friends and tell them now about how I am feeling. I do not suppress my emotions anymore*" (Archana, 23 years of age, Female, Personal Communication, January 21, 2021).

The "Cognitive reconstruction" theme emerged from the subthemes of emotion regulation, planning, distress tolerance, and hopefulness. Participants revealed how they learnt to regulate their emotions better, particularly during extreme stress. "*Whenever I am angry or anxious, I take a few deep breaths and have a glass of water. I have also removed myself from the situation to avoid aggression*" (Sheila, 19 years of age, Female, Personal Communication, August 18, 2021). Participants have used distress tolerance as a skill to cope with difficult situations. "*My suicidal thoughts have not stopped. But I know how I can cope with them. I manage my emotions first, and then sometimes the thoughts get distracted. I am not troubled any more*" (Rajiv, 22 years of age, Male, Personal Communication, January 8, 2022). Participants used mindful planning to deal with their stressors. "*I am now focussing on developing my career; it is a big distraction from the breakup and the abusive relationship. It helps to keep the 'dark thoughts' at bay*" (Naina, 21 years of age, Female, Personal Communication, July 8, 2021).

Discussion

The study discussion depends on the quantitative and qualitative findings obtained from the results after statistical analysis.

Effect of the Intervention on Suicide Ideators

Cognitive-based intervention is an evidence-based practice for depression and has been cited in academic literature for years in support of reducing depressive symptoms (Gautam et al., 2020). The intervention focuses on the maladaptive cognitions involved in affecting the emotional state of the individual, as this affective state often causes maladaptive coping methods (Boden et al., 2013). The maladaptive cognitions in the form of negative automatic thoughts, assumptions, and beliefs about the self, world, and future are challenged logically in cognitive-based intervention to induce rational thoughts and motives (Boden et al., 2013). In the present study, the intervention also brought a change in the level of depression after the eight sessions, where, among the suicidal ideation group, the level of depression lessened, suggesting an improved emotionality among the participants. This change is attributed to the challenging of the negative cognitions, especially the beliefs around worthlessness, helplessness, and hopelessness, which a crucial factors in depression. Challenging negative cognitions involves focusing on automatic thoughts, which are a result of belief systems that have been formed over time. These are a product of early childhood experiences and negative self-beliefs, which often develop through constant exposure to critical comments, hostile behaviour, and an invalidating environment (Lee & Schafer, 2021). The narrative one is exposed to later on developments into their own self-talk (Schauer et al., 2005). A critical self-talk creates a self-belief of not being good enough, which automatically creates a low mood by reducing feelings of being rewarded and, in turn, creating a reduction in dopamine levels (The Daily Telegraph, 2024). The cognitive-based intervention here helps suicide ideators to develop a better sense of self, which is free from critical narratives of the self (Büscher et al., 2020), which in turn improves mood by improving dopamine levels.

Suicidality in an individual is their risk of suicide. This is determined by suicidal ideation, suicidal planning, and suicidal attempts (both active and passive). Among individuals with suicidal ideation, suicidality is at a cognitive level and has not ever been executed behaviourally. Their suicidal cognitions often get activated during moments of extreme psychological triggers, which bring about a sense of entrapment (Williams & Pollock, 2001). This also triggers helplessness, ultimately leading to seeing suicide as a way out (Williams & Pollock, 2001). The intervention shows a reduction in suicidality among individuals with suicidal ideation. The challenge of negative automatic thoughts helps in reversing the sense of entrapment, where the participants do not see "suicide" as the only way out anymore. The reduction in suicidality constitutes a reduction in active suicidal desire, suicidal planning, and passive suicidal desire as well. This suggests that exposure to eight sessions of intervention has helped the participants challenge their thoughts to an extent where they experience a reduction in suicidality at the ideation level.

Planning is identifying and organising the steps necessary to execute an intention or achieve a goal (Lezak et al., 2004). In certain mental health conditions, the planning ability, along with other cognitive functions, often gets impaired (Trivedi, 2006). Planning has been implicated to be affected in suicide (Yin et al., 2020), which eventually culminates in suicidality. The participants see suicide as the only option, which is a result of poor planning. Often, perspectives are ignored, and cognitive biases influence a person to undergo cognitive deconstruction (Baumeister, 1990). This is a state where irrationality takes over, and the beliefs of self, the world, and the future get clouded. Cognitive deconstruction state is attained as a coping mechanism against extremely distressing emotional states. In the current study, the total time taken to solve the task of planning and success is significantly better after eight sessions of intervention. The intervention helps improve the distressing emotional state and brings about a reduction in cognitive rigidity (Rudd, 2015), which is a result of "cognitive deconstruction", thus bringing about better planning. The intervention involved challenging negative automatic thoughts, identifying underlying assumptions, and using the downward arrow technique to determine core beliefs related to suicide. It also focused on the advantage–disadvantage analysis of illogical thoughts and addressing cognitive distortions. This helped the participants to bring logicality into their thoughts, and with the help of behavioural exercises introduced in the sessions, they opted for adaptive coping strategies. This helped them to plan better, and that resulted in a significant reduction in suicidal ideation.

Verbal fluency is the quantity of words produced within a restricted category or in response to a stimulus, and usually within a time limit (Lezak et al., 2004). It is a cognitive function that facilitates information retrieval from memory as it involves processes like selective attention, selective inhibition, set-shifting, internal response generation, and self-monitoring (Patterson, 2011). Verbal fluency is a neuropsychological function that has been implicated in suicide (Audenert et al., 2002). The deficits suggest the inability of a person to retrieve relevant information in a heightened emotional state during suicidal ideation or the inability to engage in mental set-shifting, which prevents one from taking other perspectives. The results show a significant improvement in verbal fluency among suicide ideators post-eight sessions of intervention. This suggests that challenging negative cognitions and identifying cognitive bias in the form of distortion have aided in perspective-taking (Hudson et al., 2023). The participants could pull themselves out of cognitive deconstruction (Baumeister, 1990), where they were not comparing themselves to set societal standards, and showed improved cognitive flexibility. This helps them not to internalise the pressure and no longer associate themselves with a sense of inadequacy. This ultimately reverses the emotional numbness that the participants felt during suicidal ideation.

Response inhibition refers to the suppression of actions that are inappropriate in each context and that interfere with goal-driven behaviour (Mostofsky & Simmonds, 2008). Individuals often associate stimuli in the environment with rewards, and when a behaviour does not lead to successful reward attainment, they inhibit the behaviour ("Response Inhibition", 2015). In the current study, the participants with suicide-related ideation have a significant reduction in response inhibition after their exposure to the eight sessions of intervention. This indicates that post-exposure to cognitive behavioural therapy, the participants are better able to inhibit their impulses, particularly concerning suicidal ideation. The suicidal thoughts brought about a sense of escape, or "a way out" of the traumatic emotional pain most of the participants talked of. The sense of escape brings them a momentary sense of relief. Suicidal ideation is often seen as a cry of pain (Williams & Pollock, 2001), suggesting a coping mechanism against emotional pain (Baryshnikov & Isometsa, 2022). This sense of relief acts as a reward, which reinforces suicidal ideation. The connection between distressing times, leading to having suicide as an option, is a bond that is very strongly formed and consolidated as a part of information processing. The intervention attempts to break this bond by reinforcing a different bond. The new bond is the connection between adaptive coping, which incorporates emotion regulation skills, distress tolerance, and suicidal ideation. The participant now gets the reward and relief from adaptive coping and not from suicidal thoughts and ideation. The participants can now regulate their emotions, leading to the management of impulsivity.

Effect of the Intervention on Suicide Attempters

The cognitive-based intervention is known to be a chosen therapy for suicide prevention, and literature shows that it has most suicide attempters who have found it to be helpful (Stanley et al., 2009). Particularly for adolescents, a 12-week exposure to the intervention has been proven to be most effective (Stanley et al., 2009). In the present study, eight sessions of intense cognitive-based intervention have been effective in reducing depression and suicidality among suicide attempters, where there was a significant reduction in active suicidal desire and suicidal planning among the participants. This coincides with the role of the intervention in reducing depression and suicidality among adolescents (Spirito et al., 2012).

The eight sessions of the intervention have also proved effective in improving verbal fluency and response inhibition. The post-intervention scores reveal a statistically significant increase in verbal fluency and a statistically significant reduction in the Stroop score, suggesting better response inhibition.

Verbal fluency suggests the ability of the individual to retrieve information from long-term memory storage that can be utilised for executive functioning. The results suggest that the intervention has helped suicide attempters explore the long-term beliefs they had developed to help navigate around the suicidal thoughts and prevent them from being executed behaviourally. Challenging negative automatic thoughts and awareness of cognitive biases in the form of distortions helps one to reassess the beliefs one has about the "self", world, and the future. The findings are supported by a study conducted in 2023 by Amano et al., which suggests that the intervention improves reaction time for future thinking among depressed individuals, indicating the intervention's role in influencing future thinking among individuals with depression. The dearth of studies showing the connection between the intervention and verbal fluency among suicide attempters points out a research gap that is yet to be bridged. However, increased verbal fluency is related to improved communication, where verbal fluency is known to be connected to lexical access (ability to retrieve words from memory) and cognitive control (Shao et al., 2014), both of which are integral to effective communication. Effective communication involves expressive speech (del Hoyo Soriano et al., 2022), which is linked to a reduction in suicidal behaviour (Costanza et al., 2021).

Low response inhibition is a poor prognostic factor for an individual with a history of suicide attempts and has been associated with impulsivity in the individual (Meza et al., 2017). Low response inhibition contributes to the execution of suicidal ideation through an attempt or an act of self-harm. In the present study, in the post-intervention condition, there was a significant improvement in response inhibition, which suggests that the intervention was effective in helping individuals show better impulse control in response to suicidal urges and attempts. The intervention has been effective in improving inhibitory control in patients with Posttraumatic stress disorder (PTSD) (Falconer et al., 2013). The intervention taps the negative cognitions and beliefs to change the person's emotional state, ultimately. The exploration of cognitive distortions and assumptions aims to create a rational mindset and facilitates the frontal functioning of the brain. In research by Yoshimura et al. (2014), the intervention improved depressive symptoms related to changes in the Medial Prefrontal and Ventral Anterior Cingulate Cortex. Response inhibition is a product of the interactive effects of frontal systems, basal ganglia, and motor output regions (Dillon & Pizzagalli, 2008). Activation of the frontal systems through the intervention helps to inhibit subcortical systems, bringing about better response inhibition. Studies also suggest that deficits in response inhibition arise from dysfunction in frontal-basal ganglia circuits, especially in conditions like schizophrenia (Raemaekers et al., 2002) and Attention deficit- hyperactive disorder (ADHD) (Aron, 2007).

Dysfunction in planning ability is observed among suicide attempters (Yin et al., 2020), which is contributed to by the state of cognitive deconstruction (Baumeister, 1990). The heightened emotionality resulting from self-blame and falling short of societal standards leads to a negative affective state, creating an emotional numbness (Baumeister, 1990). This marks the onset of cognitive deconstruction when all meaning and purpose of life are lost (Baumeister, 1990). In the present study, planning ability is measured in terms of successful planning and the time taken to execute successful planning. The total time taken to execute planning ability is significantly less, suggesting improved time management related to planning after eight sessions of the intervention. However, regarding successful planning, there is no statistically significant difference between pre- and post-intervention results, suggesting the inefficiency of the intervention in this regard. Studies showed a three-month-long intervention of intervention on adolescents with ADHD brought marginal improvement to planning (Boyer et al., 2015). The eight-session intervention planned for the current study may not be enough to bring about a change in successful planning for suicidal ideation. A dearth of studies related to planning ability among suicide attempters in the context of the intervention creates a research gap that needs further exploration.

Conclusion

There were limited studies conducted in the field of psychology that involved cognitive-based intervention on suicide ideators and suicide attempters to explore the changes in the executive functions of planning, verbal fluency, and response inhibition. The current study was done to bridge the research gap. The intervention was effective in improving planning, verbal fluency, and response inhibition among suicide ideators. The intervention was effective in improving verbal fluency and response inhibition among suicide attempters but was not effective in improving planning among the attempters among suicide attempters but was not effective in improving planning among the attempters. The intervention aimed to reverse the sense of entrapment, and the transition from cognitive deconstruction to reconstruction, or the facilitation of executive functioning, helped manage the suicidal thoughts. The intervention also focuses on cortical arousal, leading to reduced impulsivity and better response inhibition. This helped arrest the transformation of suicidal ideation into a suicidal attempt.

References

American Psychological Association. (2025). Suicidality. Retrieved on January 4th, 2025 from https://dictionary.apa.org/suicidality

Armstrong, G., Vijayakumar, L., Pirkis, J., Jayaseelan, M., Cherian, A., Soerensen, J. B., ... & Niederkrotenthaler, T. (2019). Mass media representation of suicide in a high suicide state in India: an epidemiological comparison with suicide deaths in the population. *BMJ Open*, *9*(7), e030836.

Aron, A. R. (2007). The neural basis of inhibition in cognitive control. *The Neuroscientist*, *13*(3), 214–228.

Audenaert, K., Goethals, I., Van Laere, K., Lahorte, P., Brans, B., Versijpt, J., Vervaet, M., Beelaert, L., Van Heeringen, K., & Dierckx, R. (2002). SPECT neuropsychological activation procedure with the Verbal Fluency Test in attempted suicide patients. *Nuclear Medicine Communications*, *23*(9), 907–916. https://doi.org/10.1097/00006231-200209000-00015

Aviram, R. B., Hellerstein, D. J., Gerson, J., & Stanley, B. (2004). Adapting supportive psychotherapy for individuals with borderline personality disorder who self-injure or attempt suicide. *Journal of Psychiatric Practice*, *10*(3), 145–155. https://doi.org/10.1097/00131746-200405000-00002

Banerjee, T., & Reddy, K. J. (2023). Suicide and youth: Positive psychology perspective. In: Deb, S., Deb, S. (eds) *Handbook of Youth Development*. Singapore: Springer. https://doi.org/10.1007/978-981-99-4969-4_19

Baryshnikov, I., & Isometsä, E. (2022). Psychological pain and suicidal behaviour: A review. *Frontiers in Psychiatry*, *13*, 981353. https://doi.org/10.3389/fpsyt.2022.981353

Baumeister R. F. (1990). Suicide as escape from self. *Psychological Review*, *97*(1), 90–113. https://doi.org/10.1037/0033-295x.97.1.90

Beck, A. T., & Steer, R. A. (1991). *Beck Scale for Suicide Ideation Manual*. San Antonio, TX: Psychological Corporation.

Beck, A. T., Steer, R. A., & Brown, G. K. (1996). *Manual for the Beck Depression Inventory–II*. San Antonio, TX: Psychological Corporation.

Benton, A. L. & Hamsher, K. S. (1989). *Multilingual Aphasia Examination*. New York: Oxford University Press.

Boden, M. T., John, O. P., Goldin, P. R., Werner, K., Heimberg, R. G., & Gross, J. J. (2013). The Role of maladaptive beliefs in cognitive behavioural therapy: Evidence from social anxiety disorder. *Behavioural Research and Therapy*, *50*(5), 287–291. https://doi.org/10.1016/j.brat.2012.02.007

Boyer, B. E., Geurts, H. M., Prins, P. J., & Van der Oord, S. (2015). Two novel CBTs for adolescents with ADHD: the value of planning skills. *European Child & Adolescent Psychiatry*, *24*(9), 1075–1090. https://doi.org/10.1007/s00787-014-0661-5

Bryan, C. J., May, A. M., Rozek, D. C., Williams, S. R., Clemans, T. A., Mintz, J., Leeson, B., & Burch, T. S. (2018). Use of crisis management interventions among suicidal patients: Results of a randomized controlled trial. *Depression and Anxiety*, *35*(7), 619–628. https://doi.org/10.1002/da.22753

Büscher, R., Torok, M., & Sander, L. (2020). Internet-based cognitive behavioral therapy to reduce suicidal ideation: A systematic review and meta-analysis. *JAMA Network Open*, *3*(4), e203933.

Centre for Disease Control and Prevention. (2025). Suicide Prevention. Retrieved on January 14, 2025 from www.cdc.gov/suicide/index.html

Colman, A. M. (2009). *Oxford Dictionary of Psychology*. United Kingdom: Oxford University Press.

Costanza, A., Amerio, A., Aguglia, A., Magnani, L., Serafini, G., Amore, M., ... & Berardelli, I. (2021). "Hard to say, hard to understand, hard to live": Possible associations between neurologic language impairments and suicide risk. *Brain Sciences*, *11*(12), 1594.

Creswell, J. W. (2015). *A Concise Introduction to Mixed Methods Research*. Thousand Oaks, CA: SAGE Publications.

del Hoyo Soriano, L., Villarreal, J. C., Sterling, A., Edgin, J., Berry-Kravis, E., Hamilton, D. R., & Abbeduto, L. (2022). The association between expressive language skills and adaptive behavior in individuals with Down syndrome. *Scientific Reports*, *12*(1), 20014.

Dillon, D. G., & Pizzagalli, D. A. (2008). Inhibition of action, thought, and emotion: A selective neurobiological review. *Applied and Preventive Psychology*, *12*(3), 99–114. https://doi.org/10.1016/j.appsy.2007.09.004

Exner, J. E. (2003). *The Rorschach: A Comprehensive System*. USA: John Wiley & Sons.

Falconer, E., Allen, A., Felmingham, K. L., Williams, L. M., & Bryant, R. A. (2013). Inhibitory neural activity predicts response to cognitive-behavioural therapy for posttraumatic stress disorder. *The Journal of Clinical Psychiatry*, *74*(9), 895–901 https://doi.org/10.4088/JCP.12m08020

Fink, M., Kellner, C. H., & McCall, W. V. (2014). The role of ECT in suicide prevention. *The Journal of ECT*, *30*(1), 5–9. https://doi.org/10.1097/YCT.0b013e3182a6ad0d

Gautam, M., Tripathi, A., Deshmukh, D., & Gaur, M. (2020). Cognitive behavioural therapy for depression. *Indian Journal of Psychiatry*, *62*(2), 223–229. https://doi.org/10.4103/psychiatry.IndianJPsychiatry_772_19

Harmer, B., Lee, S., Duong, T. V. H., & Saadabadi, A. (2024). *Suicidal Ideation*. St. Petersburg, FL: StatPearls Publishing.

Hassan, A., De Luca, V., Dai, N., Asmundo, A., Di Nunno, N., Monda, M., & Villano, I. (2021). Effectiveness of antipsychotics in reducing suicidal ideation: Possible physiologic mechanisms. *Healthcare (Basel, Switzerland)*, *9*(4), 389. https://doi.org/10.3390/healthcare9040389

Hausman, C., Meffert, B. N., Mosich, M. K., & Heinz, A. J. (2019). Impulsivity and cognitive flexibility as neuropsychological markers for suicidality: A multi-modal investigation among military veterans with alcohol use disorder and PTSD. *Archives of Suicide Research*, *0*(0), 1–14. https://doi.org/10.1080/13811118.2019.1635930

Hudson, C. C., Bowers, E. M., Björgvinsson, T., & Beard, C. (2023) Changes in perspective-taking during cognitive behavioural therapy in a partial hospital setting. *Cognitive Therapy and Research*. https://doi.org/10.1007/s10608-023-10422-4

Keilp, J. G., Gorlyn, M., Russell, M., Oquendo, M. A., Burke, A. K., Harkavy-Friedman, J., & Mann, J. J. (2012). Neuropsychological function and suicidal behaviour: Attention control, memory and executive dysfunction in suicide attempt. *Psychological Medicine*, *43*(3), 539–551.

Kinderman, P. (2005). A psychological model of mental disorder. *Harvard Review of Psychiatry*, *13*(4), 206–217.

Kumar, S. & Pathare, S. (2024, February 7). Data is the new oil, but not when it comes to suicides in India. Retrieved on February 7, 2024 from https://cmhlp.org/blogs/data-is-the-new-oil-but-not-when-it-comes-to-suicides-in-india/

Lee, H., & Schafer, M. (2021). Are positive childhood experiences linked to better cognitive functioning in later life?: Examining the role of life course pathways. *Journal of Aging and Health*, *33*(3–4), 217–226.

Lezak, M. D. (2004) *Neuropsychological Assessment* (4th ed.). New York: Oxford University Press.

Lin, T. J., Ko, H. C., Wu, J. Y., Oei, T. P., Lane, H. Y., & Chen, C. H. (2019). The effectiveness of dialectical behaviour therapy skills training group vs. cognitive therapy group on reducing depression and suicide attempts for borderline personality disorder in Taiwan. *Archives of Suicide Research: Official Journal of the International Academy for Suicide Research*, *23*(1), 82–99. https://doi.org/10.1080/13811118.2018.1436104

McCauley, E., Berk, M. S., Asarnow, J. R., Adrian, M., Cohen, J., Korslund, K., Avina, C., Hughes, J., Harned, M., Gallop, R., & Linehan, M. M. (2018). Efficacy of dialectical behaviour therapy for adolescents at high risk for suicide: A randomized clinical trial. *JAMA Psychiatry*, *75*(8), 777–785. https://doi.org/10.1001/jamapsychiatry.2018.1109

Mewton, L., & Andrews, G. (2016). Cognitive behavioural therapy for suicidal behaviours: improving patient outcomes. *Psychology Research and Behaviour Management*, *9*, 21–29.

Meza, J. I., Owens, E. B., & Hinshaw, S. P. (2017). Response inhibition, peer preference and victimization, and self-harm: Longitudinal associations in young adult women with and without ADHD. *Journal of Abnormal Child Psychology*, *44*(2), 323–334. https://doi.org/10.1007/s10802-015-0036-5

Montgomery S. A. (1993). Suicide prevention and serotonergic drugs. *International Clinical Psychopharmacology*, *8* Suppl 2, 83–85. https://doi.org/10.1097/00004850-199311002-00012

Morrow, C. E., Bryan, C. J., & Apollonio, K. K. (2010). Empirically based assessment of suicide risk. In: Kumar U., Mandal M. K. (eds.) *Suicidal Behaviour: Assessment of People at Risk* (pp. 20-41) .New Delhi: Sage Publications.

Mostofsky, S. H., & Simmonds, D. J. (2008). Response inhibition and response selection: Two sides of the same coin. *Journal of Cognitive Neuroscience*, *20*(5), 751–761. https://doi.org/10.1162/jocn.2008.20500

Patterson, J. (2011). Verbal fluency. In: Kreutzer, J. S., DeLuca, J., Caplan, B. (eds) *Encyclopedia of Clinical Neuropsychology*. New York: Springer. https://doi.org/10.1007/978-0-387-79948-3_1423
Phillips, D. P. (1974). The influence of suggestion on suicide: Substantive and theoretical implications of the Werther effect. *American Sociological Review*, *39*(3), 340–354.
Raemaekers, M., Jansma, J. M., Cahn, W., Van der Geest, J. N., van der Linden, J. A., Kahn, R. S., & Ramsey, N. F. (2002). Neuronal substrate of the saccadic inhibition deficit in schizophrenia investigated with 3-dimensional event-related functional magnetic resonance imaging. *Archives of General Psychiatry*, *59*(4), 313–320.
Rao, S. L., Subbakrishna, D. K., & Gopukumar, K. (2004). *NIMHANS Neuropsychology Battery–2004 Manual* (Publication No. 60). Bangalore: National Institute of Mental Health and Neurosciences (NIMHANS). SAGE Journals+6IJIP+6Cambridge University Press & Assessment+6
Response Inhibition. (2015). Retrieved from: www.apa.org/pubs/highlights/peeps/issue-46
Rudd, M. D. (2015). Brief cognitive behavioral therapy (BCBT) for suicidality in military populations. *Military Psychology*, *27*(4), 233–241. https://doi.org/10.1037/mil0000086
Sadock, B. J., & Sadock, V. A. (2007). *Kaplan & Sadock's Synopsis of Psychiatry: Behavioral Sciences/ Clinical Psychiatry* (10th ed.). Philadelphia, PA: Lippincott Williams & Wilkins.
Saffer, B. Y., & Klonsky, E. D. (2017). Do neurocognitive abilities distinguish suicide attempters from suicide ideators? A systematic review of an emerging research area. *Clinical Psychology Science and Practice*, *25*(1), 1–12.
Schauer, M., Neuner, F., & Elbert, T. (2005). Narrative Exposure Therapy (NET): Reorganizing memories of traumatic stress, fear, and violence. Retrieved on January 13, 2025 from www.researchgate.net/publication/300919740_Narrative_Exposure_Therapy_NET_Reorganizing_Memories_of_Traumatic_Stress_Fear_and_Violence/citations
Shao, Z., Janse, E., Visser, K., & Meyer, A. S. (2014). What do verbal fluency tasks measure? Predictors of verbal fluency performance in older adults. *Frontiers in Psychology*, *5*, 772.
Spirito, A., Esposito-Smythers, C., Wolff, J., & Uhl, K. (2012). Cognitive-behavioural therapy for adolescent depression and suicidality. *Child and Adolescent Psychiatric Clinics of North America*, *20*(1), 191–204. https://doi.org/10.1016/j.chc.2011.01.012
Spirito, A., & Overholser, J. C. (Eds.). (2003). *Evaluating and Treating Adolescent Suicide Attempters: From Research to Practice*. Amsterdam, Netherlands: Elsevier.
Stanley B., Brown G., Brent D., Wells K., Poling K., Curry J., et al. (2009). Cognitive-behavioural therapy for suicide prevention (CBT-SP): Treatment model, feasibility, and acceptability. *Journal of American Academy of Child & Adolescent Psychiatry*, *48*(10), 1005–1013. Retrieved from www.ncbi.nlm.nih.gov/pmc/articles/PMC2888910/pdf/nihms148473.pdf
Stewart, J. G., Shields, G. S., Esposito, E. C., Cosby, E. A., Allen, N. B., Slavich, G. M., & Auerbach, R. P. (2019). Life stress and suicide in adolescents. *Journal of Child and Adolescent Psychopathology*, *47*(10). 1707–1722.
Suicide. (2019). Retrieved from www.apa.org/topics/suicide/
Swanson, S. A., & Colman, I. (2013). Association between exposure to suicide and suicidality outcomes in youth. *CMAJ: Canadian Medical Association Journal = journal de l'Association medicale canadienne*, *185*(10), 870–877. https://doi.org/10.1503/cmaj.121377
Tarrier, N. (2009). Broadening the cognitive behaviour therapy base: The importance of considering contextual, cognitive, emotional and behavioural factors. *Behaviour Research and Therapy*, *47*(6), 518–528. https://doi.org/10.1016/j.brat.2009.03.005
The Daily Telegraph. (2024). Does your self-talk need an upgrade? Retrieved January 13, 2025, from www.dailytelegraph.com.au/lifestyle/does-your-selftalk-need-an-upgrade/news-story/d1e0d227df45e1eb0097eddccb6b4558
Trivedi, J. K. (2006). Cognitive deficits in psychiatric disorders: Current status. *Indian Journal of Psychiatry*, *48*(1), 10–20. https://doi.org/10.4103/0019-5545.31613
Tsujii, N., Mikawa, W., Tsujimoto, E., Adachi, T., Niwa, A., Ono, H., & Shirakawa, O. (2017). Reduced left precentral regional responses in patients with major depressive disorder and history of suicide attempts. *PLoS One*, *12*(4), e0175249. https://doi.org/10.1371/journal.pone.0175249
Wenzel, A., Brown, G. K., Beck, A. T. (2015). *Cognitive Therapy for Suicidal Patients: Scientific and Clinical Applications*. Washington: American Psychological Association.

Williams, J. M. G., & L. R. Pollock (2001). Psychological aspects of the suicidal process. In: van Heeringen, K. (ed) *Understanding Suicidal Behaviour* (pp. 76–93). Chichester: John Wiley & Sons.

World Health Organisation. (2024). Suicide. Retrieved on July 7, 2024 from www.who.int/news-room/fact-sheets/detail/suicide

Yin, H. (2020). Cognitive distortion in suicidal behaviour. *Scholarly Journal of Psychology and Behavioural Sciences*, *3*(4). https://doi.org/10.32474/sjpbs.2020.03.000168

Yoshimura, S., Okamoto, Y., Onoda, K., Matsunaga, M., Okada, G., Kunisato, Y., Yoshino, A., Ueda, K., Suzuki, S. & Yamawaki, S. (2014). Cognitive behavioural therapy for depression changes medial prefrontal and ventral anterior cingulate cortex activity associated with self-referential processing. *Social Cognitive and Affective Neuroscience*, *9*(4), 487–493. https://doi.org/10.1093/scan/nst009

PART IV

Ethical, Social, and Practical Considerations

16

ETHICS OF TECHNOLOGY-ASSISTED NEUROCOGNITIVE REHABILITATION

Ajay Chandra, Arkalgud Ramaprasad, and Madhavi Rangaswamy

Introduction

Integration of technology into healthcare has significantly transformed neurocognitive rehabilitation. Artificial intelligence (AI), machine learning, and aligned technologies have rapidly reshaped the domain of neurocognitive rehabilitative care (Calderone et al., 2024). Transformative technologies enables early disease detection, personalized cognitive training programs, and intelligent virtual therapists to help patients recover brain functions lost to injury, stroke, or degenerative disorders (Al Kuwaiti et al., 2023; Calderone et al., 2024; Loya et al., 2025).

The ethics of integration of technology into neurocognitive rehabilitation is a complex problem. The problem needs to be defined systematically and understood systemically. For example, AI is reshaping neurocognitive rehabilitation through personalized insights and precise monitoring. While it enhances therapy, it also raises concerns around data privacy, consent, and the replication of human qualities like empathy and trust. As AI becomes more integrated into care, ethical and human-centric approaches are essential to ensure its responsible and compassionate use. Similarly, other transformative technologies like wearable devices and brain interfaces provide effective and real-time insights and efficacy through predictive capabilities in neurocognitive rehabilitation (Reddy, 2025). The integration of the new technologies in neurocognitive rehabilitation has facilitated the development of effective treatments, compensatory strategies, and real-world simulations (Charchat-Fichman et al., 2014).

Technology-based interventions ensure in delivering personalized neurocognitive rehabilitation and improving patient outcomes. However, challenges like lack of standardization, limited generalizability, integration hurdles, and ethical concerns ranging from data privacy to equitable access must be carefully addressed before these solutions can be widely adopted (Morris et al., 2025). They present novel challenges to balancing the four pillars of beneficence, non-malfeasance, justice, and autonomy (Kelly & Leff, 2023). The World Health Organization (2024) guidance on "Ethics and Governance of Artificial Intelligence for Health: Large Multi-Modal Models" is narrowly focused on one technology – AI, and

DOI: 10.4324/9781003646662-20

broadly focused on health – not just rehabilitation. It is a consensus recommendation from 20 experts but does not provide a logical framework that can be tailored to the challenge at hand.

Ontology of Ethics of Technology-Assisted Neurocognitive Rehabilitation

We present a framework to describe the challenges, explain the logic, set the expectations, and regulate the outcome of the ethics of technology-assisted neurocognitive rehabilitation. The framework will help address the issue systemically and systematically. In the absence of such a framework, the present approaches are selective, siloed, and segmented and do not address the challenges holistically as they should be. The "Ontology of Ethics of Technology-Driven Neurocognitive Rehabilitation" encapsulates the core logic of the problem clearly, concisely, and comprehensively. The complex interplay of ethics, efficiency, and effectiveness of technology-assisted neurocognitive rehabilitation can be visualized and analyzed using the framework which is presented in Figure 16.1 and described below.

The application and development of the ontology follow the description of the logic and process as put forth by Ramaprasad and Syn (Ramaprasad & Syn, 2014, 2015). The ontological method has been used to study mental healthcare (Chandra et al., 2024; Martínez et al., 2022), COVID-19 strategies (Sreeganga et al., 2021), public health informatics research (Ramaprasad et al., 2017), and mobile-health research (Cameron et al., 2017).

Neurocognitive Rehabilitation

Neurocognitive rehabilitation is the choice and implementation of suitable interventions based on clinical assessment of the response stage of the patient. The Response Stage, Clinical Interaction, and Intervention dimensions are defined below. The choice and implementation will be by a set of agents, based on the technologies and the objectives of rehabilitation. These three dimensions are defined subsequently.

Response Stage

A patient's neurocognitive rehabilitation intervention must be based on the clinical assessment of the response stage of the patient. The patient's response stage may range from null (or no) neurocognitive response to an appropriate purposeful response. The response scale is based on the RanchoScale (Lin & Wroten, 2022) and appropriately represents the progressive improvement of a neurocognitive rehabilitation patient. The stages of the scale are listed in the column labeled Response Stage in Figure 16.1. They are as follows: null, generalized, localized, confused (agitated, inappropriate, appropriate), and appropriate (automatic, purposeful). The purpose of rehabilitation is to move the patient from an upper stage to a lower stage and eventually to independence.

Clinical Interaction

The determination of the present neurocognitive stage of the patient, the desired stage, and the appropriate intervention to bridge the gap are based on the clinical interaction between the patient and the rehabilitating agent. The clinical interaction must include engagement with the patient, assessment of the patient, documentation of the interaction, diagnosis of

Objective		*Technology*		*Neurocognitive Rehabilitation*							
				*Stage**		*Clinical*		*Intervention**		*Agent*	
Ethics	[of]	Invasive	[technology for]	Null	[clinical]	Engagement	[for]	Total	[interevention by]	Care Provider	[agent]
Personal		Implant		Generalized		Assessment		Maximum		Physician	
Social		Deep brain stimulation		Localized		Documentation		Moderate		Nurse	
Professional		Electrocarticogrphy		Confused		Diagnosis		Minimum		Therapist	
Efficiency		Noninvasive		Agitated		Communication		Standby		Psycho-	
Cost		Cognitive training		Inappropriate		Feedback		Standby on request		Occupational	
Time		Virtual reality		Appropriate				Modified independent		Physio-	
Personnel		Artificial Intelligence		Appropriate						Speech & Hearing	
Informational		Wearable		Automatic						Caretaker	
Effectiveness		Robotic		Purposeful						Self	
Cognitive		Brain interface								Family/Friend	
Conative		Assistive								Professional	
Affective		Prosthetic								Community	
		External								Digital	
		Gaming								Handheld	
		Telerehabilitation								Standalone	
		eHealth									
		Communication									
		Monitoring									

Figure 16.1 Ontology of Ethics of Technology-Driven Neurocognitive Rehabilitation.

*Based on Racnhos scale [Lin K, Wroten M. Ranchos Los Amigos. (2022). In: StatPearls [Internet]. Treasure Island (FL): StatPearls Publishing. Available from: https://www.ncbi.nlm.nih.gov/books/NBK448151/]

the stage and determination of the appropriate intervention and its target stage, communication of the diagnosis and intervention with the patient and the rehabilitating agents, and feedback for correction and adaptation to bridge the gap. The steps may be implemented in one appointment or several appointments over an extended period with breaks of several days or weeks. All the appointments must be tracked in the healthcare information system assiduously. These steps are listed in the column labeled Clinical.

Intervention

An intervention may range from being total (maximum) to modified independent (minimal), a step short of the patient becoming independent (no intervention needed). These levels too are based on the RanchoScale (Lin & Wroten, 2022) and appropriately represent the progress of a neurocognitive rehabilitation patient. The intermediate levels of intervention, in descending order of coverage, are maximum, moderate, minimum, standby, and standby on request. The level of intervention for a patient is based on: (a) his/her clinical interaction with the rehabilitating agent to determine the present and desired stages of the patient, (b) the gap between the two, and (c) the associated diagnosis and recommendation of the rehabilitating agent.

Technology for Neurocognitive Rehabilitation

Technology can assist neurocognitive rehabilitation at all levels of intervention and steps of clinical interaction. The choice of technology must depend on the present response stage of the patient, the desired stage, the gap to be bridged, the choice of intervention to bridge the gap, and the clinical interaction. The technology may be invasive, penetrating the skin of the patient; it may be non-invasive, in contact with the skin of the patient but not penetrating it; or external, removed from the skin of the patient. The three types of technology and their subtypes are listed in the column labeled Technology (Figure 16.1). These technologies can perform specialized functions to drive interventions at different response stages and assist in the steps of clinical interaction, and the general functions to support the rehabilitation. For instance, integration of technology into neurocognitive rehabilitation intervention been effective in facilitating goal performance and satisfaction for individuals with memory problems (Grewal et al., 2025).

Agents of Neurocognitive Rehabilitation

The choice of assistance technologies must be determined by the agents delivering the rehabilitation in consultation with the patient. The agents, listed in the last column labeled Agent, may be care providers, caretakers, and digital tools. The care providers are physicians, nurses, therapists (psychotherapists, occupational therapists, physiotherapists, and speech and hearing therapists); the caretakers are the self, families/friends, professionals, and the community; and the digital tools may be handheld or standalone. These agents may act individually, as a team, and in coordination for neurocognitive rehabilitation.

The objective in choosing a technology for assistance is to balance the ethics of the choice with its efficiency and effectiveness in rehabilitation. The ethics determining the choice must include the following: (a) the personal ethics of the agent and the recipient, (b) the social ethics in the context of rehabilitation, and (c) the professional ethics of the rehabilitating agents. The ethics of a choice can affect the cost, time, personnel, and informational efficiency of rehabilitation; it may also affect the cognitive, conative, and affective effectiveness of the same. These choices must be guided by personal, social, and professional guidelines.

Objectives of Neurocognitive Rehabilitation

Technology – invasive, non-invasive, and external – to assist neurocognitive rehabilitation is advancing rapidly. It can improve the efficiency and effectiveness of rehabilitation. Yet, because of the very novelty and nascency of application of many of these technologies, their unintended, uncommon, and potentially long-term effects on the body and mind of the rehabilitated patient and the rehabilitating agent are unknown. The challenge, choice, and application of a technology may be further compounded by the neurocognitive state of the patient, and his/her ability to provide informed consent. In the following we discuss the tradeoffs between ethics of technology-assisted neurocognitive rehabilitation and the effect of the assistance on the efficiency and effectiveness of rehabilitation.

Ethical Challenges of Technology-Assisted Rehabilitation

The ethical challenge of technology-assisted neurocognitive rehabilitation of a patient by an agent can be defined using the ontology to:

- Specify the rehabilitation requirements of a patient in terms of:
  - The clinical steps;
  - The present and desired response stages, and the gap between them; and
  - The recommended interventions.
- Specify the technology assistance for the interventions.
- Specify the agent(s) for the interventions.
- Specify the ethical requirements of the technology-assisted interventions, and the tradeoff between the ethics, efficiency, and effectiveness of the interventions.

Invasive, non-invasive, and external technologies pose different sets of ethical issues that are distinctive and overlapping. The following discussion is organized around the three types. Within each section we discuss the ethical issues for care providers, caretakers, and digital agents.

Ethics of Technology-Assisted Neurocognitive Rehabilitation

The ethics of technology-assisted neurocognitive rehabilitation is complex because of the following characteristics of the interventions:

- The interventions act physiologically on the neurology of the patient, sometimes deep within the brain. Invasive technologies penetrate the deepest, non-invasive technologies far less, and external technologies not at all.
- The interventions act psychologically on the cognition of the patient, sometimes in fundamental ways. Invasive technologies can affect the psychology most fundamentally, non-invasive technologies less so, and external technologies the least.

- All three types of intervention technologies are advancing rapidly and developing new capabilities to intervene neurologically, cognitively, and externally. It takes time to establish their effectiveness. While external technology may be preferable to a non-invasive technology, and a non-invasive technology to an invasive technology, the trade-off between their effectiveness and efficiency may not be clearcut and hence the choice is difficult. The ethics of the choice will compound the difficulty.
- While the rehabilitation intervention may be narrowly focused, it can affect the patient's body and mind in their entirety. The ethics must consider these larger effects.
- Many of the technology's effects may be irreversible and unpredictable, especially in the long-term. The irreversibility/unpredictability is likely to be greatest for the invasive technologies, less for the non-invasive ones, and least for the external ones.
- The agents of intervention and the patients must decide upon the ethics of an intervention based on incomplete knowledge of its efficiency and effectiveness. The choice between the promise of technology must be weighed against the perils of ignorance about the outcomes must be made ethically by the agent and the recipient. Incorporation of patient feedback to enhance the overall effectiveness of and personal relevance of rehabilitation programs is critical (Reddy, 2025).

We discuss the implications of the above characteristics for the invasive, non-invasive, and external technologies for the ethical considerations of the care providers, caretakers, and digital agents in the following sections.

Ethics of Invasive Technology-Assisted Neurocognitive Rehabilitation

Invasive technologies for assisting neurocognitive rehabilitation include implants, deep brain stimulation, and electrocorticography. ICT-based programs for neurocognitive rehabilitation show promising results in improving cognitive functions such as attention, memory, and executive functions (Geraldo et al., 2018). While Brain–Computer Interface (BCI)–AI integration has witnessed greater expansion, there is a need standardized evaluation protocols, ethical global governance frameworks, formulation of guidelines for engaging in cutting-edge research, and an interdisciplinary collaboration (Qiu et al., 2024; Rudroff, 2025; Wang et al., 2024). For instance, there is the need for newer neuro-rights as part of ethical governance principles for integration of AI and BCI (Astobiza et al., 2019). Mental privacy and data security are crucial ethical concerns in integration of AI with BCIs, especially when there are risks of accessing intimate neural data without consent. There is a need for robust security measures to protect cognitive liberty and personal information as these technologies progress (Rudroff, 2025).

Care Provider, Caretaker, and Digital Agent Ethics for Invasive Technology

The care providers, caretakers, and digital agents of invasive technology-assisted neurocognitive rehabilitation bear a heavy responsibility in balancing the four pillars of beneficence, non-maleficence, justice, and autonomy (Kelly & Leff, 2023). Their challenge is compounded by: (a) unclear and emerging guidance on the personal, social, and professional ethics, (b) uncertainty and lack of trusted information about the cognitive, conative, and affective outcomes of the intervention, and (c) unpredictability and lack of adequate evidence about the cost, time, personnel, and other resource requirements regarding the technology. Further, there is a significant asymmetry in the information possessed by the care providers, caretakers, and digital agent – the care providers have the most trusted information, the caretakers less, and the digital agents the least. The open-source information available to the caretakers to make their choice is likely to be limited, and when available technical and difficult to understand for the layperson. Thus, the care providers have the primary responsibility for assuring the ethics of invasive technology-assisted neurocognitive rehabilitation. In addition to assuring their own ethics, they must assist the caretakers in their choice, and in guiding the digital agents' design to be ethical. Thus, the ethical requirements are as follows:

- The care providers must describe and explain to the patient and the caretakers the clinical steps, their purpose, and timeline.
- They must agree on the present and desired response stages, and the gap between them in consultation with the patient and the caretakers.
- They must agree on the recommended interventions, again in consultation with the patient and the caretakers.
- They must arrive at a consensus on the technology assistance for the interventions, and the agents that will be administering the interventions.
- In the above consultation, consensus generation, and implementation and the trade-off between the ethics, efficiency, and effectiveness of the interventions must be clarified amongst the concerned agents. The beneficence of the intervention and its non-maleficence must be considered, considering the likely diminished autonomy of the patient requiring the intervention.

Further, technology-assisted invasive neurocognitive rehabilitation will require super specialists; it is likely to be expensive, and not widely accessible. Consequently, ethically affirming justice in the delivery of such rehabilitation will be a challenge.

Ethics of Non-invasive Technology-Assisted Neurocognitive Rehabilitation

Non-invasive technologies for assisting neurocognitive rehabilitation include cognitive training, virtual reality, AI, wearable devices, robotics, brain interfaces, assistive technologies, and prosthetics.

The need for integrative approaches to enhance cognitive rehabilitation through assistive technology, telerehabilitation, and virtual reality has been well articulated (Stasolla et al., 2023). Robotic therapies have proven significant improvements cognitive and motor domains of neurocognitive rehab care (Morris et al., 2025). The need for advanced robotic assessment of neurologic impairments in case of traumatic brain injury cases are prerequisites for measuring motor and cognitive assessments (Debert et al., 2012). Healthcare robots can benefit mental health, physical rehabilitation, and general well-being (Boch & de Clermont-Tonnerre, 2024). Nonetheless, the challenges for robot-assisted cognitive trainings include ethical issues, user-centric design, reliability concerns, trust, and cost-effectiveness (Yuan et al., 2021).

Ethical conundrums in the use of robotics and the appropriateness of its use for children and elderly population are specifically discussed. Novel smart home care models for elderly population require robotic designs that are optimized, improved flexibility, feasibility, better sensor networks, and assistance from diverse technologies (Zhao & Guo, 2023).

Reference and adherence to ethical principles for devising treatment modalities could be considered acceptable (Majeed, 2016).

Data security and privacy are critical components of assistive technologies. For instance, the principles outlined in iManus offer a foundation for ethically designing robotics in remote healthcare (Maddahi et al., 2022). Further, the advantages of having smart mobility aids like the cloud-connected prosthetics with robotic exoskeletons are in terms of effective communications, remote customization, real-time data analysis, and adaptive learning (Zhang et al., 2024). However, limited research and attention have been directed toward the ethical considerations involved in the development and clinical translation of prosthetic technology (Gavette et al., 2023).

Care Provider, Caretaker, and Digital Agent Ethics for Non-invasive Technology

The requirements for assuring the ethics for non-invasive technology are like that of invasive technology but potentially less stringent. This is due to the easier reversibility and predictability of the effects of these interventions.

The care providers, caretakers, and digital agents of non-invasive technology-assisted neurocognitive rehabilitation bear a lighter responsibility in balancing the four pillars of ethics because of: (a) clearer and documented guidance on the personal, social, and professional ethics, (b) greater certainty and more trusted information about the cognitive, conative, and affective outcomes of the intervention, and (c) greater predictability and better evidence about the cost, time, personnel, and other resource requirements regarding the technology. Further, there is a lesser asymmetry in the information possessed by the care providers, caretakers, and the digital agent. The open-source information available to the caretakers to make their choice is less limited, and easier to understand for the layperson. Thus, the care providers share the responsibility for assuring the ethics of non-invasive technology-assisted neurocognitive rehabilitation. As with invasive technology, the care providers in addition to assuring their own ethics must assist the caretakers in their choice, and in guiding the digital agents' design to be ethical.

Thus, the procedure for fulfilling the ethical requirements is like that with the invasive technology. The steps are likely to be easier because of greater information symmetry, its availability, its patient centricity and ease of understanding, its availability, and greater patient autonomy. Further, technology-assisted non-invasive neurocognitive rehabilitation will require specialists available in larger numbers, is likely to be less expensive, and more widely accessible. Consequently, ethically affirming justice in the delivery of such rehabilitation will be easier than with non-invasive technology.

Ethics of External Technology-Assisted Neurocognitive Rehabilitation

External technologies for assisting neurocognitive rehabilitation include gaming, telerehabilitation, eHealth, communication, and monitoring technologies. Portable, scalable rehabilitation devices featuring clinical intelligence, along with capabilities for remote monitoring and coaching, are key enablers for providing feasible and acceptable distant neurorehabilitation (Lambercy et al., 2021). Remotely deployable digital technologies with remote coaching provide highly individualized training and could strengthen the goal-directed functioning of specific neurocognitive abilities (Loya et al., 2025).

Technological integration in neurocognitive home-rehabilitation enables standardized measures (e.g., response recording), efficiency (i.e., quick and accurate recording), and cost-effectiveness (Charchat-Fichman et al., 2014). Significant improvements in areas of attention, working memory, semantic fluency, emotional functioning, and information processing speed were evident following the use of cognitive telerehabilitation combined with the Transcranial Direct Current Stimulation (tDCS) on a 29-year-old man who had a traumatic brain injury (Eilam-Stock et al., 2021). The need for a well-conceptualized ethical framework of neuro-engineering and rehabilitation technology encompasses user-centric, value-driven design – built on fairness, biomedical ethics, and social responsibility (Ienca et al., 2017).

Care Provider, Caretaker, and Digital Agent Ethics for External Technology

External technologies provide the platform for clinical interactions and support for interventions. They do not play a direct role in either invasive or non-invasive interventions. Hence, their role in assuring beneficence, non-malfeasance, justice, and autonomy (Kelly & Leff, 2023) of an intervention is to assure that the platform fulfils these requirements.

The performance of these external technologies is central to the efficiency and effectiveness of the interventions. These technologies are also the repositories of all the information about the patients, the interventions, and the outcomes. They must be accessible to the care providers, caretakers, and the digital agents as needed. Yet, they must assure the privacy, confidentiality, anonymity, and quality of the information. Further, since many interventions are long-term, these external systems must maintain continuity of support for an extended period.

The ethics challenge of external technology is compounded today by: (a) the large volumes of information, (b) the variety of types of information, (c) the electronification of information, (d) the demand for ready accessibility of information by all the agents, (e) the legal requirements of protection of information, and (f) the exponentially increasing risk of information hacking by mal-intentioned actors.

Conclusion

Technology-assisted neurocognitive rehabilitation is advancing very rapidly propelled by the rapid advances in invasive, non-invasive, and external technologies. However, academic research, government policies, and professional practice guidelines on the ethics of these technologies and their application have not kept pace. There is very little academic research, few policy statements, and only general practice guidelines on the subject. Technologies that assist neurocognitive rehabilitation can deeply affect the body, mind, and spirit of the patients and their caretakers. The ethics of use of these technologies pose unique challenges the answers to which cannot be extended, extrapolated, or generalized from other applications. They require specific focus and attention.

The Ontology of Ethics of Technology-Driven Neurocognitive Rehabilitation is a framework to develop roadmaps for research, policies, and practices on the subject. It can be used to map the states-of-research, -policies, and -practices on the subject. The mapping will help determine the emphases and gaps in the three domains, and the translational gaps between them. Thus, it can be used to highlight the current biases and blind spots and correct them.

Last, the framework is modular. It can be extended to accommodate new technologies, agents, and objectives. Its elements can be refined to study the ethics at a finer level of granularity, or they can be aggregated to study at a higher level. Thus, it can anchor the emergence and evolution of the scientific body of knowledge on the ethics of technology-driven neurocognitive rehabilitation.

References

Al Kuwaiti, A., Nazer, K., Al-Reedy, A., Al-Shehri, S., Al-Muhanna, A., Subbarayalu, A. V., Al Muhanna, D., & Al-Muhanna, F. A. (2023). A Review of the Role of Artificial Intelligence in Healthcare. *Journal of Personalized Medicine*, *13*(6). Scopus. https://doi.org/10.3390/jpm13060951

Astobiza, A. M., Ausín, T., Toboso, M., Ferrer, R. M., Payá, M. A., & López, D. (2019). Translating Thought into Action: Brain-Computer Interfaces and the Ethical Problem of Agency. *Revista de Bioetica y Derecho*, *46*, 29–46. Scopus. https://doi.org/10.1344/rbd2019.0.26862

Boch, A., & de Clermont-Tonnerre, L. D. C. (2024). Introduction to Social Robotics Ethics: An Empirical Perspective. In *The Elgar Companion to Applied AI Ethics* (pp. 344–396). Edward Elgar Publishing Ltd.; Scopus. https://doi.org/10.4337/9781803928241.00020

Calderone, A., Latella, D., Bonanno, M., Quartarone, A., Mojdehdehbaher, S., Celesti, A., & Calabrò, R. S. (2024). Towards Transforming Neurorehabilitation: The Impact of Artificial Intelligence on Diagnosis and Treatment of Neurological Disorders. *Biomedicines*, *12*(10), Article 10. https://doi.org/10.3390/biomedicines12102415

Cameron, J. D., Ramaprasad, A., & Syn, T. (2017). An Ontology of and Roadmap for mHealth Research. *International Journal of Medical Informatics*, *100*, 16–25. https://doi.org/10.1016/j.ijmedinf.2017.01.007

Chandra, A., Sreeganga, S. D., & Ramaprasad, A. (2024). Mental Healthcare Systems Research during COVID-19: Lessons for Shifting the Paradigm Post COVID-19. *Urban Governance*, S2664328624000056. https://doi.org/10.1016/j.ugj.2024.01.005

Charchat-Fichman, H., Uehara, E., & Santos, C. F. (2014). New Technologies in Assessment and Neuropsychological Rehabilitation. *Temas Em Psicologia*, *22*(3), 539–553. https://doi.org/10.9788/TP2014.3-01

Debert, C. T., Herter, T. M., Scott, S. H., & Dukelow, S. (2012). Robotic Assessment of Sensorimotor Deficits After Traumatic Brain Injury. *Journal of Neurologic Physical Therapy*, *36*(2), 58–67. Scopus. https://doi.org/10.1097/NPT.0b013e318254bd4f

Eilam-Stock, T., George, A., & Charvet, L. E. (2021). Cognitive Telerehabilitation with Transcranial Direct Current Stimulation Improves Cognitive and Emotional Functioning Following a Traumatic Brain Injury: A Case Study. *Archives of Clinical Neuropsychology*, *36*(3), 442–453. https://doi.org/10.1093/arclin/acaa059

Gavette, H., McDonald, C. L., Kostick-Quenet, K., Mullen, A., Najafi, B., & Finco, M. G. (2023). Advances in Prosthetic Technology: A Perspective on Ethical Considerations for Development and Clinical Translation. *Frontiers in Rehabilitation Sciences*, *4*. Scopus. https://doi.org/10.3389/fresc.2023.1335966

Geraldo, A., Dores, A. R., Coelho, B., Ramião, E., Castro-Caldas, A., & Barbosa, F. (2018). Efficacy of ICT-Based Neurocognitive Rehabilitation Programs for Acquired Brain Injury: A Systematic Review on Its Assessment Methods. *European Psychologist*, *23*(3), 250–264. https://doi.org/10.1027/1016-9040/a000319

Grewal, K. S., Grewal, E. S., Cammer, A., McWilliams, L. A., Spiteri, R. J., & O'Connell, M. E. (2025). Case Reports in the Integration of Technology with Cognitive Rehabilitation for Individuals with Memory Concerns and Their Care Partners. *Archives of Clinical Neuropsychology*, *40*(4), 878–906. https://doi.org/10.1093/arclin/acae115

Ienca, M., Kressig, R. W., Jotterand, F., & Elger, B. (2017). Proactive Ethical Design for Neuroengineering, Assistive and Rehabilitation Technologies: The Cybathlon Lesson. *Journal of NeuroEngineering and Rehabilitation*, *14*(1). Scopus. https://doi.org/10.1186/s12984-017-0325-z

Kelly, E. M., & Leff, M. (2023, July 13). Ethical Issues Commonly Managed During Rehabilitation. *PM&R Knowledge Now*. https:// now.aapmr.org/ethical-issues-commonly-managed-during-rehabilitation/

Lambercy, O., Lehner, R., Chua, K., Wee, S. K., Rajeswaran, D. K., Kuah, C. W. K., Ang, W. T., Liang, P., Campolo, D., Hussain, A., Aguirre-Ollinger, G., Guan, C., Kanzler, C. M., Wenderoth, N., & Gassert, R. (2021). Neurorehabilitation from a Distance: Can Intelligent Technology Support Decentralized Access to Quality Therapy? *Frontiers in Robotics and AI*, *8*. https://doi.org/10.3389/frobt.2021.612415

Lin, K., & Wroten, M. (2022). *Ranchos Los Amigos*. StatPearls Publishing. www.ncbi.nlm.nih.gov/books/NBK448151/

Loya, F., Binder, D., Rodriguez, N., Buchanan, B., Novakovic-Agopian, T., & Chen, A. J.-W. (2025). Fostering Experiential Learning of Neurocognitive Skills in Brain Injury Tele-Rehabilitation: Bridging Gaps in Remote Training by Integrating Scenario-Based Digital Experiences with Coaching. *Frontiers in Human Neuroscience*. https://doi.org/10.3389/fnhum.2025.1593246

Maddahi, A., Leach, T. R., Saeedi, M., Dhannapuneni, P. R., Maddahi, Y., Choukou, M.-A., & Zareinia, K. (2022). Roboethics in Remote Human Interactions and Rehabilitative Therapeutics. *Applied Sciences (Switzerland)*, *12*(12). Scopus. https://doi.org/10.3390/app12126033

Majeed, A. B. A. (2016). Roboethics—Making Sense of Ethical Conundrums. In Yussof H. (Ed.), *Procedia Computer Science*, *105*, 310–315. Elsevier B.V.; Scopus. https://doi.org/10.1016/j.procs.2017.01.227

Martínez, V., Crockett, M. A., Chandra, A., Suwasrawala, S. S., Ramaprasad, A., Núñez, A., & Gómez-Rojas, M. (2022). State of Mental Health Research of Adolescents and Youth in Chile: An Ontological Analysis. *International Journal of Environmental Research and Public Health*, *19*(16), Article 16. https://doi.org/10.3390/ijerph19169889

Morris, K., Parmar, J., Torres, A., & Kamal, T. T. (2025). Balancing Ethics and Efficacy: Examining the Scope of Artificial Intelligence in Neurocognitive Rehabilitation (P4-4.004). *Neurology*, *104*(7_Supplement_1), 3037. https://doi.org/10.1212/WNL.0000000000210752

Qiu, W.-C., Ma, L., Guo, H.-Y., Yang, J.-J., & Li, X.-J. (2024). The Implementation, Clinical Progress and Technical Challenges of Implantable Brain-Computer Interface Systems. *Progress in Biochemistry and Biophysics*, *51*(10), 2478–2497. Scopus. https://doi.org/10.16476/j.pibb.2024.0276

Ramaprasad, A., & Syn, T. (2014). Design Thinking and Evaluation Using an Ontology. In M. Helfert, B. Donnellan, & J. Kenneally (Eds.), *Design Science: Perspectives from Europe* (Vol. 447, pp. 63–74). Springer International Publishing. https://doi.org/10.1007/978-3-319-13936-4_6

Ramaprasad, A., Sastry, N. K. B., & Syn, T. (2017). Envisioning Precision Healthcare Informatics: A Unified Framework. *MEDINFO 2017: Precision Healthcare through Informatics*, *245*, 564–568. https://doi.org/10.3233/978-1-61499-830-3-564

Ramaprasad, A., & Syn, T. (2015). Ontological Meta-Analysis and Synthesis. *Communications of the Association for Information Systems*, *37*, 138–153. Scopus. https://doi.org/10.17705/1cais.03707

Reddy, K. J. (2025). Patient Perspectives and Outcomes. https://doi.org/10.1007/978-3-031-88117-6_7

Rudroff, T. (2025). Decoding Thoughts, Encoding Ethics: A Narrative Review of the BCI-AI Revolution. *Brain Research*, 1850. Scopus. https://doi.org/10.1016/j.brainres.2024.149423

Sreeganga, S. D., Chandra, A., & Ramaprasad, A. (2021). Ontological Analysis of COVID-19 Vaccine Roll Out Strategies: A Comparison of India and the United States of America. *International Journal of Environmental Research and Public Health*, *18*(14), 7483. https://doi.org/10.3390/ijerph18147483

Stasolla, F., Lopez, A., Akbar, K., Vinci, L. A., & Cusano, M. (2023). Matching Assistive Technology, Telerehabilitation, and Virtual Reality to Promote Cognitive Rehabilitation and Communication Skills in Neurological Populations: A Perspective Proposal. *Technologies*, *11*(2), Article 2. https://doi.org/10.3390/technologies11020043

Wang, X.-Q., Sun, H.-Q., Si, J.-Y., Lin, Z.-Y., Zhai, X.-M., & Lu, L. (2024). Challenges and Suggestions of Ethical Review on Clinical Research Involving Brain-Computer Interfaces. *Chinese Medical Sciences Journal*, *39*(2), 131–139. Scopus. https://doi.org/10.24920/004377

World Health Organization. (2024). *Ethics and Governance of Artificial Intelligence for Health: Large Multi-Modal Models. WHO Guidance* (1st ed). World Health Organization.

Yuan, F., Klavon, E., Liu, Z., Lopez, R. P., & Zhao, X. (2021). A Systematic Review of Robotic Rehabilitation for Cognitive Training. *Frontiers in Robotics and AI*, *8*. Scopus. https://doi.org/10.3389/frobt.2021.605715

Zhang, R., Zhou, Y., Zhang, J., & Zhao, J. (2024). Cloud-Integrated Robotics: Transforming Healthcare and Rehabilitation for Individuals with Disabilities. *Proceedings of the Indian National Science Academy*, *90*(3), 752–763. Scopus. https://doi.org/10.1007/s43538-024-00264-3

Zhao, L., & Guo, Y. (2023). Development of Rehabilitation and Assistive Robots in China: Dilemmas and Solutions. *Journal of Shanghai Jiaotong University (Science)*, *28*(3), 382–390. Scopus. https://doi.org/10.1007/s12204-023-2596-9

17

HARNESSING VIRTUAL REALITY FOR ADVANCING INNOVATIONS IN NEUROCOGNITIVE REHABILITATION

Virtual Reality in Neurocognitive Rehabilitation

Navkiran Kalsi, Tara Chand, Rajanikant Panda, and Neeraj Upadhyay

17.1 Introduction: The Evolving Landscape of Neurorehabilitation

During the past 20 years, technology has revolutionized healthcare by introducing new tools to help diagnose, monitor, and treat numerous physical and psychological conditions. From these new developments, virtual reality (VR) shows a great deal of potential in helping neurocognitive rehabilitation (Buele et al., 2023; Maggio et al., 2019). Because VR simulates real-world experiences, clinicians and researchers can design projects in VR that are clinically useful and ecologically true to life (Devos et al., 2023; Hummel et al., 2023). Therefore, this approach has major relevance for neurorehabilitation, mainly in those who have cognitive difficulties due to a neurological injury, mental illness, or developmental disability (Moreno et al., 2019).

Most often, patients in neurocognitive rehabilitation practice structured tasks, fill out worksheets or participate in simulations supervised by healthcare staff. While these methods achieve some success, they usually fail to offer the important natural or personalized aspects for difficult cases. VR provides a unique way of offering therapy by placing cognitive training in surroundings modeled after everyday environments (Buele et al., 2023; Dermody et al., 2020; Lin et al., 2023). Paired with tools like teletherapy, wearable sensors, or artificial intelligence (AI)-driven feedback, VR also enables remote or home-based rehab, making it more accessible, especially in rural or underserved areas (Anghelescu, 2022; Edwards et al., 2022; Pagliari et al., 2025). As a result, people under treatment can try things such as buying groceries online, managing their virtual budgets or meeting others for social purposes in a safe and adjustable environment.

What makes VR different is its ability to involve users in a digital world and let them influence what takes place by making decisions (Chatterjee et al., 2022). Such interactivity

DOI: 10.4324/9781003646662-21

is useful because it stimulates learning and helps carry over skills developed in therapy to real life (Wen et al., 2024). Since these VR platforms are so adaptable, it is possible to track performance, modify task difficulty, and gather information on responses and bodily processes (Norrholm et al., 2016).

Neurorehabilitation is moving toward personalized care and VR is notable because its interventions can be customized for any patient's particular needs (Miltiadis et al., 2025). For individuals affected by stroke, traumatic brain injury, or living with Alzheimer's or schizophrenia, VR can help by offering tasks that test memory, attention, planning, making decisions, and social abilities (Chen et al., 2022; Ren et al., 2024; Riches et al., 2021; Shahmoradi & Rezayi, 2022). So, a patient with executive problems might use technology to complete a task that requires skills such as planning, switching between activities and checking for errors. Lately, many experiments have shown that VR boosts different areas of thinking, yet these positive outcomes are diverse (Lim et al., 2020). Many studies have pointed to clear improvements in visuospatial skills and executive functions after participants use VR for treatment (Riva et al., 2020). These cognitive tasks are best supported in environments where users can naturally explore space, handle objects, and make tough decisions. VR navigation has been trialed with patients having mild cognitive impairments (MCIs) and it demonstrates better results for learning spatial skills when compared to older, nonvirtual approaches (Tortora et al., 2024). Multisensory features like visual, auditory, or haptic feedback also open new possibilities for conditions like autism or sensory integration issues (Brugada-Ramentol et al., 2022; Gabrielli et al., 2023). Advanced VR systems are even exploring ways to boost emotional regulation and neuroplasticity (Gangemi et al., 2023; Parisi et al., 2022).

Although VR has a less regular effect on memory, paying attention and using language. Some studies claim important improvements, yet others say there are only small variances in favor of standard therapies (Cappadona et al., 2023; Optale et al., 2010; Coleman et al., 2019; Kourtesis & MacPherson, 2023). There is likely some variability because patient populations differ, technology varies from VR platform to platform and a lack of consistent methods is used. In rehab for language loss, VR can respond to people talking, but it is still difficult to get it to recognize tone and complex social-like interactions (Bryant et al., 2020; Brassel et al., 2023). Using VR in clinics is currently limited due to costs and its technical requirements (Glegg & Levac, 2018). Simple mistakes in creating VR systems may make therapy difficult for people already at risk, by frustrating them or causing too much disorientation or motion sickness (Balloufaud et al., 2025; Firdaus et al., 2024; Rouhani et al., 2024). Patient-friendly, function wise, design is necessary for success.

To provide a comprehensive understanding of VR in neurocognitive rehabilitation, this chapter outlines things that are important for using VR in neurorehab. First, it looks at the mental skills that VR trains memory, attention, executive functions, language, and visuospatial skills and checks if they help. It then explains why VR is effective, looking at the technology and ideas behind its interactive experience and how it helps the brain function. It goes on to discuss using VR in a clinic, covering tools, technology, and plans that suit each user's needs. It also looks at issues involving the costs of equipment, unreliable results, and the safety of users. In addition, it points out that VR may complement regular therapies and serve as a means to reach more people with rehab online. Subsequently, the chapter discusses the key challenges and limitations that hinder the broader adoption of VR in neurorehabilitation. These include issues such as equipment variability, lack of standardization, limited generalizability of findings, and ethical or usability concerns, especially among

Table 17.1 Virtual reality systems used in neurocognitive rehabilitation

VR System	*Platform Type*	*Degrees of Freedom*	*Immersion Level*	*Neurocognitive Rehab Applications*
Oculus Rift	High-end PC VR	6 DoF	High	Used in stroke and TBI rehab; supports immersive 3D environments
HTC Vive	High-end PC VR	6 DoF	High	Applied in memory and executive function rehab; offers precise tracking
PlayStation VR	Console-based VR	3 DoF	Moderate	Less common in rehab; used in research for attention tasks
Samsung Gear VR	Mobile VR	3 DoF	Moderate	Used in early cognitive training studies for MCI
Google Cardboard	Mobile VR	3 DoF	Low	Low-cost solution for basic cognitive exercises
Oculus Quest/ Quest 2	Standalone VR	6 DoF	High	Widely used in clinical trials; balance, memory, attention rehab
CAVE systems	Multi-wall projection VR	Varies	Very high	Used in lab settings for spatial navigation and executive function training
VR TierOne	Medical VR	6 DoF	High	Specialized in neurocognitive rehab for stroke and elderly patients
NeuroVR	Desktop-based non-immersive	N/A	Moderate	Software for customizable virtual tasks; used in clinical and research settings

vulnerable populations. Finally, the chapter concludes with a summary of its key insights and reflections on the evolving role of VR in shaping the future of cognitive rehabilitation (Table 17.1).

17.2 A Domain-Specific Exploration: VR's Impact on Cognitive Functions

VR has emerged as a promising tool in neurocognitive rehabilitation, offering interactive, immersive environments that closely replicate real-world settings. These simulated scenarios provide safe and controlled conditions in which patients can perform tasks that challenge and enhance various cognitive functions. This section delves into the primary cognitive domains targeted by VR interventions: memory, attention, executive functions, language, and visuospatial abilities. For each domain, we explore the underlying mechanisms, recent empirical findings, and assess the overall effectiveness of VR-based therapies.

17.2.1 Memory

Patients' conditions, including traumatic brain injury (TBI), stroke, dementia, and various neurodegenerative disorders, may badly affect a person's memory. VR helps memory rehabilitation by mimicking common scenes where patients can perform activities that improve their memory. Such tasks consist of navigating places, identifying the locations of things and organizing habits inside VR. VR-based projects have shown that they can increase a person's ability to remember things both quickly and for a long time. Canty et al. (2014) found that people with TBI improved their episodic memory after training using virtual environments that resembled activities in daily life, such as getting around a city and handling tasks such as errands. Elderly participants with MCI undergoing VR activities were found to have better memory recall because the activities stimulated the hippocampus.

Despite these positive findings, the variability in task design, duration of intervention, and patient demographics complicates the generalization of outcomes. Moreover, some studies report limited transfer of VR-induced memory gains to real-world settings, emphasizing the need for integrating VR with traditional memory rehabilitation strategies.

17.2.2 Attention

Many individuals with neurological disorders such as attention deficit hyperactivity disorder (ADHD), stroke, and TBI often have problems focusing. With multisensory and individually adjusted input, VR environments help people practice better attention. It can reproduce crowded places like streets, classrooms, and offices, allowing individuals to exercise their focus on important things. Studies have found that VR attention training helps people a lot. An analysis of scientific studies found that VR attention training produced medium-to-strong effect sizes, mainly in tasks that require people to keep their attention on one thing for a long time (Parsons et al., 2019). One widely used app in VR is the virtual classroom, which helps children with ADHD concentrate while facing many possible distractions. There is now the ability in some VR sets to track eye movement and measure brain activity so that attention levels can be observed live. These different innovations make it easier to understand and measure how much attention a user is giving, which can help guide doctors.

However, VR-based attention training also faces challenges. Technological limitations, such as latency or low-resolution displays, can hinder attentional engagement. Additionally, cognitive fatigue may set in more quickly in immersive environments, potentially reducing the effectiveness of prolonged sessions.

17.2.3 Executive Functions

Executive functions involve a wide range of skills, for example, planning, reasoning things out, making choices, holding things back from reacting too quickly, and being flexible with thinking. Many patients with damage to the frontal lobe, schizophrenia, and neurodegenerative diseases experience problems in these functions. Thanks to its ability to set up real-life tasks, VR is a key tool for developing executive skills. People taking part in VR-based tasks may be asked to organize their shopping, meet time deadlines, or answer new challenges. For instance, the virtual multiple errands test (VMET) offers a digital shopping mall to check and improve executive functioning, as users work on different tasks within given rules and methods. There is evidence from studies that using VR can improve people's executive functions. The researchers found that schizophrenia patients who did VR exercises involving decision-making and rule-changing improved their ability to adjust to new ideas and block distractions (Amado et al., 2016). Similarly, those who took part in VR training after their strokes mentioned they could organize daily tasks and make independent choices more easily (Raspelli et al., 2011).

Despite these advances, some researchers caution that VR tasks may not always isolate specific executive processes, making it difficult to attribute improvements to particular components of executive function. Standardized protocols and task analyses are needed to ensure the validity of such interventions.

17.2.4 Language

Language impairments can result from conditions such as aphasia, developmental disorders, or neurodegenerative diseases. While rehabilitation programs for motor skills and thinking have used VR a lot, its involvement in language therapy is not yet well established. Nonetheless, VR has the potential to help language rehabilitation because it offers environments rich with events that encourage people to speak.

On some occasions, VR has been used in early interventions to help children name things, understand them, and talk more often. To help with language skills, those with aphasia have walked through virtual grocery stores and been asked to understand or carry out instructions. Because these tasks fit naturally into social activities, the sessions are more useful in everyday life. So far, VR seems to help increase motivation and interest in language work, which plays a key role in a successful recovery. Marshall et al. (2016) reported that using VR in language training helped post-stroke aphasics name objects and put words into sentences more easily than with traditional approaches.

However, the use of VR for recovering language after harm remains mostly untested due to the lack of large-scale research. Many current applications have not added advanced natural language processing or speech recognition, which reduces how people interact with language.

17.2.5 Visuospatial Abilities

Visuospatial abilities mean the ability to detect, process, and work with visual and spatial data. By using these skills, we can manage our way around objects and space each day. Many cases of Alzheimer's, right hemisphere stroke and some forms of dementia include problems with visuospatial skills.

VR stands out in helping with training visuospatial skills because it features interactive 3D spaces. Navigating virtual mazes, arranging artifacts or estimating distances are activities that patients can practice in a safe environment. They call for using your eyes and motor skills, giving children a chance to learn using multiple senses at the same time. Studies agree that VR helps improve visuospatial skills. In their study, Caglio et al. (2012) found that TBI patients who received VR navigation training greatly improved in remembering and orienting themselves. Taking part in VR spatial training seemed to help elderly people reduce their sense of disorientation and become better at reading maps.

In this field, VR improves the experience by recording accurate performances and setting adjustable difficulty settings. However, some users, usually older adults or those severely impaired, might feel uncomfortable or struggle with interfaces in VR, which can limit what VR offers them.

17.2.6 Comparative Effectiveness and Integration across Domains

VR looks promising for all five mental abilities, but the amount of proof for each varies. Results from studies focused on executive functioning and visuospatial abilities are persistent and strong because VR environments use interactive and spatial aspects. Numerous VR games can help improve attention and memory, but results vary depending on what is being played and who the user is. Although language rehabilitation offers several benefits, it is the least developed and more research and new ideas are needed. Many studies point out that including VR in regular rehabilitation therapies could improve outcomes the most. When we include VR together with cognitive-behavioral strategies, occupational therapy, or medicines, we can achieve much better results and see these strategies applied in everyday life. It also allows therapists to keep checking patients and change treatment options quickly to fit what each person requires.

In summary, VR offers a versatile and engaging platform for cognitive rehabilitation, targeting a wide range of cognitive domains (Riva et al., 2020) with varying degrees of success in various neurocognitive disorders (see Figure 17.1). Its ability to simulate real-world scenarios, provide immediate feedback, and adapt to user performance makes it a powerful tool in therapeutic contexts. However, challenges remain, including the need for standardized protocols, better integration of technological advancements such as natural language processing (NLP), and greater attention to accessibility issues. As research advances, VR is likely to play an increasingly central role in the future of neurocognitive rehabilitation, especially when designed in conjunction with clinical best practices and empirical evidence.

17.3 The Mechanics of Immersion: How VR Functions in Neurocognitive Therapy

VR has become a valuable tool in therapy, owing to its capability to immerse, interact, and provide real-time feedback to its users. These characteristics are not just mere elements of technology; they also facilitate cognitive engagement and support neuroplasticity. Understanding how these VR features work helps us understand why VR is such a promising tool for treating mental illness, rehabilitation, and cognitive enhancement.

17.3.1 Immersion: Creating a Sense of Presence and Modulating Attention

Immersion refers to the extent to which a VR system indulges the user in a virtually created reality, substituting the real world. High immersion is created with the help of multisensory integration, blending visual, auditory, and sometimes haptic inputs. The immersion quality in VR creates a sense of "presence" – the subjective experience of being in the artificial environment rather than in the real world. The presence matters in psychology because it captures the user's attention and improves their emotional engagement.

From a neuroscientific perspective, immersion experience of a VR activates brain areas involved in sensorimotor integration and spatial navigation, including the parietal cortex, precuneus, and hippocampus (Bohil et al., 2011). Furthermore, neuroimaging studies have shown that the dorsolateral prefrontal cortex (PFC), which is connected to executive function and attentional control, is more active when people use high-immersion VR (Lorentz et al., 2023). Immersion also activates the anterior insula, leading to better interoceptive awareness of physical and emotional states, creating a salient experience for better therapeutic engagement (Owens & Duncan, 2025). The ability to shift attention can be very helpful in treating anxiety disorders, chronic pain, and post traumatic stress disorder (PTSD), since these problems are linked to maladaptive attentional bias. With immersive VR, people can channel their attention from stressful internal states towards beneficial goal-activities. Such attention modulation helps correct the maladaptive brain networks and encourages better adaptive patterns.

17.3.2 Interactivity: Enhancing Engagement through Agency and Sensorimotor Contingency

Interactivity in VR refers to the ability to actively participate and manipulate the simulated environment using actions or choices. It is beyond simply viewing the simulated environment and hence fosters a sense of agency to the users that their actions are crucial to make a difference and keep them involved. Working on agency in therapy can help people with mood and anxiety disorders recover from learned helplessness.

When we interact with media, neural networks involving the motor cortex, premotor cortex, basal ganglia, and cerebellum are activated, depending on the details of the task (Mishra et al., 2021). The frontoparietal control network and the salience network, which support goal-directed actions and cognitive resource allocation, both connect closely to these regions. Interactivity also activates the ventral tegmental area (VTA) and the nucleus accumbens, which form the brain's dopaminergic reward network (Matar et al., 2014). These brain regions reinforce engagement by reward-based learning. Interactive VR sets up a system that instantly provides feedback to the user's actions, which is important for successful cognitive-behavioral therapy (CBT) and exposure therapy.

17.3.3 Real-Time Feedback: Closing the Loop between Action and Consequence

Real-time feedback is information that is instantly updated to the user based on their performance. Users may receive this feedback in visual, auditory, or haptic information that can include performance measurements, virtual rewards, or tracking heart rate and skin conductance changes. Real-time feedback to users creates a system where actions lead to instant results, enhancing their learning experiences.

Neuroscientific research suggests that feedback in real time interacts with the anterior cingulate cortex, a brain region responsible for monitoring error and conflict, and is thus implicated in adaptive behaviors (Lorenzetti et al., 2018). When users receive feedback regarding their performance, the anterior cingulate cortex (ACC) in combination with the medial PFC processes this information and decides what actions to take in the future (Garrison et al., 2021). Such feedback-associated learning reinforces neural pathways and supports neural plasticity by strengthening existing relevant neural circuits. Biofeedback with VR, such as neurofeedback or heart rate variability training, increases the therapeutic potential of VR by providing the user with the ability to control some of their autonomic functions (Wilhelm et al., 2005). This enables learners to actively modulate and control their physiological states, allowing them to reprogram neural networks involving brain regions like the amygdala, hypothalamus, and ventromedial PFC, which are particularly involved in stress and emotion regulation.

17.3.4 Cognitive Engagement: Sustaining Mental Effort and Promoting Active Learning

Cognitive engagement refers to the ability to invest in mental activity or goal-directed effort that results in learning. For VR, cognitive engagement is a byproduct of immersion, interaction, and feedback received. These elements of VR result in a simulated environment that is dynamic, goal-oriented, and emotionally salient. Users, when engaged in such environments, tend to process information intensively, make new associations, and retain therapeutic gains better (Amin et al., 2024; Lin et al., 2024).

From a neurocognitive perspective, cognitive engagement is linked to the activation of the default mode network (DMN) and central executive network (CEN) (Cao et al., 2016; Vuper et al., 2021). Engagement of the two networks is essential in higher cognitive functions like self-referential thought, working memory, and problem-solving. The simultaneous engagement of both these networks can facilitate the seamless integration of therapeutic insights into one self-concept. Furthermore, VR interventions often have narratives and gamified features (Amin et al., 2024) that activate the hippocampus and amygdala, improving emotion-based memory encoding. This leads to increased therapeutic value as it reinforces the consolidation of healthy, adaptable beliefs and behaviors.

17.3.5 Neural Plasticity: Mechanisms of Long-Term Change

The primary mechanism for achieving therapeutic success is neural change, or reorganization of the brain and its interconnections. Experience-dependent plasticity can be achieved through immersion, interactivity, and feedback-laden environments of VR. Such environments provide ideal conditions essential for changing maladaptive neural circuits. The repetitive and substantial interactions in VR can induce changes in gray matter density and white matter integrity. For instance, VR-based rehabilitation for stroke patients is known to improve the functioning of the motor cortex and inter-hemispheric communication, thus enhancing motor recovery (Lohse et al., 2014; Tieri et al., 2018). Similarly, VR exposure therapy is known to decrease hyperactivity of the amygdala as well as gain increased top-down control of the PFC in patients suffering from PTSD (Malejko et al., 2017).

Furthermore, emotionally rewarding experiences, repetition, and salient contexts in VR can enhance long-term potentiation (LTP), the underlying mechanism of learning and memory. By repetitive engagement of neural networks in personally meaningful and emotionally charged contexts, VR can aid in forming new cognitive and behavioral patterns. The consolidation of new patterns can drive meaningful and lasting changes in behavior (Albus et al., 2021).

To sum up, VR, through its core elements of immersion, interactivity, and real-time feedback, serves as a neurocognitive interface through which modulations can occur in the brain function as well as behavior. These core elements interact with each other to foster cognitive engagement and enable neural plasticity: Immersion redirects the attention of users and enhances emotional engagement; interactivity promotes a sense of authority and learning based on rewards; real-time feedback supports regulation, error monitoring, and adaptive learning. The dynamic nature of VR allows clinicians to customize interventions while tracking progress in real time and maintaining user motivation over long periods. The treatment potential of VR goes above mere play or relaxation; it is rather a scientifically grounded tool capable of reshaping brain function and behavior (Bailenson et al., 2025). The inclusion of VR in therapeutic protocols is a significant advancement in the world of mental health and neurorehabilitation. VR offers new, groundbreaking opportunities to harness neuroplasticity in neurocognitive rehabilitation.

17.4 Practical Implementation of Virtual Reality in Neurocognitive Rehabilitation

There is a growing need to explore the pragmatic issues, including practical considerations and inherent challenges of integrating VR technology in neurocognitive rehabilitation processes (Rathinam, 2021). The practical aspects of bringing VR technology into the clinic, as opposed to practicing it in a lab setting, pose a multitude of issues that require attention, despite the technological and theoretical merits being well documented. This chapter covers the basic requirements of implementing VR, including hardware and software components, user interface (UI) design, system-level access, and specific plan customization for therapies, along with more tailored ones aimed at specific goals. The chapter also explores the key challenges that restrict the wide adoption of VR, such as variability of equipment, lack of standardization, limited scope of generalizability of empirical evidence, and ethical or usability barriers, especially with more vulnerable populations (Nieto-Escamez et al., 2023).

17.4.1 Hardware Considerations in VR Implementation

The foundation of any VR-based therapeutic intervention is the choice of suitable hardware. The VR hardware can include accessible and portable platforms like Google Cardboard or smartphone-based systems, as well as fully immersive head-mounted displays (HMDs) like the Oculus Rift, HTC Vive, and PlayStation VR.

Clinical settings often require prolonged equipment usage; hence, VR systems with high-resolution screens, reliable motion tracking, and ergonomic comfort are ideal to reduce any discomfort or cybersickness. Additionally, to improve the realism and interactivity of therapy sessions, peripheral devices like treadmills, haptic gloves, motion sensors, and electroencephalography (EEG) or biosignal monitors can be incorporated. However, installing these sophisticated setups can be expensive and require skilled technicians for maintenance and calibration purposes (Leeb & Pérez-Marcos, 2020). The balance between clinical viability and technological sophistication is still a crucial point of concern, especially for institutions with limited budgets or technical resources.

17.4.2 Software Platforms and Therapeutic Content

Like hardware, the software architecture of VR interventions holds great importance. Therapeutic software should be tailored to target particular cognitive domains, providing a range of settings that can automatically adapt to fit the user's achievements. Because most available VR games cannot be used medically, experts often build their own software with input from clinicians, neuroscientists, and software developers (Kouijzer et al., 2023). Tools like Virtual Reality Rehabilitation System (VRRS), NeuroVR, and Reh@City are VR platforms specifically developed as cognitive rehabilitation tools. Most cognitive rehabilitation tools are designed to target memory, attention, executive function, and visuospatial skills in a planned way. Essential components are goal-oriented tasks, different scenario options and real-time feedback.

Besides, the ability to work with devices like EEG and heart rate monitors greatly improves how therapeutic VR applications function by adding biofeedback (Leeb & Pérez-Marcos, 2020). Having sensors that track physiology makes therapy personalized to individual needs while retaining objectivity, allowing therapists to assess responses in real-time and adapt accordingly.

17.4.3 User Interface Design and Human-Centered Approaches

The effectiveness of a VR system often depends on the way the UI is created. Issues with the UI can make it harder to use, which may impact patient engagement and decrease the success of treatment. Interfaces designed for intuitive and accessible use for patients with neurocognitive impairment are crucial for their adoption as a therapeutic tool. The main focus when designing should be simplicity, readability, and creating minimal cognitive load.

Aiming for a strong user experience, design teams must repeatedly test prototypes with their end users (patients and therapists) (Cucinella et al., 2025). A good VR system must support individuals with poor motor control, poor vision, or hearing and those with limited abilities due to aging. It is now possible to improve accessibility by using gesture, voice or eye-tracking controls. Also, the inclusion of VR spaces with themes and languages that are culturally appropriate can greatly enhance the adaptability of VR systems in clinical settings. Personalizing elements on the screen, such as picking an avatar or selecting settings, can help users feel more involved and motivated.

17.4.4 System Accessibility and Infrastructure Readiness

Accessibility extends beyond making something usable at physical levels; it also includes organizational, financial, and infrastructure-based usability. Many hospitals in developing countries cannot incorporate VR systems because they lack the needed infrastructure. It includes fast internet, space for VR equipment, and a team to take care of this sophisticated system. Another major obstacle to the wider accessibility of VR in health technologies is their high cost (Cano-de-la-Cuerda et al., 2024; O'Neil et al., 2018). The cost for necessary hardware, software, and worker training at the start can be large. However, there are affordable options for VR, such as mobile VR or reused equipment. For VR-based interventions to be fair, it is required to have more public and private agencies to join forces, more funding agencies to come forward with specific grants and the use of effective, scalable strategies to mitigate the cost-related challenges. In addition, being prepared on the institutional level is very important for achieving outcomes. This includes clinician training programs and ensuring data privacy and security.

17.4.5 Development of Personalized Therapy Plans

Personalization is one of the biggest strengths of VR. By using personalized plans, therapists can focus the treatment content on each patient's profile. This personalization can be achieved by evaluating at the beginning level, continuous performance monitoring and adjusting its level of difficulty as needed. Experiences in VR can be customized to help patients overcome real situations like moving through a crowded street, preparing for an interview, or managing their daily routines. Because the learning is contextually relevant, the skills learned have real-world applicability. Personalization can also extend to be simple feature adjustments like sensory preferences of visual or auditory cues, to adjustments as per emotional comfort zone or motor abilities.

AI and machine learning (ML) allow websites to analyze user data and make instant suggestions based on the user's behavior. Inclusion of AI-ML helps practitioners to rely on research and in continuous refinement of therapeutic protocols.

17.5 Challenges and Limitations in VR-Based Neurocognitive Rehabilitation

Despite its promise, the broader adoption of VR in clinical neurorehabilitation faces several formidable challenges. These can be broadly categorized into technological, methodological, and ethical domains (Burdea, 2003).

17.5.1 Equipment Variability and Compatibility

As mentioned above there are many VR headsets available and everyone has specifications, features, and its own kind of proprietary operating system. Having so much variation among patients slows the process of using one treatment for all and comparing the findings of different research studies. Scaling up these interventions to other clinical environments is often blocked by the systems' inability to communicate with each other. Besides, outdated technology is a problem that keeps recurring. With how fast VR technology improves, equipment may become outdated fairly quickly and will need to be replaced often. Problems linking old hardware to newer software make it harder for investments to last longer and be practical.

17.5.2 Lack of Standardization and Regulatory Oversight

Because there are no set rules for VR-based therapy, its use as a clinical method is sometimes questionable. Because the length, contents, outcomes, and therapy aim of sessions are varied, it's challenging to compare experiments or create guidelines. Because research findings vary so much, it is difficult to use them in meta-analyses, which slows the way VR becomes part of mainstream practice. Regulatory rules for digital therapeutics are undergoing changes. Not a lot of VR applications have been tested broadly by the medical community or approved by agencies like the Food and Drug Administration (FDA) or European medical agencies (EMAs). As a consequence, clinicians could choose not to use VR tools when it is not proven that they can be used safely or are sure to work.

17.5.3 Limited Generalizability of Empirical Findings

The vast majority of VR-based research happens in the controlled conditions of labs with only a few, similar participants. Consequently, the results are only relevant to narrowly selected samples and lack use in wider, true-to-life situations. Socio-cultural differences, age, health problems, and familiarity with technology can have a major effect on how effective and comfortable a solution is for users. More research needs to focus on diverse populations, run studies at many sites, and continue to evaluate effects after a long period (Nieto-Escamez et al., 2023). Partnerships among clinicians, technologists, and social scientists can contribute to closing the gap between what is proven effective and what is useful in everyday life.

17.5.4 Ethical and Usability Concerns for Vulnerable Populations

When children, the elderly, or people with serious cognitive problems use VR interventions, extra attention must be given to ethics (Kellmeyer et al., 2019). Problems include people not fully knowing the risks, keeping confidential data private, and the risk of conditions such as cybersickness or dissociation. It is also true that people with physical or neurodevelopmental issues often have a greater challenge when using websites. Designers ought to recognize these problems and ensure their systems address different requirements while still keeping treatment effective. A second issue to worry about is the protection of data. Since VR gathers private biometric and activity data, it must be securely encrypted, anonymized, and satisfy the requirements of rules like General Data Protection Regulation (GDPR) or Health Insurance Protection Accountability Act (HIPAA). Losing this data may result in unauthorized access and earn users' distrust.

17.5.5 Conclusion: Toward Scalable and Inclusive VR Therapy

The practical implementation of VR in neurocognitive rehabilitation presents both exciting opportunities and considerable challenges. While advances in hardware, software, and interface design have made immersive therapy more feasible, systemic issues such as equipment variability, lack of standardization, limited generalizability, and ethical concerns must be addressed to ensure sustainable adoption.

Personalization remains a cornerstone of effective VR therapy, offering tailored experiences that can adapt to diverse user needs. Yet, for VR to become a standard tool in neurorehabilitation, it must be scalable, affordable, and supported by robust empirical evidence. This requires coordinated efforts among technologists, clinicians, policymakers, and researchers to develop inclusive, evidence-based, and ethically grounded frameworks for implementation. By critically engaging with these practical aspects, stakeholders can move beyond the experimental allure of VR and toward its meaningful integration into clinical practice. In doing so, they unlock the potential of this transformative technology to reshape the landscape of mental health and cognitive rehabilitation for years to come.

17.6 Future Directions and Conclusions: Advancing VR in Neurocognitive Rehabilitation

As VR becomes more used in clinics, it is becoming clear that it can greatly shape neurocognitive rehabilitation (Moon & Han, 2022). We shed light by explaining how they can support conventional therapies and ensure more people can take part in rehabilitation no matter their distance from a care center. Moreover, further studies are crucial, particularly underlining the importance of teamwork across different areas and the use of varied intervention methods (Pelosin et al., 2020; Sánchez-Cuesta et al., 2021). Finally, the section pulls together the central learnings of the chapter and reflects on how VR changes the field of cognitive rehabilitation.

17.6.1 *Potential of VR to Supplement Traditional Therapies*

Usually, the traditional approach to neurocognitive rehabilitation involves therapists organizing and supervising specific daily exercises. Nevertheless, such strategies might not provide enough engagement, consideration, or suitability to achieve great recovery. Using VR, users can interact as they wish in spaces created to appear and feel realistic. If people engage their senses and see new virtual environments, their therapy becomes fuller and easier to apply outside the clinic.

A variety of research shows that VR can be a helpful extra in regular rehabilitation. For instance, patients who took part in VR stroke rehabilitation achieved better or similar results for executive function and visuospatial skills than those who used standard therapy methods (Tomczuk et al., 2025). Using VR encourages people with dementia to remember their past that can slow down changes in their minds (Papaioannou et al., 2022). They show that you can improve existing therapy methods by using VR technology. Besides, therapists and patients can interact in more creative ways thanks to VR. During treatment, therapists notice what patients do online, assess the results quickly, and alter the therapy if possible (Corti et al., 2022). Because it can be adjusted for any purpose, this tool supplies important cognitive information used in evidence-based medicine.

17.6.2 *Remote and Home-Based Rehabilitation Models*

VR helps make cognitive rehabilitation accessible to far more people by allowing remote and at-home treatments. Individuals in rural and non-served areas can now get specialist rehabilitation for their mental conditions. Thanks to VR, individuals can now begin and follow cognitive training from wherever they are. Portable VR systems designed for home use and mHealth apps are being introduced at an increasing rate now. They can give people exercises for their minds, answer their actions quickly, and connect through cloud technology to clinicians (Matsumoto et al., 2021). New technology in medicine can be especially helpful for people with Parkinson's disease, multiple sclerosis, and traumatic brain injury.

Remote VR research takes some of the strain off healthcare systems by reducing the need for patients to go to new centers or keep visiting the doctor often. Being flexible about rehabilitation timing encourages patients to stay involved, feel committed, and heal better. For VR at home to be successful, we must concentrate on how easy it can be used, the safety of patients, and technical challenges. Assistive technology should support users of all abilities and those responsible for them are commonly taught ways to help (Giles, 2018). When used correctly under guidance, remote VR may change how patients receive neurocognitive rehabilitation.

17.6.3 *The Need for Interdisciplinary Collaboration*

For VR to keep improving cognitive rehabilitation, different disciplines must work together consistently. VR interventions are most effective when designed by researchers working in clinical psychology as well as neurology, development, engineering, and human–computer interaction (HCI) fields (Lei et al., 2023). It is important for clinicians that VR programs follow cognitive and behavioral theories to make them useful for therapy. Clinicians need to explain the neurocognitive goals, build tasks to measure them, and find suitable assessment measures. At the same time, engineers and developers must design virtual settings that are both functional, engaging and open to as many users as possible.

It is equally necessary for HCI researchers to improve the way users interact with interfaces. Problems with usability for elderly or cognitively impaired users may cause a mobile app to be less effective for therapy (Chiang et al., 2017). Involving users from the beginning in the design process allows systems to be easy to use, interesting, and fit for use by a wide range of people. Government and organizational support is needed to help interdisciplinary work succeed. All stakeholders, including research bodies, healthcare groups, and academic institutions, need to design ideas that support different sectors agreeing to partner, relying on each other's facilities, and sharing knowledge. Training and its benefits can be maximized only if the approach is coordinated.

17.6.4 Multimodal and Adaptive Interventions

VR can advance neurocognitive rehabilitation through its immersive environment as well as by pairing multiple forms of intervention. If developers add EEG, functional near-infrared spectroscopy (fNIRS), eye-tracking, motion sensors, and haptic feedback to VR, the environments for therapy can be more detailed and precise. Joining VR with neurophysiological tests provides a live report of the brain's activity, while subjects perform cognitive tasks (Plotnik et al., 2021). So, swift changes can be made to support students depending on how engaged or mentally busy they are. Likewise, to support patients with neurodegenerative illnesses, motion capture can be used to check motor performance and support exercises that improve minds and bodies at the same time. The use of biofeedback with VR appears to offer a great deal of promise. Games that connect heart rate variability or galvanic skin response to the game experience help patients control their stress, regulate their emotions, and pay better attention (Cano et al., 2024).

Using all these methods can lead to the creation of highly individual therapy plans. With ML, patient data can be studied to tailor the difficulty of tasks, the order in which they are done and the type of feedback, all to increase treatment benefits and lower the risk of patients getting bored or overloaded (Wankhede et al., 2025). With improvements in technology, it's more necessary to consider ethical and data privacy. Any research or development carried out in homes or remotely calls for key protocols in data handling and careful protection of personal data.

17.7 Key Areas for Future Research

Even though VR has great potential, a lot of research areas need to be explored further. Future research should try to accomplish the following:

- Standard method development for VR-assisted care, determining approaches for intervention, assessments of performance, and sharing outcomes.
- Set up major randomized controlled trials (RCTs): Most existing studies are too small and lack the proper controls.
- Implementation of robust randomized controlled trials is needed to test its effectiveness.
- Very few looked at the stability of cognitive changes in the long term after the end of therapy.

- Modify the research to examine how age, gender, existing illness or disability, and skills with technology influence student learning.
- Research should look at how VR compares with common digital training tools for the brain.
- Further research is needed to study whether the improvements in VR transfer to functional benefits in the real world.

Answering these questions calls for ongoing research using multiple academic disciplines and diverse groups of study participants. The synthesis of information and guidance for practice will come mainly from conducting meta-analyses and systematic reviews.

17.8 Summary and Reflections

This chapter examined the ways VR is reshaping neurocognitive therapy. The beginning involved pinpointing memory, attention, executive function, language, and visuospatial abilities within VR interventions and examining research on how these abilities were supported. After that, it analyzed the key factors that make VR valuable for therapy such as immersion, interaction, and real-time information. The rest of the chapter focused on practical issues, describing hardware, software, designs for UIs, and whether systems are accessible. How having a personalized treatment strategy can boost user involvement and success was also considered. Thereafter, the discussion centered on obstacles such as different tools, a lack of set standards, and problems with using the equipment.

The chapter also discussed how VR may be used in the future for rehabilitation. It suggested that VR could strengthen standard treatments, increase reach by being accessible remotely, and grow with innovation and collaboration between various experts. With each new technology, VR is poised to drive important progress in mental health and cognitive treatments. Recognizing the strengths and weaknesses of VR, researchers, therapists, and policymakers can use it fully in helping people recover and strengthen their mental abilities.

References

Albus, P., Vogt, A., & Seufert, T. (2021). Signaling in Virtual Reality Influences Learning Outcome and Cognitive Load. *Computers & Education*, 166, 104154. https://doi.org/10.1016/j.compedu.2021.104154

Amado, I., Brénugat-Herné, L., Orriols, E., Desombre, C., Dos Santos, M., Prost, Z., Krebs, M. O., & Piolino, P. (2016). A Serious Game to Improve Cognitive Functions in Schizophrenia: A Pilot Study. *Frontiers Psychiatry*, 7, 64. doi: 10.3389/fpsyt.2016.00064

Amin, F., Waris, A., Syed, S., Amjad, I., Umar, M., Iqbal, J., & Omer Gilani, S. (2024). Effectiveness of Immersive Virtual Reality-Based Hand Rehabilitation Games for Improving Hand Motor Functions in Subacute Stroke Patients. *IEEE Transactions on Neural Systems and Rehabilitation Engineering: A Publication of the IEEE Engineering in Medicine and Biology Society*, 32, 2060–2069. https://doi.org/10.1109/TNSRE.2024.3405852

Anghelescu, A. (2022). 39799756. Parkinson's Disease, 2022, 4370712. https://doi.org/10.1155/2022/4370712

Bailenson, J. N., DeVeaux, C., Han, E., Markowitz, D. M., Santoso, M., & Wang, P. (2025). Five Canonical Findings from 30 Years of Psychological Experimentation in Virtual Reality. *Nature Human Behaviour*. https://doi.org/10.1038/s41562-025-02216-3

Balloufaud, M., Boujut, A., Marie, R., Guinaldo, A., Fourcade, L., Hamonet-Torny, J., & Perrochon, A. (2025). Augmented Reality Exergames for Upcoming Cognitive-Motor

Rehabilitation: User-Centered Design Approach and User Experience of Healthy Children. *JMIR Rehabilitation and Assistive Technologies*, 12, e69205. https://doi.org/10.2196/69205
Bohil, C. J., Alicea, B., & Biocca, F. A. (2011). Virtual Reality in Neuroscience Research and Therapy. *Nature Reviews. Neuroscience*, 12(12), 752–762. https://doi.org/10.1038/nrn3122
Brassel, S., Brunner, M., Power, E., Campbell, A., & Togher, L. (2023). Speech-Language Pathologists' Views of Using Virtual Reality for Managing Cognitive-Communication Disorders Following Traumatic Brain Injury. *American Journal of Speech-Language Pathology*, 32(2S), 907–923. https://doi.org/10.1044/2022_AJSLP-22-00077
Brugada-Ramentol, V., Bozorgzadeh, A., & Jalali, H. (2022). Enhance VR: A Multisensory Approach to Cognitive Training and Monitoring. *Frontiers in Digital Health*, 4, 916052. https://doi.org/10.3389/fdgth.2022.916052
Bryant, L., Brunner, M., & Hemsley, B. (2020). A Review of Virtual Reality Technologies in the Field of Communication Disability: Implications for Practice and Research. *Disability and Rehabilitation: Assistive Technology*, 15(4), 365–372. https://doi.org/10.1080/17483107.2018.1549276
Buele, J., Varela-Aldás, J. L., & Palacios-Navarro, G. (2023). Virtual Reality Applications Based on Instrumental Activities of Daily Living (iADLs) for Cognitive Intervention in Older Adults: A Systematic Review. *Journal of Neuroengineering and Rehabilitation*, 20(1), 168. https://doi.org/10.1186/s12984-023-01292-8
Burdea, G. C. (2003). Virtual Rehabilitation—Benefits and Challenges. *Methods of Information in Medicine*, 42(5), 519–523.
Caglio, M., Latini-Corazzini, L., D'Agata, F., Cauda, F., Sacco, K., Monteverdi, S., Zettin, M., Duca, S., & Geminiani, G. (2012) Virtual Navigation for Memory Rehabilitation in a Traumatic Brain Injured Patient. *Neurocase*, 18(2), 123–131. doi: 10.1080/13554794.2011
Cano, N., Gómez-Hernández, J., Ariza, M., Mora, T., Roche, D., Porras-Garcia, B., & Garolera, M. (2024). A Multimodal Group-Based Immersive Virtual Reality Intervention for Improving Cognition and Mental Health in Patients with Post-Covid-19 Condition. A Quasi-Experimental Design Study. *Frontiers in Psychology*, 15, 1441018. https://doi.org/10.3389/fpsyg.2024.1441018
Cano-de-la-Cuerda, R., Blázquez-Fernández, A., Marcos-Antón, S., Sánchez-Herrera-Baeza, P., Fernández-González, P., Collado-Vázquez, S., Jiménez-Antona, C., & Laguarta-Val, S. (2024). Economic Cost of Rehabilitation with Robotic and Virtual Reality Systems in People with Neurological Disorders: A Systematic Review. *Journal of Clinical Medicine*, 13(6), 1531. https://doi.org/10.3390/jcm13061531
Canty, A. L., Fleming, J., Patterson, F., Green, H. J., Man, D., & Shum, D. H. K. (2014). Evaluation of a Virtual Reality Prospective Memory Task for Use with Individuals with Severe Traumatic Brain Injury. *Neuropsychological Rehabilitation*, 24(2), 238–265. https://doi.org/10.1080/09602011.2014.881746
Cao, W., Cao, X., Hou, C., Li, T., Cheng, Y., Jiang, L., Luo, C., Li, C., & Yao, D. (2016). Effects of Cognitive Training on Resting-State Functional Connectivity of Default Mode, Salience, and Central Executive Networks. *Frontiers in Aging Neuroscience*, 8, 70. https://doi.org/10.3389/fnagi.2016.00070
Cappadona, I., Ielo, A., La Fauci, M., Tresoldi, M., Settimo, C., De Cola, M. C., Muratore, R., De Domenico, C., Di Cara, M., Corallo, F., Tripodi, E., Impallomeni, C., Quartarone, A., & Cucinotta, F. (2023). Feasibility and Effectiveness of Speech Intervention Implemented with a Virtual Reality System in Children with Developmental Language Disorders: A Pilot Randomized Control Trial. *Children (Basel, Switzerland)*, 10(8), 1336. https://doi.org/10.3390/children10081336
Chatterjee, K., Buchanan, A., Cottrell, K., Hughes, S., Day, T. W., & John, N. W. (2022). Immersive Virtual Reality for the Cognitive Rehabilitation of Stroke Survivors. *IEEE Transactions on Neural Systems and Rehabilitation Engineering: A Publication of the IEEE Engineering in Medicine and Biology Society*, 30, 719–728. https://doi.org/10.1109/TNSRE.2022.3158731
Chen, X., Liu, F., Lin, S., Yu, L., & Lin, R. (2022). Effects of Virtual Reality Rehabilitation Training on Cognitive Function and Activities of Daily Living of Patients With Poststroke Cognitive Impairment: A Systematic Review and Meta-Analysis. *Archives of Physical Medicine and Rehabilitation*, 103(7), 1422–1435. https://doi.org/10.1016/j.apmr.2022.03.012
Chiang, V. C.-L., Lo, K.-H., & Choi, K.-S. (2017). Rehabilitation of Activities of Daily Living in Virtual Environments with Intuitive User Interface and Force Feedback. *Disability and*

Rehabilitation: Assistive Technology, 12(7), 672–680. https://doi.org/10.1080/17483107.2016.1218554

Coleman, B., Marion, S., Rizzo, A., Turnbull, J., & Nolty, A. (2019). Virtual Reality Assessment of Classroom – Related Attention: An Ecologically Relevant Approach to Evaluating the Effectiveness of Working Memory Training. *Frontiers in Psychology*, 10, 1851. https://doi.org/10.3389/fpsyg.2019.01851

Corti, C., Oprandi, M. C., Chevignard, M., Jansari, A., Oldrati, V., Ferrari, E., Martignoni, M., Romaniello, R., Strazzer, S., & Bardoni, A. (2022). Virtual-Reality Performance-Based Assessment of Cognitive Functions in Adult Patients with Acquired Brain Injury: A Scoping Review. *Neuropsychology Review*, 32(2), 352–399. https://doi.org/10.1007/s11065-021-09498-0

Cucinella, S. L., de Winter, J. C. F., Grauwmeijer, E., Evers, M., & Marchal-Crespo, L. (2025). Towards Personalized Immersive Virtual Reality Neurorehabilitation: A Human-Centered Design. *Journal of Neuroengineering and Rehabilitation*, 22(1), 7. https://doi.org/10.1186/s12984-024-01489-5

Dermody, G., Whitehead, L., Wilson, G., & Glass, C. (2020). The Role of Virtual Reality in Improving Health Outcomes for Community-Dwelling Older Adults: Systematic Review. *Journal of Medical Internet Research*, 22(6), e17331. https://doi.org/10.2196/17331

Devos, H., Ng, C., Santos, F. H., Sood, P., Hu, X., Zanwar, P., Ogawa, E., & Heyn, P. (2023). Virtual Reality for Cognitive Rehabilitation: A Beginner's Guide for Clinicians. *Archives of Physical Medicine and Rehabilitation*, 104(2), 355–358. https://doi.org/10.1016/j.apmr.2022.10.006

Edwards, D., Williams, J., Carrier, J., & Davies, J. (2022). Technologies Used to Facilitate Remote Rehabilitation of Adults with Deconditioning, Musculoskeletal Conditions, Stroke, or Traumatic Brain Injury: An Umbrella Review. *JBI Evidence Synthesis*, 20(8), 1927–1968. https://doi.org/10.11124/JBIES-21-00241

Farrell, K., & MacDougall, D. (2023). An Overview of Clinical Applications of Virtual and Augmented Reality: CADTH Horizon Scan. *Canadian Agency for Drugs and Technologies in Health*. www.ncbi.nlm.nih.gov/books/NBK596298/

Firdaus, R., Tantri, A. R., & Manggala, S. K. (2024). Factors Influencing Virtual Reality Sickness in Emergency Simulation Training. *Medical Science Educator*, 34(6), 1309–1315. https://doi.org/10.1007/s40670-024-02102-z

Gabrielli, S., Cristofolini, M., Dianti, M., Alvari, G., Vallefuoco, E., Bentenuto, A., Venuti, P., Mayora Ibarra, O., & Salvadori, E. (2023). Co-Design of a Virtual Reality Multiplayer Adventure Game for Adolescents With Autism Spectrum Disorder: Mixed Methods Study. *JMIR Serious Games*, 11, e51719. https://doi.org/10.2196/51719

Gangemi, A., De Luca, R., Fabio, R. A., Lauria, P., Rifici, C., Pollicino, P., Marra, A., Olivo, A., Quartarone, A., & Calabrò, R. S. (2023). Effects of Virtual Reality Cognitive Training on Neuroplasticity: A Quasi-Randomized Clinical Trial in Patients with Stroke. *Biomedicines*, 11(12), 3225. https://doi.org/10.3390/biomedicines11123225

Garrison, J. R., Saviola, F., Morgenroth, E., Barker, H., Lührs, M., Simons, J. S., Fernyhough, C., & Allen, P. (2021). Modulating Medial Prefrontal Cortex Activity Using Real-Time fMRI Neurofeedback: Effects on Reality Monitoring Performance and Associated Functional Connectivity. *NeuroImage*, 245, 118640. https://doi.org/10.1016/j.neuroimage.2021.118640

Giles, G. M. (2018). Neurocognitive Rehabilitation: Skills or Strategies? *The American Journal of Occupational Therapy: Official Publication of the American Occupational Therapy Association*, 72(6), 7206150010p1–7206150010p16. https://doi.org/10.5014/ajot.2018.726001

Glegg, S. M. N., & Levac, D. E. (2018). Barriers, Facilitators and Interventions to Support Virtual Reality Implementation in Rehabilitation: A Scoping Review. *PM & R: The Journal of Injury, Function, and Rehabilitation*, 10(11), 1237–1251.e1. https://doi.org/10.1016/j.pmrj.2018.07.004

Hummel, E., Cogne, M., Lange, M., Lecuyer, A., Joly, F., & Gouranton, V. (2023). VR for Vocational and Ecological Rehabilitation of Patients With Cognitive Impairment: A Survey. *IEEE Transactions on Neural Systems and Rehabilitation Engineering: A Publication of the IEEE Engineering in Medicine and Biology Society*, 31, 4167–4178. https://doi.org/10.1109/TNSRE.2023.3324131

Kellmeyer, P., Biller-Andorno, N., & Meynen, G. (2019). Ethical Tensions of Virtual Reality Treatment in Vulnerable Patients. *Nature Medicine*, 25(8), 1185–1188. https://doi.org/10.1038/s41591-019-0543-y

Kouijzer, M. M. T. E., Kip, H., Bouman, Y. H. A., & Kelders, S. M. (2023). Implementation of Virtual Reality in Healthcare: A Scoping Review on the Implementation Process of Virtual Reality in

Various Healthcare Settings. *Implementation Science Communications*, 4(1), 67. https://doi.org/10.1186/s43058-023-00442-2

Kourtesis, P., & MacPherson, S. E. (2023). An Ecologically Valid Examination of Event-Based and Time-Based Prospective Memory Using Immersive Virtual Reality: The Influence of Attention, Memory, and Executive Function Processes on Real-World Prospective Memory. *Neuropsychological Rehabilitation*, 33(2), 255–280. https://doi.org/10.1080/09602011.2021.2008983

Leeb, R., & Pérez-Marcos, D. (2020). Brain-Computer Interfaces and Virtual Reality for Neurorehabilitation. *Handbook of Clinical Neurology*, 168, 183–197. https://doi.org/10.1016/B978-0-444-63934-9.00014-7

Lei, Y., Deng, Y., Dong, L., Li, X., Li, X., & Su, Z. (2023). A Novel Sensor Fusion Approach for Precise Hand Tracking in Virtual Reality-Based Human-Computer Interaction. *Biomimetics (Basel, Switzerland)*, 8(3), 326. https://doi.org/10.3390/biomimetics8030326

Lim, J. E., Wong, W. T., Teh, T. A., Lim, S. H., Allen, J. C., Quah, J. H. M., Malhotra, R., & Tan, N. C. (2020). A Fully-Immersive and Automated Virtual Reality System to Assess the Six Domains of Cognition: Protocol for a Feasibility Study. *Frontiers in Aging Neuroscience*, 12, 604670. https://doi.org/10.3389/fnagi.2020.604670

Lin, C., Ren, Y., & Lu, A. (2023). The Effectiveness of Virtual Reality Games in Improving Cognition, Mobility, and Emotion in Elderly Post-Stroke Patients: A Systematic Review and Meta-Analysis. *Neurosurgical Review*, 46(1), 167. https://doi.org/10.1007/s10143-023-02061-w

Lin, X. P., Li, B. B., Yao, Z. N., Yang, Z., & Zhang, M. (2024). The Impact of Virtual Reality on Student Engagement in the Classroom-a Critical Review of the Literature. *Frontiers in Psychology*, 15, 1360574. https://doi.org/10.3389/fpsyg.2024.1360574

Lohse, K. R., Hilderman, C. G. E., Cheung, K. L., Tatla, S., & Van Der Loos, H. F. M. (2014). Virtual Reality Therapy for Adults Post-Stroke: A Systematic Review and Meta-Analysis Exploring Virtual Environments and Commercial Games in Therapy. *PLoS One*, 9(3), e93318. https://doi.org/10.1371/journal.pone.0093318

Lorentz, L., Schüppen, A., Suchan, B., & Binkofski, F. (2023). Neural Correlates of Virtual Reality-Based Attention Training: An fMRI Study. *NeuroImage*, 284, 120454. https://doi.org/10.1016/j.neuroimage.2023.120454

Lorenzetti, V., Melo, B., Basílio, R., Suo, C., Yücel, M., Tierra-Criollo, C. J., & Moll, J. (2018). Emotion Regulation Using Virtual Environments and Real-Time fMRI Neurofeedback. *Frontiers in Neurology*, 9, 390. https://doi.org/10.3389/fneur.2018.00390

Maggio, M. G., Maresca, G., De Luca, R., Stagnitti, M. C., Porcari, B., Ferrera, M. C., Galletti, F., Casella, C., Manuli, A., & Calabrò, R. S. (2019). The Growing Use of Virtual Reality in Cognitive Rehabilitation: Fact, Fake or Vision? A Scoping Review. *Journal of the National Medical Association*, 111(4), 457–463. https://doi.org/10.1016/j.jnma.2019.01.003

Malejko, K., Abler, B., Plener, P. L., & Straub, J. (2017). Neural Correlates of Psychotherapeutic Treatment of Post-Traumatic Stress Disorder: A Systematic Literature Review. *Frontiers in Psychiatry*, 8, 85. https://doi.org/10.3389/fpsyt.2017.00085

Marshall, J., Booth, T., Devane, N., Galliers, J., Greenwood, H., Hilari, K., Talbot, R., Wilson, S., & Woolf, C. (2016). Evaluating the Benefits of Aphasia Intervention Delivered in Virtual Reality: Results of a Quasi-Randomised Study. *PLoS One*, 11(8), e0160381. doi: 10.1371/journal.pone.0160381

Matar, E., Shine, J. M., Naismith, S. L., & Lewis, S. J. G. (2014). Virtual Reality Walking and Dopamine: Opening New Doorways to Understanding Freezing of Gait in Parkinson's Disease. *Journal of the Neurological Sciences*, 344(1–2), 182–185. https://doi.org/10.1016/j.jns.2014.06.054

Matsumoto, M. E., Wilske, G. C., & Tapia, R. (2021). Innovative Approaches to Delivering Telehealth. *Physical Medicine and Rehabilitation Clinics of North America*, 32(2), 451–465. https://doi.org/10.1016/j.pmr.2020.12.008

Miltiadis, I., Skarlis, A., & Burko, P. (2025). The Role of Virtual Reality in Personalized Medicine: Advancing Prediction, Prevention, and Participation. *Journal of Medical Systems*, 49(1), 56. https://doi.org/10.1007/s10916-025-02191-2

Mishra, S., Kumar, A., Padmanabhan, P., & Gulyás, B. (2021). Neurophysiological Correlates of Cognition as Revealed by Virtual Reality: Delving the Brain with a Synergistic Approach. *Brain Sciences*, 11(1), 51. https://doi.org/10.3390/brainsci11010051

Moon, H.-J., & Han, S. (2022). Perspective: Present and Future of Virtual Reality for Neurological Disorders. *Brain Sciences*, 12(12), 1692. https://doi.org/10.3390/brainsci12121692

Moreno, A., Wall, K. J., Thangavelu, K., Craven, L., Ward, E., & Dissanayaka, N. N. (2019). A Systematic Review of the Use of Virtual Reality and Its Effects on Cognition in Individuals with Neurocognitive Disorders. *Alzheimer's & Dementia (New York, NY)*, *5*, 834–850. https://doi.org/10.1016/j.trci.2019.09.016

Nieto-Escamez, F., Cortés-Pérez, I., Obrero-Gaitán, E., & Fusco, A. (2023). Virtual Reality Applications in Neurorehabilitation: Current Panorama and Challenges. *Brain Sciences*, 13(5), 819. https://doi.org/10.3390/brainsci13050819

Norrholm, S. D., Jovanovic, T., Gerardi, M., Breazeale, K. G., Price, M., Davis, M., Duncan, E., Ressler, K. J., Bradley, B., Rizzo, A., Tuerk, P. W., & Rothbaum, B. O. (2016). Baseline Psychophysiological and Cortisol Reactivity as a Predictor of PTSD Treatment Outcome in Virtual Reality Exposure Therapy. *Behaviour Research and Therapy*, 82, 28–37. https://doi.org/10.1016/j.brat.2016.05.002

O'Neil, O., Fernandez, M. M., Herzog, J., Beorchia, M., Gower, V., Gramatica, F., Starrost, K., & Kiwull, L. (2018). Virtual Reality for Neurorehabilitation: Insights from 3 European Clinics. *PM & R: The Journal of Injury, Function, and Rehabilitation*, 10(9 Suppl 2), S198–S206. https://doi.org/10.1016/j.pmrj.2018.08.375

Optale, G., Urgesi, C., Busato, V., Marin, S., Piron, L., Priftis, K., Gamberini, L., Capodieci, S., & Bordin, A. (2010). Controlling Memory Impairment in Elderly Adults Using Virtual Reality Memory Training: A Randomized Controlled Pilot Study. *Neurorehabilitation and Neural Repair*, 24(4), 348–357. https://doi.org/10.1177/1545968309353328

Owens, E. A., & Duncan, R. O. (2025). Evidence of a Hierarchical Representation in Bodily Self-Consciousness: The Neural Correlates of Embodiment and Presence in Virtual Worlds. *Frontiers in Human Neuroscience*, 19, 1468947. https://doi.org/10.3389/fnhum.2025.1468947

Pagliari, C., Tella, S. D., Bonanno, C., Cacciante, L., Cioeta, M., De Icco, R., Jonsdottir, J., Federico, S., Franceschini, M., Goffredo, M., Rainoldi, F., Rovaris, M., Springhetti, I., Calabrò, R. S., Tassorelli, C., Rossini, P. M., Baglio, F., & for RIN_TeleSM_Group. (2025). Enhancing the Effect of Rehabilitation on Multiple Sclerosis: A Randomized Clinical Trial Investigating the Impact of Remotely-Supervised Transcranial Direct Current Stimulation and Virtual Reality Telerehabilitation Training. *Multiple Sclerosis and Related Disorders*, 94, 106256. https://doi.org/10.1016/j.msard.2024.106256

Papaioannou, T., Voinescu, A., Petrini, K., & Stanton Fraser, D. (2022). Efficacy and Moderators of Virtual Reality for Cognitive Training in People with Dementia and Mild Cognitive Impairment: A Systematic Review and Meta-Analysis. *Journal of Alzheimer's Disease: JAD*, 88(4), 1341–1370. https://doi.org/10.3233/JAD-210672

Parisi, A., Bellinzona, F., Di Lernia, D., Repetto, C., De Gaspari, S., Brizzi, G., Riva, G., & Tuena, C. (2022). Efficacy of Multisensory Technology in Post-Stroke Cognitive Rehabilitation: A Systematic Review. *Journal of Clinical Medicine*, 11(21), 6324. https://doi.org/10.3390/jcm11216324

Parsons, T. D., Duffield, T., & Asbee, J. (2019). A Comparison of Virtual Reality Classroom Continuous Performance Tests to Traditional Continuous Performance Tests in Delineating ADHD: A Meta-Analysis. *Neuropsychology Review*, 29(3), 338–356. https://doi.org/10.1007/s11065-019-09407-6

Pelosin, E., Cerulli, C., Ogliastro, C., Lagravinese, G., Mori, L., Bonassi, G., Mirelman, A., Hausdorff, J. M., Abbruzzese, G., Marchese, R., & Avanzino, L. (2020). A Multimodal Training Modulates Short Afferent Inhibition and Improves Complex Walking in a Cohort of Faller Older Adults with an Increased Prevalence of Parkinson's Disease. *The Journals of Gerontology: Series A, Biological Sciences and Medical Sciences*, 75(4), 722–728. https://doi.org/10.1093/gerona/glz072

Plotnik, M., Ben-Gal, O., Doniger, G. M., Gottlieb, A., Bahat, Y., Cohen, M., Kimel-Naor, S., Zeilig, G., & Beeri, M. S. (2021). Multimodal Immersive Trail Making-Virtual Reality Paradigm to Study Cognitive-Motor Interactions. *Journal of Neuroengineering and Rehabilitation*, 18(1), 82. https://doi.org/10.1186/s12984-021-00849-9

Raspelli, S., Pallavicini, F., Carelli, L., Morganti, F., Poletti, B., Corra, B., Silani, V., & Riva, G. (2011). Validation of a Neuro Virtual Reality-Based Version of the Multiple Errands Test for the Assessment of Executive Functions. *Studies in Health Technology and Informatics*, 167, 92–97.

Rathinam, C. (2021). Virtual Reality for Neurorehabilitation: Hopes and Challenges for Therapists. *Developmental Medicine and Child Neurology*, 63(4), 370–371. https://doi.org/10.1111/dmcn.14791

Ren, Y., Wang, Q., Liu, H., Wang, G., & Lu, A. (2024). Effects of Immersive and Non-Immersive Virtual Reality-Based Rehabilitation Training on Cognition, Motor Function, and Daily Functioning in Patients with Mild Cognitive Impairment or Dementia: A Systematic Review and Meta-Analysis. *Clinical Rehabilitation*, 38(3), 305–321. https://doi.org/10.1177/02692155231213476

Riches, S., Pisani, S., Bird, L., Rus-Calafell, M., Garety, P., & Valmaggia, L. (2021). Virtual Reality-Based Assessment and Treatment of Social Functioning Impairments in Psychosis: A Systematic Review. *International Review of Psychiatry (Abingdon, England)*, 33(3), 337–362. https://doi.org/10.1080/09540261.2021.1918648

Riva, G., Mancuso, V., Cavedoni, S., & Stramba-Badiale, C. (2020). Virtual Reality in Neurorehabilitation: A Review of Its Effects on Multiple Cognitive Domains. *Expert Review of Medical Devices*, 17(10), 1035–1061. https://doi.org/10.1080/17434440.2020.1825939

Rouhani, R., Umatheva, N., Brockerhoff, J., Keshavarz, B., Kruijff, E., Gugenheimer, J., & Riecke, B. E. (2024). Towards Benchmarking VR Sickness: A Novel Methodological Framework for Assessing Contributing Factors and Mitigation Strategies through Rapid VR Sickness Induction and Recovery. *Displays*, 84, 102807. https://doi.org/10.1016/j.displa.2024.102807

Sánchez-Cuesta, F. J., Arroyo-Ferrer, A., González-Zamorano, Y., Vourvopoulos, A., Badia, S. B. I., Figuereido, P., Serrano, J. I., & Romero, J. P. (2021). Clinical Effects of Immersive Multimodal BCI-VR Training after Bilateral Neuromodulation with rTMS on Upper Limb Motor Recovery after Stroke. A Study Protocol for a Randomized Controlled Trial. *Medicina (Kaunas, Lithuania)*, 57(8), 736. https://doi.org/10.3390/medicina57080736

Shahmoradi, L., & Rezayi, S. (2022). Cognitive Rehabilitation in People with Autism Spectrum Disorder: A Systematic Review of Emerging Virtual Reality-Based Approaches. *Journal of Neuroengineering and Rehabilitation*, 19(1), 91. https://doi.org/10.1186/s12984-022-01069-5

Tieri, G., Morone, G., Paolucci, S., & Iosa, M. (2018). Virtual Reality in Cognitive and Motor Rehabilitation: Facts, Fiction and Fallacies. *Expert Review of Medical Devices*, 15(2), 107–117. https://doi.org/10.1080/17434440.2018.1425613

Tomczuk, D., Poleszak, J., Kowalska, A., Jankowska, N., Trubalski, M., & Turżańska, K. (2025). Virtual Reality as a Supplement to Traditional Rehabilitation after Strokes: A Preliminary Study on Therapy Effectiveness. *Wiadomosci Lekarskie* (Warsaw, Poland: 1960), 78(3), 506–512. https://doi.org/10.36740/WLek/202428

Tortora, C., Di Crosta, A., La Malva, P., Prete, G., Ceccato, I., Mammarella, N., Di Domenico, A., & Palumbo, R. (2024). Virtual Reality and Cognitive Rehabilitation for Older Adults with Mild Cognitive Impairment: A Systematic Review. *Ageing Research Reviews*, 93, 102146. https://doi.org/10.1016/j.arr.2023.102146

Vuper, T. C., Philippi, C. L., & Bruce, S. E. (2021). Altered Resting-State Functional Connectivity of the Default Mode and Central Executive Networks Following Cognitive Processing Therapy for PTSD. *Behavioural Brain Research*, 409, 113312. https://doi.org/10.1016/j.bbr.2021.113312

Wankhede, N. L., Koppula, S., Ballal, S., Doshi, H., Kumawat, R., Raju, Ss., Arora, I., Sammeta, S. S., Khalid, M., Zafar, A., Taksande, B. G., Upaganlawar, A. B., Gulati, M., Umekar, M. J., Kopalli, S. R., & Kale, M. B. (2025). Virtual Reality Modulating Dynamics of Neuroplasticity: Innovations in Neuro-Motor Rehabilitation. *Neuroscience*, 566, 97–111. https://doi.org/10.1016/j.neuroscience.2024.12.040

Wen, Y., Shen, X., & Shen, Y. (2024). Improving Immersive Experiences in Virtual Natural Setting for Public Health and Environmental Design: A Systematic Review and Meta-Analysis of Randomized Controlled Trials. *PLoS One*, 19(4), e0297986. https://doi.org/10.1371/journal.pone.0297986

Wilhelm, F. H., Pfaltz, M. C., Gross, J. J., Mauss, I. B., Kim, S. I., & Wiederhold, B. K. (2005). Mechanisms of Virtual Reality Exposure Therapy: The Role of the Behavioral Activation and Behavioral Inhibition Systems. *Applied Psychophysiology and Biofeedback*, 30(3), 271–284. https://doi.org/10.1007/s10484-005-6383-1

PART V

Future Directions

18

NEUROPLASTICITY UNLEASHED

Receiving the Brain for Recovery

Sruthy Nair and K. Jayasankara Reddy

19.1 Introduction

Neuroplasticity, the brain's capacity to reorganize by forming new neural connections, has redefined how we understand healing. Once seen as a fixed organ, the brain is now understood as dynamic, responsive to experience, and capable of significant adaptation. This realization has profound implications for recovery from neurological injuries and psychological trauma.

Historically, rehabilitation focused on compensating for lost function rather than restoring it. Early neurological models emphasized fixed functions assigned to specific brain regions, limiting therapeutic expectations (Minds Behind the Brain, n.d.). However, findings by Merzenich et al. (1983) demonstrated that the adult brain could rewire itself after injury. Subsequent research confirmed that structural and functional plasticity occurs across the lifespan (Draganski et al., 2004).

Modern rehabilitation harnesses neuroplasticity through targeted, repetitive, and engaging interventions. Mechanisms such as synaptogenesis, dendritic branching, and neurogenesis in areas like the hippocampus underpin recovery processes (Kempermann et al., 1997). Behavioral therapies, enriched environments, and physical activity enhance these processes, partly by increasing brain-derived neurotrophic factor (BDNF), a key molecule in neural growth (Cotman, 2002).

In practice, therapies like constraint-induced movement therapy (CIMT) improve motor recovery by promoting cortical reorganization (Taub et al., 1999). Similarly, psychological interventions—like cognitive–behavioral therapy and EMDR—can induce observable brain changes, challenging the separation between "mental" and "neural" treatments (Goldapple et al., 2004; Pagani et al., 2012).

While promising, the application of neuroplasticity in therapy is still evolving. Individual differences, timing, and comorbidities affect outcomes. Nevertheless, tools such as neuroimaging and non-invasive stimulation now allow clinicians to monitor and influence brain plasticity in real time (Dayan et al., 2013). As neuroscience integrates with psychology and technology, a new era of personalized, adaptive rehabilitation emerges—one where recovery is not limited to restoration but opens doors to transformation.

DOI: 10.4324/9781003646662-23

19.2 Biological Foundations of Neuroplasticity

Understanding the biological underpinnings of neuroplasticity is essential for grasping how the brain transforms in response to experience, injury, and intervention. This section delves into the molecular and structural mechanisms that support neuroplastic changes, explores the temporal dynamics of critical versus lifelong plasticity, and examines how the brain reorganizes itself after damage through complex compensatory processes.

19.2.1 Mechanisms: Synaptic Plasticity, Neurogenesis, Dendritic Pruning

At the cellular level, neuroplasticity reflects the brain's ability to adapt structurally and functionally in response to internal and external influences. This adaptability is governed by three central mechanisms: synaptic plasticity, neurogenesis, and dendritic pruning.

Synaptic plasticity involves the strengthening or weakening of synaptic connections based on their activity levels. Long-term potentiation (LTP), a persistent increase in synaptic strength following high-frequency stimulation, is a foundational mechanism underlying learning and memory (Malenka & Bear, 2004). Long-term depression (LTD), its counterpart, weakens synapses, ensuring that neural pathways remain efficient and responsive to changing demands.

Neurogenesis, once thought to be confined to early development, continues into adulthood—particularly in the hippocampus and subventricular zone. Adult neurogenesis is influenced by environmental enrichment, physical exercise, and cognitive challenges (Eriksson et al., 1998; Kempermann et al., 1997). Newly formed neurons can integrate into existing circuits, contributing to memory formation, emotional regulation, and recovery from injury.

Dendritic pruning is the selective elimination of less active synaptic connections, a process vital during development to refine neural circuits. It continues, albeit at a slower rate, into adulthood, balancing network efficiency with plastic potential (Huttenlocher, 1979).

19.2.2 Critical Period versus Lifelong Plasticity

Plasticity is not evenly distributed across the lifespan. Critical periods, developmental windows of heightened neuroplastic sensitivity, are marked by rapid synaptogenesis followed by experience-dependent pruning. Classic experiments by Hubel and Wiesel (1963) demonstrated that visual deprivation during these periods leads to irreversible deficits, highlighting the necessity of early sensory input for normal cortical development.

These windows are not limited to sensory systems. Language acquisition, emotional regulation, and social cognition also rely on timely environmental inputs (Knudsen, 2004). The closure of critical periods involves molecular brakes such as perineuronal nets, which stabilize circuits but also limit malleability.

Yet, lifelong plasticity remains a central tenet of modern neuroscience. While the capacity for change may diminish with age, the adult brain retains significant plastic potential. Interventions such as cognitive training, mindfulness practices, and physical activity can induce plastic changes in adult brains, particularly in the prefrontal cortex, hippocampus, and sensorimotor regions (Buonomano & Merzenich, 1998; Holzel et al., 2010).

19.2.3 Brain Network Reorganization after Injury

When injury disrupts neural networks, the brain initiates compensatory strategies to restore function. These include contralesional compensation, perilesional reorganization, functional substitution, and LTP.

Perilesional reorganization occurs when regions adjacent to the damaged site assume lost functions. For example, after a stroke affecting the motor cortex, therapy can help surrounding areas reorganize to support motor output (Nudo et al., 1996).

Contralesional compensation involves the recruitment of homologous areas in the opposite hemisphere. In post-stroke aphasia, the right hemisphere may support language recovery, although this is sometimes associated with reduced efficiency due to hemispheric imbalance (Saur et al., 2006).

Functional substitution engages alternate brain regions to support tasks previously managed by the damaged area. Cross-modal plasticity, where visual cortices in blind individuals activate during tactile processing (e.g., Braille reading), illustrates this phenomenon (Sadato et al., 1996).

LTP is instrumental in embedding these reconfigurations. Enhanced synaptic efficacy through LTP supports the reinforcement of newly formed pathways, especially under the influence of targeted rehabilitation strategies such as CIMT or neuromodulation (Liepert et al., 2000). Together, these processes demonstrate that neuroplasticity is not only a reactive mechanism but a proactive, integrative force—linking biology, experience, and recovery.

19.3 Clinical Evidence Supporting Neuroplastic Recovery

Historically, recovery was believed to plateau within the first six months post-injury, with residual impairments considered largely permanent (Katz et al., 2004). However, a growing body of interdisciplinary research challenges this paradigm by offering compelling clinical and neuroimaging evidence of neuroplastic recovery well into the chronic phase. This section synthesizes clinical case studies, longitudinal neuroimaging data, and experimental findings to illuminate how neuroplasticity—modulated by factors such as task specificity, intervention intensity, and timing—can facilitate meaningful functional recovery, even years after injury.

19.3.1 Case Study of Late-Stage Recovery in TBI

Clinical case reports provide humanized, contextual insights into the extended timeline of neuroplastic potential. For instance, a landmark case documented by Taub et al. (2006) illustrated significant motor recovery in a patient with chronic hemiparesis nearly a decade after a severe traumatic brain injury (TBI), facilitated by CIMT. The patient, previously unable to perform daily tasks with the affected limb, demonstrated not only improved functionality but also cortical reorganization, as evidenced by functional magnetic resonance imaging (fMRI) scans.

In another illustrative case, Wilson and Betteridge (2019) described the neurocognitive and emotional rehabilitation of a middle-aged adult who sustained diffuse axonal injury from a motor vehicle accident. Despite minimal progress during the first year post-injury, intensive integrative cognitive rehabilitation initiated 24 months post-injury resulted in notable improvements in memory, executive function, and emotional regulation. These changes were corroborated by both neuropsychological testing and improved neural connectivity as revealed by diffusion tensor imaging (DTI).

A particularly compelling case comes from the work of Coetzer and Balchin (2014), who reported on a young adult male who experienced a severe TBI following an industrial accident. After nearly three years with minimal documented improvement, he entered a structured neuropsychological rehabilitation program tailored around vocational tasks, social re-engagement, and personalized cognitive–behavioral strategies. Over the subsequent two years, he regained independence in daily living, secured part-time employment, and reported enhanced emotional well-being. These improvements were mirrored by enhanced activation in frontoparietal networks on fMRI, suggesting that late-stage re-engagement with meaningful activities can catalyze neural change.

Moreover, rehabilitation involving assistive technology and immersive environments has also led to unexpected gains. A case reported by Cicerone et al. (2011) involved a female TBI survivor with persistent language and attentional deficits, who underwent therapy via virtual reality (VR)-enhanced cognitive training. Initiated nearly four years post-injury, her therapy emphasized realistic simulations of multitasking environments, such as virtual shopping and scheduling. Improvements were seen in task-switching ability and verbal fluency, with parallel increases in prefrontal activation patterns.

These cases collectively emphasize the potential for recovery long after initial injury and highlight the importance of personalized, engaging, and meaningful rehabilitative environments. The human capacity for change, both neural and functional, extends far beyond early expectations if the right combination of interventions and motivations is in place.

19.3.1 Longitudinal Neuroimaging: Visualizing Plastic Change

Longitudinal neuroimaging studies provide critical insights into how brain structure and function evolve post-injury, validating the biological substrates of observed clinical improvements.

For example, Bonnelle et al. (2011) followed patients with moderate to severe TBI over a two-year period using resting-state fMRI (rs-fMRI). They found that sustained cognitive rehabilitation led to increases in connectivity within the default mode network (DMN)—an effect associated with improvements in memory and attention. These findings provided evidence that cognitive improvements were not merely compensatory but accompanied by reorganization within key brain networks.

Similarly, Palacios et al. (2020) employed DTI to track changes in white matter integrity at three intervals across a year post-injury. Patients who received targeted interventions continued to show gains in fractional anisotropy, even beyond the 12-month mark, indicating ongoing myelin repair and axonal integrity. While individual differences were notable, the trend toward prolonged structural change reinforces the case for long-term rehabilitation.

These studies, while fewer in number relative to early-phase trials, lend biological credibility to the extended timeline of recovery observed in case studies and point to neuroplasticity as a dynamic, experience-driven process.

19.3.3 Role of Intensity, Timing, and Task-Specific Practice

The modulation of neuroplasticity hinges significantly on the principles of experience-dependent plasticity, as outlined by Kleim and Jones (2008). Three particularly salient factors emerge in the context of clinical recovery: intensity, timing, and task specificity.

Intensity: Multiple studies demonstrate a dose–response relationship between the intensity of intervention and the magnitude of neuroplastic change. For instance, Lang et al. (2009) found that patients with chronic stroke receiving high-dose motor training showed significantly greater improvements in both motor function and sensorimotor cortex activation compared to those receiving conventional therapy. Although this study focused on stroke, the principles extend to TBI, where increased engagement in rehabilitative activities has been associated with enhanced outcomes (Nakase-Richardson et al., 2011).

Timing: While early intervention remains critical, delayed interventions are not necessarily futile. Evidence from longitudinal rehabilitation research indicates that neuroplastic gains can be achieved even when therapy is initiated years post-injury. As discussed earlier, studies such as that by Wilson and Betteridge (2019) reveal that the brain retains a capacity for reorganization far beyond the acute period. This realization shifts the therapeutic framework from one constrained by rigid timelines to one informed by personalized readiness and responsiveness.

Task specificity: Plasticity is inherently use-dependent. This principle is exemplified in motor rehabilitation strategies such as CIMT and robotic-assisted gait training, which target specific impaired functions. A study by Lotze et al. (2003) revealed that stroke patients who practiced task-specific motor actions displayed increased activation in the ipsilateral motor cortex, suggesting recruitment of alternate neural pathways for functional compensation. Comparable outcomes have been documented in TBI populations, particularly when interventions focus on ecological tasks relevant to daily life (Cicerone et al., 2011).

In essence, recovery is optimized when therapeutic activities are tailored to the individual's deficits, are sufficiently challenging, and are practiced repetitively in a contextually meaningful manner.

19.3.4 Psychological and Environmental Modulators of Neuroplastic Recovery

Beyond neurobiological mechanisms, recovery trajectories are also shaped by psychological and environmental variables. Emotional engagement, social support, and the therapeutic alliance have been identified as critical enablers of neuroplastic change (Coetzer & Balchin, 2014). For example, patients who demonstrate greater intrinsic motivation and resilience often show superior functional outcomes, even when matched for injury severity and rehabilitation intensity.

Moreover, enriched environments—characterized by cognitive stimulation, physical activity, and social interaction—have been shown to promote synaptogenesis and dendritic branching in animal models (Nithianantharajah & Hannan, 2006). Translating this to human populations, structured day programs and community reintegration initiatives may extend the gains made in clinical settings and foster long-term recovery.

These psychosocial factors are not merely adjuncts but core components of a truly biopsychosocial model of neuroplastic rehabilitation.

The accumulated evidence compels a reconsideration of how clinicians conceptualize prognosis and design intervention timelines. Rather than viewing chronicity as a barrier, it may be more productive to see it as a different phase of opportunity—one that requires tailored strategies and sustained support. This shift has practical implications:

- Policy: Funding models should accommodate long-term rehabilitation, not just early post-acute care.
- Clinical guidelines: Protocols should incorporate flexible timelines for assessment and re-intervention.
- Therapist training: Emphasis should be placed on dynamic, individualized approaches that evolve with the patient's recovery trajectory.

Such changes will enable healthcare systems to align more closely with the plastic potential of the human brain.

In sum, the clinical and neurobiological evidence increasingly supports a dynamic model of recovery—one in which neuroplasticity remains an active and modifiable process long after the initial injury. Case studies reveal profound recoveries previously deemed impossible; neuroimaging confirms the brain's capacity for structural and functional reorganization; and behavioral research identifies the conditions that best support this recovery. The implication is clear: we must continue to challenge static notions of prognosis and embrace therapeutic strategies that honor the brain's enduring capacity for change.

19.4 Technology-Driven Plasticity Interventions

Advancements in neuroscience have underscored the brain's remarkable capacity for neuroplasticity—the ability to reorganize, adapt, and repair itself across the lifespan. In the context of rehabilitation, this capacity opens a gateway to transformative interventions. The integration of technology in neurorehabilitation has accelerated the development of methods aimed at promoting plasticity. This section explores six such interventions: transcranial magnetic stimulation (TMS) and transcranial direct current stimulation (tDCS), neurofeedback and real-time functional MRI (rt-fMRI), brain–computer interfaces (BCIs), artificial intelligence (AI)-enhanced cognitive training platforms, and VR with gamification strategies. These modalities, individually and in combination, signify a paradigmatic shift in how clinicians and researchers conceptualize, induce, and sustain plasticity.

19.4.1 Non-Invasive Brain Stimulation: TMS and tDCS

TMS and tDCS are non-invasive neuromodulation techniques that aim to alter cortical excitability and modulate synaptic plasticity. These methods have gained traction in both research and clinical settings for their potential to restore function following neurological injury or to ameliorate psychiatric symptoms.

TMS involves the generation of magnetic pulses that penetrate the skull and induce electric currents in targeted brain regions. Depending on the frequency and pattern of stimulation, TMS can either excite or inhibit neuronal activity (Hallett, 2007). Repetitive TMS (rTMS) has shown efficacy in treating depression, stroke-related motor deficits, chronic pain, and even auditory hallucinations in schizophrenia (George, 2010; Slotema et al., 2010). For instance, Lefaucheur et al. (2014) conducted a meta-analysis confirming the efficacy of rTMS over the dorsolateral prefrontal cortex in alleviating treatment-resistant depression. Furthermore, theta burst stimulation, a patterned form of TMS, has demonstrated enhanced efficacy due to its ability to mimic endogenous firing patterns (Huang et al., 2005).

In contrast, tDCS delivers a low-intensity, constant electric current via electrodes placed on the scalp. Anodal stimulation generally enhances cortical excitability, while cathodal stimulation reduces it (Nitsche & Paulus, 2000). tDCS has been explored for improving working memory, motor function, and language recovery post-stroke (Elsner et al., 2019). Notably, when applied during cognitive or motor tasks, tDCS appears to facilitate activity-dependent plasticity through mechanisms akin to LTP. For example, Baker et al. (2010) found that coupling tDCS with task-specific language therapy in post-stroke aphasia patients led to improved naming and comprehension skills compared to sham stimulation.

Clinical trials have also investigated the use of tDCS in mood disorders, addiction, and chronic pain. In patients with fibromyalgia, daily tDCS over the primary motor cortex resulted in significant pain relief and functional improvement (Fregni et al., 2006). However, the heterogeneity in individual response remains a major challenge. Ongoing efforts are aimed at identifying biomarkers that predict stimulation responsiveness and personalizing stimulation protocols.

Despite their promise, both TMS and tDCS face challenges related to interindividual variability, optimal stimulation parameters, dosage scheduling, and long-term efficacy. Yet, when integrated into personalized therapeutic protocols and combined with other interventions like pharmacotherapy or behavioral therapy, these methods hold considerable potential for promoting functional plasticity and enhancing rehabilitation outcomes.

19.4.2 Neurofeedback and Real-Time fMRI

Neurofeedback involves training individuals to regulate their own brain activity using real-time feedback from electroencephalography (EEG) or fMRI. Rooted in the operant conditioning framework, neurofeedback seeks to reinforce desirable neural patterns while inhibiting maladaptive ones.

EEG-based neurofeedback has been successfully used in treating ADHD, epilepsy, anxiety, depression, insomnia, and PTSD (Thibault et al., 2015; Gruzelier, 2013). For example, a landmark study by Lubar and Shouse (1976) demonstrated that children with ADHD could learn to increase their sensorimotor rhythm (SMR), leading to reduced hyperactivity. More recently, neurofeedback has gained attention in the context of autism spectrum disorders (ASD), with some evidence suggesting improvements in social cognition and attentional control.

rt-fMRI extends the capabilities of neurofeedback by offering high spatial resolution of brain activity. This allows for targeting deep brain structures implicated in affective and cognitive regulation. DeCharms et al. (2005) used rt-fMRI to enable participants to modulate anterior cingulate cortex activity, resulting in decreased chronic pain perception. Other studies have shown that individuals can learn to control amygdala activity, with potential implications for mood and anxiety disorders (Zotev et al., 2013).

Importantly, neurofeedback supports metacognitive development by allowing individuals to consciously engage in neural self-regulation. This has profound implications for psychoanalytic practice, where self-awareness and introspection are central to therapeutic change. From a neuropsychoanalytic perspective, the capacity to observe and modulate one's neural processes represents an externalized encounter with the unconscious, thus bridging technology and intrapsychic dynamics.

Despite its promise, challenges remain. The standardization of protocols, placebo effects, training duration, and generalizability of gains are ongoing concerns. Furthermore, ethical considerations arise when neurofeedback is used for enhancement rather than therapeutic purposes, raising questions about agency, consent, and societal expectations.

19.4.3 Brain–Computer Interfaces (BCIs)

BCIs translate brain signals into commands that can control external devices or computer applications. BCIs bypass traditional neuromuscular pathways, offering alternative communication channels for individuals with severe motor impairments such as locked-in syndrome, amyotrophic lateral sclerosis (ALS), and spinal cord injury.

BCI systems typically use EEG, electrocorticography (ECoG), or implanted electrodes to detect neural activity. The signals are decoded using machine learning (ML) algorithms and translated into actions, such as cursor movements, text input, or robotic arm control (Wolpaw & Wolpaw, 2012).

In neurorehabilitation, BCIs allow patients to engage in motor imagery tasks that activate sensorimotor pathways, facilitating Hebbian plasticity and promoting cortical reorganization (Biasiucci et al., 2018).

A study by Ramos-Murguialday et al. (2013) demonstrated that stroke patients who engaged in motor-imagery-based BCI training coupled with functional electrical stimulation showed greater improvement in motor function than controls. This underscores the synergistic potential of combining intention (neural signals) with action (feedback). Beyond motor recovery, BCIs are being explored for emotional regulation, communication in disorders of consciousness, and even artistic expression. Researchers are developing effective BCIs that respond to emotional states, paving the way for emotionally adaptive environments and devices.

The future of BCIs likely lies in their integration with wearable technologies, neural implants, and AI. However, ethical concerns loom large, including issues of identity, cognitive liberty, and surveillance. As the boundary between self and machine blurs, interdisciplinary dialogue is vital to ensure that technological progress remains human-centered.

19.4.4 AI-Enhanced Cognitive Training Platforms

AI is revolutionizing cognitive rehabilitation by enabling adaptive, personalized training modules that evolve in real time based on user performance. These platforms utilize big data, reinforcement learning, and neural network models to optimize task difficulty, duration, and content alignment with individual cognitive profiles.

AI-driven tools have been used to target executive functions, memory, attention, and reasoning. For instance, Lumosity and BrainHQ offer gamified exercises supported by research on neurocognitive enhancement. While commercial claims have often outpaced empirical support, some randomized controlled trials suggest modest but measurable benefits, particularly when interventions are long-term and embedded in multimodal rehabilitation (Hardy et al., 2015).

More robust platforms are emerging in clinical neuroscience. The Adaptive Cognitive Evaluation (ACE) platform uses Bayesian algorithms to assess and train cognitive functions with high precision (O'Laughlin et al., 2025). These systems enable real-time feedback loops that align with the Hebbian principle of "fire together, wire together," facilitating synaptic reinforcement and network remodeling. AI also supports predictive analytics, identifying individuals at risk for cognitive decline and recommending early interventions.

An emerging frontier in AI-enhanced rehab is natural language processing (NLP)-based cognitive therapies. Chatbots and conversational agents can simulate therapeutic dialogues, provide memory prompts, or engage users in reflective activities. These platforms offer scalability in mental health care, especially in low-resource settings.

In neurodegenerative disorders such as Alzheimer's disease, personalized cognitive stimulation integrated with neuroimaging biomarkers may slow cognitive deterioration by harnessing compensatory plasticity (Klimova & Maresova, 2017). AI models trained on large datasets may eventually guide clinical decisions and suggest interventions based on real-time monitoring of brain activity and behavior. However, the integration of AI also necessitates caution. Biases in training data, lack of transparency in decision-making algorithms, and the potential for over-reliance on machine-generated outcomes must be addressed. The goal must remain to augment human capacities, not replace them.

19.4.5 Virtual Reality and Gamification in Rehabilitation

VR provides immersive, interactive environments that simulate real-world contexts. In rehabilitation, VR is used to engage patients in task-specific exercises that promote motor and cognitive recovery. When combined with gamification—the application of game design elements such as points, levels, and feedback—VR can enhance motivation, adherence, and neuroplastic engagement.

A meta-analysis by Laver et al. (2017) found that VR-based interventions improved upper limb function post-stroke compared to conventional therapy. These systems enable repetition, variability, and real-time performance feedback—key principles of motor learning. For example, systems like the CAREN (Computer Assisted Rehabilitation Environment) offer multi-sensory integration through treadmills, motion capture, and VR goggles to simulate dynamic tasks like crossing a street or climbing stairs.

Gamification has also proven effective in cognitive rehabilitation. Programs like Cogmed and RehaCom incorporate game mechanics to train working memory and attention in individuals with TBI, ADHD, or mild cognitive impairment. The competitive and rewarding aspects of gamification stimulate the dopaminergic system, facilitating learning and engagement (Anguera et al., 2013).

Therapeutic VR is also being applied in the treatment of phobias, PTSD, and pain management. By creating controlled yet immersive environments, VR enables graded exposure therapy or distraction from aversive stimuli, capitalizing on emotional regulation and attentional control mechanisms. Moreover, social VR platforms offer opportunities for patients with social anxiety or autism to rehearse social interactions in safe virtual settings.

Nevertheless, challenges remain in standardizing VR protocols, addressing cybersickness, and ensuring long-term transfer of skills to real-life functioning. Cultural relevance and accessibility must also be considered, especially in diverse or underserved populations.

19.4.6 Interdisciplinary Integration and Ethical Considerations

The convergence of these technological interventions marks a shift from compartmentalized to integrated rehabilitation paradigms. Combining tDCS with VR, neurofeedback with AI analytics, or BCIs with gamified environments can yield synergistic effects that maximize plasticity. Such hybrid approaches demand interdisciplinary collaboration among neuroscientists, clinicians, engineers, psychologists, and ethicists.

Moreover, as these interventions become more personalized and data-intensive, ethical considerations regarding privacy, equity, and agency must be addressed. Questions around cognitive enhancement versus therapeutic application, long-term neuroethics, and algorithmic biases in AI systems necessitate a framework that balances innovation with responsibility (Ienca & Andorno, 2017).

Training professionals to interpret and ethically apply these technologies will be crucial. Ethical frameworks should not be post hoc but embedded in the design, deployment, and regulation of these tools. Inclusive policymaking, public dialogue, and ongoing research into social implications will be central to aligning technological advances with human values.

Technology-driven interventions have opened new frontiers in our understanding and application of neuroplasticity. From non-invasive brain stimulation and neurofeedback to AI-enhanced platforms and immersive VR environments, these tools offer unprecedented potential for tailoring rehabilitation to individual needs. As we move toward integrative and ethically informed models, the fusion of technology and neuroscience stands poised to redefine the landscape of recovery and cognitive transformation. These interventions not only advance clinical outcomes but also invite us to rethink the boundaries of mind, machine, and meaning in the context of healing and growth.

19.5 Personalized and Predictive Neurorehabilitation

As neurorehabilitation evolves into a more integrative and individualized field, the focus has shifted from standardized protocols to patient-specific models that consider the brain's dynamic, self-organizing capabilities. In the wake of rapid technological and computational advances, the landscape of neurorehabilitation is becoming increasingly personalized and predictive. This transformation is anchored in three critical developments: (1) the identification and utilization of biomarkers of plasticity such as EEG rhythms and fMRI connectivity; (2) the implementation of adaptive interventions tailored to neural responses; and (3) the deployment of ML algorithms to optimize therapy planning and monitor therapeutic progress.

Building on the foundational concepts of technology-driven plasticity interventions, this section explores how neuroscientific precision can be harnessed to guide real-time therapeutic decisions, anticipate outcomes, and individualize care. These personalized frameworks not only improve efficacy but also align with the psychoanalytic imperative to treat the individual as a whole, honoring both biological specificity and subjective experience.

19.5.1 Biomarkers of Plasticity: Quantifying the Brain's Readiness to Change

Biomarkers are quantifiable biological indicators that offer insights into an organism's physiological state. In neurorehabilitation, biomarkers of plasticity provide a window into the brain's capacity to reorganize and recover. Among the most promising are those derived from EEG and fMRI.

EEG Biomarkers: EEG is a cost-effective and temporally precise method for monitoring neural activity. Oscillatory patterns in EEG—such as alpha (8–12 Hz), beta (13–30 Hz), and theta (4–8 Hz) rhythms—are increasingly viewed as neurophysiological markers of cognitive states and neural flexibility. Event-related potentials (ERPs), such as the P300 component, have been linked to attention, decision-making, and recovery prognosis following stroke or brain injury (Polich, 2007).

The concept of "EEG reactivity"—the brain's ability to modulate its rhythms in response to sensory input—has gained traction as a predictive marker. For example, increased theta-band activity in frontal regions has been associated with enhanced attentional plasticity during cognitive training (Enriquez-Geppert et al., 2017). Similarly, SMR amplitude is often used in neurofeedback protocols aimed at motor rehabilitation (Pichiorri et al., 2015).

fMRI Connectivity Patterns: While EEG provides excellent temporal resolution, fMRI offers superior spatial resolution. rs-fMRI and task-based fMRI are used to assess functional connectivity—how different brain regions communicate during rest and activity. Changes in connectivity within the DMN, salience network, and frontoparietal control network have been linked to cognitive and emotional impairments in neurological patients.

Connectivity disruptions are often observed in stroke, TBI, and neurodegenerative disorders. Importantly, longitudinal fMRI studies have shown that rehabilitative training can enhance connectivity in motor and cognitive networks (Grefkes & Fink, 2014). Moreover, individual differences in baseline connectivity can predict the effectiveness of specific interventions. For instance, greater prefrontal–motor connectivity has been linked to better motor recovery after VR-based training (Riva et al., 2020). Biomarker-driven rehabilitation thus allows clinicians to "listen" to the brain and tailor interventions accordingly, maximizing the potential for meaningful plastic change.

19.5.2 Adaptive Interventions: Real-Time Responsiveness to Neural States

Traditional rehabilitation protocols often follow rigid schedules and intensity levels, assuming homogeneity among patients. However, this one-size-fits-all model fails to capture the dynamism of neural systems. Adaptive interventions—those that respond in real time to an individual's physiological or behavioral state—embody a new paradigm that respects the fluctuating readiness and capacity for plastic change.

Neurofeedback and Closed-Loop Systems: Adaptive interventions frequently utilize closed-loop systems, where real-time brain activity is monitored and used to adjust ongoing stimulation or task parameters. Neurofeedback exemplifies this model. EEG-based neurofeedback enables individuals to observe and modulate their brain rhythms, leading to improved outcomes in attention disorders, epilepsy, and mood regulation (Thibault et al., 2015).

Similarly, closed-loop deep brain stimulation (DBS) and TMS have been developed to synchronize stimulation with endogenous neural oscillations, thus enhancing efficacy and reducing side effects (Widge et al., 2018). For example, synchronizing TMS pulses with alpha-phase activity in the prefrontal cortex has shown promise in improving working memory performance (Zrenner et al., 2016).

Task Adaptation and Cognitive Load Regulation: Adaptive systems also monitor cognitive or emotional workload using physiological markers (e.g., pupil dilation, heart rate variability, EEG), adjusting task difficulty in real time. This is particularly relevant in BCI training and cognitive remediation programs. For instance, adaptive BCI frameworks modulate feedback sensitivity based on the participant's neural fatigue levels, ensuring sustained engagement and optimizing learning (Roy et al., 2019).

19.5.3 Machine Learning in Therapy Planning and Progress Tracking

The advent of ML offers unprecedented opportunities for pattern recognition, predictive modeling, and decision-making in neurorehabilitation. These algorithms can detect subtle, non-linear relationships among large datasets—relationships that would be imperceptible to human analysis.

Predictive Modeling: Supervised learning algorithms, such as support vector machines (SVMs), random forests, and deep neural networks, are being employed to predict therapy outcomes based on pre-treatment neuroimaging and behavioral data. For example, Pustina et al. (2017) used SVMs to predict post-stroke language recovery using fMRI connectivity and lesion mapping, achieving high predictive accuracy.

Similarly, ML models trained on EEG data can differentiate responders from non-responders in TMS therapy for depression, enabling personalized protocol selection (Drysdale et al., 2016). These predictive capabilities enhance treatment planning, allowing clinicians to prioritize interventions with the highest likelihood of success.

Real-Time Progress Monitoring: ML also facilitates real-time monitoring of therapeutic progress. By continuously analyzing EEG or behavioral data during training, algorithms can detect plateaus or regressions and recommend adaptive changes. This continuous feedback loop creates a living therapeutic model that evolves with the patient.

Unsupervised learning techniques, such as clustering and principal component analysis (PCA), are useful for discovering subtypes within patient populations, leading to better stratification and individualized care. Reinforcement learning models, which optimize decision-making through trial and error, are particularly suited for guiding adaptive therapy schedules.

While ML offers powerful tools, it also introduces concerns about interpretability and transparency. The "black box" nature of deep learning raises ethical questions, especially when these models are used in clinical decision-making. Therefore, explainable AI (XAI) approaches are being developed to make model predictions interpretable by clinicians and patients alike (Doshi-Velez & Kim, 2017).

Ethical AI in neurorehabilitation must also address data privacy, informed consent, and bias mitigation. Equitable access to AI-enhanced care demands attention to the social and cultural contexts in which these technologies are deployed.

19.5.4 Integrative Perspectives: Toward a Neuro-Psycho-Technological Model

The integration of biomarkers, adaptive systems, and ML embodies a systemic approach to neurorehabilitation. It reframes therapy as a co-evolving relationship between patient and environment, mediated by technology but grounded in humanistic care.

Adaptive technologies, when implemented ethically and reflectively, become instruments of attunement—responsive not just to neural activity, but to the lived experience of recovery. They enable a model of care that is not only data-informed but meaning-driven.

19.6 Ethical and Practical Considerations in Neurotechnology-Driven Rehabilitation

As the field of neurorehabilitation grows more sophisticated through the incorporation of technologies like EEG-guided feedback, real-time fMRI, and AI-powered adaptive interfaces, its promise is undeniably compelling. Yet, alongside its scientific and clinical advancements come complex ethical and practical questions. Who gets access to these technologies? How are data privacy and autonomy protected? What are the risks of overstimulation or maladaptive plasticity? These questions are not peripheral—they strike at the core of equitable and humane care. Addressing them is essential if neurorehabilitation is to remain not only effective but ethically grounded.

This section integrates the sociotechnical challenges posed by neurotechnology-driven rehabilitation. It examines inequalities in access, global disparities in health infrastructure, privacy and surveillance concerns, and the potential unintended consequences of neuroplastic interventions. These dimensions connect conceptually to the personalization models previously discussed, demanding that innovation be guided by principles of justice, transparency, and person-centered care.

19.6.1 Access and Equity: Who Benefits from Neurotechnology?

The deployment of advanced neurotechnologies has the potential to exacerbate existing disparities in healthcare access. While EEG neurofeedback and AI-enhanced cognitive training platforms may be available in high-resource settings, their accessibility in rural, low-income, or underdeveloped regions remains limited (Ienca & Andorno, 2017).

Socioeconomic barriers: Many cutting-edge interventions—such as closed-loop DBS or individualized machine learning protocols—require expensive equipment, specialized personnel, and long-term follow-up. These resource-intensive features create systemic barriers to access. Even within wealthy countries, neurotechnological interventions are often restricted to private clinics, leaving marginalized populations underserved. For example, a study by Illes and Racine (2005) noted that neurological interventions often benefit those who are already socioeconomically privileged, while global health systems struggle to implement even basic rehabilitation services.

Global health disparities: Low- and middle-income countries (LMICs) face additional challenges, including lack of infrastructure, inconsistent electricity, limited internet access, and scarcity of trained clinicians. As a result, the global distribution of neurotechnological care is skewed. Without intentional investment and cross-border collaboration, neurorehabilitation risks becoming a luxury of the Global North (Patel et al., 2018).

To address these disparities, some researchers advocate for the development of scalable, low-cost solutions, such as portable EEG devices or smartphone-based neurofeedback applications. While promising, these innovations must be culturally adapted and ethically evaluated to ensure they do not replicate exploitative models of tech transfer.

19.6.2 Data Privacy and Autonomy: Neurotechnology in the Age of Surveillance

With neurotechnological interventions comes an exponential increase in data—brain activity recordings, behavioral profiles, and AI-generated predictions. This data is deeply personal, raising urgent questions about ownership, consent, and surveillance.

Neurodata as Identity: Unlike traditional health data, neurodata carries intimate insights into a person's cognition, mood, and behavioral tendencies. It may reveal not just health status but preferences, vulnerabilities, and psychological traits. The European Commission has proposed framing such data under the category of "mental privacy"—a subset of cognitive liberty that protects individuals from unauthorized brain data collection (Ienca & Andorno, 2017).

However, commercial entities are increasingly involved in neurotechnology development, often without clear regulatory oversight. Consumer-grade neurodevices, such as EEG headbands marketed for focus and relaxation, collect data that can be sold or shared. The use of AI to interpret these datasets introduces additional opacity: users may not know how their data is being used or how decisions are made based on it (Yuste et al., 2017).

Informed Consent and Algorithmic Opacity: True informed consent requires that users understand what data is being collected, how it is stored, and for what purpose it will be used. Yet the complexity of ML models often renders this transparency difficult. Many algorithms are "black boxes," making predictions or recommendations without explainable reasoning (Doshi-Velez & Kim, 2017).

This is particularly problematic in therapeutic contexts, where trust in the clinician–device–patient relationship is paramount. Ethical implementation demands XAI models, strong consent protocols, and robust data governance frameworks.

19.6.3 Neuroplastic Risks: When Plasticity Goes Away

While neuroplasticity is the foundation of rehabilitation, its capacity for change also introduces vulnerability. Repeated stimulation, unregulated feedback, or overly intensive protocols can lead to maladaptive plasticity.

Overstimulation and Sensory Fatigue: TMS, tDCS, and high-intensity cognitive training can overstimulate neural circuits if not carefully calibrated. Reports have linked excessive or improperly targeted stimulation with negative outcomes such as anxiety, irritability, cognitive fatigue, or emotional blunting (Davis & van Koningsbruggen, 2013).

In children or individuals with neurodevelopmental conditions, the risk of overtraining is especially concerning. Because their brains are still developing, the unintended reinforcement of compensatory or pathological pathways can have long-lasting consequences.

Maladaptive Rewiring: Plasticity can also reinforce harmful behaviors or maladaptive strategies. For instance, in phantom limb pain, cortical reorganization that once served an adaptive function may become pathological. Similarly, in PTSD, hyperconnectivity in threat-related networks may resist extinction (Pitman et al., 2012). Interventions must therefore be guided by comprehensive understanding of individual brain states and psychological contexts.

This risk underscores the need for continuous monitoring, interdisciplinary oversight, and careful balancing of intensity and rest. Interventions must be personalized not just for efficacy but for safety and long-term neuropsychological integrity.

19.6.4 Ethical Frameworks for the Future

Ethical neurorehabilitation requires more than institutional review boards (IRBs) and patient forms. It requires a paradigm shift—an integrative framework that honors both the technological complexity and the subjective depth of each individual.

Approaches such as the four principles of biomedical ethics—autonomy, beneficence, non-maleficence, and justice (Beauchamp & Childress, 2001)—remain foundational. However, neurotechnology demands further elaboration. Scholars advocate for a fifth principle: neurojustice—the fair and inclusive distribution of cognitive enhancement and rehabilitation technologies (Chatterjee, 2006).

19.7 The Future Vision

As the science of neuroplasticity becomes increasingly nuanced, its translation into clinical care must evolve from reactive compensation toward proactive restoration. Historically, neurorehabilitation has emphasized compensatory strategies—training individuals to work around lost function. However, advances in neuroscience, neurotechnology, and AI are ushering in a transformative vision: one that prioritizes the regeneration and restoration of neural networks through targeted, personalized interventions embedded within standard clinical pathways.

19.7.1 From Compensation to Restoration

Traditionally, patients recovering from stroke or TBI have been taught to rely on unaffected brain regions to compensate for damaged areas. While this approach can improve function, it does not necessarily promote recovery at the neural level. Recent evidence suggests that timely, intensive, and neurotechnology-supported interventions can stimulate activity in perilesional and even ipsilesional areas, enhancing true functional restoration (Stinear et al., 2017). For instance, CIMT, when combined with non-invasive brain stimulation, has been shown to increase corticospinal excitability and motor performance (Ward et al., 2019).

19.7.2 Embedding Neuroplastic Principles in Clinical Pathways

To fully leverage neuroplasticity, clinical systems must shift from episodic to continuous models of care. This means integrating biomarkers of plasticity—such as EEG coherence or fMRI network connectivity—into diagnostic and progress-monitoring routines (Grefkes &

Fink, 2020). Early identification of individuals with high plastic potential could guide intervention timing and intensity, optimizing outcomes while avoiding resource waste.

Moreover, digital tools such as AI-based progress tracking platforms can continuously adapt rehabilitation tasks based on real-time neural and behavioral feedback. These platforms not only offer scalability but also align with the move toward precision medicine in neurology and psychiatry (Kourtis et al., 2019).

19.7.2 The Power of Cross-Disciplinary Convergence

The future of neurorehabilitation lies in collaboration. Neuroscience provides the empirical foundation, psychology brings insights into motivation and behavior, rehabilitation sciences ensure clinical viability, and AI facilitates precision and scalability. For example, BCIs paired with VR can simulate real-life environments for patients to rehearse cognitive or motor tasks. Such integrations have shown promise in enhancing engagement and promoting cortical reorganization.

Importantly, psychological principles—such as self-efficacy, therapeutic alliance, and affect regulation—must be considered in designing neurorehabilitative technologies. Devices that ignore the lived emotional experience of recovery risk disengagement and reduced efficacy. In essence, the future vision of neurorehabilitation is not confined to novel devices or protocols. It is a systemic, human-centered evolution in care—where science, ethics, and lived experience converge to restore not only function but also dignity and autonomy.

19.8 Conclusion

Neuroplasticity, once a contested concept, now stands as a central tenet in our understanding of brain recovery. Yet, the most profound shift is not scientific but conceptual: reframing plasticity not as a fixed window of opportunity but as a dynamic, lifelong capacity for change. This perspective invites a new kind of hope—one grounded in ongoing adaptation rather than finite restoration. As we look toward the next decade, the implications of this reframing ripple across clinical practice, research agendas, and health policy.

Clinicians are increasingly called to treat not only deficits but potential (Doidge, 2008).

Rehabilitation protocols must move beyond static task repetitions to embrace adaptive, feedback-informed, and person-centered approaches. Technologies like EEG-based neurofeedback, AI-curated interventions, and VR therapies are tools—but they are only as effective as the frameworks guiding their use. Understanding plasticity as a continuous, context-sensitive process empowers clinicians to personalize care trajectories, reassess goals dynamically, and support long-term engagement. Moreover, emotional, motivational, and relational dimensions must be integrated into neurorehabilitation. Hope is not merely an outcome; it is also a therapeutic resource. Cultivating resilience and agency through therapy—rather than treating recovery as a return to a prior baseline—represents a paradigm shift aligned with current neuroscientific insights.

For researchers, the future lies in translational and interdisciplinary work. We must move from describing plasticity to harnessing it—through biomarkers, predictive models, and cross-modal interventions. Longitudinal studies exploring the interplay of neural, behavioral, and psychosocial variables will be essential. There is also growing recognition that lived experience matters: subjective accounts of recovery, identity change, and adaptation are critical data points, not just anecdotal ephemera (Doidge, 2008). In addition, ethical engagement with the risks and possibilities of neurotechnology must remain at the forefront (Ienca & Andorno, 2017). Researchers have a responsibility to co-develop technologies with users, ensure explainability in AI systems, and prioritize equity in access and outcomes.

Policymakers face the task of embedding neuroplastic insights into healthcare infrastructure. This includes investing in scalable neurorehabilitation services, supporting interdisciplinary training, and enacting regulations that protect mental privacy. Health systems must shift from acute care models to chronic, adaptive support frameworks that account for the long tail of recovery (Patel et al., 2018).

Reframing neuroplasticity as a long-term, dynamic opportunity reorients the entire ecosystem of brain health. It challenges deterministic models of decline and introduces a more hopeful, albeit complex, vision—one that honors both biology and human variability. As we step into the next decade, the task is not only to innovate, but to integrate: translating neuroplasticity from a scientific concept into a lived, supported reality.

References

Anguera, J. A., Boccanfuso, J., Rintoul, J. L., Al-Hashimi, O., Faraji, F., Janowich, J., Kong, E., Larraburo, Y., Rolle, C., Johnston, E., & Gazzaley, A. (2013). Video game training enhances cognitive control in older adults. *Nature*, *501*(7465), 97–101. https://doi.org/10.1038/nature12486

Baker, J. M., Rorden, C., & Fridriksson, J. (2010). Using transcranial Direct-Current stimulation to treat stroke patients with aphasia. *Stroke*, *41*(6), 1229–1236. https://doi.org/10.1161/strokeaha.109.576785

Beauchamp, T. L., & Childress, J. F. (2001). *Principles of Biomedical Ethics*. Oxford University Press.

Biasiucci, A., Leeb, R., Iturrate, I., Perdikis, S., Al-Khodairy, A., Corbet, T., Schnider, A., Schmidlin, T., Zhang, H., Bassolino, M., Viceic, D., Vuadens, P., Guggisberg, A., & Millán, J. D. R. (2018). Brain-actuated functional electrical stimulation elicits lasting arm motor recovery after stroke. *Nature Communications*, *9*(1). https://doi.org/10.1038/s41467-018-04673-z

Bonnelle, V., Leech, R., Kinnunen, K. M., Ham, T. E., Beckmann, C. F., De Boissezon, X., Greenwood, R. J., & Sharp, D. J. (2011). Default mode network connectivity predicts sustained attention deficits after traumatic brain injury. *Journal of Neuroscience*, *31*(38), 13442–13451. https://doi.org/10.1523/jneurosci.1163-11.2011

Buonomano, D. V., & Merzenich, M. M. (1998). CORTICAL PLASTICITY: From synapses to maps. *Annual Review of Neuroscience*, *21*(1), 149–186. https://doi.org/10.1146/annurev.neuro.21.1.149

Chatterjee, A. (2006). The promise and predicament of cosmetic neurology. *Journal of Medical Ethics*, *32*(2), 110–113. https://doi.org/10.1136/jme.2005.013599

Cicerone, K. D., Langenbahn, D. M., Braden, C., Malec, J. F., Kalmar, K., Fraas, M., Felicetti, T., Laatsch, L., Harley, J. P., Bergquist, T., Azulay, J., Cantor, J., & Ashman, T. (2011). Evidence-based cognitive rehabilitation: Updated review of the literature from 2003 through 2008. *Archives of Physical Medicine and Rehabilitation*, *92*(4), 519–530. https://doi.org/10.1016/j.apmr.2010.11.015

Coetzer, R., & Balchin, R. (2014). *Working with Brain Injury*. Psychology Press eBooks. https://doi.org/10.4324/9781315816432

Cotman, C. (2002). Exercise: A behavioral intervention to enhance brain health and plasticity. *Trends in Neurosciences*, *25*(6), 295–301. https://doi.org/10.1016/s0166-2236(02)02143-4

Davis, N. J., & Van Koningsbruggen, M. G. (2013). "Non-invasive" brain stimulation is not non-invasive. *Frontiers in Systems Neuroscience*, *7*. https://doi.org/10.3389/fnsys.2013.00076

Dayan, E., Censor, N., Buch, E. R., Sandrini, M., & Cohen, L. G. (2013). Noninvasive brain stimulation: From physiology to network dynamics and back. *Nature Neuroscience*, *16*(7), 838–844. https://doi.org/10.1038/nn.3422

deCharms, R. C., Maeda, F., Glover, G. H., Ludlow, D., Pauly, J. M., Soneji, D., Gabrieli, J. D. E., & Mackey, S. C. (2005). Control over brain activation and pain learned by using real-time functional MRI. *Proceedings of the National Academy of Sciences*, *102*(51), 18626–18631. https://doi.org/10.1073/pnas.0505210102

Doidge, N. (2008). *The Brain That Changes Itself: Stories of Personal Triumph from the Frontiers of Brain Science*. Penguin UK.

Doshi-Velez, F., & Kim, B. (2017, February 28). Towards a *Rigorous Science* of *Interpretable Machine Learning*. arXiv.org. https://arxiv.org/abs/1702.08608

Draganski, B., Gaser, C., Busch, V., Schuierer, G., Bogdahn, U., & May, A. (2004). Changes in grey matter induced by training. *Nature*, *427*(6972), 311–312. https://doi.org/10.1038/427311a

Drysdale, A. T., Grosenick, L., Downar, J., Dunlop, K., Mansouri, F., Meng, Y., Fetcho, R. N., Zebley, B., Oathes, D. J., Etkin, A., Schatzberg, A. F., Sudheimer, K., Keller, J., Mayberg, H. S., Gunning, F. M., Alexopoulos, G. S., Fox, M. D., Pascual-Leone, A., Voss, H. U., …, & Liston, C. (2016). Resting-state connectivity biomarkers define neurophysiological subtypes of depression. *Nature Medicine*, *23*(1), 28–38. https://doi.org/10.1038/nm.4246

Elsner, B., Kugler, J., Pohl, M., & Mehrholz, J. (2019). Transcranial direct current stimulation (tDCS) for improving aphasia in adults with aphasia after stroke. *Cochrane Library*, *2019*(5). https://doi.org/10.1002/14651858.cd009760.pub4

Enriquez-Geppert, S., Huster, R. J., & Herrmann, C. S. (2017). EEG-neurofeedback as a tool to modulate cognition and behavior: A review tutorial. *Frontiers in Human Neuroscience*, *11*. https://doi.org/10.3389/fnhum.2017.00051

Eriksson, P. S., Perfilieva, E., Björk-Eriksson, T., Alborn, A., Nordborg, C., Peterson, D. A., & Gage, F. H. (1998). Neurogenesis in the adult human hippocampus. *Nature Medicine*, *4*(11), 1313–1317. https://doi.org/10.1038/3305

Fregni, F., Gimenes, R., Valle, A. C., Ferreira, M. J. L., Rocha, R. R., Natalle, L., Bravo, R., Rigonatti, S. P., Freedman, S. D., Nitsche, M. A., Pascual-Leone, A., & Boggio, P. S. (2006). A randomized, sham-controlled, proof of principle study of transcranial direct current stimulation for the treatment of pain in fibromyalgia. *Arthritis & Rheumatism*, *54*(12), 3988–3998. https://doi.org/10.1002/art.22195

George, M. S. (2010). Transcranial magnetic stimulation for the treatment of depression. *Expert Review of Neurotherapeutics*, *10*(11), 1761–1772. https://doi.org/10.1586/ern.10.95

Goldapple, K., Segal, Z., Garson, C., Lau, M., Bieling, P., Kennedy, S., & Mayberg, H. (2004). Modulation of Cortical-Limbic pathways in major depression. *Archives of General Psychiatry*, *61*(1), 34. https://doi.org/10.1001/archpsyc.61.1.34

Grefkes, C., & Fink, G. R. (2014). Connectivity-based approaches in stroke and recovery of function. *The Lancet Neurology*, *13*(2), 206–216. https://doi.org/10.1016/s1474-4422(13)70264-3

Grefkes, C., & Fink, G. R. (2020). Recovery from stroke: Current concepts and future perspectives. *Neurological Research and Practice*, *2*(1). https://doi.org/10.1186/s42466-020-00060-6

Gruzelier, J. H. (2013). EEG-neurofeedback for optimising performance. I: A review of cognitive and affective outcome in healthy participants. *Neuroscience & Biobehavioral Reviews*, *44*, 124–141. https://doi.org/10.1016/j.neubiorev.2013.09.015

Hallett, M. (2007). Transcranial magnetic stimulation: A primer. *Neuron*, *55*(2), 187–199. https://doi.org/10.1016/j.neuron.2007.06.026

Hardy, J. L., Nelson, R. A., Thomason, M. E., Sternberg, D. A., Katovich, K., Farzin, F., & Scanlon, M. (2015). Enhancing cognitive abilities with comprehensive training: A large, online, randomized, active-controlled trial. *PLoS One*, *10*(9), e0134467. https://doi.org/10.1371/journal.pone.0134467

Hölzel, B. K., Carmody, J., Vangel, M., Congleton, C., Yerramsetti, S. M., Gard, T., & Lazar, S. W. (2010). Mindfulness practice leads to increases in regional brain gray matter density. *Psychiatry Research Neuroimaging*, *191*(1), 36–43. https://doi.org/10.1016/j.pscychresns.2010.08.006

Huang, Y., Edwards, M. J., Rounis, E., Bhatia, K. P., & Rothwell, J. C. (2005). Theta burst stimulation of the human motor cortex. *Neuron*, *45*(2), 201–206. https://doi.org/10.1016/j.neuron.2004.12.033

Hubel, D. H., & Wiesel, T. N. (1963). Receptive fields of cells in striate cortex of very young, visually inexperienced kittens. *Journal of Neurophysiology*, *26*(6), 994–1002. https://doi.org/10.1152/jn.1963.26.6.994

Huttenlocher, P. R. (1979). Synaptic density in human frontal cortex—Developmental changes and effects of aging. *Brain Research*, *163*(2), 195–205. https://doi.org/10.1016/0006-8993(79)90349-4

Ienca, M., & Andorno, R. (2017). Towards new human rights in the age of neuroscience and neurotechnology. *Life Sciences Society and Policy*, *13*(1). https://doi.org/10.1186/s40504-017-0050-1

Illes, J., & Racine, E. (2005). Imaging or imagining? A neuroethics challenge informed by genetics. *The American Journal of Bioethics*, *5*(2), 5–18. https://doi.org/10.1080/15265160590923358

Katz, D. I., White, D. K., Alexander, M. P., & Klein, R. B. (2004). Recovery of ambulation after traumatic brain injury. *Archives of Physical Medicine and Rehabilitation*, *85*(6), 865–869. https://doi.org/10.1016/j.apmr.2003.11.020

Kempermann, G., Kuhn, H. G., & Gage, F. H. (1997). More hippocampal neurons in adult mice living in an enriched environment. *Nature*, *386*(6624), 493–495. https://doi.org/10.1038/386493a0

Kleim, J. A., & Jones, T. A. (2008). Principles of experience-dependent Neural Plasticity: Implications for rehabilitation after brain damage. *Journal of Speech, Language, and Hearing Research*, *51*(1), S225–S239. https://doi.org/10.1044/1092-4388(2008/018)

Klimova, B., & Maresova, P. (2017). Computer-based training programs for older people with mild cognitive impairment and/or dementia. *Frontiers in Human Neuroscience*, *11*. https://doi.org/10.3389/fnhum.2017.00262

Knudsen, E. I. (2004). Sensitive periods in the development of the brain and behavior. *Journal of Cognitive Neuroscience*, *16*(8), 1412–1425. https://doi.org/10.1162/0898929042304796

Kourtis, L. C., Regele, O. B., Wright, J. M., & Jones, G. B. (2019). Digital biomarkers for Alzheimer's disease: The mobile/wearable devices opportunity. *npj Digital Medicine*, *2*(1). https://doi.org/10.1038/s41746-019-0084-2

Lang, C. E., MacDonald, J. R., Reisman, D. S., Boyd, L., Kimberley, T. J., Schindler-Ivens, S. M., Hornby, T. G., Ross, S. A., & Scheets, P. L. (2009). Observation of amounts of movement practice provided during stroke rehabilitation. *Archives of Physical Medicine and Rehabilitation*, *90*(10), 1692–1698. https://doi.org/10.1016/j.apmr.2009.04.005

Laver, K. E., Lange, B., George, S., Deutsch, J. E., Saposnik, G., & Crotty, M. (2017). Virtual reality for stroke rehabilitation. *Cochrane Library*, *2018*(1). https://doi.org/10.1002/14651858.cd008349.pub4

Lefaucheur, J., André-Obadia, N., Antal, A., Ayache, S. S., Baeken, C., Benninger, D. H., Cantello, R. M., Cincotta, M., De Carvalho, M., De Ridder, D., Devanne, H., Di Lazzaro, V., Filipović, S. R., Hummel, F. C., Jääskeläinen, S. K., Kimiskidis, V. K., Koch, G., Langguth, B., Nyffeler, T., ..., & Garcia-Larrea, L. (2014). Evidence-based guidelines on the therapeutic use of repetitive transcranial magnetic stimulation (rTMS). *Clinical Neurophysiology*, *125*(11), 2150–2206. https://doi.org/10.1016/j.clinph.2014.05.021

Liepert, J., Bauder, H., Miltner, W. H. R., Taub, E., & Weiller, C. (2000). Treatment-induced cortical reorganization after stroke in humans. *Stroke*, *31*(6), 1210–1216. https://doi.org/10.1161/01.str.31.6.1210

Lotze, M. (2003). Motor learning elicited by voluntary drive. *Brain*, *126*(4), 866–872. https://doi.org/10.1093/brain/awg079

Lubar, J. F., & Shouse, M. N. (1976). EEG and behavioral changes in a hyperkinetic child concurrent with training of the sensorimotor rhythm (SMR). *Biofeedback and Self-Regulation*, *1*(3), 293–306. https://doi.org/10.1007/bf01001170

Malenka, R. C., & Bear, M. F. (2004). LTP and LTD. *Neuron*, *44*(1), 5–21. https://doi.org/10.1016/j.neuron.2004.09.012

Merzenich, M., Kaas, J., Wall, J., Nelson, R., Sur, M., & Felleman, D. (1983). Topographic reorganization of somatosensory cortical areas 3b and 1 in adult monkeys following restricted deafferentation. *Neuroscience*, *8*(1), 33–55. https://doi.org/10.1016/0306-4522(83)90024-6

Minds behind the Brain. (n.d.). *Stanley Finger* – Oxford University Press. https://global.oup.com/academic/product/minds-behind-the-brain-9780195181821

Nakase-Richardson, R., Whyte, J., Giacino, J. T., Pavawalla, S., Barnett, S. D., Yablon, S. A., Sherer, M., Kalmar, K., Hammond, F. M., Greenwald, B., Horn, L. J., Seel, R., McCarthy, M., Tran, J., & Walker, W. C. (2011). Longitudinal outcome of patients with disordered consciousness in the NIDRR TBI model systems programs. *Journal of Neurotrauma*, *29*(1), 59–65. https://doi.org/10.1089/neu.2011.1829

Nithianantharajah, J., & Hannan, A. J. (2006). Enriched environments, experience-dependent plasticity and disorders of the nervous system. *Nature Reviews Neuroscience*, *7*(9), 697–709. https://doi.org/10.1038/nrn1970

Nitsche, M. A., & Paulus, W. (2000). Excitability changes induced in the human motor cortex by weak transcranial direct current stimulation. *The Journal of Physiology*, *527*(3), 633–639. https://doi.org/10.1111/j.1469-7793.2000.t01-1-00633.x

Nudo, R., Milliken, G., Jenkins, W., & Merzenich, M. (1996). Use-dependent alterations of movement representations in primary motor cortex of adult squirrel monkeys. *Journal of Neuroscience*, *16*(2), 785–807. https://doi.org/10.1523/jneurosci.16-02-00785.1996

O'Laughlin, K. D., Cheng, B. H., Volponi, J. J., Lorentz, J. D. A., Obregon, S. A., Younger, J. W., Gazzaley, A., Uncapher, M. R., & Anguera, J. A. (2025). Validation of an adaptive assessment of executive functions (ACE-X): Longitudinal and cross-sectional analysis of cognitive task performance (PREPRINt). *Journal of Medical Internet Research*, *27*, e60041. https://doi.org/10.2196/60041

Pagani, M., Di Lorenzo, G., Verardo, A. R., Nicolais, G., Monaco, L., Lauretti, G., Russo, R., Niolu, C., Ammaniti, M., Fernandez, I., & Siracusano, A. (2012). Neurobiological correlates of EMDR monitoring – An EEG study. *PLoS One*, *7*(9), e45753. https://doi.org/10.1371/journal.pone.0045753

Palacios, E. M., Owen, J. P., Yuh, E. L., Wang, M. B., Vassar, M. J., Ferguson, A. R., Diaz-Arrastia, R., Giacino, J. T., Okonkwo, D. O., Robertson, C. S., Stein, M. B., Temkin, N., Jain, S., McCrea, M., MacDonald, C. L., Levin, H. S., Manley, G. T., & Mukherjee, P. (2020). The evolution of white matter microstructural changes after mild traumatic brain injury: A longitudinal DTI and NODDI study. *Science Advances*, *6*(32). https://doi.org/10.1126/sciadv.aaz6892

Patel, V., Saxena, S., Lund, C., Thornicroft, G., Baingana, F., Bolton, P., Chisholm, D., Collins, P. Y., Cooper, J. L., Eaton, J., Herrman, H., Herzallah, M. M., Huang, Y., Jordans, M. J. D., Kleinman, A., Medina-Mora, M. E., Morgan, E., Niaz, U., Omigbodun, O., ..., & UnÜtzer, J. (2018). The Lancet Commission on global mental health and sustainable development. *The Lancet*, *392*(10157), 1553–1598. https://doi.org/10.1016/s0140-6736(18)31612-x

Pichiorri, F., Morone, G., Petti, M., Toppi, J., Pisotta, I., Molinari, M., Paolucci, S., Inghilleri, M., Astolfi, L., Cincotti, F., & Mattia, D. (2015). Brain–computer interface boosts motor imagery practice during stroke recovery. *Annals of Neurology*, *77*(5), 851–865. https://doi.org/10.1002/ana.24390

Pitman, R. K., Rasmusson, A. M., Koenen, K. C., Shin, L. M., Orr, S. P., Gilbertson, M. W., Milad, M. R., & Liberzon, I. (2012). Biological studies of post-traumatic stress disorder. *Nature Reviews Neuroscience*, *13*(11), 769–787. https://doi.org/10.1038/nrn3339

Polich, J. (2007). Updating P300: An integrative theory of P3a and P3b. *Clinical Neurophysiology*, *118*(10), 2128–2148. https://doi.org/10.1016/j.clinph.2007.04.019

Pustina, D., Avants, B., Faseyitan, O. K., Medaglia, J. D., & Coslett, H. B. (2017). Improved accuracy of lesion to symptom mapping with multivariate sparse canonical correlations. *Neuropsychologia*, *115*, 154–166. https://doi.org/10.1016/j.neuropsychologia.2017.08.027

Ramos-Murguialday, A., Broetz, D., Rea, M., Läer, L., Yilmaz, Ö., Brasil, F. L., Liberati, G., Curado, M. R., Garcia-Cossio, E., Vyziotis, A., Cho, W., Agostini, M., Soares, E., Soekadar, S., Caria, A., Cohen, L. G., & Birbaumer, N. (2013). Brain–machine interface in chronic stroke rehabilitation: A controlled study. *Annals of Neurology*, *74*(1), 100–108. https://doi.org/10.1002/ana.23879

Riva, G., Mancuso, V., Cavedoni, S., & Stramba-Badiale, C. (2020). Virtual reality in neurorehabilitation: A review of its effects on multiple cognitive domains. *Expert Review of Medical Devices*, *17*(10), 1035–1061. https://doi.org/10.1080/17434440.2020.1825939

Roy, Y., Banville, H., Albuquerque, I., Gramfort, A., Falk, T. H., & Faubert, J. (2019). Deep learning-based electroencephalography analysis: A systematic review. *Journal of Neural Engineering*, *16*(5), 051001. https://doi.org/10.1088/1741-2552/ab260c

Sadato, N., Pascual-Leone, A., Grafman, J., Ibañez, V., Deiber, M., Dold, G., & Hallett, M. (1996). Activation of the primary visual cortex by Braille reading in blind subjects. *Nature*, *380*(6574), 526–528. https://doi.org/10.1038/380526a0

Saur, D. (2006). Dynamics of language reorganization after stroke. *Brain*, *129*(6), 1371–1384. https://doi.org/10.1093/brain/awl090

Slotema, C. W., Blom, J. D., Hoek, H. W., & Sommer, I. E. C. (2010). Should we expand the toolbox of psychiatric treatment methods to include repetitive transcranial magnetic stimulation (RTMS)? *The Journal of Clinical Psychiatry*, *71*(07), 873–884. https://doi.org/10.4088/jcp.08m04872gre

Stinear, C. M., Byblow, W. D., Ackerley, S. J., Barber, P. A., & Smith, M. (2017). Predicting recovery potential for individual stroke patients increases rehabilitation efficiency. *Stroke*, *48*(4), 1011–1019. https://doi.org/10.1161/strokeaha.116.015790

Taub, E., Uswatte, G., Mark, V. W., & Morris, D. M. M. (2006). The learned nonuse phenomenon: Implications for rehabilitation. *PubMed*, *42*(3), 241–256. https://pubmed.ncbi.nlm.nih.gov/17039223

Taub, E., Uswatte, G., & Pidikiti, R. (1999). Constraint-Induced Movement Therapy: A new family of techniques with broad application to physical rehabilitation – A clinical review. *PubMed*, *36*(3), 237–251. https://pubmed.ncbi.nlm.nih.gov/10659807

Thibault, R. T., Lifshitz, M., Birbaumer, N., & Raz, A. (2015). Neurofeedback, self-regulation, and brain imaging: Clinical science and FAD in the service of mental disorders. *Psychotherapy and Psychosomatics*, *84*(4), 193–207. https://doi.org/10.1159/000371714

Ward, N. S., Brander, F., & Kelly, K. (2019). Intensive upper limb neurorehabilitation in chronic stroke: Outcomes from the Queen Square programme. *Journal of Neurology Neurosurgery & Psychiatry*, *90*(5), 498–506. https://doi.org/10.1136/jnnp-2018-319954

Widge, A. S., Malone, D. A., & Dougherty, D. D. (2018). Closing the loop on deep brain stimulation for treatment-resistant depression. *Frontiers in Neuroscience*, *12*. https://doi.org/10.3389/fnins.2018.00175

Wilson, B. A., & Betteridge, S. (2019). *Essentials of Neuropsychological Rehabilitation*. Guilford Publications.

Wolpaw, J., & Wolpaw, E. W. (2012). *Brain-Computer interfaces: Principles and Practice*. Oxford University Press.

Yuste, R., Goering, S., Arcas, B. A. Y., Bi, G., Carmena, J. M., Carter, A., Fins, J. J., Friesen, P., Gallant, J., Huggins, J. E., Illes, J., Kellmeyer, P., Klein, E., Marblestone, A., Mitchell, C., Parens, E., Pham, M., Rubel, A., Sadato, N., …, & Wolpaw, J. (2017). Four ethical priorities for neurotechnologies and AI. *Nature*, *551*(7679), 159–163. https://doi.org/10.1038/551159a

Zotev, V., Phillips, R., Yuan, H., Misaki, M., & Bodurka, J. (2013). Self-regulation of human brain activity using simultaneous real-time fMRI and EEG neurofeedback. *NeuroImage*, *85*, 985–995. https://doi.org/10.1016/j.neuroimage.2013.04.126

Zrenner, C., Belardinelli, P., Müller-Dahlhaus, F., & Ziemann, U. (2016). Closed-loop neuroscience and non-invasive brain stimulation: A tale of two loops. *Frontiers in Cellular Neuroscience*, *10*. https://doi.org/10.3389/fncel.2016.00092

19

NEUROPLASTICITY UNLEASHED

Rewiring the Brain through BCI and HEG Neurofeedback

Mehulkumar B. Dave

Introduction

Far from a monolithic organ, the human brain is dynamic, flexible, and always changing in response to both internal and external stimuli. For most of the 20th century, neuroscience assumed that the brain's neuronal architecture was essentially unchangeable after a certain age, a hard network with fixed paths, much like railroad tracks. But decades of scientific study have disproved this theory and replaced convincing proof of the brain's capacity for lifelong rewiring and reorganizing (Doidge, 2007; Kolb & Gibb, 2011). We call this amazing capacity neuroplasticity.

More than a scientific phenomenon, neuroplasticity reflects the brain's resilience and ability for rejuvenation. It is fundamental to our recovery from strokes and severe brain injuries, our capability to acquire new languages and musical instruments, our capacity to adjust following psychological stress, and even our attempts to break rooted habits and establish new behaviors. Far from being locked into pre-defined neuronal configurations, the brain shows a lifetime capacity for change, growing, pruning, reorganizing, and optimizing itself in response to experience (Draganski et al., 2006).

Neuroplastic alterations are functional as well as morphological. While functional changes might involve the reorganizing of cortical areas following damage or through training, anatomical alterations include neurogenesis and synaptogenesis. Technologies like Hemoencephalography (HEG) neurofeedback (NFB), one of the types of brain–computer interfaces (BCIs), are revealing new ways to enhance and direct plastic changes as neuroscience tools get more sophisticated, so providing hitherto unheard-of paths for therapy, cognitive training, and human augmentation (Cramer et al., 2011; Sitaram et al., 2017).

The scientific underpinnings of neuroplasticity and their ramifications for many disciplines, including rehabilitation, education, aging, creativity, and personal growth, will be discussed in this chapter. Whether your field of expertise is neurobiology, medicine, education, or just curiosity about releasing the latent potential of your own brain, this trip

 DOI: 10.4324/9781003646662-24

into the realm of neuroplasticity promises to challenge conventional wisdom—and maybe change your perspective of human potential itself.

Types of Neuroplasticity

Much of human learning, memory, and damage recovery in the brain depends on neuroplasticity—that amazing ability of the brain to adapt and reorganize itself. It speaks of the dynamic modifications in the structure and operation of the brain in response to both internal and external stimuli. Two generally related groups can be distinguished from broad neuroplasticity: structural plasticity and functional plasticity.

Structural Plasticity

Structural plasticity is the physical modification in the architecture of the brain brought about by development, damage, or learning. These developments comprise synaptic pruning—the deletion of weaker connections to increase network coherence and efficiency, neurogenesis, the birth of new neurons, and synaptogenesis, the synthesis of new synaptic connections (Kolb & Whishaw, 2015). Although its degree changes with age and environmental enrichment, structural flexibility is clearly present throughout the lifespan.

Functional Plasticity

Functional plasticity, on the other hand, is the brain's capacity to move activities from damaged to undamaged areas. Following a catastrophic brain injury or stroke, this type of plasticity is especially important. By the use of alternative neural networks, functional restructuring helps the brain to compensate for lost functions, hence supporting ongoing cognitive and motor tasks (Cramer et al., 2011).

LTP and LTD: Synaptic Mechanisms

Neuroplasticity at the cellular level is under control by synaptic strength variations. Key processes in this system are two well-defined ones: long-term potentiation (LTP) and long-term depression (LTD). Widely documented in the hippocampal area vital for memory encoding, LTP is the strengthening of synapses by repeated activation (Bliss & Collingridge, 1993). LTD, on the other hand, lowers synaptic efficacy, so honing neuronal circuits by eliminating redundant or obsolete connections (Malenka & Bear, 2004). Learning and synaptic homeostasis are built on the interaction between LTP and LTD.

Uses of Neuroplasticity Therapeutically

Recent technical developments have opened the path for using neuroplasticity in therapeutic environments. The creation of brain–computer interfaces (BCIs) is one notable improvement. By means of a direct communication link between the brain and outside devices, BCIs enable real-time conversion of neural activity into useful directives. For patients with motor problems, such as those with amyotrophic lateral sclerosis or spinal cord injuries (Lebedev & Nicolelis, 2006), these devices have great promise. Engaging neuroplastic ideas helps BCIs to not only restore function but also possibly encourage brain restructuring and healing.

Learning about HEG Neurofeedback and BCI Technology

Two new technologies—brain–computer interfaces (BCIs) and HEG NFB—are proven rather helpful in the developing field of applied neuroscience in converting the ideas of neuroplasticity into active treatments and augmentative tactics. Although they work on essentially different ideas and are appropriate for different uses in clinical, educational, and rehabilitative environments, both instruments help interaction with brain function in real time.

Brain–Computer Interface (BCI)

Usually avoiding normal neuromuscular outputs like speech or movement, a brain–computer interface (BCI) creates a direct communication link between the brain and an external device. Usually using electroencephalography (EEG), BCIs decode brain activity and transform it into commands able to operate prostheses, computers, or other electronic systems (Wolpaw & Wolpaw, 2012). BCIs not only let users engage with their surroundings but also drive neuroplastic remodeling by engaging the endogenous electrical patterns of the brain, therefore supporting functional recovery in situations of paralysis, stroke, or neurodegenerative diseases.

Affordable, adaptable, and non-invasive modern BCI devices abound. In clinical trials, they are helping locked-in patients communicate, regain motor ability, and even provide NFB training for disorders such as ADHD and depression (Lebedev & Nicolelis, 2006; Daly & Wolpaw, 2008). Crucially, BCI's potential is enhanced when combined with virtual reality and artificial intelligence (AI) to create immersive environments encouraging brain involvement and learning.

HEG or Hemoencephalography Neurofeedback

HEG is a form of NFB method measuring cerebral blood flow and oxygenation, usually in the prefrontal cortex, while BCIs mainly monitor and decode electrical activity. It is grounded in the same physiological basis as fMRI: the idea that brain activity corresponds with variations in regional blood volume and oxygenation. HEG detects these fluctuations by the use of either passive infrared thermography (pIR HEG) or infrared (near-IR HEG) sensors positioned on the scalp (Toomim et al., 2004).

HEG training, or HEG NFB, raises blood levels and blood oxygenation, so enhancing the metabolic capacity of specific areas of the brain. It mostly occurs in the executive center of the brain the prefrontal cortex, behind the forehead, also known as the seat of civilization (Figure 19.1).

HEG offers one of the best advantages over EEG training in that it does not track electrical signals. Training with EEG in the front of the head places the electrodes/sensors very near to the eye muscles, which generate significant electrical signals when a client blinks or moves his/her eyes (muscles work with bio-electric signals just like the heart and brain). Since the EEG amplifier and/or software cannot distinguish brain signals from eye blink signals, it includes these eye blinks in the EEG we are training. It is known as an "artifact"

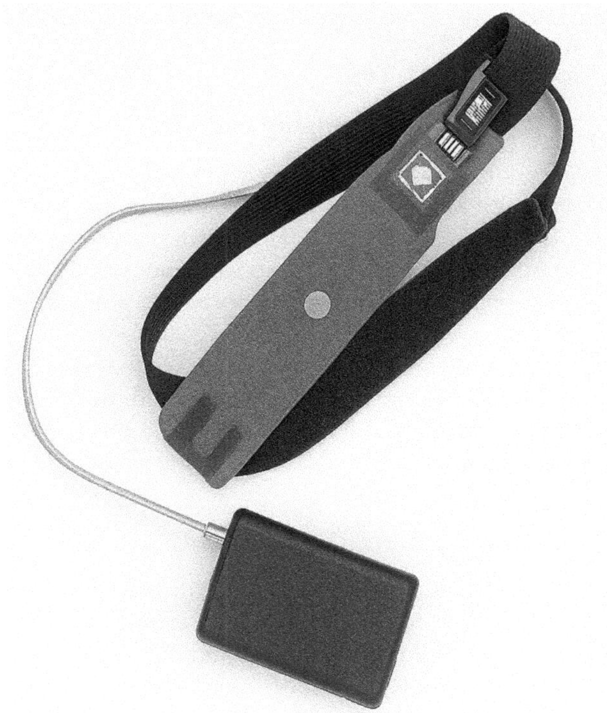

Figure 19.1 HEG belt with peanut.
Source: Photograph by the Author.

since it is not essentially from the brain. Avoiding this eye blink artifact from electrodes placed on the front of the head—and especially from those placed on the forehead/prefrontal area—is challenging.

Since HEG gauges blood perfusion or oxygen levels, it is unaffected by electrical activity; thus, you can blink as much as you want without influencing the signal. HEG can thus be a very useful training method for the many individuals who seek brain training and have a lot of slowing in their prefrontal areas, affecting planning, organizing, attention, impulse control, or other executive functions. Brain cells have more oxygen and glucose available to them, so the more readily their blood supply can be obtained, the speeds their firing when necessary.

There is a lot to learn about locating sites, placing electrodes to get a good signal, all the several training options, performing assessments, etc. EEG has a rather steep learning curve. Starting HEG is quite easy, and making many mistakes in working with it is difficult. For a new trainer, it can thus be a great starting point or a great addition for an experienced EEG trainer.

For training self-regulation of attention and executive skills, HEG NFB is especially helpful; it has also been used to treat diseases like migraine, ADHD, autism, and mood disorders (Carmen, 2004). In therapeutic environments, HEG NFB is frequently simpler to use and offers a more consistent signal for tracking long-term prefrontal activity than conventional EEG-based NFB. Strong connection of HEG with prefrontal activity helps to support treatments meant to improve working memory, emotional control, focus, and overall cognitive functions.

Complementarity and Uses

BCI and HEG NFB technologies show combined promise when included into training systems. HEG NFB can be utilized, for example, to track blood flow variations linked with cognitive effort during BCI-mediated activities, hence generating a deeper feedback loop for performance enhancement. Both devices act as strong modulators of neuroplastic change since they depend on real-time neural feedback to promote positive brain states (Sitaram et al., 2017).

Moreover, the mobility and accessibility of these technologies are extending their application outside of medical settings. From home-based BCI devices helping stroke recovery to classrooms adopting HEG NFB for attention training, the democratization of neurotechnology is materializing.

Functions in Neurorehabilitation in Clinical Settings

Beyond compensatory paradigms of treatment, clinical neurorehabilitation has evolved significantly in recent years to encompass treatments grounded in neuroplasticity, that is, the natural potential of the brain to restructure and heal itself. New treatments have surfaced to not only control symptoms but also actively encourage healing and functional restoration as scientific knowledge of plasticity deepens. Among these, HEG NFB and brain–computer interventions (BCIs) show great promise in enhancing cognitive and motor rehabilitation, hence changing the way doctors interact with injured brain systems.

Therapeutic Basis: Neuroplasticity

Many neuro-rehabilitative treatments are based on the fundamental concept of activity-dependent neuroplasticity, that is, the theory that particular, meaningful tasks can cause brain restructuring. Evidence indicates that stimulating impaired neural networks can help to partially or even totally recover motor and cognitive abilities in stroke, traumatic brain injury (TBI), cerebral palsy, and neurodegenerative illnesses (Kleim & Jones, 2008).

Clinical treatments seek to strengthen adaptive neuronal pathways by means of LTP and LTD, hence undermining dysfunctional ones. Personalized, interesting therapies enhanced with feedback criteria well matched by neurotechnological approaches increase these effects.

Motor Rehabilitation BCIs

Particularly in spinal cord injuries and post-stroke, brain–computer interfaces (BCIs) have become rather popular in motor rehabilitation. BCIs provide a means to restore broken motor paths in these populations by converting motor intent, e.g., imagined hand movement, into real-world movement via robotic devices, exoskeletons, or visual feedback (Ang et al., 2014; Cervera et al., 2018). This strengthens motor pathways in the sensorimotor cortex, therefore fostering cortical remodeling and voluntary control throughout time.

Combining physiotherapy with BCI-based motor training greatly increases upper limb motor performance according to randomized controlled trials (Pichiorri et al., 2015). Furthermore, improving learning efficiency and neuroplastic alterations is a closed-loop system with real-time feedback (Soekadar et al., 2015). BCIs have also been used to control prostheses, wheelchairs, and computer interfaces for those with severe paralysis, restoring some degree of independence and enhancing quality of life.

HEG Neurofeedback for Cognitive Reconstruction

Targeting prefrontal cortex activity, HEG NFB provides a non-invasive, user-friendly approach to treat cognitive and affective dysfunctions. In populations with ADHD, autistic spectrum disorder, post-concussion syndrome, and TBI (Carmen, 2004; Toomim et al., 2004), HEG has especially helped with attention deficiencies, working memory problems, and emotional dysregulation.

Unlike EEG, which gauges electrical activity, HEG provides a consistent surrogate for continuous cognitive work by emphasizing blood flow and oxygenation. HEG NFB is used clinically to teach patients to self-regulate their brain activity using visual or auditory cues. Regular practice can result in long-lasting behavioral improvements and help executive function networks to be strengthened (Thatcher, 2010). Research also implies that in older populations, HEG NFB helps with cognitive augmentation, mood stabilization, and migraine treatment.

Combining with Multimodal Rehabilitation Programs

Increasingly, BCI and HEG NFB technologies are being included in multimodal rehabilitation programs, which can call for pharmaceutical treatment, virtual reality, occupational therapy, and physical therapy. Their inclusion enables tailored exercises in real time and ongoing brain activity monitoring, therefore enabling individualized and adaptive rehabilitation.

A stroke rehabilitation program might combine, for instance, task-specific physiotherapy for functional motor tasks; cognitive–behavioral therapy (CBT) to address psychosocial aspects; BCI-driven robotic arm training to restore motor coordination; HEG-based NFB to improve attention and working memory.

Such integrated models fit extremely nicely with modern ideas of rehabilitation as an interactive, brain–environment conversation in which recovery depends critically on motivation, feedback, and repetition (Dobkin, 2005).

Challenges and Future Pathways

Although the use of BCI and HEG NFB in regular clinical practice presents various difficulties, even if the results seem great. These include variability in patient responsiveness, high costs and complexity of various BCI systems; limited clinical training in neurotechnology; and the need for defined procedures and outcome measurements.

By means of wearable, reasonably priced, and AI-enhanced devices that can adjust to user performance and provide precise therapy (Ruffini et al., 2019), future studies seek to fill in these voids. Furthermore, growing interest in home-based NFB systems would enable patients to keep their recovery outside of medical environments, therefore fostering long-term neuroplastic changes.

Particularly with an eye toward the efficacy of neuroplasticity-driven interventions, including brain–computer interventions (BCIs) and HEG NFB, this section offers case studies and empirical research to validate the theoretical foundations and technological advancements previously discussed. These narratives highlight outcomes in therapeutic spheres, including traumatic brain injury (TBI), autism spectrum disorder (ASD), stroke rehabilitation, and ADHD.

Ang et al. (2014) Robotic Hand Operated by brain–computer interface in Post-Stroke Motor Rehabilitation Case Study 27 chronic stroke patients undergoing rehabilitation using a BCI-driven robotic hand device made up Ang and colleagues' (2014) randomized controlled experiment. Members of the BCI group used motor imagery to control hand movements using EEG inputs.

Case Studies

Case Study-1: HEG Neurofeedback in a 78-Year-Old Stroke Survivor with Deep White Matter and Basal Ganglia Lesions

Mr. T, a 78-year-old retired engineer, was admitted to the hospital following an acute ischemic stroke. Neuroimaging revealed multiple infarcts in the deep white matter and basal ganglia, accompanied by generalized cerebral atrophy. The neurological damage resulted in left-sided hemiplegia, characterized by a complete loss of voluntary motor movement, marked spasticity in the left hand and leg, and severe functional impairment. At the same time, Mr. T experienced expressive aphasia, loss of speech, and a twisted mouth consistent with facial paresis. These deficits profoundly impacted his independence, communication, and quality of life.

Given the extent of his neurological impairments and limited response to conventional rehabilitation, Mr. T was suggested to start HEG (HEG) NFB intervention. Over a period of six months, he completed 50 sessions of HEG NFB, targeting regions implicated in motor control and cognitive function.

Each session involved the application of a near-infrared (NIR) sensor to the forehead, providing real-time feedback on cerebral blood flow. The goal was to train Mr. T to self-regulate his brain activity, thereby promoting neuroplasticity and functional reorganization in the affected neural circuits. The intervention was integrated with ongoing physiotherapy and speech therapy, but the HEG NFB protocol constituted the primary experimental approach during this period.

Throughout the intervention, Mr. T's progress was monitored using standardized motor and cognitive assessments, as well as clinical observation. The HEG NFB sessions were structured to encourage active engagement and self-regulation, with visual and auditory feedback provided to reinforce optimal brain states.

- *Motor Function:* Over time, Mr. T demonstrated gradual improvement in voluntary movement of his left hand and leg. Spasticity decreased, and he regained the ability to perform simple self-care tasks, such as grasping objects and shifting position in bed.
- *Speech and Facial Function:* While speech remained limited, Mr. T began to articulate short words and phrases with increasing clarity. The asymmetry of his mouth improved, facilitating better oral intake and social interaction.

- *Cognitive Function:* Positive changes were noted in working memory, planning, and overall cognitive performance. Mr. T became more alert, engaged, and capable of following multi-step instructions, reflecting enhanced executive function.
- *Functional Ability:* There was a noticeable improvement in overall functional independence. Mr. T required less assistance for mobility and daily activities, and his mood and motivation for rehabilitation improved.

Follow-up CT scans provided objective evidence supporting the clinical improvements. While the structural damage remained, there were subtle changes in cortical activation patterns, suggesting functional reorganization and increased cerebral blood flow in previously underactive regions. These findings align with the principles of neuroplasticity and the expected effects of HEG NFB (Figures 19.2–19.5):

Comparative City Scan Images

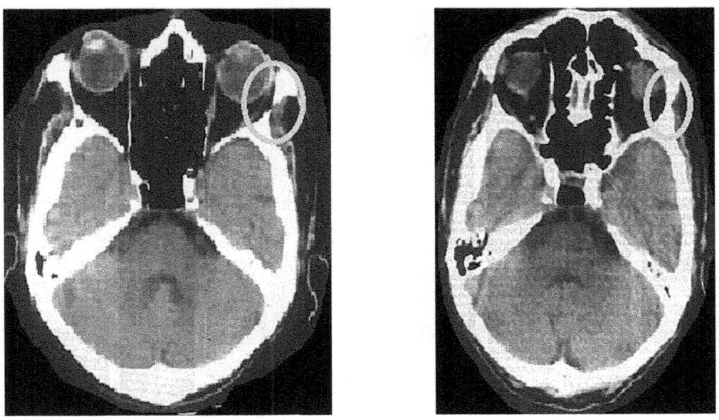

Figure 19.2 Increased white matter density at Sylvian fissure.
Source: Photographs by the Author.

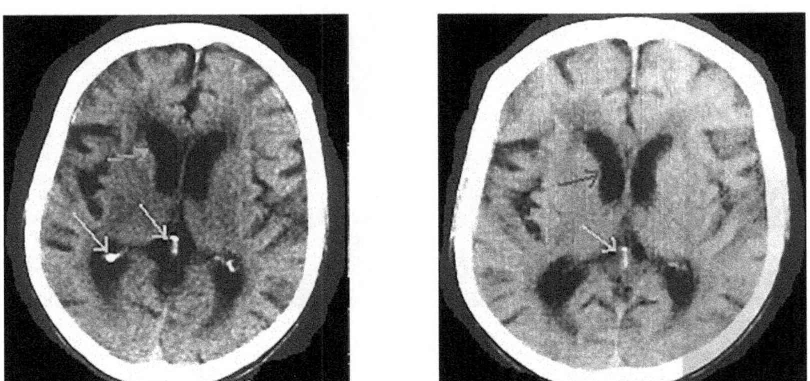

Figure 19.3 Lateral ventricles in proper size and shape, prominence of basal cisterns.
Source: Photographs by the Author.

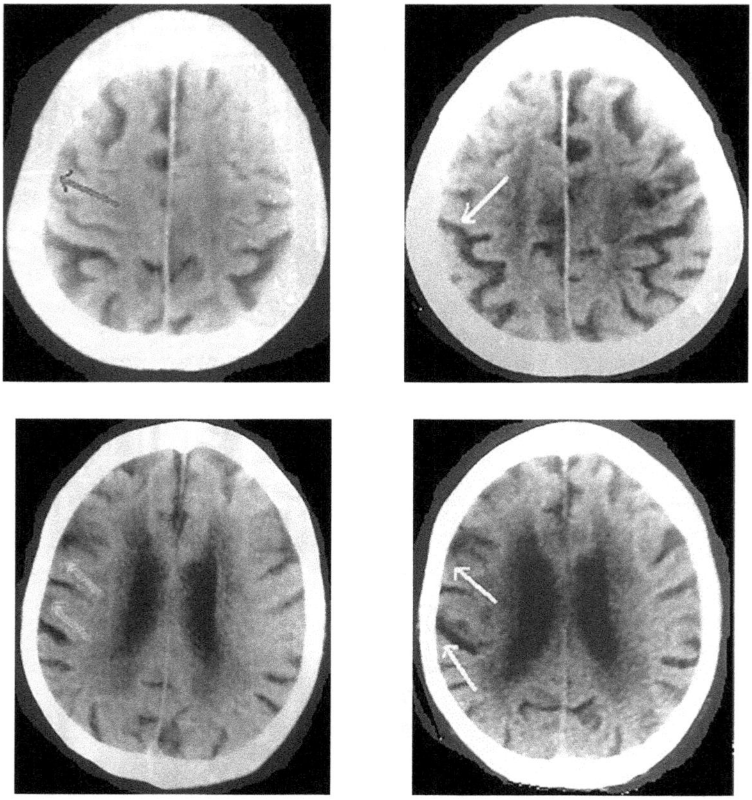

Figure 19.4 Increased white matter densities at sulcal spaces.
Source: Photographs by the Author.

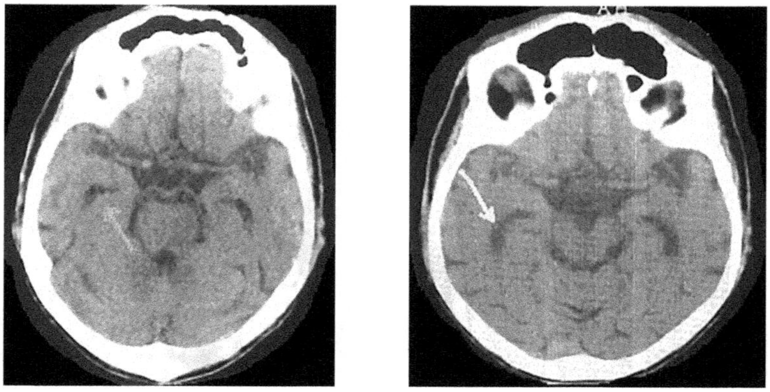

Figure 19.5 Healing of chronic ischemic lesion at the basal ganglia.
Source: Photographs by the Author.

Case Study-2: Efficacy of HEG NFB Training in Autism Spectrum Disorder

This case study explores the journey of a young child with ASD who underwent HEG NFB training, highlighting the clinical process, observed changes, and broader implications for the intervention in autism.

Neel was an 8-year-old boy diagnosed with ASD. Neel's early years were marked by challenges in social interaction, notably poor eye contact, limited emotional expressiveness, and pronounced hyperactivity. His parents described him as a bright but quite restless child, often struggling to engage in conversations or maintain attention during structured activities at school and socially. Neel's difficulties extended to language production, where he would frequently use short, repetitive phrases and show limited spontaneous communication.

Before starting HEG NFB training, Neel underwent comprehensive neuropsychological and behavioral assessments. These evaluations confirmed deficits in executive functioning, attentional control, and social responsiveness. The clinical team, aware of emerging research on HEG NFB and its potential to influence prefrontal brain function, recommended a course of HEG NFB tailored to Neel's specific needs.

Neel's HEG training protocol was based on the methods described by Toomim et al. (2004). Sessions were conducted twice weekly, each lasting approximately 30 minutes. During each session, a sensor was placed on Neel's forehead, monitoring changes in cerebral blood flow and oxygenation in the prefrontal cortex using NIR technology. During training, three areas were targeted—F7, Fp2, and Fpz. Real-time temperature feedback was provided, allowing Neel to learn how to self-regulate his brain activity.

The training environment was designed to be supportive and engaging. Neel was encouraged to watch animated videos sync with instrumental music in the background, with the feedback signal guiding him toward optimal brain states. His parents and a trained therapist were present throughout, offering gentle prompts and positive reinforcement.

Over the course of several weeks, Neel's parents and teachers began to notice subtle but meaningful changes. His eye contact improved, and he started to initiate interactions with peers and adults more frequently. Hyperactivity decreased, and Neel was better able to sit through classroom activities without becoming agitated.

Speech production showed notable gains. Neel began to use longer sentences and express a wider range of emotions. His attention span lengthened, and he demonstrated increased persistence in tasks that had previously frustrated him.

Behavioral checklists and parent/teacher reports documented improvements in social responsiveness, emotional expressiveness, and attentional control. Standardized measures of executive function and language skills reflected these positive trends, with Neel's scores moving closer to age-appropriate norms.

Neel's case, along with similar reports from Toomim et al. (2004) and other researchers, suggests that HEG training can favorably influence prefrontal brain activity in children with ASD. The observed improvements in social responsiveness, language, and attention underscore the potential of NFB as a complementary intervention for ASD.

Challenges, Limitations, and Emerging Trends in Neuroplasticity-Based Interventions

The integration of neuroplasticity-driven approaches such as brain–computer interfaces (BCIs) and HEG NFB into clinical neurorehabilitation represents a transformative advance.

However, despite their promise, these technologies face a range of practical, technical, scientific, and ethical challenges that currently limit their widespread adoption. Recognizing these barriers is essential for guiding future research and development toward more effective, accessible, and personalized neurorehabilitation therapies.

Technical and Methodological Challenges

A primary obstacle in the clinical application of BCI and HEG NFB is the complexity and variability of underlying neurophysiological signals. Non-invasive BCIs, particularly those relying on EEG, are often hampered by low signal-to-noise ratios, susceptibility to artifacts (such as those caused by muscle or eye movements), and considerable variability among users (Lotte et al., 2015). These issues necessitate advanced signal processing and individualized calibration, which can be both technically demanding and time-consuming.

HEG NFB, while less affected by electrical interference, generally offers lower temporal resolution than EEG and remains less extensively studied. Most HEG systems monitor changes in cerebral blood flow as a proxy for neural activity, which may not accurately reflect rapid cognitive or emotional shifts. Additionally, the absence of standardized training protocols and methodologies across studies undermines the ability to compare results and replicate findings (Thibault et al., 2016).

Clinical and User-Specific Variability

Individual differences in brain anatomy, cognitive profile, motivation, and the nature of neurological injury can significantly influence the effectiveness of BCI and HEG interventions. For example, some individuals, especially those with significant cortical damage, may be "BCI-illiterate," unable to generate reliable control signals even with intact cognition (Blankertz et al., 2010). This highlights the need for more adaptable and inclusive interface designs.

Furthermore, neurorehabilitation populations often experience cognitive fatigue, anxiety, and attention deficits, which can reduce engagement and undermine the benefits of NFB or BCI training. Sustaining motivation and participation over the long term remains a considerable challenge in clinical practice.

Training, Cost, and Accessibility

The high costs associated with advanced neurotechnology, such as multi-channel EEG systems, robotic BCI platforms, and commercial HEG devices, restrict their availability in many rehabilitation settings. Effective implementation also requires a skilled workforce with expertise spanning clinical therapy, engineering, and neuroscience. The scarcity of such professionals, particularly in resource-limited regions, further constrains access to these interventions.

Efforts to develop wearable, wireless, and consumer-grade neurotechnology devices offer potential solutions to these barriers. However, these innovations must still meet rigorous standards for clinical accuracy and data integrity (Ruffini et al., 2019).

Regulatory and Ethical Considerations

The application of neurotechnology in vulnerable populations, including individuals with cognitive or communication impairments, raises important ethical questions regarding autonomy, informed consent, and data privacy. Inaccurate or misinterpreted brain signals in BCI systems, for example, could inadvertently disadvantage users, especially those in locked-in states.

As NFB systems increasingly incorporate cloud-based storage and artificial intelligence, safeguarding the security and ownership of neurological data becomes paramount. Robust regulatory frameworks are needed to protect users and ensure the efficacy and safety of these technologies as they move toward commercialization and home use (Ienca & Andorno, 2017).

Innovations and Future Directions

Despite these challenges, emerging trends in machine learning, wearable biosensors, neuroadaptive interfaces, and home-based systems are poised to significantly enhance the utility and personalization of BCI and HEG NFB applications.

- *Hybrid Systems:* Integrating EEG with other modalities, such as functional near-infrared spectroscopy (fNIRS) or electromyography (EMG), can improve accuracy and adaptability (Fazli et al., 2012).
- *Closed-Loop and AI-Driven Feedback:* Real-time AI algorithms enable dynamic adjustment of feedback based on performance or neurophysiological state, optimizing learning and therapeutic outcomes (Soekadar et al., 2015).
- *Gamification and Virtual Reality:* Incorporating game-like elements and virtual reality into NFB systems can boost engagement and neuroplasticity, especially in pediatric and cognitive rehabilitation (Georgiev et al., 2021).
- *Remote and Mobile Platforms:* Cloud-based remote monitoring and mobile neurotechnology solutions offer promising avenues for home-based rehabilitation, potentially reducing costs and maintaining therapeutic continuity (Ruffini et al., 2019).

Looking ahead, the integration of biomarkers from EEG, HEG, and behavioral data is expected to enable precision rehabilitation, allowing clinicians to tailor interventions based on real-time brain activity and longitudinal performance trends.

Final Reflections

Advances in neuroplasticity research and rapid bio signal processing have positioned BCI and HEG NFB as compelling innovations in neurorehabilitation. However, technical limitations, individual variability, financial and access barriers, and ethical complexities continue to constrain their clinical impact. Overcoming these challenges will require interdisciplinary collaboration, regulatory clarity, and user-centered design. As these technologies evolve, they hold the potential to revolutionize the treatment of brain disorders by delivering more personalized, adaptive, and accessible neurotherapies.

References

Ang, K. K., Guan, C., Phua, K. S., Wang, C., Zhou, L., Tang, K. Y., …, & Kuah, C. W. K. (2014). Brain-computer interface-based robotic end effector system for wrist and hand rehabilitation: Results of a three-armed randomized controlled trial for chronic stroke. *Frontiers in Neuroengineering, 7*, 30. https://doi.org/10.3389/fneng.2014.00030

Birbaumer, N., et al. (1999). A spelling device for the paralysed. *Nature, 398*(6725), 297–298. https://doi.org/10.1038/18581

Blankertz, B., Sannelli, C., Halder, S., Hammer, E. M., Kübler, A., Müller, K. R., …, & Schalk, G. (2010). Neurophysiological predictor of SMR-based BCI performance. *NeuroImage, 51*(4), 1303–1309. https://doi.org/10.1016/j.neuroimage.2010.03.022

Bliss, T. V. P., & Collingridge, G. L. (1993). A synaptic model of memory: Long-term potentiation in the hippocampus. *Nature, 361*(6407), 31–39. https://doi.org/10.1038/361031a0

Carmen, J. A. (2004). Passive Infrared Hemoencephalography: Four years and 100 migraines. *Journal of Neurotherapy, 8*(3), 23–51. https://doi.org/10.1300/J184v08n03_03

Cervera, M. A., et al. (2018). Brain-computer interfaces for post-stroke motor rehabilitation: A meta-analysis. *Annals of Clinical and Translational Neurology, 5*(5), 651–663. https://doi.org/10.1002/acn3.544

Cramer, S. C., Sur, M., Dobkin, B. H., O'Brien, C., Sanger, T. D., Trojanowski, J. Q., …, & Vinogradov, S. (2011). Harnessing neuroplasticity for clinical applications. *Brain, 134*(6), 1591–1609. https://doi.org/10.1093/brain/awr039

Daly, J. J., & Wolpaw, J. R. (2008). Brain–computer interfaces in neurological rehabilitation. *The Lancet Neurology, 7*(11), 1032–1043. https://doi.org/10.1016/S1474-4422(08)70223-0

Dobkin, B. H. (2005). Rehabilitation after stroke. *New England Journal of Medicine, 352*(16), 1677–1684. https://doi.org/10.1056/NEJMcp043511

Doidge, N. (2007). *The brain that changes itself: Stories of personal triumph from the frontiers of brain science*. Penguin Books.

Draganski, B., Gaser, C., Busch, V., Schuierer, G., Bogdahn, U., & May, A. (2006). Neuroplasticity: Changes in grey matter induced by training. *Nature, 427*(6972), 311–312. https://doi.org/10.1038/nature02132

Fazli, S., Mehnert, J., Steinbrink, J., Curio, G., Villringer, A., Müller, K. R., & Blankertz, B. (2012). Enhanced performance by a hybrid NIRS–EEG brain computer interface. *NeuroImage, 59*(1), 519–529. https://doi.org/10.1016/j.neuroimage.2011.07.084

Ienca, M., & Andorno, R. (2017). Towards new human rights in the age of neuroscience and neurotechnology. *Life Sciences, Society and Policy, 13*, 5. https://doi.org/10.1186/s40504-017-0050-1

Kleim, J. A., & Jones, T. A. (2008). Principles of experience-dependent neural plasticity: Implications for rehabilitation after brain damage. *Journal of Speech, Language, and Hearing Research, 51*(1), S225–S239. https://doi.org/10.1044/1092-4388(2008/018)

Kolb, B., & Gibb, R. (2011). Brain plasticity and behaviour in the developing brain. *Journal of the Canadian Academy of Child and Adolescent Psychiatry, 20*(4), 265–276.

Kolb, B., & Whishaw, I. Q. (2015). *An introduction to brain and behavior* (4th ed.). Worth Publishers.

Lebedev, M. A., & Nicolelis, M. A. L. (2006). Brain–machine interfaces: Past, present and future. *Trends in Neurosciences, 29*(9), 536–546. https://doi.org/10.1016/j.tins.2006.07.004

Lotte, F., Bougrain, L., & Clerc, M. (2015). Electroencephalography (EEG)-based brain–computer interfaces. In *Handbook of neuroengineering* (pp. 295–320). Springer. https://doi.org/10.1007/978-1-4614-6434-1_45

Malenka, R. C., & Bear, M. F. (2004). LTP and LTD: An embarrassment of riches. *Neuron, 44*(1), 5–21. https://doi.org/10.1016/j.neuron.2004.09.012

Mekyska, J., Mzik, M., Galaz, Z., & Eliasova, I. (2021). Gamification in neurorehabilitation: Toward a personalized and engaging therapy. *Frontiers in Neurology, 12*, 640444. https://doi.org/10.3389/fneur.2021.640444

Pichiorri, F., Morone, G., Petti, M., Toppi, J., Pisotta, I., Molinari, M., …, & Mattia, D. (2015). Brain-computer interface boosts motor imagery practice during stroke recovery. *Annals of Neurology, 77*(5), 851–865. https://doi.org/10.1002/ana.24390

Ruffini, G., Dunne, S., Farres, E., Marco-Pallares, J., Ray, C., & Mendoza, E. (2019). A transcranial current stimulation system for home use: Feasibility study for personalized neuromodulation. *Frontiers in Neuroscience, 13*, 671. https://doi.org/10.3389/fnins.2019.00671
Sitaram, R., Ros, T., Stoeckel, L., Haller, S., Scharnowski, F., Lewis-Peacock, J., …, & Birbaumer, N. (2017). Closed-loop brain training: The science of neurofeedback. *Nature Reviews Neuroscience, 18*(2), 86–100. https://doi.org/10.1038/nrn.2016.164
Soekadar, S. R., Birbaumer, N., Slutzky, M. W., & Cohen, L. G. (2015). Brain–machine interfaces in neurorehabilitation of stroke. *Neurobiology of Disease, 83*, 172–179. https://doi.org/10.1016/j.nbd.2014.11.025
Thatcher, R. W. (2010). Validity and reliability of quantitative electroencephalography. *Journal of Neurotherapy, 14*(2), 122–152. https://doi.org/10.1080/10874201003773500
Thibault, R. T., Lifshitz, M., & Raz, A. (2016). The self-regulating brain and neurofeedback: Experimental science and clinical promise. *Cortex, 74*, 247–261. https://doi.org/10.1016/j.cortex.2015.10.024
Toomim, H., Carmen, J., & Toomim, M. (2004). Hemoencephalography: A new therapy for attention deficit hyperactivity disorder. *Journal of Neurotherapy, 8*(3), 5–21. https://doi.org/10.1300/J184v08n03_02
Wolpaw, J. R., & Wolpaw, E. W. (Eds.). (2012). *Brain-computer interfaces: Principles and practice.* Oxford University Press.

20
ADVANCING THE FRONTIERS OF NEUROCOGNITIVE REHABILITATION
Research and Practice Ahead

Aparna Prashanth and K. Jayasankara Reddy

20.1 Introduction

Over the past several decades, neurocognitive rehabilitation has emerged as a cornerstone of clinical neuroscience, psychology, and allied health disciplines, offering structured pathways for individuals recovering from brain injuries, neurological illnesses, developmental disorders, and cognitive aging. The evolution of this field, from relatively rudimentary behavioral interventions to sophisticated, integrative programs leveraging neuroscience, psychology, and technology, marks a profound shift in how clinicians and researchers understand and facilitate cognitive recovery. As scientific understanding of brain plasticity has expanded, so too has our capacity to develop interventions that not only compensate for deficits but also actively stimulate neuroregeneration and functional reorganization (Nagappan et al., 2020).

Reflecting on the progress achieved, the field has benefited from foundational contributions that established the efficacy of domain-specific retraining (e.g., attention, memory, executive functions), cognitive-behavioral interventions for functional adaptation, and environmental modifications to support daily living. Frameworks such as the International Classification of Functioning, Disability and Health (ICF) have also reframed rehabilitation in terms of participation and contextual factors, shifting the focus beyond symptoms to the broader lived experience of individuals with cognitive impairments. More recently, the integration of neuroimaging, electrophysiological assessments, and computational modeling has deepened our capacity to assess treatment responses and personalize interventions. Despite these advancements, however, the field now faces a critical inflection point, one that demands a renewed vision grounded in innovation, equity, and sustainability.

This forward-looking synthesis is rooted in the recognition that current approaches, while valuable, often fall short of addressing the heterogeneity and complexity of neurocognitive challenges in real-world settings. Traditional rehabilitation programs are frequently

 DOI: 10.4324/9781003646662-25

constrained by limited access, lack of cultural contextualization, and challenges in long-term implementation. Furthermore, rapid shifts in population demographics, particularly the global increase in aging populations, the rising incidence of stroke and neurodegenerative diseases, and the long-term cognitive sequelae of conditions such as traumatic brain injury and post-COVID neurological syndromes, demand urgent advancements in both scientific research and clinical infrastructure (Maas et al., 2022). The question, then, is not only how far we have come, but also how we can advance the field to meet the evolving cognitive health needs of the 21st century.

Central to this endeavor is the growing call for interdisciplinary collaboration. Neurocognitive rehabilitation is no longer the domain of neurologists or psychologists alone. Today's landscape involves neuroscientists, rehabilitation physicians, occupational therapists, computer scientists, bioengineers, and even ethicists working together to co-design interventions that are technologically innovative and clinically meaningful. The rise of hybrid models that combine cognitive training with neuromodulatory techniques, such as neurofeedback training (NFT), transcranial direct current stimulation (tDCS), or virtual reality (VR)-based environments, illustrates the expanding toolbox available to rehabilitation scientists. Importantly, these technologies are not merely adjuncts; they have the potential to redefine the principles of rehabilitation altogether by making therapy more adaptive, engaging, and precise.

The introduction of artificial intelligence (AI) and machine learning into cognitive rehabilitation platforms presents additional transformative possibilities. AI-powered systems can analyze vast streams of behavioral and neural data to predict treatment responsiveness, personalize cognitive tasks, and support real-time adjustments in therapeutic protocols. Similarly, wearable biosensors and mobile health (mHealth) platforms are enabling continuous cognitive monitoring outside clinical settings, thereby bridging the longstanding gap between episodic clinic-based care and the dynamic realities of patients' daily lives. These innovations point toward a future where neurocognitive rehabilitation becomes not only more effective but also more accessible and sustainable.

Yet, technological advancement alone is not sufficient. As the field progresses, it must reckon with critical systemic challenges. These include the uneven distribution of rehabilitation services across socioeconomic and geographic contexts, the limited availability of trained professionals in many regions, and the need for policies that support long-term rehabilitation coverage. Moreover, as emerging technologies become embedded within care systems, ethical considerations around data security, algorithmic bias, and informed consent grow increasingly complex. Thus, advancing the frontiers of neurocognitive rehabilitation is as much about redefining clinical practices as it is about rethinking the structures that govern them.

The urgency of innovation is particularly pressing in the context of an aging global population. According to the World Health Organization, by 2050, the number of people aged 60 and over will double, reaching over 2 billion. With this demographic shift comes a surge in age-related cognitive impairments, including mild cognitive impairment (MCI), Alzheimer's disease, and other dementias. These conditions not only strain families and health systems but also call for scalable, preventive, and rehabilitative strategies that can be deployed across diverse cultural and healthcare contexts. Neurocognitive rehabilitation, therefore, must evolve from being predominantly reactive, focused on mitigating deficits, to becoming proactively embedded in lifelong cognitive health promotion.

Simultaneously, neurodevelopmental and acquired conditions in younger populations, from attention-deficit/hyperactivity disorder (ADHD) and learning disabilities to traumatic brain injuries and perinatal hypoxia, require age-sensitive, developmentally informed interventions that bridge medical, educational, and psychosocial domains. The future of the field thus lies in its ability to cater to the full lifespan, harnessing neuroplasticity at every stage while recognizing the unique recovery trajectories of individuals across age, culture, and context.

In response to these multifaceted challenges and opportunities, this chapter aims to chart a roadmap for the future of neurocognitive rehabilitation. It will explore key areas of innovation, including the integration of NFT and other neuromodulation strategies, the role of digital technologies and AI, and the imperative of cultural adaptation and ethical governance. It will examine how translational research can bridge the gap between laboratory findings and real-world clinical impact. It will also reflect on how emerging practices are reshaping professional training, interdisciplinary teamwork, and patient-centered care models.

In essence, this chapter seeks to push the boundaries of what is currently possible in neurocognitive rehabilitation by asking bold questions: How can we make rehabilitation more precise, inclusive, and accessible? What does it mean to truly personalize therapy in a way that honors the neurobiological, psychological, and social dimensions of cognition? How can we ensure that innovation in this field contributes not only to individual recovery but also to broader health equity?

By reflecting on past achievements and envisioning future possibilities, this chapter positions neurocognitive rehabilitation at the nexus of science, technology, and human-centered care, laying the groundwork for a more integrated, anticipatory, and inclusive approach to cognitive well-being in the years ahead.

20.2 Key Challenges in Current Practice

Despite the significant advancements in neurocognitive rehabilitation over the past few decades, several enduring challenges continue to hinder its consistent and equitable application across populations. These challenges are not merely technical or clinical in nature but are deeply embedded in the ways in which systems are designed, interventions are delivered, and outcomes are measured. As the field strives to advance toward a more responsive and person-centered model of care, it is crucial to critically examine the limitations in current practice. Addressing these limitations is essential for creating scalable, culturally sensitive, and sustainable models of neurocognitive rehabilitation that serve diverse individuals across the lifespan.

One of the most pressing issues is the lack of long-term outcome tracking. Much of the existing research in neurocognitive rehabilitation focuses on short-term gains, often measured within a span of weeks or months following intervention (Kumar et al., 2017). While these studies provide valuable insight into initial efficacy, they rarely account for the maintenance of functional gains, the generalization of skills to real-world environments, or the impact on quality of life over time (Carruthers et al., 2020). This gap limits our understanding of the true effectiveness of interventions and undermines efforts to develop comprehensive, evidence-based protocols. Without robust longitudinal data, it becomes difficult to determine which therapies yield durable benefits, for whom, and under what conditions. This limitation also makes it challenging to advocate for policy changes or insurance coverage that support extended rehabilitation services beyond the acute recovery phase.

A second critical concern involves the inconsistent implementation of evidence-based practices (EBPs) in routine clinical settings. While a wealth of scientific literature supports specific cognitive remediation strategies, executive function training programs, and compensatory techniques, there is a notable gap between research and practice (Bahar-Fuchs et al., 2013). Factors contributing to this gap include insufficient clinician training, limited time and resources, competing treatment priorities, and the lack of accessible implementation guidelines. Moreover, variability in how EBPs are interpreted and delivered across institutions leads to inconsistent patient outcomes. In many regions, especially low-resource or rural settings, clinicians may be unaware of or unable to access current best practices (Gopal et al., 2021). As a result, clients often receive fragmented or outdated interventions, diminishing the potential impact of rehabilitation.

The cultural and contextual mismatch of standardized interventions presents another formidable challenge. Many of the neurocognitive assessment tools and therapeutic approaches in use today have been developed and validated in Western, high-income countries. These interventions, while grounded in empirical rigor, often fail to account for cultural values, linguistic nuances, educational backgrounds, and familial structures prevalent in non-Western contexts (Sue et al., 2008). Applying such tools without appropriate cultural adaptation can result in inaccurate diagnoses, misinterpretation of cognitive strengths or deficits, and reduced engagement in therapy. Furthermore, standardized interventions may overlook local conceptions of health, healing, and disability, leading to interventions that are misaligned with clients' lived realities. This lack of cultural sensitivity not only impacts treatment effectiveness but also risks marginalizing entire populations who may not see their needs reflected in the rehabilitation process.

Finally, the field continues to struggle with fragmentation across service delivery systems. Neurocognitive rehabilitation often requires input from multiple professionals, like clinical psychologists, speech and occupational therapists, neurologists, social workers, and caregivers. However, in practice, these professionals frequently operate in silos, with limited communication or shared treatment planning. This lack of integration results in duplicated efforts, contradictory recommendations, or gaps in care. For example, a patient recovering from a traumatic brain injury may receive cognitive retraining from one clinician, vocational counseling from another, and emotional support from yet another, each using different frameworks and goals (Matney et al., 2022). The absence of coordinated care not only places an undue burden on clients and their families but also impedes the development of unified, outcome-driven rehabilitation models. Additionally, systemic barriers, such as disconnected health records, rigid insurance structures, and institutional hierarchies, further complicate the ability to provide seamless, interdisciplinary care.

These challenges are exacerbated in low- and middle-income countries, where resources are limited, and infrastructure for neurorehabilitation is often underdeveloped. In such contexts, even foundational services such as neuropsychological assessment, case management, and family education may be inaccessible. The lack of a unified policy framework for neurocognitive rehabilitation across healthcare systems globally results in wide disparities in care quality and availability. Consequently, while some individuals may benefit from cutting-edge interventions, others may be left with minimal or no support, reinforcing health inequities and social disparities.

To move forward, it is imperative for the field to adopt a systems-level perspective that recognizes and responds to these interrelated challenges. Longitudinal research must become a standard component of outcome evaluation. Training programs should emphasize real-world implementation of EBPs and foster interdisciplinary collaboration. Cultural adaptation of interventions must be prioritized, not as an afterthought but as a foundational step in treatment planning (Marsiglia & Booth, 2014). Most importantly, integrated care models must be supported through both policy initiatives and institutional practices that promote communication, coordination, and person-centered care across service delivery systems.

In sum, while neurocognitive rehabilitation has achieved substantial progress, its full potential will remain unrealized unless these foundational challenges are addressed. As we look toward the future of the field, acknowledging and confronting these limitations is not an admission of failure, but rather a commitment to growth, equity, and excellence in cognitive care. This chapter builds on this critical reflection to explore emerging solutions, technological innovations, and collaborative models that can bridge these gaps and usher in the next era of neurocognitive rehabilitation.

20.3 Emerging Research Priorities

As neurocognitive rehabilitation continues to evolve, the research agenda guiding its future must be equally dynamic, inclusive, and translational. In an era marked by rapid technological advancement and increasing cognitive health demands, the need to deepen our scientific understanding of recovery trajectories has never been more urgent. Emerging research priorities in the field reflect a collective recognition that the path to cognitive restoration is highly individualized, shaped by an interplay of neural, psychological, environmental, and sociocultural variables (Rakesh et al., 2024). In particular, NFT stands out as a promising avenue for exploring these complex interactions, offering a non-invasive, self-regulatory method that directly targets brain function to facilitate neuroplasticity. Aligning these research priorities with NFT can not only enhance our mechanistic understanding of brain–behavior links but also pave the way for more personalized, precise, and accessible rehabilitation strategies (Gkintoni et al., 2025).

One of the foremost research imperatives is the implementation of longitudinal studies on neurocognitive recovery. Short-term evaluations provide insight into immediate treatment effects, but long-term studies are essential for understanding the sustainability and real-life translation of cognitive gains (Wu et al., 2016). In the context of NFT, longitudinal designs can illuminate whether self-regulation of neural activity leads to enduring improvements in attention, executive function, or emotional regulation. Furthermore, such studies can help determine how NFT interacts with life transitions, comorbid conditions, and environmental stressors across time (Orndorff-Plunkett et al., 2017). By tracking individuals across multiple time points post-intervention, researchers can also examine booster effects, relapse patterns, and the need for maintenance protocols, thus informing best practices for long-term cognitive support (Shallcross et al., 2018).

A second critical area is mechanistic research linking brain changes with behavioral outcomes. While many interventions report functional improvements, the underlying neural mechanisms often remain poorly understood (Dutcher & Creswell, 2018). NFT is uniquely positioned to address this gap, as it allows researchers to monitor and train specific neural oscillations or regions in real time. Investigations using electroencephalography

(EEG), functional magnetic resonance imaging (fMRI), or other neuroimaging tools can help delineate how targeted NFT protocols modulate brain networks associated with attention, working memory, impulse control, or mood regulation. For example, up-training frontal midline theta rhythms or down-training beta overactivity in sensorimotor regions may yield observable changes in cognitive flexibility or emotional regulation. By mapping these neural shifts to behavioral data, researchers can develop more targeted NFT protocols grounded in neurobiological precision, strengthening both theoretical models and clinical applications (Viviani & Vallesi, 2021).

The optimization of intervention dosing, timing, and sensitivity to neuroplasticity windows is another emerging frontier. Neurocognitive rehabilitation, particularly in neurological and psychiatric populations, is not a one-size-fits-all process. The frequency, duration, and timing of interventions can significantly influence outcomes, especially in relation to critical periods of neuroplasticity, windows during which the brain is more malleable and responsive to change. NFT protocols must be calibrated to these sensitive periods, which may differ by age, injury severity, or neurodevelopmental stage. For instance, early post-injury periods or developmental phases such as adolescence may offer enhanced responsiveness to neuromodulation. Research into optimal dosing schedules (e.g., session length, frequency per week) and tapering strategies will be instrumental in maximizing the impact of NFT while minimizing fatigue, dropout, or overtraining (Jiang et al., 2021).

Lastly, there is an urgent need to expand research inclusion across diverse populations. Historically, neurocognitive rehabilitation research, including studies on NFT, has been disproportionately centered in Western, high-income, and often homogenous samples (Whalley et al., 2004). This limits generalizability and contributes to the underrepresentation of ethnically, linguistically, and socioeconomically diverse groups. Inclusion of diverse participants is not merely a matter of equity; it is a scientific necessity. Variables such as cultural beliefs about mental health, baseline familiarity with technology, and differing educational or occupational contexts may influence how individuals engage with NFT and how outcomes manifest (Sun et al., 2022). Moreover, tailoring NFT protocols to linguistic, cultural, and neuropsychological profiles relevant to non-Western contexts can enhance efficacy and adherence. Future research must actively engage underrepresented communities through participatory design, community partnerships, and the co-development of culturally informed intervention materials (Marques et al., 2018).

Taken together, these four research priorities, longitudinal tracking, mechanistic exploration, optimization of timing and dosing, and inclusive study design, form the backbone of a forward-looking, ethically grounded, and evidence-rich future for neurocognitive rehabilitation. When aligned with the capabilities of NFT, these priorities offer a framework for innovation that is not only technologically advanced but also deeply responsive to human variability (Zakaria et al., 2025).

As the field moves ahead, NFT has the potential to serve not just as a tool for intervention, but as a research platform in itself. Its real-time data generation, individual-level customization, and potential for at-home application make it uniquely suited to address the current limitations in research and practice. Prioritizing rigorous, diverse, and theory-driven studies will be essential to unlocking NFT's full potential and integrating it meaningfully into the next generation of neurocognitive rehabilitation strategies (Gkintoni et al., 2025).

20.4 Innovations in Assessment and Intervention

The future of neurocognitive rehabilitation is being redefined by a surge of technological innovations that are transforming both how we assess cognitive functioning and how we deliver interventions. These innovations mark a shift from static, one-size-fits-all models to dynamic, personalized, and ecologically valid approaches. Advances in wearable technology, neurophysiological monitoring, virtual environments, and genomics are expanding the frontiers of what is possible in cognitive assessment and recovery. Importantly, NFT serves as both a central tool and a conceptual bridge across many of these emerging domains, enabling real-time brain monitoring, tailored intervention delivery, and individualized feedback that can accelerate and sustain cognitive change (Hao et al., 2022). This section explores how these innovations are reshaping the neurocognitive rehabilitation landscape and enhancing the precision, accessibility, and impact of treatment protocols.

One of the most transformative developments is the real-time cognitive and neural monitoring enabled by wearable devices and EEG-based tools. Traditional neuropsychological assessments provide valuable insights but are often limited to clinic-based, time-bound snapshots of functioning (Schaefer et al., 2023). In contrast, wearable biosensors and portable EEG systems allow continuous tracking of neural and cognitive activity in naturalistic settings. For example, EEG-based neurofeedback systems can monitor attention-related brainwave patterns (such as theta/beta ratios) in real time, enabling individuals to gain greater control over their cognitive states. This has profound implications for rehabilitation, as therapists can now move beyond subjective self-reports to objective, ongoing data streams that reflect actual cognitive engagement, stress levels, and executive functioning in daily life. Such feedback can inform both moment-to-moment adjustments and long-term treatment planning, significantly enhancing the granularity and relevance of intervention.

Closely aligned with this capability is the emergence of adaptive digital therapeutics and gamified interventions. These platforms use real-time data to adjust the difficulty, duration, and type of cognitive tasks based on the user's current performance and neural activity. Neurofeedback interfaces are increasingly being embedded within gamified environments that promote engagement and motivation while targeting specific neural circuits. For instance, attention training games integrated with EEG neurofeedback can reinforce sustained focus in individuals with ADHD or traumatic brain injury (Tosti et al., 2024). These platforms not only enhance compliance but also support neuroplastic changes by creating rich, repetitive, and rewarding learning contexts. Their adaptability allows for continuous recalibration, ensuring that the intervention remains challenging yet achievable, a key determinant of neurorehabilitation success.

Another frontier in innovation is the use of VR and augmented reality (AR) in ecological training. Unlike traditional lab-based tasks, VR and AR platforms simulate real-life scenarios, such as navigating a busy street, preparing a meal, or participating in a conversation, allowing users to rehearse cognitive and behavioral skills in immersive, controlled environments (Tan et al., 2022). When integrated with EEG or neurofeedback systems, these environments can provide neuroadaptive feedback, adjusting virtual stimuli based on the user's brain activity. For example, a VR module designed for social cognition training could become more complex as frontal lobe engagement increases or slow down in response to signs of cognitive overload (Nijman et al., 2020). This combination of immersion, feedback, and interactivity not only enhances ecological validity but also facilitates the transfer of trained skills to real-world contexts, addressing a longstanding gap in traditional neurocognitive interventions.

Perhaps the most groundbreaking development is the integration of neurogenetics and personalized neurorehabilitation, a paradigm that aligns with the broader movement toward precision medicine. Individual differences in genetic makeup, such as polymorphisms related to dopamine or serotonin regulation, can influence neural plasticity, treatment responsiveness, and recovery trajectories (Stewart & Cramer, 2017). As our understanding of gene–brain–behavior relationships deepens, interventions like NFT can be tailored not only to behavioral symptoms or EEG profiles, but also to individual neurobiological predispositions. For instance, individuals with specific brain-derived neurotrophic factor (BDNF) gene variants may exhibit different responses to neurofeedback targeting memory-related networks, necessitating customized protocols or adjunctive strategies (Zhang et al., 2006). While the field of neurogenetic rehabilitation is still in its infancy, its integration with real-time brain monitoring and adaptive intervention systems holds tremendous potential for crafting personalized, biologically informed rehabilitation pathways.

These innovations also present opportunities for broader health system transformation. Cloud-based platforms can aggregate and analyze large datasets from wearables and NFT devices, enabling population-level insights and predictive modeling. Remote delivery of NFT and cognitive training via tele-neurorehabilitation platforms can address access barriers in underserved or rural communities. Meanwhile, cross-platform interoperability between EEG systems, VR environments, and digital health records can foster integrated care and seamless data sharing among multidisciplinary teams.

However, innovation must be tempered with ethical foresight and critical reflection. The increasing granularity of brain data, the use of AI in decision-making, and the genetic stratification of treatment raise important questions about privacy, consent, and health equity. As neurofeedback becomes more embedded within these sophisticated ecosystems, safeguarding user autonomy and ensuring equitable access to these technologies will be vital.

In conclusion, innovations in assessment and intervention are propelling neurocognitive rehabilitation into a new era, one defined by personalization, real-time feedback, ecological relevance, and neurobiological precision. NFT, with its capacity to translate brain activity into actionable change, is central to this transformation. By integrating NFT with wearable technology, gamified digital therapeutics, immersive environments, and genetic profiling, clinicians and researchers can design interventions that are not only effective but also deeply aligned with the needs, contexts, and capacities of each individual. These converging innovations offer a vision of rehabilitation that is responsive, scalable, and empowering, one that redefines what it means to recover, adapt, and thrive in the face of cognitive challenges.

20.5 Artificial Intelligence and Predictive Models

The integration of AI into neurocognitive rehabilitation represents one of the most transformative developments in the field to date. AI technologies are rapidly redefining how clinicians assess, diagnose, and intervene in cognitive impairments, offering new levels of precision, scalability, and efficiency. As neurocognitive disorders grow in prevalence, particularly among aging populations and individuals recovering from brain injuries or neurological illnesses, AI provides a critical opportunity to augment human expertise with data-driven insights. In tandem with NFT, which offers real-time access to brain-based data, AI-driven models can help bridge the gap between massive, multidimensional datasets and individualized, clinically relevant decisions. This section explores how AI is being utilized in neurocognitive rehabilitation, the promises and limitations of machine learning for outcome prediction, the ethical considerations surrounding AI use, and real-world examples from ongoing trials and pilot programs that illustrate its evolving potential.

One of the primary applications of AI in neurorehabilitation is diagnostic support and treatment planning. AI systems can analyze large-scale data, ranging from neuroimaging and EEG signals to behavioral metrics and patient-reported outcomes, to identify diagnostic patterns that might be missed by traditional assessments. For example, deep learning algorithms can detect subtle EEG biomarkers associated with attention deficits, executive dysfunction, or early-stage dementia, helping clinicians refine diagnostic accuracy and intervene earlier (Chudzik et al., 2024). In the context of NFT, AI can automate the selection of target brainwave frequencies or regions based on individualized baseline data, thus optimizing protocol design. This is especially valuable in complex cases where comorbidities or atypical presentations make standard interventions less effective. Moreover, AI-assisted planning tools can dynamically adjust NFT protocols in response to real-time data, ensuring that sessions remain responsive to the patient's evolving neural and behavioral profile.

Complementing this is the development of machine learning models for outcome prediction. These models use historical and ongoing patient data to forecast likely treatment trajectories, functional outcomes, and relapse risks. In neurofeedback, machine learning can track neural plasticity markers, such as changes in spectral power, coherence, or event-related potentials, to estimate the likelihood of sustained cognitive gains or identify the optimal number of sessions for a particular individual. Predictive models can also support differential treatment planning by identifying which individuals are most likely to benefit from NFT versus alternative or adjunctive therapies. For instance, a supervised learning algorithm might predict that adolescents with ADHD and high frontal theta activity will respond more favorably to NFT targeting attentional regulation than to stimulant medication alone (Cao et al., 2023). These insights can help clinicians allocate resources more efficiently, minimize trial-and-error approaches, and personalize care in ways that were previously impossible.

However, with the increasing integration of AI into clinical workflows comes the imperative for ethical AI use in clinical decision-making. AI systems must be transparent, explainable, and accountable. Clinicians need to understand how and why an algorithm arrived at a specific prediction or recommendation in order to interpret and contextualize the results within the broader therapeutic framework. There are also concerns about bias, particularly if training data do not reflect the diversity of real-world populations. For instance, an AI model trained primarily on data from high-income, urban settings may not generalize to rural or culturally diverse communities, thereby risking inequitable outcomes (Celi et al., 2022). Furthermore, privacy concerns are heightened when sensitive neural and behavioral data are collected through wearable or NFT devices and fed into AI systems. Robust data governance frameworks, informed consent processes, and clinician oversight are essential to ensure that AI enhances rather than undermines ethical, person-centered care.

Despite these challenges, early examples from current trials and pilot programs demonstrate the powerful potential of AI in neurocognitive rehabilitation. In several ongoing studies, AI-enabled NFT systems are being tested for their ability to dynamically adjust feedback thresholds based on user performance, attention levels, and EEG variability (Adolphe et al., 2025). One pilot trial in Europe is exploring an AI-enhanced NFT platform for stroke patients, which tailors cognitive training exercises based on real-time motor cortex activity and recovery metrics (Blaznik & Marusic, 2025). Another study in the U.S. is using AI to predict treatment adherence and engagement in adolescent populations undergoing NFT for emotional dysregulation, allowing clinicians to intervene proactively when disengagement risk is high. Similarly, VR-based rehabilitation programs integrated with AI analytics are being used to adaptively modify cognitive tasks in response to user fatigue, stress indicators, and brain signal fluctuations, pushing the boundary of individualized neurorehabilitation.

In educational and research contexts, AI is also being used to generate personalized learning curves from NFT data, highlighting when a user reaches a plateau, when reinforcement is most effective, or when a protocol should be modified to avoid overtraining. Such insights would be nearly impossible to derive manually and illustrate AI's role not just in scaling services, but also in deepening our understanding of how different brains respond to therapy.

In sum, AI offers unprecedented opportunities to enhance the efficiency, precision, and adaptability of neurocognitive rehabilitation, particularly when integrated with interventions like NFT. By supporting diagnosis, predicting outcomes, personalizing protocols, and identifying risk patterns, AI stands to revolutionize both how care is delivered and how success is measured. Yet, this revolution must be accompanied by a strong ethical framework, ongoing human oversight, and a commitment to inclusivity. When thoughtfully applied, AI can transform neurofeedback from a powerful tool into a fully intelligent system, capable of learning, adapting, and healing alongside the human mind it seeks to serve.

20.6 Scaling and Global Access

As the burden of cognitive impairments continues to rise globally, driven by aging populations, neurological disorders, and developmental conditions, scaling neurocognitive rehabilitation in an equitable, accessible, and sustainable manner has become a pressing public health priority. While many evidence-based interventions, including NFT, have demonstrated significant promise in improving attention, executive function, and emotional regulation, their reach remains largely confined to urban centers and high-income populations. To ensure that the benefits of these innovations are not limited to a privileged few, the field must embrace new models of service delivery, build professional capacity in underserved areas, and advocate for systemic reforms that address the structural barriers to care. This section explores the opportunities and challenges of scaling neurocognitive rehabilitation through tele-rehabilitation, mHealth technologies, training expansion in low-resource settings, and policy reforms, all while emphasizing the potential of NFT as a portable, low-cost, and adaptable tool in the global neurorehabilitation ecosystem.

One of the most transformative strategies for expanding access is the rise of **tele-rehabilitation and hybrid care models**. By leveraging digital platforms, clinicians can deliver cognitive training, psychoeducation, and real-time NFT sessions to individuals in remote or underserved locations. Hybrid models, which combine in-person assessment with remote intervention delivery, offer flexibility and continuity, particularly for populations with mobility challenges or limited access to transportation. Recent developments in wearable EEG devices and cloud-based neurofeedback platforms have made it possible for patients to engage in NFT from home, with therapists monitoring progress and adjusting protocols remotely. Such models have proven especially valuable during crises like the COVID-19 pandemic, where traditional clinic-based rehabilitation was disrupted. As these systems become more robust, tele-neurofeedback may emerge as a mainstream option for cognitive rehabilitation across geographical and economic boundaries.

Parallel to tele-rehabilitation is the growing potential of mHealth solutions and low-cost interventions. Smartphones, tablets, and other digital devices are becoming ubiquitous, even in low- and middle-income countries, creating new pathways for scalable, low-cost care delivery. NFT protocols can now be embedded into mobile applications that use simple EEG headsets to train users in attention regulation, stress management, and emotion control. These mobile tools can be paired with gamified interfaces to promote engagement and sustained use, particularly among children and adolescents. Additionally, app-based platforms can integrate behavioral tracking, progress visualization, and AI-powered feedback, enabling users to self-monitor and self-regulate over time. For low-resource settings, this technological simplification lowers the cost barrier and circumvents infrastructure constraints, allowing even modestly equipped health centers to offer neurocognitive rehabilitation services.

However, technology alone is not enough. Effective scaling depends on training needs and capacity building in low-resource settings. A key barrier to implementation is the lack of trained professionals, neuropsychologists, cognitive therapists, EEG technicians, who can deliver, supervise, and adapt interventions like NFT. Many regions lack formal training programs in neurorehabilitation, and even fewer incorporate emerging modalities such as neurofeedback. Capacity building must therefore include curriculum development, professional training initiatives, remote supervision models, and the creation of regional centers of excellence that can support decentralized learning and mentorship. Community health workers and non-specialist providers can also be trained to deliver basic components of NFT or assist in monitoring sessions under remote supervision, further expanding the reach of services in underserved populations. Building this infrastructure is not only a logistical requirement but also a moral imperative if the field seeks to address the global inequities in cognitive healthcare.

These efforts must be supported by robust policy and funding frameworks that promote equitable care. Despite growing evidence of its efficacy, neurocognitive rehabilitation, including NFT, is rarely integrated into national health systems or insurance schemes. This exclusion perpetuates disparities, particularly for individuals in marginalized communities or those with long-term, non-acute cognitive conditions. Advocating for the inclusion of NFT and related services in public health policies, reimbursement models, and disability support programs is essential for mainstreaming these interventions. At the same time, policies must prioritize ethical safeguards, quality assurance, and standardization across service providers to ensure consistency and safety in scaled delivery. Multilateral collaborations, between governments, NGOs, academic institutions, and technology developers, can catalyze the development of open-access tools, localized intervention manuals, and implementation toolkits to guide large-scale rollout.

In recent years, several pilot programs have demonstrated the feasibility and impact of scalable NFT-based interventions. For instance, portable EEG neurofeedback kits have been deployed in rural India and sub-Saharan Africa as part of pilot studies targeting children with attention difficulties, showing positive trends in engagement and preliminary outcomes (Amaral et al., 2017). Other initiatives have used mobile EEG headsets to train disaster-affected populations in emotional regulation, contributing to post-trauma resilience. These examples underscore that when carefully adapted and thoughtfully implemented, neurofeedback and other neurocognitive interventions can be both scientifically robust and socially inclusive.

In conclusion, scaling neurocognitive rehabilitation globally requires a paradigm shift, from niche, clinic-bound models to inclusive, tech-enabled, and culturally grounded systems of care. NFT, with its real-time feedback, adaptability, and portability, is well-positioned to be a central player in this transformation. However, success will depend on more than technological innovation; it will require investments in human capital, policy reform, and global solidarity. By embracing these principles, the field can move toward a future where access to cognitive rehabilitation is not a privilege but a universal right, anchored in equity, scientific rigor, and a commitment to human dignity.

20.7 Ethical, Legal, and Social Implications (ELSI)

As neurocognitive rehabilitation enters an era defined by AI, wearable technologies, mHealth platforms, and NFT, it becomes increasingly important to critically engage with the ethical, legal, and social implications (ELSI) that accompany these innovations. While the integration of advanced technologies into assessment and intervention holds tremendous promise for personalization, accessibility, and effectiveness, it also introduces new layers of complexity regarding consent, data stewardship, accountability, and clinical integrity. Innovations in neurocognitive care, particularly those involving real-time brain monitoring, AI-driven decision-making, and remote digital platforms, raise fundamental questions about autonomy, privacy, equity, and trust (Lee et al., 2021). This section examines the most pressing ELSI issues in the context of digital and AI-enhanced neurorehabilitation, with a particular focus on NFT, and calls for a proactive, rights-based framework to guide innovation responsibly.

A primary concern is the evolving nature of informed consent in digital and AI-driven interventions. Unlike traditional therapies that are confined to face-to-face sessions, digital NFT platforms and AI-assisted tools often operate continuously, gather passive data, and adapt protocols dynamically. This presents challenges to conventional models of consent, which typically involve a one-time agreement based on a static understanding of treatment. In neurofeedback, for example, participants may not fully grasp how their brain data are being used to train algorithms or optimize session feedback. Consent, therefore, must be reimagined as an ongoing, iterative process that provides users with regular updates, real-time explanations of AI-based changes, and the ability to pause or opt out of specific functions. Special considerations are also needed for vulnerable populations, such as children, individuals with cognitive impairments, or those with limited digital literacy, who may require additional supports to understand and exercise their rights in technologically mediated settings.

Equally critical is the issue of data security and ownership, especially given the highly sensitive nature of neurocognitive data. EEG recordings, behavioral performance metrics, and predictive models derived from NFT or AI tools offer intimate insights into an individual's mental states, vulnerabilities, and cognitive trajectories. When stored or transmitted across platforms, this data is subject to risks including unauthorized access, breaches, commercial exploitation, or unintended algorithmic profiling. Legal frameworks around data ownership remain fragmented across jurisdictions, and many digital neurotechnologies operate in regulatory gray zones (Ochang et al., 2022). It is essential that developers and clinicians implement end-to-end encryption, anonymization protocols, and transparent data governance policies that clearly articulate who owns the data, how it will be used, and who has the right to access or delete it. Moreover, users must be empowered with meaningful control over their data, not just through dense legal documents, but through user-friendly dashboards and consent management tools that foster true data sovereignty.

Another vital consideration is balancing innovation with clinical responsibility. As NFT and AI-powered tools become more autonomous, automatically adjusting protocols, suggesting diagnoses, or tracking engagement patterns, there is a risk that human oversight may be diminished or devalued. Clinicians may begin to rely excessively on algorithmic recommendations without fully understanding their basis, leading to overconfidence in systems that may contain hidden biases or errors. Additionally, the allure of technological sophistication may drive premature adoption of tools that have not undergone rigorous clinical validation. While innovation should be encouraged, it must not come at the expense of safety, transparency, or therapeutic rigor. Every new NFT protocol, AI feature, or mobile application must be subject to the same standards of evidence, peer review, and ethical scrutiny as traditional interventions (Gupta, 2023). Moreover, clinicians must retain ultimate accountability for treatment decisions, using technology to augment, not replace, their professional judgment and therapeutic relationship with the client.

Importantly, ELSI considerations extend beyond the individual to encompass broader social justice and equity concerns. Technologies that are not designed with cultural sensitivity, linguistic accessibility, or inclusivity in mind risk exacerbating existing disparities in access to care. For example, NFT platforms that require high-cost devices or rely on high-bandwidth internet connectivity may be inaccessible to users in low-resource settings, further marginalizing them from cutting-edge interventions (Orndorff-Plunkett et al., 2017). Similarly, AI models trained predominantly on homogeneous datasets may underperform for individuals from underrepresented populations, leading to inaccurate predictions or suboptimal treatment recommendations. To ensure fairness and inclusivity, developers and researchers must prioritize diversity in data collection, engage with communities during the design process, and subject technologies to equity audits that identify and mitigate potential biases (Chen et al., 2023).

Several ongoing initiatives are attempting to address these ELSI issues through multidisciplinary collaboration. Research ethics boards are beginning to require digital-specific informed consent procedures for NFT and AI studies. International working groups are developing ethical guidelines for neurotechnology, including the Organisation for Economic Co-operation and Development (OECD) Recommendation on Responsible Innovation in Neurotechnology (O'Shaughnessy et al., 2023). Meanwhile, advocacy groups are pushing for stronger legal protections around neurodata, including proposals to classify brain data as a special category of health information under privacy legislation. These collective efforts point to a growing recognition that ethical foresight must evolve in tandem with technological innovation.

In summary, the rapid integration of NFT, AI, and digital tools into neurocognitive rehabilitation demands an equally rapid evolution in our ethical, legal, and social frameworks. Informed consent must become dynamic and user-centric; data security and ownership must be protected through transparent and enforceable governance; and the role of clinicians must be safeguarded against unchecked algorithmic authority. At the same time, equity, access, and social justice must remain central considerations in the design and deployment of new technologies. If these challenges are met with intentionality and care, the future of neurocognitive rehabilitation will not only be innovative but also ethical, inclusive, and grounded in the fundamental dignity and rights of every individual it serves.

20.8 Future Clinical Pathways

As the science and technology of neurocognitive rehabilitation continue to advance, so too must the clinical models that deliver these interventions to individuals across the lifespan. The future of neurorehabilitation will be defined not only by the sophistication of tools such as NFT, brain–computer interfaces, and AI-driven platforms, but also by how well these tools are embedded into everyday clinical practice in ways that are personalized, inclusive, and sustainable. To truly realize the potential of these innovations, the field must move beyond fragmented, episodic care toward integrated, multimodal pathways that prioritize prevention, early intervention, and accessibility across diverse settings. This section outlines key directions for future clinical models, including the importance of multimodal, team-based care rooted in personalization, the rise of preventive neurorehabilitation for at-risk populations, and the integration of cognitive rehabilitation into primary and community-based healthcare systems.

The foundation of future clinical pathways lies in multimodal, team-based approaches that emphasize individualized care. No single modality, whether pharmacological, behavioral, or neuromodulatory, can adequately address the complex interplay of neural, psychological, and social factors that contribute to cognitive impairment. Future models will increasingly combine interventions such as neurofeedback, cognitive-behavioral therapy, pharmacological support, occupational and speech therapy, and psychoeducation into cohesive, person-centered treatment plans (Wright et al., 2020). These approaches require close collaboration among neurologists, neuropsychologists, rehabilitation therapists, AI analysts, and social workers, working as an integrated team to develop, monitor, and adapt care in real time. NFT will play a central role in this ecosystem, not just as a tool for attention or executive function training, but as a mechanism for tracking neural responses to a range of interventions and adjusting strategies accordingly (Watanabe et al., 2017). The integration of multimodal protocols grounded in individualized neurocognitive profiles will lead to more nuanced, efficient, and effective rehabilitation outcomes.

A second critical shift is the growing emphasis on preventive neurorehabilitation for at-risk populations. Traditionally, cognitive rehabilitation has been reactive, initiated after injury, illness, or noticeable decline. However, emerging research suggests that early intervention, especially among populations genetically or environmentally predisposed to cognitive decline, can significantly mitigate future impairment. This includes individuals with a family history of neurodegenerative diseases, children with perinatal complications, adolescents with early signs of ADHD or learning difficulties, and adults with chronic stress or lifestyle-related cognitive risk factors (Cecil & Nigg, 2022). NFT, with its capacity for non-invasive, early detection of atypical brainwave patterns and its potential to promote self-regulation, is uniquely suited for preventive applications. For example, neurofeedback protocols targeting stress reduction and attentional control in adolescents may buffer against future mental health challenges, while executive function training in aging adults may delay the onset of MCI (Trambaiolli et al., 2021). Embedding preventive strategies into school settings, wellness programs, and community health initiatives represents a paradigm shift, from managing disease to cultivating lifelong brain health.

To scale these benefits equitably, the field must also embrace the integration of cognitive rehabilitation into primary and community care. Cognitive impairments are often first noticed not in specialty clinics but in everyday settings, during school evaluations, routine doctor visits, or family caregiving observations (McCollum & Karlawish, 2020). Yet, primary care providers and community health workers frequently lack the tools or training to identify cognitive risks or refer patients to appropriate services. The future of neurorehabilitation lies in equipping these first-line providers with screening tools, digital assessment platforms, and training in basic cognitive support techniques (Iosa et al., 2023). NFT devices, particularly those designed for mobile use, can be introduced into these settings to facilitate early detection and short-term interventions. For instance, a community health worker trained in basic NFT protocols could support attention training in rural schools, while a general practitioner could use a digital dashboard to track a patient's cognitive profile over time and make timely referrals (Abdulmalek et al., 2022). This model not only democratizes access but also alleviates the burden on overextended specialty services by creating a tiered, scalable system of care.

To implement these future pathways, several enabling conditions must be met. Clinical training programs need to adapt, equipping professionals with interdisciplinary competencies and technological fluency. Healthcare systems must invest in interoperable data infrastructure, allowing for seamless communication across teams and care settings. Policies must support integration through appropriate reimbursement models, regulatory clarity, and investment in digital health infrastructure (Spreafico et al., 2025). Most importantly, patients and their families must be seen as active partners in the rehabilitation process, with care models designed to be flexible, responsive, and culturally sensitive to their lived realities.

Already, pilot programs in integrated neurorehabilitation are demonstrating promising outcomes. Initiatives in urban hospitals have combined NFT with physical and occupational therapy for stroke survivors in early recovery. School-based interventions have used NFT alongside behavioral coaching to support executive function in children with ADHD. Community-based aging programs are experimenting with preventive NFT for mild memory complaints, with therapists visiting homes or connecting via telehealth (Poon et al., 2005). These early models provide a glimpse into a future where neurocognitive rehabilitation is no longer confined to specialized centers but is woven into the fabric of daily care across the lifespan.

In conclusion, the clinical future of neurocognitive rehabilitation must be inclusive, proactive, and deeply personalized. Multimodal collaboration, preventive care, and community integration are not just aspirational goals; they are necessary evolutions if the field is to meet the growing global need for effective, equitable brain health interventions. NFT, as a flexible and scalable tool, will be instrumental in this transformation, supporting early identification, personalized therapy, and lifelong cognitive resilience (Loriette et al., 2021). By charting these new clinical pathways, neurocognitive rehabilitation stands to move from the margins to the mainstream, supporting not only recovery, but flourishing, in every corner of society.

20.9 Conclusion

As neurocognitive rehabilitation evolves into a dynamic and interdisciplinary field, its future lies at the intersection of neuroscience, technology, ethics, and community care. This chapter has mapped the contours of what lies ahead, highlighting emerging research priorities, clinical innovations, and scalable models that are poised to redefine how we understand, assess, and intervene in cognitive impairment. At the heart of this transformation is a shift from reactive, fragmented approaches toward proactive, person-centered, and technologically integrated systems of care. NFT, in particular, emerges as a central tool, uniquely positioned to serve both as an intervention and a feedback-rich platform that bridges neural data with behavioral change.

One of the most significant developments shaping the future of neurocognitive rehabilitation is the integration of real-time brain monitoring, AI, and adaptive digital therapeutics. These technologies are making it possible to personalize interventions with unprecedented precision, responding to moment-to-moment changes in brain activity and cognitive performance. Innovations such as wearable EEG systems, gamified neurofeedback interfaces, virtual and AR platforms, and AI-driven predictive models are allowing for more immersive, engaging, and effective rehabilitation experiences. Yet, alongside these technical advancements comes the need for rigorous clinical validation, ethical oversight, and inclusivity. Ensuring that these tools serve diverse populations, and not just the digitally privileged, will be essential to the field's equitable growth.

The movement toward prevention and early intervention further underscores the importance of integrating neurocognitive support into primary care, schools, and community settings. NFT, due to its portability and adaptability, is especially well-suited for these environments. Whether used to train attention in classrooms, build stress resilience among at-risk youth, or support healthy aging in community centers, neurofeedback represents a flexible, low-cost, and scalable solution. However, these applications will only be sustainable if they are supported by robust training, interdisciplinary teamwork, and the inclusion of families and caregivers in the rehabilitation process.

Moreover, as NFT and other digital tools become embedded in healthcare systems, ELSI must be actively addressed. The neurocognitive data gathered through these platforms are deeply personal, and its misuse or misinterpretation could have serious consequences. From ensuring meaningful informed consent to safeguarding data privacy and promoting algorithmic fairness, the field must develop frameworks that prioritize the dignity, rights, and autonomy of the individuals it serves. Clinical responsibility cannot be outsourced to machines; rather, technology must remain a tool in the hands of skilled and compassionate human practitioners.

Ultimately, the future of neurocognitive rehabilitation demands both innovation and intention. It will require bold investments in research, infrastructure, and education, alongside a deep commitment to inclusion, cultural sensitivity, and ethical practice. By embracing multimodal, personalized, and preventive models of care, and by centering tools like neurofeedback within those frameworks, the field can move toward a future where cognitive rehabilitation is not only more effective, but also more just, accessible, and human-centered. In doing so, we can ensure that advances in brain science translate into real-world impact, supporting individuals not only in recovering cognitive function but in reclaiming autonomy, purpose, and quality of life.

References

Abdulmalek, S., Nasir, A., Jabbar, W. A., Almuhaya, M. a. M., Bairagi, A. K., Khan, M. A., & Kee, S. (2022). IoT-based healthcare-monitoring system towards improving quality of life: A review. *Healthcare*, *10*(10), 1993. https://doi.org/10.3390/healthcare10101993

Adolphe, M., Pech, M., Sawayama, M., Maurel, D., Delmas, A., Oudeyer, P., & Sauzeon, H. (2025). Exploring the potential of artificial intelligence in individualized cognitive training: A systematic review. *PLoS One*, *20*(6), e0316860. https://doi.org/10.1371/journal.pone.0316860

Amaral, C. P., Simões, M. A., Mouga, S., Andrade, J., & Castelo-Branco, M. (2017). A novel Brain Computer Interface for classification of social joint attention in autism and comparison of 3 experimental setups: A feasibility study. *Journal of Neuroscience Methods*, *290*, 105–115. https://doi.org/10.1016/j.jneumeth.2017.07.029

Bahar-Fuchs, A., Clare, L., & Woods, B. (2013). Cognitive training and cognitive rehabilitation for persons with mild to moderate dementia of the Alzheimer's or vascular type: A review. *Alzheimer's Research & Therapy*, *5*(4), 35. https://doi.org/10.1186/alzrt189

Blaznik, L., & Marusic, U. (2025). Exploring the impact of electroencephalography-based neurofeedback (EEG NFB) on motor deficits in Parkinson's disease: A targeted literature review. *Applied Sciences*, *15*(5), 2496. https://doi.org/10.3390/app15052496

Cao, M., Martin, E., & Li, X. (2023). Machine learning in attention-deficit/hyperactivity disorder: New approaches toward understanding the neural mechanisms. *Translational Psychiatry*, *13*(1). https://doi.org/10.1038/s41398-023-02536-w

Carruthers, S., Pickles, A., Slonims, V., Howlin, P., & Charman, T. (2020). Beyond intervention into daily life: A systematic review of generalisation following social communication interventions for young children with autism. *Autism Research*, *13*(4), 506–522. https://doi.org/10.1002/aur.2264

Cecil, C. A. M., & Nigg, J. T. (2022). Epigenetics and ADHD: Reflections on current knowledge, research priorities and translational potential. *Molecular Diagnosis & Therapy*, *26*(6), 581–606. https://doi.org/10.1007/s40291-022-00609-y

Celi, L. A., Cellini, J., Charpignon, M., Dee, E. C., Dernoncourt, F., Eber, R., Mitchell, W. G., Moukheiber, L., Schirmer, J., Situ, J., Paguio, J., Park, J., Wawira, J. G., & Yao, S. (2022). Sources of bias in artificial intelligence that perpetuate healthcare disparities—A global review. *PLOS Digital Health*, *1*(3), e0000022. https://doi.org/10.1371/journal.pdig.0000022

Chen, Y., Clayton, E. W., Novak, L. L., Anders, S., & Malin, B. (2023). Human-centered design to address biases in artificial intelligence. *Journal of Medical Internet Research*, *25*, e43251. https://doi.org/10.2196/43251

Chudzik, A., Śledzianowski, A., & Przybyszewski, A. W. (2024). Machine learning and digital biomarkers can detect early stages of neurodegenerative diseases. *Sensors*, *24*(5), 1572. https://doi.org/10.3390/s24051572

Dutcher, J. M., & Creswell, J. D. (2018). Behavioral interventions in health neuroscience. *Annals of the New York Academy of Sciences*, *1428*(1), 51–70. https://doi.org/10.1111/nyas.13913

Gkintoni, E., Aroutzidis, A., Antonopoulou, H., & Halkiopoulos, C. (2025). From neural networks to emotional networks: A systematic review of EEG-based emotion recognition in cognitive neuroscience and real-world applications. *Brain Sciences*, *15*(3), 220. https://doi.org/10.3390/brainsci15030220

Gopal, D. P., Chetty, U., O'Donnell, P., Gajria, C., & Blackadder-Weinstein, J. (2021). Implicit bias in healthcare: Clinical practice, research and decision making. *Future Healthcare Journal*, *8*(1), 40–48. https://doi.org/10.7861/fhj.2020-0233

Gupta, M. (2023). Reviewing the relationship between blockchain and NFT with world famous NFT market places. *Deleted Journal*, *1*(1), 1–8. https://doi.org/10.36676/sjmbt.v1i1.01

Hao, Z., He, C., Ziqian, Y., Haotian, L., & Xiaoli, L. (2022). Neurofeedback training for children with ADHD using individual beta rhythm. *Cognitive Neurodynamics*, *16*(6), 1323–1333. https://doi.org/10.1007/s11571-022-09798-y

Iosa, M., Paolucci, S., & Morone, G. (2023). The future of neurorehabilitation: Putting the brain and body together again. *Brain Sciences*, *13*(12), 1617. https://doi.org/10.3390/brainsci13121617

Jiang, H., Stieger, J., Kreitzer, M. J., Engel, S., & He, B. (2021). Frontolimbic alpha activity tracks intentional rest BCI control improvement through mindfulness meditation. *Scientific Reports*, *11*(1). https://doi.org/10.1038/s41598-021-86215-0

Kumar, K. S., Samuelkamaleshkumar, S., Viswanathan, A., & Macaden, A. S. (2017). Cognitive rehabilitation for adults with traumatic brain injury to improve occupational outcomes. *Cochrane Library*, *2017*(6). https://doi.org/10.1002/14651858.cd007935.pub2

Lee, E. E., Torous, J., De Choudhury, M., Depp, C. A., Graham, S. A., Kim, H., Paulus, M. P., Krystal, J. H., & Jeste, D. V. (2021). Artificial intelligence for mental health care: Clinical applications, barriers, facilitators, and artificial wisdom. *Biological Psychiatry Cognitive Neuroscience and Neuroimaging*, *6*(9), 856–864. https://doi.org/10.1016/j.bpsc.2021.02.001

Loriette, C., Ziane, C., & Hamed, S. B. (2021). Neurofeedback for cognitive enhancement and intervention and brain plasticity. *Revue Neurologique*, *177*(9), 1133–1144. https://doi.org/10.1016/j.neurol.2021.08.004

Maas, A. I. R., Menon, D. K., Manley, G. T., Abrams, M., Åkerlund, C., Andelic, N., Aries, M., Bashford, T., Bell, M. J., Bodien, Y. G., Brett, B. L., Büki, A., Chesnut, R. M., Citerio, G., Clark, D., Clasby, B., Cooper, D. J., Czeiter, E., Czosnyka, M., …, & Zemek, R. (2022). Traumatic brain injury: Progress and challenges in prevention, clinical care, and research. *The Lancet Neurology*, *21*(11), 1004–1060. https://doi.org/10.1016/s1474-4422(22)00309-x

Marques, B., McIntosh, J., & Campays, P. (2018). Participatory Design for Under-Represented Communities. In *Advances in Human and Social Aspects of Technology Book Series* (pp. 1–15). https://doi.org/10.4018/978-1-5225-4197-4.ch001

Marsiglia, F. F., & Booth, J. M. (2014). Cultural adaptation of interventions in real practice settings. *Research on Social Work Practice*, *25*(4), 423–432. https://doi.org/10.1177/1049731514535989

Matney, C., Bowman, K., & Berwick, D. (2022, February 1). Rehabilitation and Long-Term Care Needs After Traumatic Brain Injury. *Traumatic Brain Injury*. NCBI Bookshelf. www.ncbi.nlm.nih.gov/books/NBK580075/

McCollum, L., & Karlawish, J. (2020). Cognitive impairment evaluation and management. *Medical Clinics of North America*, *104*(5), 807–825. https://doi.org/10.1016/j.mcna.2020.06.007

Nagappan, P. G., Chen, H., & Wang, D. (2020). Neuroregeneration and plasticity: A review of the physiological mechanisms for achieving functional recovery postinjury. *Military Medical Research*, *7*(1). https://doi.org/10.1186/s40779-020-00259-3

Nijman, S. A., Veling, W., Greaves-Lord, K., Vos, M., Zandee, C. E. R., Rot, M. A. H., Geraets, C. N. W., & Pijnenborg, G. H. M. (2020). Dynamic interactive social cognition training in virtual reality (DiSCoVR) for people with a psychotic disorder: Single-group feasibility and acceptability study. *JMIR Mental Health*, *7*(8), e17808. https://doi.org/10.2196/17808

O'Shaughnessy, M. R., Johnson, W. G., Tournas, L. N., Rozell, C. J., & Rommelfanger, K. S. (2023). Neuroethics guidance documents: Principles, analysis, and implementation strategies. *Journal of Law and the Biosciences*, *10*(2). https://doi.org/10.1093/jlb/lsad025

Ochang, P., Stahl, B. C., & Eke, D. (2022). The ethical and legal landscape of brain data governance. *PLoS One*, *17*(12), e0273473. https://doi.org/10.1371/journal.pone.0273473

Orndorff-Plunkett, F., Singh, F., Aragón, O., & Pineda, J. (2017). Assessing the effectiveness of neurofeedback training in the context of clinical and social neuroscience. *Brain Sciences*, *7*(8), 95. https://doi.org/10.3390/brainsci7080095

Poon, P., Hui, E., Dai, D., Kwok, T., & Woo, J. (2005). Cognitive intervention for community-dwelling older persons with memory problems: Telemedicine versus face-to-face treatment. *International Journal of Geriatric Psychiatry*, *20*(3), 285–286. https://doi.org/10.1002/gps.1282

Rakesh, D., McLaughlin, K. A., Sheridan, M., Humphreys, K. L., & Rosen, M. L. (2024). Environmental contributions to cognitive development: The role of cognitive stimulation. *Developmental Review*, *73*, 101135. https://doi.org/10.1016/j.dr.2024.101135

Schaefer, L. A., Thakur, T., & Meager, M. R. (2023, May 16). *Neuropsychological Assessment.* StatPearls – NCBI Bookshelf. www.ncbi.nlm.nih.gov/books/NBK513310/

Shallcross, A. J., Willroth, E. C., Fisher, A., Dimidjian, S., Gross, J. J., Visvanathan, P. D., & Mauss, I. B. (2018). Relapse/recurrence prevention in major depressive disorder: 26-month follow-up of mindfulness-based cognitive therapy versus an active control. *Behavior Therapy*, *49*(5), 836–849. https://doi.org/10.1016/j.beth.2018.02.001

Spreafico, A. M. C., Tarricone, R., & Stern, A. D. (2025). Comprehensive policies for scaling systemic and equitable integration of digital health technologies. *npj Digital Medicine*, *8*(1). https://doi.org/10.1038/s41746-025-01820-x

Stewart, J. C., & Cramer, S. C. (2017). Genetic variation and neuroplasticity: Role in rehabilitation after stroke. *Journal of Neurologic Physical Therapy*, *41*, S17–S23. https://doi.org/10.1097/npt.0000000000000180

Sue, S., Zane, N., Hall, G. C. N., & Berger, L. K. (2008). The case for cultural competency in psychotherapeutic interventions. *Annual Review of Psychology*, *60*(1), 525–548. https://doi.org/10.1146/annurev.psych.60.110707.163651

Sun, J., Shen, H., Ibn-Ul-Hassan, S., Riaz, A., & Domil, A. E. (2022). The association between digitalization and mental health: The mediating role of wellbeing at work. *Frontiers in Psychiatry, 13*. https://doi.org/10.3389/fpsyt.2022.934357

Tan, B., Shi, J., Yang, S., Loh, H., Ng, D., Choo, C., & Medalia, A. (2022). The use of virtual reality and augmented reality in psychosocial rehabilitation for adults with neurodevelopmental disorders: A systematic review. *Frontiers in Psychiatry, 13*. https://doi.org/10.3389/fpsyt.2022.1055204

Tosti, B., Corrado, S., Mancone, S., Di Libero, T., Rodio, A., Andrade, A., & Diotaiuti, P. (2024). Integrated use of biofeedback and neurofeedback techniques in treating pathological conditions and improving performance: A narrative review. *Frontiers in Neuroscience, 18*. https://doi.org/10.3389/fnins.2024.1358481

Trambaiolli, L. R., Cassani, R., Mehler, D. M. A., & Falk, T. H. (2021). Neurofeedback and the Aging Brain: A Systematic Review of Training Protocols for Dementia and Mild Cognitive Impairment. *Frontiers in Aging Neuroscience, 13*. https://doi.org/10.3389/fnagi.2021.682683

Viviani, G., & Vallesi, A. (2021). EEG-neurofeedback and executive function enhancement in healthy adults: A systematic review. *Psychophysiology, 58*(9). https://doi.org/10.1111/psyp.13874

Watanabe, T., Sasaki, Y., Shibata, K., & Kawato, M. (2017). Advances in fMRI real-time neurofeedback. *Trends in Cognitive Sciences, 21*(12), 997–1010. https://doi.org/10.1016/j.tics.2017.09.010

Whalley, L. J., Deary, I. J., Appleton, C. L., & Starr, J. M. (2004). Cognitive reserve and the neurobiology of cognitive aging. *Ageing Research Reviews, 3*(4), 369–382. https://doi.org/10.1016/j.arr.2004.05.001

Wright, J., Sohlberg, M. M., Watson-Stites, R., & McCart, M. (2020). Identification of key therapy ingredients for SLPs serving on multidisciplinary teams facilitating return to learn for students with prolonged cognitive effects after concussion. *Topics in Language Disorders, 40*(1), 6–35. https://doi.org/10.1097/tld.0000000000000198

Wu, L., Subramanian, N., Abdulrahman, M. D., Liu, C., & Pawar, K. S. (2016). Short-term versus long-term benefits: Balanced sustainability framework and research propositions. *Sustainable Production and Consumption, 11*, 18–30. https://doi.org/10.1016/j.spc.2016.09.003

Zakaria, T. M., Langi, A. Z., Sophian, N. M., & Anshori, I. (2025). Artificial Intelligence (AI) in neurofeedback therapy using electroencephalography (EEG), Heart Rate Variability (HRV), Galvanic Skin Response (GSR): review. *IEEE Access*, 1. https://doi.org/10.1109/access.2025.3582805

Zhang, H., Ozbay, F., Lappalainen, J., Kranzler, H. R., Van Dyck, C. H., Charney, D. S., Price, L. H., Southwick, S., Yang, B., Rasmussen, A., & Gelernter, J. (2006). Brain derived neurotrophic factor (BDNF) gene variants and Alzheimer's disease, affective disorders, posttraumatic stress disorder, schizophrenia, and substance dependence. *American Journal of Medical Genetics Part B Neuropsychiatric Genetics, 141B*(4), 387–393. https://doi.org/10.1002/ajmg.b.30332

INDEX

Note: Page numbers in *italics* indicate figures and in **bold** indicate tables on the corresponding pages.

accountability of AI tools 138–139
acetylcholinesterase inhibitors (ChEIs) 34
action understanding **213**
adaptation 4
Adaptive Cognitive Evaluation (ACE) platform 311
adaptive therapy: brain-computer interfaces (BCIs) in 157; real-time responsiveness to neural states using 313–314; virtual reality (VR) in 294
aerobic and resistance training 35, 60
affective computing (AC) 67, 125
aging-related cognitive impairments *see* geriatric rehabilitation
algorithmic bias 112–113, 137, 160
alpha wave 166
altered gravitational environments: effects of microgravity and, on brain function 178–179; future directions and interdisciplinary potential of HDT and 181; HDT relevance to clinical rehabilitation and 180–181; neuroplastic mechanisms in response to 179–180
Alzheimer's disease 4, 26, 27, 29, 50, 339; apolipoprotein E (APOE) ε4 allele and 53; early detection through AI-enabled wearables 129–130; P300 wave in 97; personalized cognitive stimulation for 311; virtual reality (VR) and 285
anticholinergics 63
anxiety disorders: neurofeedback training for 169; social brain in 215
aphasia and virtual reality (VR) 285
apolipoprotein E (APOE) ε4 allele 53
artificial intelligence (AI) 36, 70–71, 119–120, 140–141, 269, 339; in assessment 120–124, **124**; brain-computer interfaces (BCIs) and (*see* brain-computer interfaces (BCIs)); clinical applications and case studies on 107–113; cognitive training platforms for neuroplasticity and 310–311; as complement *vs.* replacement for psychotherapy 106–107; current role in mental health interventions 105–106; data privacy concerns with 111–112; ethical and human-centric considerations with 109–113; future directions and global perspectives on 114–115, 139–140; hybrid model with traditional therapy 114–115; in monitoring 124–132, *127*; neurocognitive rehabilitation, psychotherapy, and 103–104; in neuroimaging for assessment of cognitive disorders 122–123; potential algorithmic bias and 112–113; predictive models and 345–347; real-time data and personalized interventions using 108; therapeutic nuances difficult for 110–111; in therapy 132–137, **136**; *see also* machine learning (ML); natural language processing (NLP); technology-assisted neurocognitive rehabilitation; virtual reality (VR)
assessments: artificial intelligence (AI) in 120–124, **124**; functional 55; neuropsychological 31–32, *55–56*; in older adults 31–32, *55–56*
assistive communication, brain-computer interfaces (BCIs) for 155–157, *156*

atrophy, brain 28
attention 9; virtual reality (VR) and 284
attention deficit-hyperactivity disorder (ADHD) 167–168, 185–186, 340; Cognitive Remediation Therapy (CRT) and Behavioral Therapy (BT) for (*see* Cognitive Remediation Therapy (CRT) and Behavioral Therapy (BT)); virtual reality (VR) and 284
augmented reality (AR) 36; in AI-enhanced rehabilitation 135–137, **136**
autism spectrum disorders: hemoencephalography (HEG) neurofeedback and 333; neurofeedback training for 169; social brain in 214–215, 222–223
automated speech analysis (ASA) 121–122
autonomic regulation in long COVID 200–201
autonomy and human oversight of AI tools 138

Baddeley Working Memory Model 237
balance exercises 60
baseline tracking 56
BDNF 60
Beers Criteria 35, 65
Behavioral Therapy (BT) *see* Cognitive Remediation Therapy (CRT) and Behavioral Therapy (BT)
behaviorism 7
Behavior Rating Inventory of Executive Function, Second Edition (BRIEF- 2) 188
Ben-Yishay, Yehuda 7, 16, 18–19
benzodiazepines 63
beta wave 166–167
Biofeedback Certification International Alliance (BCIA) 172
biomarkers of plasticity 312–313
biopsychosocial model 15–16, 18
Biopsychosocial Model of Mental Health 245
Biopsychosocial Model of Suicide 245
biosensors 126
Biosound Therapy System 202
bipolar disorder, social brain in 215
borderline personality disorder (BPD) 216; suicidal ideation and attempts and 246
brain: biomarkers of plasticity in 312–313; effects of microgravity and altered gravitational environments on 178–179; hemoencephalography (HEG) neurofeedback and (*see* hemoencephalography (HEG) neurofeedback); mild cognitive impairment and dementias and 28–30, **30**; neuroplasticity of (*see* neuroplasticity); normal aging of 26–27, 50; reorganization of, after injury 305; social (*see* social brain); structural and functional changes linked to aging 27–28, 52–53; virtual reality (VR) impact on 286–287; *see also* neurofeedback (NFB)
brain-computer interfaces (BCIs) 11, 17, 135, 325–326; applications in neurorehabilitation 151–154, *152*; case study on 155–157, *156*; challenges with 158–160, 329–330, 333–335; clinical trials and technological advancements in 154–155; critical elements of 148–150, *150*; detecting neural intent with 150; effectors and feedback in 151; ethical considerations with 159–160; future innovations and global trends in 160–162, **161**; hemoencephalography (HEG) neurofeedback and 326–328, *327*; intersection of AI and 147–148, *149*; motor rehabilitation 328–329; neuroplasticity and 310; in personalized and adaptive therapy 157; translating brain signals into commands through neural decoding 150–151
Brain Gym Therapy: development of 237–238; discussion and conclusion on 240–241; findings on 239–240; research methodology on 238–239
BrainHQ 59, 310
BrainTrack 128–129
brain tumors 5

CAPE model 81, *82*
caregivers: ethics of technology-assisted neurocognitive rehabilitation and 274–275, 276, 277; of older adults 62, 66
case management and care coordination in geriatric rehabilitation (GR) 65–66
central executive network (CEN) 288
chatbots 311
cholinesterase inhibitors 63, 64
clock-drawing test 31, 121
closed-loop systems 313–314
cognitive behavioral therapy (CBT) 62, 104, 105, 107, 303, 329; for suicidal ideation and attempts 246
cognitive frailty 54
cognitive impairments, evaluation of, in older adults 30–32
Cognitive Load Theory 237
cognitive-motor interventions 18
cognitive neuropsychological model 14–15
Cognitive Remediation Therapy (CRT) and Behavioral Therapy (BT) 185–186; data analysis on 190–191, **191**; efficacy and distinctions between 191–192; interventions in 189–190; literature review on 186–187; research method on **188**, 188–190
cognitive training and stimulation 58–59; AI-driven 133–134; long COVID and 201
compensation 4, 10
computer-assisted cognitive training 17
computer-based cognitive programs 59

computer vision 119, 125
Confusion Assessment Method (CAM) 55
constraint-induced movement therapy (CIMT) 303
Constructivist Theory 237
contextual validity 14
continuous monitoring 126–129
contralesional compensation 305
Controlled Oral Word Association (COWA) test 248
COVID-19 pandemic 17; *see also* long COVID
crisis management 246
critical periods, neuroplasticity 304
cultural considerations in tele-rehabilitation 70

dance performance, EEG neurofeedback training in 170
data acquisition 125
data preprocessing 125–126
data privacy 111–112, 137, 159, 316
decision making and suicidality 245
default mode network (DMN) 288
deficit remediation 80
delta wave 166
dementia 4, 24, 25, 28–30, 50, 339; virtual reality (VR) and 283–284, 285
dendritic pruning 304
depressive disorders: P300 abnormalities in 93–94; social brain in 215, 223; suicidal ideation and attempts and 243
developmental disorders and virtual reality (VR) 285
diabetes 53
Diagnostic and Statistical Manual of Mental Disorders, Fifth Edition (DSM-5) 236
dialectic behaviour therapy (DBT) 246
digital biomarkers 129
digital cognitive therapies 67–68
digital phenotyping 125, 128
Digit Span Backward (DSB) 239
Digit Span Forward (DSF) 239
Digit Span Test (DST) 238
Diller, Leonard 7, 16, 18–19
dual-task training 18

EARLY- COGN^3 initiative 68
ecological momentary assessment (EMA) 128
ecological validity 14
electroconvulsive therapy (ECT) and suicidality 246
electroencephalography (EEG) 87, 147–148; applications of neurofeedback training and 167–170; biomarkers 313; defined 166; neurofeedback training using 165–166; non-clinical applications of neurofeedback using 169–170; procedure of neurofeedback training and 167; standard band frequencies in 166–167
EMDR 303
emotion and reward processing **213**
emotion recognition and theory of mind 220
empathy 109; schizophrenia and neuromodulation for deficits in 226–229
environmental modulators of neuroplastic recovery 307–308
epilepsy 5; neurofeedback training for 168
Episodic Buffer 237
equity in neurocognitive rehabilitation 6, 21; AI tools and 138
ethical, legal, and social implications (ELSI) 349–351
ethical issues: in AI-driven neurocognitive rehabilitation 137–139; artificial intelligence (AI) 109–113; in brain-computer interfaces (BCIs) 159–160; geriatric rehabilitation 36–37, 70; long COVID and 203–204; neuroplasticity and 312; in neurotechnology-driven rehabilitation 315–317; in Qualitative Phase Before Intervention study 253; in technology-assisted neurocognitive rehabilitation 269–278, **271**
ethics of technology-assisted neurocognitive rehabilitation 269–270, 273–274; agents of neurocognitive rehabilitation and 272–273; external technologies and 276–277; invasive technologies and 274–275; non-invasive technologies and 275–276; objectives of neurocognitive rehabilitation and 273; ontology of 270–273, **271**; technology of neurocognitive rehabilitation and 272
event-related potentials (ERPs): electroencephalography and 87; P300 wave 88–96, **91–92**; types of 88
evidence-based practice 5, 13, 341–342; geriatric rehabilitation 34–35, 58
executive functions 9–10, 54; exercises for older adults 59–62; suicidality and 245; virtual reality (VR) and 284–285
explainability and clinical trust 138
extracellular space volume fraction 171

FINGER trial 39
flexibility exercises 60
forensic populations 5, 10, 21
frontotemporal dementia (FTD) 29, 50–51
functional brain reorganization 52
functional measures 13
functional MRI 5, 11; connectivity patterns 313; neurofeedback and 309–310
functional plasticity 325
functional rehabilitation 12–13, 57
functional reorganization 11

functional skill development 16
functional substitution 305

Gage, Phineas 6, 7, 19–20
Game of Dice Task (GDT) 245
gamification 311
gamma wave 167
genetic risk factors in older adults 53
geriatric rehabilitation (GR) 24–26, 39–40; case studies on 61–62; epidemiological trends and public health significance of 51; evaluation of cognitive impairments and 30–32; evidence-based intervention strategies in 34–35; family, caregiver, cultural, ethical, and policy considerations in 36–37; future directions in 38–39, 70–71; guidelines from recent meta-analyses of 60; medications used in 34–35, 63–65; multidisciplinary team approach in 65–71; neurocognitive changes in older adults and 26–30, **30**; outcome measures and program evaluation 38, 71; overview of aging-related cognitive impairments addressed by 49–51; pharmacological adjuncts in 63; principles of 32–34, 57–59; psychosocial and behavioral interventions in 62; scope of 48–49; technological and innovative approaches in 36; *see also* older adults
gingko biloba 64
goal-directed interventions 12–13, 57

Halstead-Reitan Neuropsychological Battery 7
hardware, virtual reality (VR) 289
HAROLD pattern 52
head-down tilt (HDT) *see* altered gravitational environments
health disparities 70
heart rate variability (HRV) 126, 128
Hebbian learning theory 10
hemoencephalography (HEG) neurofeedback 324; in autism spectrum disorder 333; BCI technology 326–330, *327*; case studies on 330–335; in stroke survivor with deep white matter and basal ganglia lesions 330–332, *331–332*
hippocampal growth with exercise 60
historical development of neurocognitive rehabilitation 6–9
holistic neuropsychological rehabilitation approach 6, 16, 18–19, 21; geriatric rehabilitation 71–72
Holmes, Gordon 8
home-based rehabilitation *see* remote and home-based neurorehabilitation
HomeCoRe 36
human-centered approach with virtual reality (VR) 290
human emotions in therapy 110
hyperbaric oxygen therapy (HBOT) for long COVID 202–203
hyperlipidemia 53
hypertension 53

immersion, virtual reality (VR) 286–287
individualization 12, 58
inference and feedback loops 126, *127*
information processing model 15, 17–18
informed consent 137, 160
integrative approach 6, 13–14, 20–21; neuro-psycho-technological model 315
interactivity, virtual reality (VR) 287
interdisciplinary collaboration in neurocognitive rehabilitation 5, 8–9, 56; neuroplasticity and 312; virtual reality (VR) and 293–294
International Classification of Functioning, Disability and Health (ICF) 12, 18, 338
Internet of Things 39
Iowa Gambling Task (IGT) 245

language impairments and virtual reality (VR) 285
learning difficulties (LDs) 340; classification of 236; discussion and conclusion on Brain Gym Therapy for 240–241; factors contributing to 236; memory strategies for students with 236–237; research findings on 239–240; research methodology on 238–239
Lewy body dementia (LBD) 29, 50
lifelong plasticity 304
lifestyle-related risk factors in older adults 53
long COVID: autonomic regulation in 200–201; challenges, limitations and ethical considerations 203–204; cognitive dysfunction in 196–197; combination therapies for 201–203; future research directions and conclusions on 204; hyperbaric oxygen therapy (HBOT) for 202–203; mechanisms of action in neurocognitive recovery 199–201; neuro-endocrine modulatory systems and 201; neuroinflammation modulation in 200; neuroplasticity enhancement and 199–200; neurostimulation modalities for 197; nutrition therapy for 203; transcranial direct current stimulation (tDCS) for 198; transcranial magnetic stimulation (TMS) for 197–198; vagus nerve stimulation and transcutaneous auricular stimulation for 198–199
long-term depression (LTD) 325
long-term potentiation (LTP) 325
LTP 305
Lumosity 59, 310
Luria, Alexander 6–7, 19–20

machine learning (ML) 70–71, 119, 339; in early detection and diagnosis 120; model training 126; in neuro-psycho-technological model 315; for personalized therapy plans 133; systematic evidence supporting 120–121; in therapy planning and progress tracking 314
MediaPipe 68
memantine 63, 64
memory 9; interventions for 59–62; learning difficulties (LDs) and deficiencies in 236–237; virtual reality (VR) impact on 283–284; working 237, 238–239, 245
mental health populations 5
metacognitive training 16
microgravity *see* altered gravitational environments
mild cognitive impairment (MCI) 24, 27, 28–30, 50, 339; virtual reality (VR) and 284
Mini-Mental State Examination (MMSE) 30–31, 32, 55
mirror neuron system (MNS) 210–211, 212
mobile applications 128
model training, machine learning 126
moderate cognitive impairment (MCI) 25
monitoring, artificial intelligence (AI): barriers and potential solutions in 132; benefits of 131–132; brain-computer interfaces (BCIs) 155; clinical applications and case examples 129–131; continuous monitoring 126–129; data preprocessing 125–126; digital biomarkers 129; inference and feedback loops 126, *127*; model training 126; neurocognitive monitoring 124–125; remote 67–68, 128–129
Montreal Cognitive Assessment (MoCA) 30–31, 55
mood disorders, neurofeedback training for 169
motor behavior monitoring 125
multidisciplinary approach 13–14; geriatric rehabilitation (GR) 65–71
multimodal interventions 57–58; brain-computer interfaces (BCIs) in 158, 329; virtual reality (VR) 294
multimodal monitoring systems 128
Multiple Errands Test 14
multiple sclerosis (MS) 5
multiplexing 171
music performance, EEG neurofeedback training and 170
music therapy 35

National Institute of Mental Health and Neuro Sciences (NIMHANS) 238
natural language processing (NLP) 71, 106, 111, 119, 286, 311; leveraging speech and facial analysis for assessment 121–122; virtual therapists and 134–135
neurocognitive rehabilitation 19–22, 338–340, 352–353; adaptive interventions in 157, 294; altered gravitational environments and 180–181; applications of brain-computer interfaces (BCIs) and AI in 151–154, *152*; artificial intelligence (AI) in (*see* artificial intelligence (AI)); core principles of 11–14; current trends and future directions in 16–19; definition and purpose of 3–4; early conceptualizations of 3; emerging research priorities in 342–343; ethical, legal, and social implication (ELSI) 349–351; future clinical pathways for 351–352; historical development of 6–9; innovations in assessment and intervention 344–345; invasive 274–275; key challenges in current practice of 340–342; key concepts and terminology in 9–11; populations and conditions served by 4–5; relevance and growth in the field 5–6; remote and home-based 158, 293; scaling and global access to 347–348; technology-assisted (*see* technology-assisted neurocognitive rehabilitation); theoretical framework and models in 14–16; virtual reality (VR) in (*see* virtual reality (VR))
neurodegenerative disorders 4; brain-computer interfaces (BCIs) and 155; P300 in 94–95; virtual reality (VR) and 283–284
neuro-endocrine modulatory systems and long COVID 201
neurofeedback (NFB) 151, 165–166, 313–314; clinical applications of 167–169; critical issues and concluding remarks on 171–172; long COVID and 201–202; non-clinical applications of 169–170; procedure in 167; *see also* hemoencephalography (HEG) neurofeedback
neurogenesis 52, 171, 304
neuroimaging: AI in 122–124, **124**, 129; neuroplasticity and 306
neuroinflammation 52; exercise and 60; long COVID and 200
neuromodulation for empathy deficits 226–229
neurons 52
neurophysiological signals 125
neuroplasticity 11, 52, 60, 170–171, 303, 324–325; biological foundations of 304–305; biomarkers of 312–313; brain-computer interfaces (BCIs) and 158; clinical evidence supporting neuroplastic recovery and 305–308; critical period *versus* lifelong plasticity in 304; ethical and practical considerations in 315–317; future vision for 317–318; learning difficulties (LDs) and 236; long COVID and 199–200; in response to gravity changes 179–180; technology-driven

interventions for 308–312; therapeutic basis of 325–326, 328–329; types of 325–326; virtual reality (VR) and 288–289
neuropsychology, positive 79–80, 81
neuro-psycho-technological model 315
neurostimulation: for long COVID 197; neuromodulation for empathy deficits in schizophrenia 226–229; social brain and 216–217, 221–226, *225*
neurosurgical populations 5
neurotransmitters 52, 171
non-invasive brain stimulation (NIBS) *219*, 219–221, 308–309
non-invasive technology-assisted neurocognitive rehabilitation, ethics of 275–276
non-pharmacological treatments: geriatric rehabilitation 35; for suicidal ideation and attempts 246
non-synaptic neurotransmission 171
nootropics 64
novel biomarkers 70
nutrition therapy for long COVID 203

obsessive-compulsive disorder (OCD): P300 abnormalities in 93; social brain in 215–216
occupational therapy (OT), geriatric rehabilitation 35, 61–62
older adults 24–26; adherence and motivation in 70; caregivers of 62, 66; common risk factors in 53; diagnostic and assessment strategies for 55–56; epidemiological trends and public health significance of 51; evaluation of cognitive impairments in 30–32; importance of early detection and baseline tracking in 56; interplay between cognitive, sensory, and physical decline in 54; memory strategies and executive function exercises for 59–62; neurocognitive changes in 26–30, **30**; normal aging in 26–27; overview of aging-related cognitive impairments in 49–51; pathophysiology and risk factors of cognitive decline in 52–53; polypharmacy in 35, 64; *see also* geriatric rehabilitation
omega-3s 64
outcome measures and program evaluation, geriatric rehabilitation 38, 71
oxidative stress 52

P300 wave 88–90, **91–92**; applications in neurorehabilitation 95–96; in depressive disorders 93–94; limitations and future directions in use of 96–97; in neurodegenerative disorders 94–95; in obsessive-compulsive disorder 93; in schizophrenia and psychotic disorders 90
Parkinson's disease 4, 97
pediatric patients, brain-computer interfaces (BCIs) and 155
PENSIEVE-AI 121
perilesional reorganization 305
personality disorders: social brain in 216, 223–224; suicidality and 246
personalized rehabilitation 6, 39, 70; brain-computer interfaces (BCIs) in 157; neuroplasticity 312; virtual reality (VR) and 290–291
person-centered care 12, 57; geriatric rehabilitation 71–72
person perception 212
pharmacologic treatments: geriatric rehabilitation 34–35, 63–65; for suicidal ideation and attempts 246
Phonological Loop 237
physical decline 54
physical exercise: long COVID and 202; neuroplasticity and 60; for older adults 35, 59–60
Picture Recall Test (PRT) 238
policy and funding prioritization of neurocognitive rehabilitation 5; geriatric rehabilitation 37
polypharmacy 35, 64
positive neuropsychology 79–80, 81
post-traumatic stress disorder, neurofeedback training for 169
predictive modeling 314
predictive neurorehabilitation 312
prevention 4
psychiatric conditions, socio-cognitive deficits in 213–216
psychiatric rehabilitation, brain-computer interfaces (BCIs) in 155
psychological modulators of neuroplastic recovery 307–308
psychosocial support 16
psychosocial therapies 35
psychotherapy 104; AI as complement *vs.* replacement for 106–107; hybrid AI and 114–115; for suicidal ideation and attempts 246; traditional and current landscape of 104–105
psychotic disorders 90

Qualitative Phase Before Intervention study: cognitive-based intervention used in **249**; discussion of 258–261; ethical considerations in 253; qualitative findings of 255–258; qualitative questions used in 252; research approach and paradigm in 246–247; research hypothesis on 247; research questions on 247; results of **253–255**; sampling in 248–250; study objectives on 247; study tools used in 250–251; variables and operational definitions in 247–248

randomized controlled trials (RCTs) 4, 7
real-time adaptive interventions 128
real-time feedback, virtual reality (VR) 287
real-time progress monitoring 314
regeneration and neurogenesis 171
regulation of AI tools 138–139
reminescence therapy 35, 62
remote and home-based neurorehabilitation: brain-computer interfaces (BCIs) in 158; virtual reality (VR) in 293
remote monitoring 67–68, 128–129
response inhibition and suicidality 245
restorative rehabilitation 4, 10, 57
robotics 119, 134
rural and underserved populations 68–69

safety 109
sarcopenia 54
scalability 21
schizophrenia 90, 97; neuromodulation for empathy deficits in 226–229; social brain in 214, 221–222; virtual reality (VR) 284–285; virtual reality (VR) and 284
selective serotonin reuptake inhibitors (SSRIs) 34–35
self-perception 212
sensorimotor rhythm 166
sensory decline 54
siloed model 20–21
SMART (specific, measurable, achievable, relevant, and time-bound) goals 12
social brain 210–211, *211*; emerging applicability in other psychiatric disorders 224–226, *225*; functional domains of 212–213, **213**; neurostimulation and 216–217, 220–221; non-invasive brain stimulation and *219*, 219–221; social processing and 220–221; transcranial direct current stimulation (tDCS) and 218–219; transcranial magnetic stimulation and 217–218, *218*
social cognition 210
Social Learning Theory 237
social perception **213**
social processing and social brain 220–221
socio-cognitive deficits 213–216
socioeconomic status (SES), aging-related cognitive impairments and 24–25
software platforms, virtual reality (VR) 289–290
solution-focused brief therapy (SFBT): application of CAPE model with 81, *82*; applications in different clinical settings 78; assumptions and tenets of 79; challenges, benefits, and future directions in 83–84; defined 78; neuropsychological insights and mechanisms in 82–83; positive neuropsychology and 79–80, 81; reimagining therapeutic relationship in 81; strength-based empowerment in 80
space exploration *see* altered gravitational environments
speech analysis 125
sports performance, EEG neurofeedback training in 170
standardized neuropsychological assessments 13
STOPP/START tools 65
strength-based empowerment 80
stroke 4, 53; brain-computer interfaces (BCIs) and 154; hemoencephalography (HEG) neurofeedback and 330–332, *331–332*; recovery and rehabilitation through AI and wearable sensors 130; virtual reality (VR) and 283–284
structural plasticity 325
substance use disorders (SUD), neurofeedback training for 168–169
suicidal ideation and attempts: decision making, response inhibition, and working memory in 245; definitions in 243–245; depression and 243; executive functions and 245; management of suicide and 245; Qualitative Phase Before Intervention and (*see* Qualitative Phase Before Intervention study); statistics on 243; suicidality and 245
suicide contagion 244
supplements 64
support vector machines (SVMs) 314
synapses 52; exercise and 60; LTP and LTD 325
synaptic plasticity 304

talk therapy 104
tasks of mentalization 212
technology-assisted neurocognitive rehabilitation 5, 6, 17, 20; brain-computer interfaces (BCIs) 154–155; ethics of 269–278, **271**; in geriatric rehabilitation 36, 66–67; neuroplasticity and 308–312; remote monitoring and digital cognitive therapies 67–68; wearables 39, 68, 126, 128–130; *see also* artificial intelligence (AI); virtual reality (VR)
telehealth 8, 34
tele-rehabilitation 17, 21, 36; geriatric rehabilitation 66–67; health disparities and cultural considerations in 70; for rural or underserved populations 68–69
theoretical framework and models in neurocognitive rehabilitation 14–16

theory of mind (ToM) 210, **213**, 214; emotion recognition and 220; neurostimulation and 221–226
theta wave 166
Tower of London (TOL) Test 248
transcranial direct current stimulation (tDCS) 11, 17, 308–309, 339; for long COVID 198; social brain and 218–219
transcranial magnetic stimulation (TMS) 11, 308–309; for long COVID 197–198; social brain and 217–218, *218*
transcutaneous auricular stimulation for long COVID 198–199
traumatic brain injury (TBI) 4; brain-computer interfaces (BCIs) and 154–155; monitoring cognitive and emotional function via AI systems for 131; neurofeedback training for 168; neuroplastic recovery and 305–306; virtual reality (VR) and 283–284
T-Rehab System 67–68

user interface design, virtual reality (VR) 290

vagus nerve stimulation (VNS) for long COVID 198–199
validation 109
vascular dementia (VaD) 29, 50
vascular risk factors in older adults 53
Veterans Administration 8
virtual multiple errands test (VMET) 284
virtual reality (VR) 4, 8, 17, 19, 36, 68, 105, 339; in AI-enhanced rehabilitation 135–137, **136**; attention and 284; challenges and limitations with 291–292; in cognitive assessment and stimulation 122; cognitive engagement with 288; comparative effectiveness and integration across domains 286; development of personalized therapy plans for 290–291; in evolving landscape of neurorehabilitation 281–283, **283**; executive functions and 284–285; future directions and conclusions on 292–294; immersion with 286–287; impact on cognitive functions 283–286; interactivity with 287; key areas for future research in 294–295; language impairments and 285; mechanics in neurocognitive therapy 286–289; memory and 283–284; neuoplasticity and 311; neural plasticity and 288–289; practical implementation of 289–291; real-time feedback with 287; system accessibility and infrastructure readiness with 290; visuospatial abilities and 285
virtual therapists 134–135
Visual-Spatial Sketchpad 237
visuospatial abilities and virtual reality (VR) 285
vitamin E 64

wartime influence on brain injury treatment 7–8, 19–20
wearable technology 39, 68, 126, 128–129; Alzheimer's disease early detection using 129–130; stroke recovery and rehabilitation through 130
white matter 28
Word Recall Test (WRT) 238–239
working memory 237, 238–239; Brain Gym Therapy effects on 239–240; suicide and 245

For Product Safety Concerns and Information please contact our EU
representative GPSR@taylorandfrancis.com
Taylor & Francis Verlag GmbH, Kaufingerstraße 24, 80331 München, Germany

www.ingramcontent.com/pod-product-compliance
Ingram Content Group UK Ltd.
Pitfield, Milton Keynes, MK11 3LW, UK
UKHW052224190926
475702UK00007B/1194
* 9 7 8 1 0 4 1 0 8 7 1 1 3 *